Robert

Marc

Scott Whitfield

THE THEORY OF ATOMIC SPECTRA

THE THEORY OF ATOMIC SPECTRA

by

E. U. CONDON, PH.D.
PROFESSOR OF PHYSICS, WASHINGTON UNIVERSITY
ST LOUIS, MISSOURI

and

G. H. SHORTLEY, PH.D.
ASSOCIATE DIRECTOR RESEARCH CENTER
BORG WARNER CORPORATION

CAMBRIDGE
AT THE UNIVERSITY PRESS
1957

PUBLISHED BY
THE SYNDICS OF THE CAMBRIDGE UNIVERSITY PRESS
Bentley House, 200 Euston Road, London, N.W. 1
American Branch: 32 East 57th Street, New York 22, N.Y.

First printed 1935
Reprinted with corrections 1951
Reprinted 1953
1957

First printed in Great Britain at the University Press, Cambridge
Reprinted by offset-litho by
Bradford & Dickens

To

HENRY NORRIS RUSSELL

CONTENTS

CHAPTER III (*continued*)

CHAPTER IV

CHAPTER V

CHAPTER VI

CHAPTER VII

CHAPTER VIII

CHAPTER IX

CHAPTER X

CHAPTER XI

CHAPTER XII

CHAPTER XIII

CHAPTER XIV

CHAPTER XV

CHAPTER XVI

CHAPTER XVII

CHAPTER XVIII

APPENDIX

PREFACE

In this monograph we have undertaken a survey of the present status of the problem of interpreting the line spectra due to atoms. This interpretation seems to us to be in a fairly closed and highly satisfactory state. All known features of atomic spectra are now at least semi-quantitatively explained in terms of the quantum-mechanical treatment of the nuclear-atom model.

This does not mean that the period of fruitful research in atomic spectra is at an end. Fundamental questions are still outstanding in regard to the relativistic theory of the many-electron problem and also in regard to the theory of the interaction of radiation and matter. In addition every reader will see many places where more experimental information and better or more detailed theoretical calculations are desirable. It is our hope that the book will be useful in stimulating progress along these lines. With this end in view we have aimed to give explicitly an example of the use of each of the general types of calculation involved and an adequate survey of the literature of the more specialized calculations. The literature is covered approximately to the summer of 1934 although a few later references have been incorporated.

There exists confusion in the original literature about two matters which we have made every effort to clear up in this book:

In the first place it has been too little recognized that a matrix is not fully useful in the transformation theory unless the relative phases of the states to which the components refer are in some way specified, since two matrices cannot be added or multiplied unless these phases are the same in both. The phase choice is arbitrary, just like the troublesome sign conventions in geometrical optics, but one choice does have to be made and adhered to throughout a given set of calculations. To facilitate the use of the formulas and tables of the book in other calculations, we have attempted in every case to make explicit the specification of the phase choice employed.

In the second place there occurs, particularly in the theoretical literature, a great diversity of spectroscopic terminology. We have attempted to adhere as closely as possible to the original meanings of the nouns which denote energy levels and spectral lines, and find that this gives a nomenclature that is convenient and unambiguous. Briefly, our usage is the following: a *component* (of a line) results from a radiative transition between two *states* of an atom; a *line* results from the totality of transitions between two *levels*; in Russell-Saunders coupling, a *multiplet* is the totality of lines connecting the levels of two *terms*, and a *supermultiplet* is the totality of multiplets connecting the terms of two *polyads*; a *transition array* is the totality of lines connecting the levels of two *configurations*.

We have defined in § 7[4] a quantity, called the *strength* of a line, which we find to give a more convenient theoretical specification of the radiation intensity than either of the Einstein transition probabilities. We hope that this new usage will find favour among spectroscopists.

We take pleasure in making acknowledgment here of the valued help of many friends with whom we have discussed various parts of the theory, and who have criticized portions of the manuscript. In particular we are indebted to Prof. H. P. Robertson of Princeton for most of the elegant new treatment of spherical harmonics given in § 4[3]; to Dr F. Seitz of Princeton and to Prof. C. W. Ufford of Allegheny College for special calculations; and to Mr B. Napier for assistance in preparing the figures. Completion of the work was greatly facilitated by generous arrangements by the senior author's colleagues at Princeton to relieve him of most of his teaching duties for a term. The junior author is appreciative of courtesies shown by the physics faculties of the University of Minnesota, where he spent the summer of 1933, and of the Massachusetts Institute of Technology, where he spent the year 1933–1934 as a National Research Fellow. The greater part of his work on the book was done while he was a Fellow in Princeton University in the years 1931–1933.

Finally we wish to record our great enthusiasm for the beautiful typographical work of the Cambridge University Press as exemplified once more in the following pages.

E. U. C.
G. H. S.

March 1935

We are naturally gratified at the reception that our work has had and regret very much that other duties have prevented our giving it the thorough revision which it needs. The present printing is essentially a reprint of the 1935 edition except that a number of errors and misprints have been corrected.

E. U. C.
G. H. S.

May 1950

CHAPTER I

INTRODUCTION

"And so the true Cause of the Length of that Image was detected to be no other, than that *Light* is not similar or Homogenial, but consists of Difform Rays, some of which are more Refrangible than others." NEWTON.

Spectroscopy is that branch of physics which is the direct outgrowth of a classic experiment of Newton's which led him to the conclusion which we have placed at the head of this chapter. Newton's experimental arrangement is shown in Fig. 1[1], which is taken from Voltaire's *Elémens de la Philosophie de Newton* (Amsterdam, 1738, p. 116). The beam of sunlight enters a hole in the window shutters and traverses the prism, falling on the screen *P*. The image is not round like the hole in the shutter, but long in the direction perpendicular to the axis of the prism and coloured.

Fig. 1[1]. Newton's discovery of dispersion.

This arrangement is the proto-type of the modern spectroscope. As we are here interested in the theory of spectra we shall not concern ourselves with the technique of spectroscopy as it has developed from Newton's time to the present. Our principal interest will be the information concerning the nature of the atom which is obtained from a study of the characteristic radiations emitted by monatomic vapours.

After the discovery of sharp dark lines in the solar spectrum and sharp emission lines in spectra of flames, arcs, and sparks, the physicists of the nineteenth century seized upon spectroscopy as a valuable tool for qualitative chemical analysis. At that stage the experimental problem consisted in correlating the various lines and bands seen in the spectroscope with the chemical nature of the emitting substance. This task in itself was by no means simple, for the spectroscope is extraordinarily sensitive to small impurities and it was difficult to deal with sources pure enough to make certain the correlation of the observed lines to the substances in the source.

Another thing which made for difficulty is the fact that the spectra do not depend simply on the chemical elements present but on their state of chemical combination, which in turn is usually altered by the conditions which render the substance luminescent. To this period also belongs the beginning of the great task of setting up accurate standards of wave-length.

It early became clear that the observed spectra are of three general types. *Continuous*, that is, having no line structure in the spectroscopes of greatest resolving power. These are emitted by incandescent solids, but also under some circumstances by molecules and even by single atoms. *Banded*, having a special form of line structure in which close groups of many lines occur so densely packed that in smaller instruments they appear continuous. These are characteristic of the spectra of molecules and arise from the many possible changes in the rotational state of the molecule during the radiation process. *Line*, where the lines are well separated and generally show no obvious simple arrangement although in many cases they are grouped into small related groups of a few lines. Such spectra are due to individual atoms.

The line spectrum due to a single chemical element in the form of a monatomic vapour shows still another complication. It was early learned that quite different spectra are obtained from the same element according to the energetic violence with which it is excited to luminescence. In the electric spark more energy is put into the emitting atoms than in the electric arc and generally quite different lines result from the same element for these two modes of excitation. These differences are now known to arise from different degrees of ionization of the same element.

Evidently the next step is that of trying to understand the structural nature of an atom which enables it to emit its characteristic radiations. For the purposes of the classical kinetic theory of gases it is not necessary to assume anything more about atoms than that they behave something like hard elastic balls. As a consequence the range of phenomena which that theory embraces is not in a position to tell us more about the structure of atoms than an estimate of their size. In the spectrum of one element we are given a vast amount of data which is measurable with great precision. Evidently it is somehow determined by the structure of the atom, so spectroscopy stood out clearly in the minds of physicists as an important means for studying that structure.

In the latter part of the nineteenth century Maxwell developed his electromagnetic wave theory which received experimental confirmation in the experiments of Hertz and Oliver Lodge. Because the velocity of light agreed with the velocity of electric waves, the theory of electric waves was early applied as a theory of light. The wave theory of light, hitherto developed as a theory of elastic vibrations in the universal medium, the ether, was re-written in terms of the theory of electromagnetic waves. A little later the electron was discovered through researches on cathode rays and the chemical theory of ions was developed in connection with the electrolytic dissociation hypothesis. The view became current that atoms are structures built out of electrons and positive ions. A branch of physics called 'the electron theory of matter' came into being whose programme called for the explanation of the properties of matter in terms of this picture with the aid of the laws of the electromagnetic field.

During this period empirical regularities in line spectra were being found. The best known of these was Balmer's simple formula (1885) for the wave-lengths of the visible lines of the hydrogen spectrum.

Although experimentalists still prefer to express their measurements in terms of wave-lengths, Hartley showed (1883) that there are regularities in the spacing of related doublet or triplet lines which are more simply expressed in terms of the reciprocal of the wave-length, that is, the wave-number or number of waves in unit length. This discovery is of the greatest theoretical importance. There is no logical reason to-day for dealing with wave-lengths at all, and they are seldom mentioned in this book. But the custom of thinking in terms of them in the laboratory is probably too firmly entrenched to be shaken off for a long time. After the work of Balmer comes the important researches of Rydberg and of Kayser and Runge, who discovered that many spectral lines in various atomic spectra, chiefly those of alkali and alkaline-earth metals, can be organized into series obeying formulas similar to the formula of Balmer.

These empirical discoveries of spectral regularities reach their culmination in the clear establishment of the *Ritz combination principle*. This came in 1908 after two decades of important work on the study of spectral series. According to this result each atom may be characterized by a set of numbers called terms, dimensionally like wave-numbers, such that the actual wave-numbers of the spectral lines are given by differences between these terms. Ritz thought that lines were associated with all possible differences between these terms, and this is in accord with modern theoretical views, except that the lines associated with some differences are millions of times weaker than others so that practically there are important selection rules needed to tell which differences give strong lines.

The principle received striking confirmation in the same year through Paschen's discovery of an infra-red series in hydrogen. The wave-numbers of the Balmer series are represented by the formula

$$\sigma = \mathsf{R}\left(\frac{1}{2^2} - \frac{1}{n^2}\right), \qquad (n = 3, 4, 5, \ldots)$$

where $\mathsf{R} = 109677\ \text{cm}^{-1}$, an empirical coefficient called the Rydberg constant. In Ritz's language this means that this particular set of lines arises as differences of the terms $\sigma_n = \mathsf{R}/n^2$ and the particular term $\sigma_2 = \mathsf{R}/2^2$. The combination principle suggests the existence of lines given more generally by

$$\sigma = \mathsf{R}\left(\frac{1}{m^2} - \frac{1}{n^2}\right).$$

For $m = 3$ and $n = 4, 5, 6, \ldots$, the lines fall in the infra-red. Paschen found them at the predicted places. Lyman also found in the ultra-violet three lines corresponding to $m = 1$ and $n = 2, 3, 4$. The principle was quickly assimilated as an important rule in the analysis of spectra. It holds in all cases, even when the individual terms cannot be represented by a simple formula as is the case with hydrogen. Its applicability is of great generality, holding for molecular as well as atomic spectra.

The first decade of the twentieth century was important as showing, through the work of Planck on black-body radiation and Einstein on the photo-electric effect, that there is much more to the laws of interaction of matter and radiation than is given by the nineteenth century electromagnetic theory. These developments mark the birth of quantum theory. The electron theory programme had led to some simple assumptions concerning atomic structure and had had some notable successes, particularly in Lorentz's calculation of the effect of a magnetic field on the spectral lines, as observed by Zeeman. Spectral lines were associated with the electromagnetic radiation coming from motion of the electrons in an atom, generally regarded simply as harmonic oscillations about an equilibrium position. Great difficulty attached to the interpretation of the enormous number of spectral lines without the introduction of unreasonable complications in the model. On the experimental side a most important step was the recognition, through experiments on scattering of alpha particles made by Rutherford, that the positive electricity in an atom is confined to a small particle which is now called the nucleus of the atom. Its linear dimensions are not greater than about 10^{-4} those of the whole atom.

The stage is now set for the great theoretical developments made by Bohr from 1913 onward. Rutherford's experiments had given a general picture of a nuclear atom—a positively charged massive nucleus surrounded by the negatively charged and much less massive electrons. The theoretical developments had given imperfect and unclear indications of the need of fundamental changes in the electron theory for the process of emission and absorption of radiation. Empirical spectroscopy was organized by means of the Ritz combination principle and the extensive study of spectral series. In 1913 Bohr's first work on atomic structure gave a theory of the spectrum of hydrogen which involved several important advances.

Most general was the idea of stationary states and the interpretation of the Ritz combination principle. It is postulated that the possible states of atoms and molecules are restricted to certain values of the total energy. These values are determined by the structure of the atom or molecule and may be continuous in some ranges, as in the classical theory, or may be restricted to a set of discrete values. Then the postulate is made that the emission or absorption of radiation is connected with a process in which the atom passes from one energy level to another. This is rendered precise by the statement that the frequency of the radiation emitted is given by the equation

$$h\nu = E_1 - E_2,$$

where h is Planck's constant, ν the frequency of the radiation, E_1 the energy of the atom before the radiative process and E_2 its energy afterwards. If $E_2 > E_1$ the frequencies come out formally negative—their numerical values are the frequencies of light which can be absorbed by atoms in the state of energy E_1. This expresses the frequencies as differences of numbers characteristic of the atom and establishes a coordination between the energy levels, E, and the terms, σ, through the relation

$$E = -hc\sigma.$$

The minus sign arises since the conventional way of measuring term values was by counting them as positive when measured from the series limit. There is no reason to adhere to this convention, so that we shall always write

$$E = +hc\sigma$$

and regard σ simply as a measure of the energy in the auxiliary unit, cm^{-1}.

The hypothesis suggests its own means of experimental verification. If atoms are excited to radiate by single electron impacts in which the electrons have known kinetic energy E, then the only spectral lines appearing should be those for which the energy of the initial state E_1 is less than E. (This assumes that the zero of energy is the energy of the lowest state and that before impact the atoms are all in this lowest state.) This is the idea underlying the experiments of Franck and Hertz and many others on 'critical potentials.' Such experiments have fully confirmed the energy level interpretation of the spectroscopic terms and have been a valuable tool in experimental work. By controlled electron impact it is possible to bring out a spectrum bit by bit as the kinetic energy of the impacting electrons is gradually increased, thereby simplifying the task of determining the energy levels which are associated with the production of the different lines.

The other part of Bohr's early work was the development of a special dynamical model for the hydrogen atom and the study of rules for the determination of the allowed energy levels. The model for this simplest atom consisted of an electron and proton describing orbits about their centre of mass according to classical mechanics under their mutual attraction as given by the Coulomb inverse-square law. The allowed circular orbits were determined simply by the requirement (an additional postulate of the quantum theory) that the angular momentum of the system be an integral multiple of $\hbar = h/2\pi$. This served to give an energy

$$E_n = -\frac{2\pi^2\mu e^4}{h^2}\frac{1}{n^2}$$

for the circular orbit whose angular momentum is $n\hbar$, whence the term values are

$$\sigma_n = \frac{E_n}{hc} = -\frac{2\pi^2\mu e^4}{h^3 c}\frac{1}{n^2}.$$

Not only is the variation as n^{-2} in accord with the scheme of terms as given by the Lyman-Balmer-Paschen series in hydrogen, but the numerical coefficient comes out correctly, so by this means the empirical Rydberg constant R was for the first time related to universal constants.

Naturally this definite accomplishment stimulated other work and in the next few years great advances were made in interpretation of finer details of the hydrogen spectrum due to relativistic effects (Sommerfeld) and the effect of an electric field on the hydrogen spectrum (Epstein, Schwarzschild). Also it gave rise to much important work in extending the model and the quantum principle to other more complicated atomic and molecular structures. These studies were eminently successful in a semi-quantitative way and gave a great impetus to the experimental study and analysis of atomic spectra. The theory called for a study of the model by means of classical mechanics. The so-called multiply periodic motions had to be sought out and of these the allowed ones determined by a rule of quantization which was an outgrowth of Bohr's requirement on the angular momentum for the circular orbits of hydrogen. We shall not trace in detail the work along these lines: for this the reader is referred to Sommerfeld's *Atombau und Spektrallinien* and to Van Vleck's *Quantum Principles and Line Spectra*.

The theory was quite incomplete, however, in regard to the details of the interaction of the atom with the electromagnetic field. It gave no definite basis for the calculation of the relative or absolute intensities of the spectral lines. It also failed to give satisfactory results when attempts were made to calculate the energy levels of atoms containing more than one electron. Numerous attempts were made to calculate the energy of the lowest state of helium but without securing agreement with the experimental result. Evidently Bohr's principles had to be regarded as provisional indications of the direction in which a more satisfactory theory was to be sought. The unsatisfactoriness of the theory came more and more into the foreground in the early part of the 1920–30 decade after almost ten years in which physicists were busy making such progress as was possible with the original Bohr theory.

This was the period just preceding the discovery of the formalism of quantum mechanics, the discovery which has been so extraordinarily fruitful for all parts of atomic physics in the past eight years. In this period much was done in the study of atomic spectra and the formulation of the results in the language of the Bohr theory. The most important theoretical development was *Bohr's correspondence principle*. This emphasizes that the laws of atomic physics must be of such a character that they agree with the classical mechanics and electromagnetic theory in the limit of large quantum numbers. This principle was able to make much more definite some of the results of the previous theory in the way of special calculations—in particular it gave an approximate method of calculating the relative intensity of spectral lines. Its successes in such special applications were sufficient to create confidence in the principle. What was much more important than the special applications was the use of the principle as a broad general guide in the attempts to formulate a more complete set of laws for atomic physics.

The important problem before theoretical physics was thus the development of a rational system of quantum mechanics. The earlier work is usually referred to as quantum theory: it consisted of a few quantum postulates patched on to the classical kinematics and dynamics. By quantum mechanics we mean the much more unified theory of atomic physics which we owe to de Broglie, Heisenberg, Schrödinger, Dirac, and others. We shall give just a sketch of the origin of the new theory. It developed rapidly from 1925 onwards, at first along two quite different lines which were quickly brought into close relationship.

De Broglie built his work on an analogy, due to Hamilton, between the laws of mechanics and the laws of geometrical optics. He was led to the formal conclusion that a wave motion is associated with the motion of a particle such that if the momentum of the particle is $\boldsymbol{p}$ in magnitude and direction, the associated wave is propagated with a wave-number $\boldsymbol{\sigma}$ related to $\boldsymbol{p}$ by the equation

$$\boldsymbol{p} = h\boldsymbol{\sigma}.$$

This was suggested by the equation $E = h\nu$ together with the fact that in relativity theory E is the time-like component of a four-vector of which $\boldsymbol{p}$ is the space-like part, while ν is the time-like component of a four-vector of which $\boldsymbol{\sigma}$ is the space-like part. This suggestion was made in 1924. Something over a year later it was taken up by Schrödinger and developed in his famous series of papers on wave mechanics. Very roughly the idea is that just as geometrical optics is adequate for phenomena in which all apertures are large compared with the wave-length of light—otherwise diffraction effects are important—so also classical mechanics is adequate where the linear dimensions involved are large compared to h/p—otherwise effects akin to diffraction are important: these are the special effects which we find in the region of the quantum phenomena of atomic physics.

At the same time Heisenberg took a decisive step in the formulation of matrix mechanics as a definite realization of the correspondence principle programme. The initial step was to regard the atom as characterized by an ensemble of quantities of the type $P_{nm}e^{2\pi i\nu_{nm}t}$, in which P_{nm} gives the amplitude of a classical harmonic oscillator whose intensity of radiation and type of polarization is the same as that of the light of frequency ν_{nm} from the actual atom. This double array of quantities, for all values of n and m, was regarded by Heisenberg as a single mathematical entity. By treating it and other quantities related to the atom's structure by the mathematical rules of matrix algebra it was found that a definite formulation of the laws of quantum mechanics could be given in accordance with the correspondence principle. In a few months the mathematical equivalence of Schrödinger's wave mechanics and Heisenberg's matrix mechanics was established. The work of Jordan and of Dirac led to a formulation of a single mathematical system of quantum mechanics. This is the formulation in terms of which all current work of importance in quantum physics is expressed to-day.

During this period the study of atomic spectra was being actively pushed on. After the study of series spectra of the alkali-like metals in terms of the Bohr theory the next important steps were the empirical discovery by Landé of the laws of the Zeeman effect and the discovery and extensive study of related groups of lines called multiplets in complex spectra. The modern study of multiplets was begun by Catalán. The multiplet structure and the problems of the anomalous Zeeman effect called for an essential generalization of the electron-orbit model which was supplied in 1925 by Uhlenbeck and Goudsmit by postulating an intrinsic magnetic moment and angular momentum for the electron. This 'spin' hypothesis quickly cleared up many difficult points, so that it at once gained acceptance.

In the pre-quantum-mechanical period the general method of working on a detailed question in the theory of atomic spectra was to calculate from an assumed model by means of classical mechanics, and then to try to alter the formulas so obtained in such a way that the change was negligible for large quantum numbers but was of such a nature that it brought about agreement with experiment for small quantum numbers. It is really remarkable how much of the modern theory of line spectra was developed in this way. Important contributions were made by Pauli, Heisenberg, Hund, and Russell. There was developed a vector-coupling model for complex atoms in which the quantization of the angular momenta of the individual electron orbits and of their vector resultant played a dominant rôle. To this period also belongs the discovery of the important exclusion principle of Pauli according to which no two electrons in an atom may have the same set of quantum numbers. This first found a rational place in the theory with the advent of quantum mechanics. As an empirical principle, however, it was of the greatest importance, especially through the work of Hund in predicting the general structure of complex spectra and extending the theory of the periodic system of the elements as begun by Bohr.

Thus it happened that by ingenious use of the correspondence principle a great deal of the modern theory of atomic spectra was worked out without the aid of quantum mechanics. But that does not mean that the new mechanics is without importance for our understanding of line spectra, since the results obtained were not part of a closed structure of definite physical principles, but were obtained in semi-empirical ways from consideration of a formulation of the theory that was only true in the limit of large quantum numbers. Moreover, not all the problems of interest could be handled in this way, so there were detailed calculations on which to test the quantum-mechanical method as well as the task of securing from definite calculations by the new methods those results which had been cleverly guessed with the aid of the correspondence principle. This application of quantum mechanics to the atomic model has been the programme of research in the theory of atomic spectra from 1926 onwards. In spite of the mathematical complexities, it became clear in the next five or six years that the quantum mechanics of Heisenberg, Schrödinger and Dirac when applied systematically to the study of the nuclear model of the atom is adequate to give an accurate and complete unification of the great amount of empirical data accumulated through analyses of atomic spectra. Naturally in so short a time it was not possible to make precise and detailed calculations of the system of energy levels of all atoms, but enough has been done to give rise to a general conviction that the theory is quite adequate for the interpretation of atomic spectra.

Much remains to be done in the way of precise and detailed calculations and it may well be that when these are made it will be found that such an estimate of the power and scope of our present theoretical knowledge is over-optimistic. Complete as the general picture seems at present, there may well be lurking somewhere important residual effects, like the advance of the perihelion of Mercury in celestial mechanics, which will necessitate essential alterations in the theory. At present the theory exists in a somewhat closed and complete form, so that it is possible to give a unified deductive treatment of atomic spectra in terms of the quantum-mechanical theory of the nuclear atom. That is the programme of this book. Future development in this field may be merely a story of the filling in of details without essential change in the present views. Or it may be that essential difficulties will appear, requiring radical revision of the present theory. In either case the time seems ripe for taking stock of the rapid developments of the past decade in order to prepare ourselves for the completion of the study of line spectra of atoms in the experimental and theoretical studies of the future.

At present there are several systematic accounts of the subject of line spectra and their theoretical interpretation. The most important are

SOMMERFELD, *Atombau und Spektrallinien*, F. Vieweg, 1931;
HUND, *Linienspektren und periodisches System der Elemente*, J. Springer, 1927;
PAULING and GOUDSMIT, *The Structure of Line Spectra*, McGraw-Hill, 1930;
RUARK and UREY, *Atoms, Molecules and Quanta*, McGraw-Hill, 1930;
GIBBS, *Line Spectra of the Elements*, Rev. Mod. Phys. **4**, 278 (1932);
WHITE, *Introduction to Atomic Spectra*, McGraw-Hill, 1934.

All except the last of these are written from the standpoint of the correspondence principle. In this book we make no attempt at following the historical order. We have confined our attention to the historical developments to the brief sketch of this introductory chapter. Likewise we make no attempt at a complete account of the formulation of the principles of quantum mechanics, as there are now in existence several works on this subject. Most important among these are

DIRAC, *Quantum Mechanics*, Oxford University Press, second edition, 1935;
VON NEUMANN, *Mathematische Grundlagen der Quantenmechanik*, J. Springer, 1932;
FRENKEL, *Wave Mechanics, Advanced General Theory*, Oxford, 1934.

More elementary, and hence more suited to a first approach to the subject, are

SOMMERFELD (translation by BROSE), *Wave Mechanics*, Methuen, 1930;
BORN and JORDAN, *Elementare Quantenmechanik*, J. Springer, 1930;
FRENKEL, *Wave Mechanics, Elementary Theory*, Oxford, 1932;
CONDON and MORSE, *Quantum Mechanics*, McGraw-Hill, 1929;
MOTT, *An Outline of Wave Mechanics*, Cambridge University Press, 1930.

In order to make this book more useful for independent reading, however, we have opened the detailed discussion with a brief account, in the next chapter, of the principles of quantum mechanics. This account is intended simply as a review of principles used throughout the rest of the work and as a repository of the theory in convenient form for reference. The succeeding two chapters are devoted to the presentation of special results from the general theory which are used throughout the book; the detailed development of the theory of atomic spectra begins with Chapter v.

We wish finally to make a few remarks concerning the place of the theory of groups in the study of the quantum mechanics of atomic spectra. The reader will have heard that this mathematical discipline is of great importance for the subject. We manage to get along without it. When Dirac visited Princeton in 1928 he gave a seminar report on his paper showing the connection of exchange energy with the spin variables of the electron. In the discussion following the report, Weyl protested that Dirac had said that he would derive the results without the use of group theory, but, as Weyl said, all of Dirac's arguments were really applications of group theory. Dirac replied, "I said I would obtain the results without previous knowledge of group theory."

That incident serves to illustrate our attitude on this point. When a physicist is desirous of learning of new theoretical developments in his subject, one of the greatest barriers is that it generally involves new mathematical techniques with which he is apt to be unfamiliar. Relativity theory brought the necessity of learning tensor calculus and Riemannian geometry. Quantum mechanics forces him to a more careful study of boundary value problems and matrix algebra. Hence if we can minimize the amount of new mathematics he must learn in order to penetrate a new field we do him a real service. Weyl's protest to Dirac is certainly also applicable to this book. But so is Dirac's answer. Many things which are done here could be done more simply if the theory of groups were part of the ordinary mathematical equipment of physicists. But as it is not, it seems like putting unnecessary obstacles in the way to treat the subject by a method which requires this equipment. On the other hand the pure mathematician studying a new branch of physics is likely to take most delight in the fact that the theory exemplifies parts of pure mathematics which have been hitherto rather devoid of physical applications. To him our plan is not as satisfactory as one which shows how the theory of the structure of the atom is related to the abstract theory of groups.

This does not mean that we underestimate the value of group theory for atomic physics nor that we feel that physicists should omit the study of that branch of mathematics now that it has been shown to be an important tool in the new theory. It is simply that the new developments bring with them so many new things to be learned that it seems inadvisable to add this additional burden to the load.

For those who wish to regard the theory of atomic spectra from the standpoint of the theory of groups there are three books available at present:

Weyl (translation by Robertson), *Group Theory and Quantum Mechanics*, Methuen, 1931;
Wigner, *Gruppentheorie und ihre Anwendung auf die Quantenmechanik der Atomspektren*, F. Vieweg, 1931;
van der Waerden, *Die gruppentheoretische Methoden in der Quantenmechanik*, J. Springer, 1932.

CHAPTER II

THE QUANTUM MECHANICAL METHOD

The formulation of the laws of quantum mechanics which is most suited to our purpose is that due to Dirac, and discussed at length in his book.* We shall review in brief such of this formulation as we shall need, giving page references to Dirac, where a fuller account may be found. This will be followed by a discussion of matrix mechanics, perturbation theory, and related subjects.

1. Symbolic algebra of states and observables.

In this section the principles of the theory will be set down in terms of the properties of certain abstract symbols in a manner corresponding to the purely symbolic treatment of vector analysis which is independent of any coordinate system.

The state of a system (Dirac, p. 11) is described by a quantity called $\boldsymbol{\psi}$ which is analogous to a unit vector in a space of a great many (in general an infinite number of) dimensions (p. 18).† $\boldsymbol{\psi}$ has as many components as the system has independent states. In any given representation the components of $\boldsymbol{\psi}$ are ordinary (in general, complex) numbers. The sum of two $\boldsymbol{\psi}$'s is another $\boldsymbol{\psi}$ whose components are the sums of the components of the two $\boldsymbol{\psi}$'s. The product of a $\boldsymbol{\psi}$ by an ordinary number c is a $\boldsymbol{\psi}$ whose components are c times those of the original $\boldsymbol{\psi}$.

Two states are not considered to be distinct unless the $\boldsymbol{\psi}$'s which describe them are linearly independent.

Since the components of $\boldsymbol{\psi}$ are allowed to be complex, and only real numbers may occur in the interpretation of the theory, we introduce $\bar{\boldsymbol{\psi}}$, the symbolic *conjugate imaginary* of $\boldsymbol{\psi}$.‡ $\bar{\boldsymbol{\psi}}$ is not a vector in the same space as $\boldsymbol{\psi}$, but is a vector in the dual space. Its components are the ordinary complex conjugates of the corresponding components of $\boldsymbol{\psi}$. Since $\boldsymbol{\psi}$ and $\bar{\boldsymbol{\psi}}$ are vectors in different spaces, there is no place in the algebra for the addition of a $\boldsymbol{\psi}$ and a $\bar{\boldsymbol{\psi}}$. The distinction between $\boldsymbol{\psi}$ and $\bar{\boldsymbol{\psi}}$ is more fundamental than that between ordinary complex conjugates; there is no sense in which we can split $\boldsymbol{\psi}$ into a real and an imaginary part.

* DIRAC, *Quantum Mechanics* (page references are to the second edition). For a mathematically more rigorous formulation see VON NEUMANN, *Mathematische Grundlagen der Quantenmechanik*.

† We shall in this chapter use bold-faced type for symbolic quantities, reserving ordinary type for functions and operators in the Schrödinger scheme and for ordinary numbers. Thus confusion between the symbolic $\boldsymbol{\psi}$ and Schrödinger's ψ function will be avoided.

‡ Dirac uses $\boldsymbol{\phi}$ in place of $\bar{\boldsymbol{\psi}}$, but we prefer the latter notation because it is more symmetrical and more easily translated into the Schrödinger representation. Furthermore, in subsequent chapters we shall not restrict ourselves to the letter $\boldsymbol{\psi}$ to represent a state, but shall at times use different letters for different states.

We now suppose that any ψ and any $\bar{\psi}$ have a product which is defined as the analogue of the scalar product of two vectors in vector analysis and is therefore an ordinary number, in general complex (p. 22). This product will by convention always be written with the $\bar{\psi}$ first, e.g. $\bar{\psi}_r\psi_s$, where the subscripts denote different states of a system. It follows, then, that $\bar{\psi}_r\psi_s$ and $\bar{\psi}_s\psi_r$ are complex conjugate numbers, and that $\bar{\psi}_r\psi_r$ is real and positive; unless otherwise stated we shall take all ψ's to be normalized. Hence, for any r and s,

$$\bar{\psi}_r\psi_s = \overline{\bar{\psi}_s\psi_r}; \quad \bar{\psi}_r\psi_r = 1. \tag{1}$$

In observing a system experimentally we build an apparatus on a macroscopic scale which interacts with the system by a certain set of operations, resulting in a scale- or pointer-reading. The essential feature of classical physics has been that we have expected to be able to formulate the laws of physics in terms of functional relationships between the pointer-readings given by various sets of observing apparatus. All of physics and exact natural science has proceeded along such lines hitherto. Quantum mechanics does not do this. Any set of experimental apparatus and operations does not appear in the theory simply as the source of certain pointer-readings which bear a direct functional relationship to other sets of pointer-readings. Instead it appears as a quantity of a more complicated sort about to be described. Thus we are dealing not merely with a new set of laws but with an entirely new mathematical canvas on which to represent these laws. In this respect the quantum mechanics is a much more far-reaching departure from classical physics than was the theory of relativity.

In an experiment we are generally concerned with determining, directly or indirectly, the particular number which expresses the value of a dynamical variable, e.g. the position or momentum of an electron, at a particular time. Any such dynamical variable will be called an *observable*, and will be represented in this theory by a linear operator α (p. 24). An observable in the mathematical theory is a rule for acting on any ψ and converting it into another ψ. In this respect, it is analogous to a tensor of the second rank, or to the dyadic of Gibbs.

We denote the result of operation of α on ψ by $\alpha\psi$, where $\alpha\psi$ is another ψ. We need make no provision for the operation of α on $\bar{\psi}$.* The conjugate imaginary to $\alpha\psi$ will be written as $\overline{\alpha\psi}$. Linearity of α means that we have $\alpha(a\psi_r + b\psi_s) = a\alpha\psi_r + b\alpha\psi_s$, where ψ_r and ψ_s are any two states and a and b are ordinary numbers. We define the sum, $\alpha_1 + \alpha_2$, of two observables by $(\alpha_1 + \alpha_2)\psi = \alpha_1\psi + \alpha_2\psi$, where ψ is arbitrary. The product, $\alpha_1\alpha_2$, of two observables is defined by $(\alpha_1\alpha_2)\psi = \alpha_1(\alpha_2\psi)$; in general $\alpha_2\alpha_1 \neq \alpha_1\alpha_2$. Since

* Dirac does provide for the operation of α on $\bar{\psi}$, writing $\bar{\psi}\alpha = \overline{\alpha^{\dagger}\psi}$; see equation (2).

$\boldsymbol{\alpha}\boldsymbol{\psi}$ is another $\boldsymbol{\psi}$, $\bar{\boldsymbol{\psi}}_r\boldsymbol{\alpha}\boldsymbol{\psi}_s = \bar{\boldsymbol{\psi}}_r(\boldsymbol{\alpha}\boldsymbol{\psi}_s)$ is an ordinary number. The observable (Hermitian-) conjugate to $\boldsymbol{\alpha}$, denoted by $\boldsymbol{\alpha}^\dagger$, is defined by the equation

$$\bar{\boldsymbol{\psi}}_s\boldsymbol{\alpha}^\dagger\boldsymbol{\psi}_r = \overline{\bar{\boldsymbol{\psi}}_r\boldsymbol{\alpha}\boldsymbol{\psi}_s} = \overline{\boldsymbol{\alpha}\boldsymbol{\psi}_s}\boldsymbol{\psi}_r, \tag{2}$$

where the two states r and s are arbitrary (p. 29). The relation between $\boldsymbol{\alpha}$ and $\boldsymbol{\alpha}^\dagger$ is the same as that between ordinary complex conjugates; $(i\boldsymbol{\alpha})^\dagger = -i\boldsymbol{\alpha}^\dagger$; there is no rule against adding $\boldsymbol{\alpha}$ and $\boldsymbol{\alpha}^\dagger$. An observable $\boldsymbol{\alpha}$ is said to be *real* if $\boldsymbol{\alpha}^\dagger = \boldsymbol{\alpha}$, i.e. if

$$\bar{\boldsymbol{\psi}}_s\boldsymbol{\alpha}\boldsymbol{\psi}_r = \overline{\bar{\boldsymbol{\psi}}_r\boldsymbol{\alpha}\boldsymbol{\psi}_s} = \overline{\boldsymbol{\alpha}\boldsymbol{\psi}_s}\boldsymbol{\psi}_r \tag{3}$$

for all states r and s. An observable $\boldsymbol{\alpha}$ is said to be *purely imaginary* if $\boldsymbol{\alpha}^\dagger = -\boldsymbol{\alpha}$; such an observable may be written as $i\boldsymbol{\beta}$, where $\boldsymbol{\beta}$ is a real observable.* It can be shown from (2) that

$$(\boldsymbol{\alpha}_1\boldsymbol{\alpha}_2)^\dagger = \boldsymbol{\alpha}_2^\dagger\boldsymbol{\alpha}_1^\dagger; \tag{4}$$

hence if $\boldsymbol{\alpha}_1$ and $\boldsymbol{\alpha}_2$ are two real observables the commutator $(\boldsymbol{\alpha}_1\boldsymbol{\alpha}_2 - \boldsymbol{\alpha}_2\boldsymbol{\alpha}_1)$ is a purely imaginary observable (p. 29), if $\boldsymbol{\alpha}_1\boldsymbol{\alpha}_2$ is also real this commutator vanishes.

We now make the important physical postulate (p. 30) that a state for which

$$\boldsymbol{\alpha}\boldsymbol{\psi}_r = a\boldsymbol{\psi}_r, \tag{5}$$

where a is an ordinary number, is characterized by the fact that the observable $\boldsymbol{\alpha}$ *has the value* a. That is, that a measurement of the observable $\boldsymbol{\alpha}$ with the system in the state $\boldsymbol{\psi}_r$ will certainly give for the result the number a. For a real observable $\boldsymbol{\alpha}$ the number a will be necessarily real, as can be shown by multiplying (5) by $\bar{\boldsymbol{\psi}}_r$ to get $\bar{\boldsymbol{\psi}}_r\boldsymbol{\alpha}\boldsymbol{\psi}_r = a\bar{\boldsymbol{\psi}}_r\boldsymbol{\psi}_r = a$ [from (1)]. Here $\bar{\boldsymbol{\psi}}_r\boldsymbol{\alpha}\boldsymbol{\psi}_r$ is real by (3), so a is real.

The possible values of an observable $\boldsymbol{\alpha}$, i.e. the possible results of an observation of $\boldsymbol{\alpha}$, are the ordinary numbers α' for which the equation in $\boldsymbol{\psi}$,

$$\boldsymbol{\alpha}\boldsymbol{\psi} = \alpha'\boldsymbol{\psi}, \tag{6}$$

has solutions (p. 30). These are called the *allowed*, *proper*, *characteristic*, or *eigenvalues* of $\boldsymbol{\alpha}$, and may form a discrete set or a continuous set of numbers.† We shall follow Dirac in calling the $\boldsymbol{\psi}$'s which satisfy this equation the *eigen-*$\boldsymbol{\psi}$'s and the states which they denote the *eigenstates* of the observable $\boldsymbol{\alpha}$. We shall speak of an eigen-$\boldsymbol{\psi}$ which satisfies this equation for a given α' as *belonging* to the eigenvalue α', and denote it by $\boldsymbol{\psi}(\alpha')$. That is,

$$\boldsymbol{\alpha}\,\boldsymbol{\psi}(\alpha') = \alpha'\,\boldsymbol{\psi}(\alpha'). \tag{7a}$$

The complex imaginary to $\boldsymbol{\psi}(\alpha')$ we denote by $\bar{\boldsymbol{\psi}}(\alpha')$. This satisfies the relation

$$\overline{\boldsymbol{\alpha}\,\boldsymbol{\psi}(\alpha')} = \overline{\alpha'}\,\bar{\boldsymbol{\psi}}(\alpha'). \tag{7b}$$

* The operators corresponding to real and purely imaginary observables are said to be *Hermitian* and *anti-Hermitian* respectively, because of the Hermitian and anti-Hermitian character of their matrices (§ 7^2). Similarly, $\boldsymbol{\alpha}^\dagger$ is called the Hermitian conjugate of $\boldsymbol{\alpha}$ because its matrix is the Hermitian conjugate of the matrix of $\boldsymbol{\alpha}$.

† The whole set of eigenvalues is often called the *spectrum* of $\boldsymbol{\alpha}$.

There may be several, say $d_{\alpha'}$, linearly independent $\boldsymbol{\psi}$'s belonging to the same eigenvalue α'. In this case the value α' will be said to be $d_{\alpha'}$-fold degenerate, and the different $\boldsymbol{\psi}(\alpha')$'s will require another index which takes on the values 1, 2, ..., $d_{\alpha'}$ for their complete specification.

Now if one makes observations of $\boldsymbol{\alpha}$ on a system in a state $\boldsymbol{\psi}$ which is not an eigenstate of $\boldsymbol{\alpha}$, one does not always obtain the same value, but observes the various proper values of $\boldsymbol{\alpha}$ with certain probabilities. We now make a second physical postulate, *viz.* that the *average* of the values of $\boldsymbol{\alpha}$ obtained in this way is the number $\bar{\boldsymbol{\psi}}\boldsymbol{\alpha}\boldsymbol{\psi}$ which is characteristic of the state $\boldsymbol{\psi}$ (p. 43). $\boldsymbol{\psi}$ is here assumed normalized in the sense of (1).

For a real observable this average value is seen from (3) to be real for any state $\boldsymbol{\psi}$. Moreover, by writing $\boldsymbol{\alpha}=\boldsymbol{\alpha}_r+i\boldsymbol{\alpha}_i$, where $\boldsymbol{\alpha}_r$ and $\boldsymbol{\alpha}_i$ are real observables, it is seen that a necessary condition for the reality of $\bar{\boldsymbol{\psi}}\boldsymbol{\alpha}\boldsymbol{\psi}$ for all $\boldsymbol{\psi}$ is the vanishing of $\boldsymbol{\alpha}_i$. Since the results of physical observation are real for all states $\boldsymbol{\psi}$, we conclude that *all physically measurable quantities are real observables*.

PROBLEMS

1. Show that the allowed values of any function $f(\boldsymbol{\alpha})$ expressible as a power series in $\boldsymbol{\alpha}$ are $f(\alpha')$. This can be generalized to other functions (p. 38). Hence show that the first physical postulate above is a special case of the second.

2. Prove that two real observables commute if and only if their product is real.

3. Show that the observable $\boldsymbol{S}\boldsymbol{\alpha}\boldsymbol{S}^{-1}$ (where $\boldsymbol{S}^{-1}\boldsymbol{S}=\boldsymbol{S}\boldsymbol{S}^{-1}=1$) has the same allowed values as $\boldsymbol{\alpha}$, and find the relation between the eigen-$\boldsymbol{\psi}$'s of $\boldsymbol{\alpha}$ and those of $\boldsymbol{S}\boldsymbol{\alpha}\boldsymbol{S}^{-1}$.

2. Representations of states and observables.

We shall first show that the eigen-$\boldsymbol{\psi}$'s of a system belonging to two different allowed values of a *real* observable are (unitary-) orthogonal in the sense that

$$\bar{\boldsymbol{\psi}}(\alpha')\boldsymbol{\psi}(\alpha'')=0 \text{ unless } \alpha'=\alpha''. \tag{1}$$

To prove this, observe that

$$\bar{\boldsymbol{\psi}}(\alpha')\boldsymbol{\alpha}\boldsymbol{\psi}(\alpha'')=\alpha''\bar{\boldsymbol{\psi}}(\alpha')\boldsymbol{\psi}(\alpha'') \qquad \text{(from } 1^27\text{a)}$$

$$=\overline{\boldsymbol{\alpha}\boldsymbol{\psi}(\alpha')}\boldsymbol{\psi}(\alpha'')=\alpha'\bar{\boldsymbol{\psi}}(\alpha')\boldsymbol{\psi}(\alpha''). \qquad \text{(from } 1^23 \text{ and } 1^27\text{b)}$$

Therefore $(\alpha'-\alpha'')\bar{\boldsymbol{\psi}}(\alpha')\boldsymbol{\psi}(\alpha'')=0$; hence the theorem (p. 33).*

We now assume that the eigen-$\boldsymbol{\psi}$'s of any real observable† form a complete system in which we can make a Fourier expansion of an arbitrary $\boldsymbol{\psi}$ (p. 34). This amounts to assuming the whole of a kind of generalized Sturm-Liouville theory at one step; hence much needs to be filled in here by a study of exactly what classes of $\boldsymbol{\psi}$'s can be so expanded.

* Note that whenever the combination $\bar{\boldsymbol{\psi}}(\alpha')\boldsymbol{\alpha}$ arises, this may be replaced by $\alpha'\bar{\boldsymbol{\psi}}(\alpha')$ if $\boldsymbol{\alpha}$ is a real observable.

† Note that these considerations hold only for real observables; the eigen-$\boldsymbol{\psi}$'s of a general observable do not form an orthogonal set. Although we shall have occasion to employ non-real observables as an aid in calculation, we shall never have occasion to determine their characteristic values or functions, since these lack physical significance in quantum mechanics.

Consider first the case in which $\boldsymbol{\alpha}$ has a purely discrete spectrum. If the value α' is $d_{\alpha'}$-fold degenerate, we shall choose the $d_{\alpha'}$ linearly independent states belonging to α' normalized and mutually orthogonal. This we may do by a Schmidt process* since any linear combination of states belonging to α' will also belong to α'. Denote these $d_{\alpha'}$ states by $\psi(\alpha' r')$ for $r' = 1, \ldots, d_{\alpha'}$. We then have, according to these requirements, the relations

$$\bar{\psi}(\alpha' r')\,\psi(\alpha'' r'') = \delta(\alpha', \alpha'')\,\delta(r', r'') \tag{2}$$

satisfied. Here $\delta(\alpha', \alpha'') = \delta_{\alpha'\alpha''}$ is a function of the discrete set of points representing the allowed values of $\boldsymbol{\alpha}$ which is zero if $\alpha' \neq \alpha''$ and unity if $\alpha' = \alpha''$ (the Kronecker delta). The assumption is, then, that any state ψ may be expanded in terms of these states $\psi(\alpha' r')$. We shall write the expansion coefficient as $(\alpha' r'|\quad)$, placing it after the eigenstate:

$$\psi = \sum_{\alpha' r'} \psi(\alpha' r')\,(\alpha' r'|\quad). \tag{3}$$

The blank space is reserved to characterize the ψ which is being expanded; for example we write

$$\psi(\beta') = \sum_{\alpha' r'} \psi(\alpha' r')\,(\alpha' r'|\beta'). \tag{4}$$

Consider the state $\boldsymbol{\beta}\,\psi(\alpha'' r'')$ which results from the action of the observable $\boldsymbol{\beta}$ on $\psi(\alpha'' r'')$. This state can be expanded in terms of the complete set $\psi(\alpha' r')$. For this expansion we adopt the notation

$$\boldsymbol{\beta}\,\psi(\alpha'' r'') = \sum_{\alpha' r'} \psi(\alpha' r')\,(\alpha' r'|\beta|\alpha'' r''). \tag{5}$$

Before proceeding further with the algebra of this expansion theory we shall show how we may obtain a significant choice and characterization of the different states belonging to the degenerate level α'. Let us choose an observable $\boldsymbol{\beta}$ which commutes with $\boldsymbol{\alpha}$. We observe that

$$\boldsymbol{\beta\alpha}\,\psi(\alpha' r') = \alpha'\boldsymbol{\beta}\,\psi(\alpha' r') = \sum_{\alpha'' r''} \alpha'\,\psi(\alpha'' r'')\,(\alpha'' r''|\beta|\alpha' r')$$

from (5), while

$$\boldsymbol{\alpha\beta}\,\psi(\alpha' r') = \sum_{\alpha'' r''} \alpha''\,\psi(\alpha'' r'')\,(\alpha'' r''|\beta|\alpha' r').$$

If $\boldsymbol{\alpha\beta} = \boldsymbol{\beta\alpha}$, the coefficients of $\psi(\alpha'' r'')$ on the right of these two equations must be equal; hence $(\alpha' - \alpha'')(\alpha'' r''|\beta|\alpha' r') = 0$, from which we draw the important conclusion that

$$(\alpha'' r''|\beta|\alpha' r') = 0 \text{ unless } \alpha' = \alpha''. \quad (\boldsymbol{\beta\alpha} - \boldsymbol{\alpha\beta} = 0) \tag{6}$$

Let us now set ourselves the problem of finding the eigenstates of $\boldsymbol{\beta}$. Denote such a state by $\psi(\beta')$ and let (4) represent its expansion—we must then determine the coefficients $(\alpha' r'|\beta')$. Expanding the allowed values equation

$$\boldsymbol{\beta}\,\psi(\beta') = \beta'\,\psi(\beta')$$

* See for example, Courant and Hilbert, *Methoden der Mathematischen Physik*, p. 34.

we obtain

$$\sum_{\alpha'' r'' \alpha' r'} \boldsymbol{\psi}(\alpha' r')\,(\alpha' r'|\boldsymbol{\beta}|\alpha'' r'')\,(\alpha'' r''|\beta') = \sum_{\alpha' r'} \beta'\, \boldsymbol{\psi}(\alpha' r')\,(\alpha' r'|\beta'),$$

whence

$$\sum_{\alpha'' r''} (\alpha' r'|\boldsymbol{\beta}|\alpha'' r'')\,(\alpha'' r''|\beta') = \beta'\,(\alpha' r'|\beta') \quad \text{for all } \alpha', r'. \tag{7}$$

This set of homogeneous linear equations for the coefficients $(\alpha' r'|\beta')$ reduces when we use the special property (6) of $\boldsymbol{\beta}$ to the simpler set

$$\sum_{r''} (\alpha' r'|\boldsymbol{\beta}|\alpha' r'')\,(\alpha' r''|\beta') = \beta'(\alpha' r'|\beta') \quad \text{for all } \alpha', r'. \tag{8}$$

Here all the coefficients in a given equation refer to the same eigenvalue of $\boldsymbol{\alpha}$. Hence we may consider independently the $d_{\alpha'}$ equations corresponding to the eigenvalue α'. These $d_{\alpha'}$ equations determine the direction* of that part of $\boldsymbol{\psi}(\beta')$ which lies in the $d_{\alpha'}$-dimensional subspace characterized by α'. Let us find the solutions which lie wholly in this subspace, i.e. let us set $(\alpha'' r''|\beta')$ equal to zero unless $\alpha'' = \alpha'$. In order that we may find a non-vanishing solution of this type, the determinant

$$\begin{vmatrix} (\alpha' 1|\boldsymbol{\beta}|\alpha' 1) - \beta' & (\alpha' 1|\boldsymbol{\beta}|\alpha' 2) & \cdots & (\alpha' 1|\boldsymbol{\beta}|\alpha' d_{\alpha'}) \\ (\alpha' 2|\boldsymbol{\beta}|\alpha' 1) & (\alpha' 2|\boldsymbol{\beta}|\alpha' 2) - \beta' & \cdots & (\alpha' 2|\boldsymbol{\beta}|\alpha' d_{\alpha'}) \\ \cdots & \cdots & \cdots & \cdots \\ (\alpha' d_{\alpha'}|\boldsymbol{\beta}|\alpha' 1) & (\alpha' d_{\alpha'}|\boldsymbol{\beta}|\alpha' 2) & \cdots & (\alpha' d_{\alpha'}|\boldsymbol{\beta}|\alpha' d_{\alpha'}) - \beta' \end{vmatrix}$$

must vanish.† Setting this determinant equal to zero gives an equation of $d_{\alpha'}{}^{\text{th}}$ degree in β' whose roots are the $d_{\alpha'}$ eigenvalues β' which are consistent with the conditions assumed. To each of these roots belongs an eigenstate of $\boldsymbol{\beta}$ lying wholly in the subspace characterized by α'. To a multiple ($d_{\alpha'\beta'}$-fold) root β' will belong $d_{\alpha'\beta'}$ linearly independent eigenstates lying in this subspace. Since all the eigenstates of $\boldsymbol{\beta}$ are given by (8), and since it is clear that we have found a complete set of solutions of (8), we have found a complete set of states $\boldsymbol{\psi}(\alpha' \beta')$ which are simultaneously eigenstates of $\boldsymbol{\alpha}$ and of $\boldsymbol{\beta}$. If there is still a degeneracy in these states, we may choose a third observable $\boldsymbol{\gamma}$ (independent of $\boldsymbol{\alpha}$ and $\boldsymbol{\beta}$ in the sense that $\boldsymbol{\gamma}$ is not a function of $\boldsymbol{\alpha}$ and $\boldsymbol{\beta}$) which commutes with both $\boldsymbol{\alpha}$ and $\boldsymbol{\beta}$. We may by a process similar to the above find a complete set of eigenstates of $\boldsymbol{\gamma}$ which are simultaneously eigenstates of $\boldsymbol{\alpha}$ and $\boldsymbol{\beta}$, i.e. of the form $\boldsymbol{\psi}(\alpha' \beta' \gamma')$. We continue to introduce independent commuting observables until we find no degeneracy in a simultaneous eigenstate of all of them. The number of such observables is the quantum-mechanical analogue of the classical number of degrees of freedom. Thus a calcium atom with 20 electrons and a fixed nucleus will be found to

* A set of linear homogeneous equations determines only the ratios of the unknowns, i.e. the direction of a vector but not its magnitude. But because of the normalization condition, an eigenstate is determined by its direction except for an arbitrary phase factor $e^{i\delta}$.

† See, e.g., Bôcher, *Higher Algebra*, p. 47.

require 80 quantum numbers for the complete description of its state. A set of observables, say $\boldsymbol{\gamma}_1, \boldsymbol{\gamma}_2, \ldots, \boldsymbol{\gamma}_k$, in terms of whose simultaneous eigenstates we can describe the state of a system completely, is said to be a *complete set.* We shall denote this set of observables by $\boldsymbol{\Gamma}$ for short, and shall write the general state characterized by the *quantum numbers* $\gamma_1', \gamma_2', \ldots, \gamma_k'$ as $\psi(\Gamma')$.* We can choose a complete set of observables in various ways since any independent commuting functions $\boldsymbol{\delta}_1, \boldsymbol{\delta}_2, \ldots, \boldsymbol{\delta}_k$ of the $\boldsymbol{\gamma}$'s will also form a complete set, which we shall denote by $\boldsymbol{\Delta}$.

Let us now return to a consideration of the algebra of representations in terms of the eigenstates of such complete sets of observables. A general state ψ has the expansion [cf. (3)]

$$\psi = \sum_{\Gamma'} \psi(\Gamma')\,(\Gamma'|\quad) \tag{10}$$

The expansion coefficients are given explicitly by multiplying with $\bar{\psi}(\Gamma'')$:

$$(\Gamma''|\quad) = \bar{\psi}(\Gamma'')\,\psi. \tag{11}$$

For the complex imaginary equation to (10) we shall use the notation

$$\bar{\psi} = \sum_{\Gamma'} (\quad|\Gamma')\,\bar{\psi}(\Gamma'), \tag{12}$$

where

$$(\quad|\Gamma') = \overline{(\Gamma'|\quad)} = \bar{\psi}\,\psi(\Gamma'). \tag{13}$$

The set of eigenstates is a set of orthogonal unit vectors. The component $(\Gamma'|\quad)$ of ψ is given by the scalar product, in the proper sense, of ψ with the unit vector $\psi(\Gamma')$.

If the ψ we are expanding is an eigen-ψ of any set of observables $\boldsymbol{\Delta}$, say $\psi(\Delta')$, we use the following convenient notation, which is seen to be completely self-consistent:

$$\psi(\Delta') = \sum_{\Gamma'} \psi(\Gamma')\,(\Gamma'|\Delta'), \tag{10'}$$

$$(\Gamma'|\Delta') = \bar{\psi}(\Gamma')\,\psi(\Delta'), \tag{11'}$$

$$\bar{\psi}(\Delta') = \sum_{\Gamma'} (\Delta'|\Gamma')\,\bar{\psi}(\Gamma'), \tag{12'}$$

$$(\Delta'|\Gamma') = \overline{(\Gamma'|\Delta')} = \bar{\psi}(\Delta')\,\psi(\Gamma'). \tag{13'}$$

These equations express the (*unitary*†) transformation from one system of eigenstates to another; hence $(\Gamma'|\Delta')$ and $(\Delta'|\Gamma')$ are known as *transformation coefficients*.

* The normalization condition satisfied by the eigenstates of such a set of observables is given by

$$\bar{\psi}(\Gamma')\,\psi(\Gamma'') = \bar{\psi}(\gamma_1' \cdots \gamma_k')\,\psi(\gamma_1'' \cdots \gamma_k'') = \delta(\gamma_1', \gamma_1'') \ldots \delta(\gamma_k', \gamma_k'') = \delta(\Gamma', \Gamma''). \tag{9}$$

As a matter of convenience we shall write this product of δ's as $\delta(\gamma_1' \cdots \gamma_k';\ \gamma_1'' \cdots \gamma_k'')$ or simply $\delta(\Gamma', \Gamma'')$.

† A unitary transformation is a transformation from one set of normalized (unitary-) orthogonal states to another such set.

The observable $\boldsymbol{\alpha}$ is characterized in the $\boldsymbol{\Gamma}$ scheme by the set of numbers $(\Gamma''|\alpha|\Gamma')$ defined [cf. (5)] by*

$$\boldsymbol{\alpha}\boldsymbol{\psi}(\Gamma') = \sum_{\Gamma''} \boldsymbol{\psi}(\Gamma'')(\Gamma''|\alpha|\Gamma'), \tag{14}$$

and given explicitly by

$$(\Gamma''|\alpha|\Gamma') = \bar{\boldsymbol{\psi}}(\Gamma'')\boldsymbol{\alpha}\boldsymbol{\psi}(\Gamma'). \tag{15}$$

The square array formed from these numbers as Γ' and Γ'' range over the allowed values of the observables $\boldsymbol{\Gamma}$ is called the *matrix* of $\boldsymbol{\alpha}$, and the numbers themselves are called *matrix elements* or *components* (cf. § 7[2]).

In the $\boldsymbol{\Gamma}$ scheme the matrix of any one of the $\boldsymbol{\gamma}$'s, say $\boldsymbol{\gamma}_i$, is *diagonal* [i.e. $(\Gamma'|\gamma_i|\Gamma'')$ is zero unless $\Gamma' \equiv \Gamma''$]. Since $\boldsymbol{\gamma}_i\boldsymbol{\psi}(\Gamma') = \gamma_i'\boldsymbol{\psi}(\Gamma')$, we have from (15) that $(\Gamma''|\gamma_i|\Gamma') = \gamma_i'\delta_{\Gamma'\Gamma''}$. *The matrix of an observable* $\boldsymbol{\alpha}$ *which commutes with* $\boldsymbol{\gamma}_i$ *is diagonal with respect to* γ_i', i.e. $(\Gamma'|\alpha|\Gamma'')$ is zero unless $\gamma_i' = \gamma_i''$. This follows at once from (6).

From the definition 1[2]2 it follows that the elements of the matrix of $\boldsymbol{\alpha}^\dagger$ are given by

$$(\Gamma'|\alpha^\dagger|\Gamma'') = \overline{(\Gamma''|\alpha|\Gamma')}. \tag{16}$$

The matrix of $\boldsymbol{\alpha}^\dagger$ is the *Hermitian conjugate* of the matrix of $\boldsymbol{\alpha}$. The matrix of a real observable, for which

$$(\Gamma'|\alpha|\Gamma'') = \overline{(\Gamma''|\alpha|\Gamma')}, \tag{17}$$

is said to be *Hermitian*.

PROBLEMS

1. Show that the relation between the components of $\boldsymbol{\psi}$ in two different representations is

$$(\Delta'|\ \ ) = \sum_{\Gamma'} (\Delta'|\Gamma')(\Gamma'|\ \ ). \tag{18}$$

2. Show that the relation between the matrix components of $\boldsymbol{\alpha}$ in two different representations is

$$(\Delta'|\alpha|\Delta'') = \sum_{\Gamma'\Gamma''} (\Delta'|\Gamma')(\Gamma'|\alpha|\Gamma'')(\Gamma''|\Delta''). \tag{19}$$

3. Derive the following identity, which is characteristic of a unitary transformation

$$\sum_{\Delta'} (\Gamma'|\Delta')(\Delta'|\Gamma'') = \delta_{\Gamma'\Gamma''}. \tag{20}$$

4. Show that the sum of the diagonal elements (the trace) of the matrix of an observable is independent of representation, i.e. that

$$\sum_{\Gamma'} (\Gamma'|\alpha|\Gamma') = \sum_{\Delta'} (\Delta'|\alpha|\Delta'), \tag{21}$$

and that this sum is just equal to the sum of the characteristic values of the observable $\boldsymbol{\alpha}$, weighted by their degeneracies. (*The Diagonal-Sum Rule*)

5. In the $\boldsymbol{\Gamma}$-scheme we can represent the action of $\boldsymbol{\alpha}$ on any $\boldsymbol{\psi}$ [cf. (10)] in the following fashion:

$$\boldsymbol{\alpha}\boldsymbol{\psi} = \sum_{\Gamma''\Gamma'} \boldsymbol{\psi}(\Gamma'')(\Gamma''|\alpha|\Gamma')(\Gamma'|\ \ ). \tag{22}$$

---

* In this notation one obtains a complete analogy between the form of a quantum-mechanical observable and Gibbs' dyadic if one writes $\boldsymbol{\alpha}$ as the operator

$$\boldsymbol{\alpha} = \sum_{\Gamma'\Gamma''} \boldsymbol{\psi}(\Gamma')(\Gamma'|\alpha|\Gamma'')\bar{\boldsymbol{\psi}}(\Gamma'').$$

Here the fact that the $\bar{\boldsymbol{\psi}}$ and the $\boldsymbol{\psi}$ appear in the order opposite to that conventional for multiplication indicates that this is not their scalar product but that they form a unit dyad.

6. Show that if α anticommutes with γ_i, the matrix element

$$(\Gamma'|\alpha|\Gamma'')=0 \text{ unless } \gamma_i'=-\gamma_i''. \qquad (\alpha\gamma_i+\gamma_i\alpha=0) \tag{23}$$

7. Show that the elements of the matrix of the product $\alpha\beta$ of two observables are given by

$$(\Gamma'|\alpha\beta|\Gamma'')=\sum_{\Gamma'''}(\Gamma'|\alpha|\Gamma''')(\Gamma'''|\beta|\Gamma''). \tag{24}$$

8. Consider a set of m states $\psi(\Gamma_a^1), \psi(\Gamma_a^2), \ldots, \psi(\Gamma_a^m)$ and a set of m states $\psi(\Delta_a^1), \ldots, \psi(\Delta_a^m)$ derived from these by a unitary transformation:

$$\psi(\Delta_a^i)=\sum_{j=1}^{m}\psi(\Gamma_a^j)(\Gamma_a^j|\Delta_a^i).$$

Consider also a set of n states $\psi(\Gamma_b^1), \ldots, \psi(\Gamma_b^n)$ and a set $\psi(\Delta_b^1), \ldots, \psi(\Delta_b^n)$ derived from these by a unitary transformation. If α is any observable, show that

$$\sum_{i=1}^{m}\sum_{j=1}^{n}|(\Gamma_a^i|\alpha|\Gamma_b^j)|^2=\sum_{i=1}^{m}\sum_{j=1}^{n}|(\Delta_a^i|\alpha|\Delta_b^j)|^2. \tag{25}$$

(*The Principle of Spectroscopic Stability*)

9. For the states of the preceding problem, show that if

$$\sum_{k=1}^{n}(\Gamma_a^i|\alpha|\Gamma_b^k)(\Gamma_b^k|\alpha|\Gamma_a^j)=K\,\delta_{ij}, \tag{26a}$$

where K is a constant independent of i and j, then also

$$\sum_{k=1}^{n}(\Delta_a^i|\alpha|\Delta_b^k)(\Delta_b^k|\alpha|\Delta_a^j)=K\,\delta_{ij}. \tag{26b}$$

3. Continuous eigenvalues and the Schrödinger representation.

These formulas must be modified in the case of an observable $\boldsymbol{w}$ whose spectrum of eigenvalues is not wholly discrete, but is wholly or partially continuous. An eigen-ψ belonging to a value w' lying in the continuous range is to be normalized according to

$$\bar{\psi}(w')\psi(w'')=\delta(w'-w''), \tag{1}$$

where $\delta(x)$ is an improper even function of x defined by

$$\delta(x)=0 \text{ for } x\neq 0; \quad \int_{-\infty}^{+\infty}\delta(x)\,dx=1 \tag{2}$$

(Dirac, Chapter IV, which see for a complete discussion of the δ function).* The expansion of an arbitrary ψ in terms of the eigen-ψ's of $\boldsymbol{w}$ will now be of the form

$$\psi=\sum\psi(w')(w'|\quad)+\int\psi(w'')\,dw''\,(w''|\quad), \tag{3}$$

where w' ranges over the discrete spectrum, w'' over the continuous spectrum of $\boldsymbol{w}$. Here

$$(w'''|\quad)=\bar{\psi}(w''')\psi$$

for all w'''. If w''' is in the continuous spectrum this follows, on multiplication of (3) with $\bar{\psi}(w''')$, from (1) and the relation

$$\int f(x)\,\delta(x-a)\,dx=f(a) \tag{4}$$

* The δ function as used in connection with continuous spectra is to be regarded as a convenient abbreviated notation which permits simplification in the handling of more complicated mathematically rigorous expressions. This treatment of the continuous spectrum, although able to lead one into trouble, is certainly the most convenient; the δ function may be in general treated as an exact symbol when integrated out according to (4).

(Dirac, p. 73). In general, a similar replacement of integration over the continuous spectrum for summation over the discrete, and of $\delta(w'-w'')$ for $\delta_{\alpha'\alpha''}$ must be made in the preceding discussion in the case of a continuous spectrum.

The most important instance of an observable with a continuous spectrum is the coordinate $\boldsymbol{x}$, which we assume to have a completely continuous range of allowed values. For a single particle in three dimensions $\boldsymbol{x}$, $\boldsymbol{y}$, and $\boldsymbol{z}$ form a set of commuting observables (complete except for the peculiarity of electron spin which we shall later introduce as a fourth commuting observable $\boldsymbol{\sigma}$ having just two eigenvalues). For n particles we have the set $\boldsymbol{x}_1, \boldsymbol{y}_1, \boldsymbol{z}_1, \ldots, \boldsymbol{x}_n, \boldsymbol{y}_n, \boldsymbol{z}_n$. We shall denote this whole set of observables by $\boldsymbol{x}$ for short, writing

$$\bar{\psi}(x')\,\psi(x'') = \delta(x'-x''), \tag{5}$$

where $\delta(x'-x'') = \delta(x_1'-x_1'')\,\delta(y_1'-y_1'')\,\delta(z_1'-z_1'')\ldots\delta(z_n'-z_n'')$.

We shall usually denote an allowed value of the above set of observables merely* by x. The expansion of an arbitrary ψ in terms of the eigen-ψ's of $\boldsymbol{x}$ is then written

$$\psi = \int \psi(x)\,(x|\quad)\,dx, \quad \bar{\psi} = \int (\quad|x)\,\bar{\psi}(x)\,dx, \tag{6}$$ †

where $(x|\quad)$ and $(\quad|x)$ are complex conjugate functions of x.

This representation in terms of the eigen-ψ's of $\boldsymbol{x}$ (the Schrödinger representation) is the most useful of all representations, since from it is derived the Schrödinger equation which has proved to be a most powerful tool in the explicit solution of quantum-mechanical problems. It will be noted that the function $(x|\quad)$ completely determines ψ. This function $(x|\quad)$ is called the Schrödinger representative of ψ and will be written as ψ (the Schrödinger ψ function), or ψ_x if we wish to indicate the independent variable explicitly. Coordinates will throughout this work be relegated to the position of subscripts when they need be written, the space on the line being reserved for the more important quantum numbers. Thus the function $(x|\Gamma')$ which is the Schrödinger representative of $\psi(\Gamma')$ will be written as $\psi_x(\Gamma')$ and called the (Schrödinger) eigen-ψ belonging to Γ'. Since ψ determines ψ, we can equally well speak of a system as being in the state ψ or the state ψ, the state $\psi(\Gamma')$ or the state $\psi(\Gamma')$. In this ψ notation the expansion (6) becomes

$$\begin{aligned} \psi &= \int \psi(x)\,\psi\,dx, \quad \psi(\Gamma') = \int \psi(x)\,\psi(\Gamma')\,dx; \\ \bar{\psi} &= \int \bar{\psi}\,\bar{\psi}(x)\,dx, \quad \bar{\psi}(\Gamma') = \int \bar{\psi}(\Gamma')\,\bar{\psi}(x)\,dx. \end{aligned} \tag{7}$$

* There will be no confusion with the Schrödinger operator x because of the omission of the prime here, since no distinction need be made between the operator x and the allowed value x in the Schrödinger scheme.

† $dx = dx_1\,dy_1\,dz_1\ldots dz_n$.

If $\boldsymbol{\alpha}$ is an observable, $\boldsymbol{\alpha\psi}$ is a $\boldsymbol{\psi}$ and will have a Schrödinger representative which we shall write as $\alpha\psi$, i.e. we shall write

$$\boldsymbol{\alpha\psi} = \int \boldsymbol{\psi}(x)\, \alpha\psi\, dx. \tag{8}$$

This equation defines the operator α in the Schrödinger sense as that functional operator which acts on the function ψ which represents $\boldsymbol{\psi}$ and turns it into the function $\alpha\psi$ which represents $\boldsymbol{\alpha\psi}$.*

We shall now prove two theorems which show the great similarity in behaviour of $\boldsymbol{\psi}$ and ψ that justifies their being called by the same letter.

I. *Any linear relation which holds between* $\boldsymbol{\psi}$*'s holds also between their representatives.* This follows from the fact that the representative of the sum of two $\boldsymbol{\psi}$'s is the sum of the representatives of the individual $\boldsymbol{\psi}$'s; that the representative of $c\boldsymbol{\psi}$ is c times the representative of $\boldsymbol{\psi}$; and that if two $\boldsymbol{\psi}$'s are equal, their representatives are equal.

II. *The scalar product of two* $\boldsymbol{\psi}$*'s equals the integral of the product of their representatives* in the sense that if $\boldsymbol{\psi}(1)$ is represented by $\psi(1)$, $\boldsymbol{\psi}(2)$ by $\psi(2)$, then

$$\bar{\boldsymbol{\psi}}(1)\boldsymbol{\psi}(2) = \int \bar{\psi}(1)\,\psi(2)\, dx. \tag{10}$$

The proof of this follows from (4), (5), and (7),

$$\begin{aligned}\bar{\boldsymbol{\psi}}(1)\boldsymbol{\psi}(2) &= \int \bar{\psi}_x(1)\,\bar{\boldsymbol{\psi}}(x)\, dx \int \boldsymbol{\psi}(x')\,\psi_{x'}(2)\, dx' \\ &= \int\int \bar{\psi}_x(1)\,\delta(x-x')\,\psi_{x'}(2)\, dx\, dx' \\ &= \int \bar{\psi}_x(1)\,\psi_x(2)\, dx.\end{aligned}$$

Now the eigenstates $\boldsymbol{\psi}(\Gamma')$ are normalized ($2^2 9$) according to

$$\bar{\boldsymbol{\psi}}(\Gamma')\boldsymbol{\psi}(\Gamma'') = \delta_{\Gamma'\Gamma''};$$

hence the $\psi(\Gamma')$ have the normalization (theorem II)

$$\int \bar{\psi}(\Gamma')\,\psi(\Gamma'')\, dx = \delta_{\Gamma'\Gamma''}.\dagger \tag{11}$$

* This notation assumes that the representative of $f(\boldsymbol{x})\,\boldsymbol{\psi}$, where $f(\boldsymbol{x})$ is any algebraic function of the set of observables $\boldsymbol{x}$, is just the function $f(x)\,\psi$. This is easily seen to be the case:

$$f(\boldsymbol{x})\,\boldsymbol{\psi} = \int f(\boldsymbol{x})\,\boldsymbol{\psi}(x)\,\psi\, dx = \int \boldsymbol{\psi}(x)\, f(x)\,\psi\, dx. \tag{9}$$

† If w is an observable which has a continuous range of eigenvalues, its eigenstates must be normalized according to (1) instead of $1^2 1$ and integrals must be substituted for sums in the expansions in terms of $\psi(w')$. In this case we obtain for the normalizing condition

$$\int \bar{\psi}_x(w')\,\psi_x(w'')\, dx = \delta(w'-w'') \tag{12}$$

in place of (11). This normalization of $\psi_x(w')$ is accomplished if we choose the normalization factor in such a way that the set of *eigendifferentials*

$$\chi_x(w') = \frac{1}{\sqrt{\Delta}} \int_{w'}^{w'+\Delta} \psi_x(w')\, dw' \tag{13}$$

Any $\boldsymbol{\psi}$ may be expanded in the Γ' scheme according to

$$\boldsymbol{\psi}=\sum_{\Gamma'}\boldsymbol{\psi}(\Gamma')\,(\Gamma'|\quad).$$

Hence (theorem I)

$$\psi=\sum_{\Gamma'}\psi(\Gamma')\,(\Gamma'|\quad), \tag{16}$$

where*

$$(\Gamma'|\quad)=\bar{\boldsymbol{\psi}}(\Gamma')\,\boldsymbol{\psi}=\int\bar{\psi}(\Gamma')\,\psi\,dx. \tag{17}$$

The state $\boldsymbol{\alpha}\boldsymbol{\psi}(\Gamma')$ has the expansion

$$\boldsymbol{\alpha}\boldsymbol{\psi}(\Gamma')=\sum_{\Gamma''}\boldsymbol{\psi}(\Gamma'')\,(\Gamma''|\alpha|\Gamma').$$

Hence the $\alpha\,\psi(\Gamma')$ of (8) will have the expansion

$$\alpha\,\psi(\Gamma')=\sum_{\Gamma''}\psi(\Gamma'')\,(\Gamma''|\alpha|\Gamma'), \tag{18}$$

where from $2^2 15$ and theorem II

$$(\Gamma''|\alpha|\Gamma')=\bar{\boldsymbol{\psi}}(\Gamma'')\,\boldsymbol{\alpha}\boldsymbol{\psi}(\Gamma')=\int\bar{\psi}(\Gamma'')\,\alpha\,\psi(\Gamma')\,dx. \tag{19}$$

Finally the allowed values equation

$$\boldsymbol{\alpha}\boldsymbol{\psi}(\alpha')=\alpha'\,\boldsymbol{\psi}(\alpha')$$

becomes the functional equation

$$\alpha\,\psi(\alpha')=\alpha'\,\psi(\alpha'), \tag{20}$$

which is the general form of Schrödinger's equation for any observable. As

is normalized to unity in the sense of (11). The χ's will then be orthogonal if $|w'-w''|>\Delta$, and will for values of w' chosen at the small intervals Δ have all the properties of a set of eigenfunctions going with discrete eigenvalues. To see this let us evaluate the integral

$$\begin{aligned}\int\bar{\chi}_x(w')\,\chi_x(w'')\,dx&=\frac{1}{\Delta}\int\int_{w'}^{w'+\Delta}\bar{\psi}_x(w')\,dw'\int_{w''}^{w''+\Delta}\psi_x(w'')\,dw''dx\\&=\frac{1}{\Delta}\int_{w'}^{w'+\Delta}\int_{w''}^{w''+\Delta}\int\bar{\psi}_x(w')\,\psi_x(w'')\,dx\,dw'dw''\\&=\frac{1}{\Delta}\int_{w'}^{w'+\Delta}\int_{w''}^{w''+\Delta}\delta(w'-w'')\,dw'dw''.\end{aligned}$$

Now the integral over w'' has the value 1 for all values of w' which lie between w'' and $w''+\Delta$, and the value 0 for all other values of w'. The whole integral then becomes

$$\begin{aligned}&1 \text{ if } w'=w'',\\&0 \text{ if } |w'-w''|\geqslant\Delta,\\&\frac{\kappa-1}{\kappa} \text{ if } |w'-w''|=\frac{\Delta}{\kappa}\leqslant\Delta.\end{aligned} \tag{14}$$

The expansion of a ψ in terms of the eigen-ψ's of w may then be written as a sum over $\chi_x(w^r)$, where w^r takes on values spaced by the interval Δ (this is rigorously true asymptotically as $\Delta\rightarrow 0$):

$$\psi_x=\int\psi_x(w')\,(w'|\quad)\,dw'\doteq\sum_{w^r}\sqrt{\Delta}\,\chi_x(w^r)\,(w^r|\quad). \tag{15}$$

We could define, in exact analogy, a symbolic $\boldsymbol{\chi}(w')$ which would have properties similar to the above, and it is in some such way that a mathematically rigorous treatment of the continuous spectrum must be given.

* The set of functions $\psi(\Gamma')$ are to be considered as a set of unit vectors in ordinary function space (Hilbert space) in terms of which any arbitrary ψ may be expanded.

yet we have defined the operator α only symbolically; the discovery of the exact functional form of α for observables that are functions of the coordinates and momenta was made by Schrödinger and will be the subject of § 6^2.

4. The statistical interpretation.

If we make a measurement of the complete set of commuting observables $\boldsymbol{\gamma}_1, \boldsymbol{\gamma}_2, \ldots, \boldsymbol{\gamma}_k$ on a system in the state $\boldsymbol{\psi}$, what is the probability $P(\Gamma')$ of finding the values $\gamma_1', \gamma_2', \ldots, \gamma_k'$? This question may be answered in the following way (Dirac, p. 65).

Consider any function $f(\boldsymbol{\Gamma})$ of $\boldsymbol{\gamma}_1, \boldsymbol{\gamma}_2, \ldots, \boldsymbol{\gamma}_k$. The average value obtained for this function when making measurements on $\boldsymbol{\psi}$ is

$$\bar{\boldsymbol{\psi}} f(\boldsymbol{\Gamma}) \boldsymbol{\psi} = \{\sum_{\Gamma'} (\quad|\Gamma')\, \bar{\boldsymbol{\psi}}(\Gamma')\}\{\sum_{\Gamma''} f(\Gamma'')\, \boldsymbol{\psi}(\Gamma'')\,(\Gamma''|\quad)\}^{*}$$
$$= \sum_{\Gamma'} f(\Gamma')\, |(\Gamma'|\quad)|^2.$$

On the other hand, this average value must equal

$$\sum_{\Gamma'} f(\Gamma')\, P(\Gamma').$$

Since the last two expressions are equal for arbitrary functions f, we can equate coefficients of each $f(\Gamma')$ to obtain

$$P(\Gamma') = |(\Gamma'|\quad)|^2. \tag{1}$$

Hence the probability that the $\boldsymbol{\gamma}$'s have the values Γ' in the state $\boldsymbol{\psi}$ is $|(\Gamma'|\quad)|^2$. Similarly the probability that the $\boldsymbol{\gamma}$'s have the values Γ' in a state in which the $\boldsymbol{\delta}$'s are known to have the values Δ' [i.e. in the state $\boldsymbol{\psi}(\Delta')$] is

$$|(\Gamma'|\Delta')|^2.$$

If we now ask the probability $P(x)\,dx$ that the set $\boldsymbol{x}$ of observables have values in dx at x, we must modify the above computation to suit the case of a continuous observable. The mean value of $f(\boldsymbol{x})$ in the state $\boldsymbol{\psi}$ is given by

$$\bar{\boldsymbol{\psi}} f(\boldsymbol{x}) \boldsymbol{\psi} = \int \bar{\psi} f(x)\, \psi\, dx$$

from $3^2 9$ and $3^2 10$. On the other hand this mean value

$$= \int f(x)\, P(x)\, dx.$$

Since $f(x)$ is an arbitrary function of the variables x, we obtain from this the physical interpretation of Schrödinger's ψ function: that *$\bar{\psi}\psi\, dx$ is the probability that the system be in the coordinate volume element between x and $x + dx$.* This interpretation is seen to be consistent with the normalization of ψ as given by ($3^2 11$).

* Since $f(\boldsymbol{\Gamma})\, \boldsymbol{\psi}(\Gamma') = f(\Gamma')\, \boldsymbol{\psi}(\Gamma')$, an obvious generalization to the case of a set of commuting observables of Problem 1, § 1^2.

PROBLEMS

1. Show directly that $$\sum_{\Gamma'} P(\Gamma') = 1.$$

2. Show that if a real observable α has no negative eigenvalues, then
$$\bar{\psi}\alpha\psi \geqslant 0$$
for all ψ; and conversely that if $\bar{\psi}\alpha\psi \geqslant 0$ for all ψ, then α has no negative eigenvalues.

3. Show that $\alpha^2 + \beta^2$, where α and β are any two real observables, is real and has no negative eigenvalues.

5. The laws of quantum mechanics.

We have now laid out the mathematical pattern in terms of which the laws of atomic physics are formulated. The remainder of the theory consists of the recognition of the properties of the operators which are to represent various observables. To a particular mode of observation with certain apparatus is to be associated a certain operator. The laws of nature are not, as before, the functional relations between the numerical values given by certain experiments, but relations between the operators that stand for various modes of observation. The recognition of what operator is to be associated with each set of experimental operations has been carried out thus far partly by appeal to the correspondence principle (as with coordinate position and conjugate momentum) and partly by appeal to experiment (as with electron spin). Of course the correspondence principle itself is a broad generalization from experiment, so the known relations between operators for physical quantities all spring from experiment.

The cartesian coordinates $\boldsymbol{x}_1, \boldsymbol{y}_1, \boldsymbol{z}_1, \ldots, \boldsymbol{x}_n, \boldsymbol{y}_n, \boldsymbol{z}_n$, or for convenience $\boldsymbol{x}_1, \boldsymbol{x}_2, \ldots, \boldsymbol{x}_{3n}$, of a system of n particles, and the conjugate momenta $\boldsymbol{p}_1, \boldsymbol{p}_2, \ldots, \boldsymbol{p}_{3n}$ satisfy the following quantum-theoretic laws of nature (Dirac, p. 91):
$$\begin{aligned} [\boldsymbol{x}_i, \boldsymbol{x}_j] &= \boldsymbol{x}_i\boldsymbol{x}_j - \boldsymbol{x}_j\boldsymbol{x}_i = 0, \\ [\boldsymbol{p}_i, \boldsymbol{p}_j] &= \boldsymbol{p}_i\boldsymbol{p}_j - \boldsymbol{p}_j\boldsymbol{p}_i = 0, \qquad (1)^* \\ [\boldsymbol{x}_i, \boldsymbol{p}_j] &= \boldsymbol{x}_i\boldsymbol{p}_j - \boldsymbol{p}_j\boldsymbol{x}_i = i\hbar\delta_{ij}. \end{aligned}$$
These equations are consistent with $\boldsymbol{x}_i$ and $\boldsymbol{p}_i$ both having a continuous range of allowed values. The first equation has been presupposed in using $\boldsymbol{x}_1, \boldsymbol{x}_2, \ldots, \boldsymbol{x}_{3n}$ as commuting observables for a Schrödinger representation in § 3[2].

Analogous to the total energy of the system is a Hamiltonian function $\boldsymbol{H}$, *which is the same function of the* $\boldsymbol{p}$*'s and* $\boldsymbol{x}$*'s as on the classical theory for the*

* We shall use the notation $[\boldsymbol{A}, \boldsymbol{B}]$ for the simple commutator $\boldsymbol{AB} - \boldsymbol{BA}$. (Dirac uses $[\boldsymbol{A}, \boldsymbol{B}]$ for $(\boldsymbol{AB} - \boldsymbol{BA})/i\hbar$ in closer analogy to the classical Poisson bracket.) The commutator of a single observable with the product of two is given by formulas similar to that for a derivative of a product:
$$\begin{aligned} [\boldsymbol{AB}, \boldsymbol{C}] &= [\boldsymbol{A}, \boldsymbol{C}]\boldsymbol{B} + \boldsymbol{A}[\boldsymbol{B}, \boldsymbol{C}] \\ [\boldsymbol{A}, \boldsymbol{BC}] &= [\boldsymbol{A}, \boldsymbol{B}]\boldsymbol{C} + \boldsymbol{B}[\boldsymbol{A}, \boldsymbol{C}]. \end{aligned} \qquad (2)$$

*analogous dynamical system.** The importance of the Hamiltonian in the classical theory lay in the fact that through Hamilton's equations of motion it determined the time variation of the state. That continues to be its importance here, the dependence of the state $\boldsymbol{\psi}$ on the time being given (Dirac, p. 115) by

$$i\hbar\frac{\partial\boldsymbol{\psi}}{\partial t}=\boldsymbol{H}\boldsymbol{\psi}, \tag{3}$$

where $\partial\boldsymbol{\psi}/\partial t$ indicates a vector whose components are the time derivatives of the components of $\boldsymbol{\psi}$ in a fixed representation. In Schrödinger's notation this becomes

$$i\hbar\frac{\partial\psi}{\partial t}=H\psi. \tag{3'}$$

The time dependence of the eigenstates of energy is particularly simple. One has for the eigenstate belonging to H',

$$\boldsymbol{H}\boldsymbol{\psi}=H'\boldsymbol{\psi},$$

and therefore such a state merely changes its phase in time:

$$\begin{aligned}\boldsymbol{\psi}_t(H')&=\boldsymbol{\psi}_0(H')\,e^{-iH't/\hbar}\\ \bar{\boldsymbol{\psi}}_t(H')&=\bar{\boldsymbol{\psi}}_0(H')\,e^{+iH't/\hbar}.\end{aligned} \tag{4}$$

Corresponding equations obtain for the Schrödinger function $\psi_t(H')$. In an eigenstate of energy the average value of any observable $\boldsymbol{\alpha}$ is independent of the time.† This is proved as follows:

$$\bar{\boldsymbol{\psi}}_t(H')\,\boldsymbol{\alpha}\,\boldsymbol{\psi}_t(H')=\bar{\boldsymbol{\psi}}_0(H')\,e^{+iH't/\hbar}\,\boldsymbol{\alpha}\,\boldsymbol{\psi}_0(H')\,e^{-iH't/\hbar}=\bar{\boldsymbol{\psi}}_0(H')\,\boldsymbol{\alpha}\,\boldsymbol{\psi}_0(H').$$

Since the probability that the set of observables $\boldsymbol{\Gamma}$ have the values Γ' depends only on the average values of all functions of the $\boldsymbol{\gamma}$'s (§ 4[2]), this probability is independent of the time in such a state. Because of these properties an eigenstate of energy is called a *stationary state*.

PROBLEM

Show that the commutator

$$f(\boldsymbol{x})\,\boldsymbol{p}_i-\boldsymbol{p}_i f(\boldsymbol{x})=i\hbar\frac{\partial f}{\partial \boldsymbol{x}_i}, \tag{5}$$

where $f(\boldsymbol{x})$ is any function of $\boldsymbol{x}_1, \boldsymbol{x}_2, \ldots, \boldsymbol{x}_{3n}$ expressible as a power series.

6. Schrödinger's equation.

We shall now discuss the question of the functional form of an observable in the Schrödinger sense, as defined by 3[2]8:

$$\boldsymbol{\alpha}\boldsymbol{\psi}=\int\boldsymbol{\psi}(x)\,\alpha\psi\,dx,$$

* However, the exact order of factors and related questions, which are immaterial in the classical theory, are of great importance in the quantum theory, and for any particular problem must be determined, in the last analysis, by experiment.

† If $\boldsymbol{\alpha}$ does not involve the time explicitly.

where $\boldsymbol{\alpha}$ is supposed to be an algebraic function $F(\boldsymbol{x}, \boldsymbol{p})$ of $\boldsymbol{x}_1, \ldots, \boldsymbol{x}_{3n}, \boldsymbol{p}_1, \ldots, \boldsymbol{p}_{3n}$.

The Schrödinger operator for $f(\boldsymbol{x})$ is merely $f(x)$, as shown by 3²9.

Schrödinger's discovery* consisted of the observation that the proper operator for the observable $\boldsymbol{p}_i$ is $-i\hbar\partial/\partial x_i$. This is easily shown to satisfy the quantum conditions 5²1; for example

$$\boldsymbol{x}_j\boldsymbol{p}_i\boldsymbol{\psi} = -i\hbar\int \boldsymbol{\psi}(x)\, x_j \frac{\partial\psi}{\partial x_i}\, dx,$$

$$\boldsymbol{p}_i\boldsymbol{x}_j\boldsymbol{\psi} = -i\hbar\int \boldsymbol{\psi}(x)\, \frac{\partial}{\partial x_i}(x_j\psi)\, dx,$$

hence

$$(\boldsymbol{x}_j\boldsymbol{p}_i - \boldsymbol{p}_i\boldsymbol{x}_j)\,\boldsymbol{\psi} = i\hbar\int \boldsymbol{\psi}(x)\left\{\frac{\partial}{\partial x_i}(x_j\psi) - x_j\frac{\partial\psi}{\partial x_i}\right\}dx = i\hbar\int \boldsymbol{\psi}(x)\,\delta_{ij}\psi\, dx = i\hbar\delta_{ij}\boldsymbol{\psi}.$$

We obtain the operator for $\boldsymbol{p}_j\boldsymbol{p}_i$ by letting $\boldsymbol{p}_j$ operate on $\boldsymbol{p}_i$:

$$\boldsymbol{p}_j\boldsymbol{p}_i\boldsymbol{\psi} = -\hbar^2\int \boldsymbol{\psi}(x)\,\frac{\partial}{\partial x_j}\left(\frac{\partial\psi}{\partial x_i}\right)dx.$$

Hence the Schrödinger operator for $\boldsymbol{p}_j\boldsymbol{p}_i$ is $(-i\hbar\partial/\partial x_j)(-i\hbar\partial/\partial x_i)$, and in general the operator for a function $g(\boldsymbol{p})$ is $g(-i\hbar\partial/\partial x)$.

From an extension of these considerations it is seen that the Schrödinger operator for a general algebraic function

$$\boldsymbol{\alpha} = F(\boldsymbol{x}, \boldsymbol{p})$$

is

$$\alpha = F(x, p) = F(x, -i\hbar\partial/\partial x). \tag{1}$$

Hence the Schrödinger equation (3²20)

$$\alpha\,\psi(\alpha') = \alpha'\,\psi(\alpha')$$

becomes

$$F\left(x, -i\hbar\frac{\partial}{\partial x}\right)\psi(\alpha') = \alpha'\,\psi(\alpha'). \tag{2}$$

This is a standard type of characteristic value differential equation whose characteristic values and functions are to be determined by the use of auxiliary boundary conditions† on $\psi(\alpha')$. The original equation given by Schrödinger‡ was this equation for the particular case $\alpha = H$, which gives the energy levels and eigenstates of total energy.

7. Matrix mechanics.

We have seen that by the use of Schrödinger's representation in terms of the continuous set of allowed values of the variables $\boldsymbol{x}_i$, we can obtain a formulation of the theory entirely independent of symbolic $\boldsymbol{\psi}$'s and $\boldsymbol{\alpha}$'s. In

* SCHRÖDINGER, Ann. der Phys. **79**, 734 (1926).

† See p. 143 of DIRAC for a discussion of boundary conditions. This discussion is not complete; the governing condition must be that one find a complete *orthogonal* set of solutions of (2). See VON NEUMANN's book, page 53 *et seq.*

‡ SCHRÖDINGER, Ann. der Phys. **79**, 361 (1926).

the same manner the theory can be formulated in a way in which we deal only with the matrices representing states and observables in terms of the discrete set of allowed values of a complete set of observables $\mathbf{\Gamma}$. This corresponds to the original matrix mechanics of Heisenberg, Born, and Jordan.*

We represent ψ (or ψ) by the array of coefficients $(\Gamma'|\quad)$ in the expansion in terms of the $\psi(\Gamma')$ [or $\psi(\Gamma')$] written as a matrix of one *column*. That is, ψ and $\psi(\Delta')$ are represented respectively by

$$\left\| \begin{matrix} (\Gamma^1|\quad) \\ (\Gamma^2|\quad) \\ (\Gamma^3|\quad) \\ \cdots \\ \cdots \end{matrix} \right\| \quad \text{and} \quad \left\| \begin{matrix} (\Gamma^1|\Delta') \\ (\Gamma^2|\Delta') \\ (\Gamma^3|\Delta') \\ \cdots \\ \cdots \end{matrix} \right\|.$$

When a ψ is represented this way, a $\bar{\psi}$ must be represented by a matrix of one *row*, in order that $\bar{\psi}_r\psi_s$ may be an ordinary number (a matrix of one row and one column). For example, the product $\bar{\psi}(\Delta')\psi(\Delta'')$ becomes

$$\|(\Delta'|\Gamma^1)\,(\Delta'|\Gamma^2)\,(\Delta'|\Gamma^3)\ldots\|\cdot\left\| \begin{matrix} (\Gamma^1|\Delta'') \\ (\Gamma^2|\Delta'') \\ (\Gamma^3|\Delta'') \\ \cdots \\ \cdots \end{matrix} \right\| = \|\sum_{\Gamma'}(\Delta'|\Gamma')\,(\Gamma'|\Delta'')\| = \delta(\Delta',\Delta'') \tag{1}$$

by the usual rule of matrix multiplication† (cf. 2^220). In this form we see clearly that we cannot add a ψ and a $\bar{\psi}$, and that we must form their product in the order $\bar{\psi}\psi$.

The observable α is represented by its matrix as defined in § 2^2, i.e. by

$$\|(\Gamma'|\alpha|\Gamma'')\| = \left\| \begin{matrix} (\Gamma^1|\alpha|\Gamma^1) & (\Gamma^1|\alpha|\Gamma^2) & (\Gamma^1|\alpha|\Gamma^3) & \cdots \\ (\Gamma^2|\alpha|\Gamma^1) & (\Gamma^2|\alpha|\Gamma^2) & (\Gamma^2|\alpha|\Gamma^3) & \cdots \\ (\Gamma^3|\alpha|\Gamma^1) & (\Gamma^3|\alpha|\Gamma^2) & (\Gamma^3|\alpha|\Gamma^3) & \cdots \\ \cdots & \cdots & \cdots & \cdots \\ \cdots & \cdots & \cdots & \cdots \end{matrix} \right\|,$$

where by convention *the first index* Γ' *in* $(\Gamma'|\alpha|\Gamma'')$ *labels the row and the second index* Γ'' *the column*. If α is real, this matrix is Hermitian, which means (cf. 2^217) that the diagonal elements are real and that the corresponding elements on opposite sides of the diagonal are complex conjugates.

* HEISENBERG, Zeits. für Phys. **33**, 879 (1925);
BORN and JORDAN, *ibid.* **34**, 858 (1925);
BORN, HEISENBERG, and JORDAN, *ibid.* **35**, 557 (1925).

† The product of a matrix $\mathscr{A}$ with m rows and k columns and a matrix $\mathscr{B}$ with n columns and k rows is defined as the matrix of m rows and n columns obtained as follows:

$$\mathscr{A}\mathscr{B} = \left\| \begin{matrix} a_{11} & a_{12} & \cdots & a_{1k} \\ a_{21} & a_{22} & \cdots & a_{2k} \\ & \cdots & & \\ a_{m1} & a_{m2} & \cdots & a_{mk} \end{matrix} \right\| \cdot \left\| \begin{matrix} b_{11} & b_{12} & \cdots & b_{1n} \\ b_{21} & b_{22} & \cdots & b_{2n} \\ & \cdots & & \\ b_{k1} & b_{k2} & \cdots & b_{kn} \end{matrix} \right\| = \left\| \begin{matrix} \sum a_{1j}b_{j1} & \sum a_{1j}b_{j2} & \cdots & \sum a_{1j}b_{jn} \\ \sum a_{2j}b_{j1} & \sum a_{2j}b_{j2} & \cdots & \sum a_{2j}b_{jn} \\ & \cdots & & \\ \sum a_{mj}b_{j1} & \sum a_{mj}b_{j2} & \cdots & \sum a_{mj}b_{jn} \end{matrix} \right\|,$$

where all the Σ's are over $j = 1, 2, \ldots, k$. For a discussion of matrix algebra see, for example, BÔCHER, *Introduction to Higher Algebra*, Chapter VI.

It is now readily seen that all of the preceding equations remain true when one substitutes for ψ's and α's their respective matrices and uses the laws of matrix algebra. The allowed-values equation becomes

$$\begin{Vmatrix} (\Gamma^1|\alpha|\Gamma^1) & (\Gamma^1|\alpha|\Gamma^2) & \cdots \\ (\Gamma^2|\alpha|\Gamma^1) & (\Gamma^2|\alpha|\Gamma^2) & \cdots \\ \cdots & \cdots & \cdots \\ \cdots & \cdots & \cdots \end{Vmatrix} \cdot \begin{Vmatrix} (\Gamma^1|\alpha') \\ (\Gamma^2|\alpha') \\ \cdots \\ \cdots \end{Vmatrix} = \begin{Vmatrix} \alpha' & 0 & 0 & \cdots \\ 0 & \alpha' & 0 & \cdots \\ 0 & 0 & \alpha' & \cdots \\ \cdots & \cdots & \cdots & \cdots \end{Vmatrix} \cdot \begin{Vmatrix} (\Gamma^1|\alpha') \\ (\Gamma^2|\alpha') \\ \cdots \\ \cdots \end{Vmatrix}. \quad (2a)$$

Carrying out the matrix multiplication, we obtain the set of simultaneous linear equations (cf. 2^27 *et seq.*)

$$\sum_{\Gamma'} [(\Gamma''|\alpha|\Gamma') - \alpha'\delta_{\Gamma'\Gamma''}](\Gamma'|\alpha') = 0 \qquad \text{for all } \Gamma''. \quad (2b)$$

The condition that this set shall have a non-trivial solution is that the determinant of the coefficients vanish, i.e. that

$$\begin{vmatrix} (\Gamma^1|\alpha|\Gamma^1) - \alpha' & (\Gamma^1|\alpha|\Gamma^2) & (\Gamma^1|\alpha|\Gamma^3) & \cdots \\ (\Gamma^2|\alpha|\Gamma^1) & (\Gamma^2|\alpha|\Gamma^2) - \alpha' & (\Gamma^2|\alpha|\Gamma^3) & \cdots \\ \cdots & \cdots & \cdots & \cdots \\ \cdots & \cdots & \cdots & \cdots \end{vmatrix} = 0. \quad (3)$$

This determinant, though usually of infinite order, will often be such that the only non-vanishing elements lie in sub-squares along the diagonal, so that the infinite determinant will factor into an infinite product of finite determinants each of which can be treated by ordinary algebraic methods. An approximation to its solution in the case in which certain of the non-diagonal elements may be neglected will be considered in the next section. The roots of (3), which is called the *secular equation*, give the allowed values of α. For a $d_{\alpha'}$-fold root, the equations (2) furnish $d_{\alpha'}$ linearly independent sets of transformation coefficients $(\Gamma'|\alpha' r')$ for $r' = 1, \ldots, d_{\alpha'}$. The values of these coefficients must be chosen in accordance with the orthonormalization condition (1):

$$\sum_{\Gamma'} (\alpha' r'|\Gamma')(\Gamma'|\alpha' r'') = \delta_{r'r''}. \quad (4)$$

These transformation coefficients determine the eigen-ψ's of α, i.e. they determine those linear combinations of the eigen-ψ's of Γ in terms of which the matrix of α is diagonal; hence this calculation is known as the *diagonalization* of the matrix of α.

While the coefficients $(\Gamma'|\Delta')$ may be arranged into single rows or columns to represent the matrix of $\psi(\Delta')$, there is a sense in which the whole square array of $(\Gamma'|\Delta')$'s for all Γ' and Δ' may conveniently be considered as the matrix which transforms the eigen-ψ's of Γ into the eigen-ψ's of Δ. Let us write

$$\|\psi(\Delta^1)\ \psi(\Delta^2) \ldots\| = \|\psi(\Gamma^1)\ \psi(\Gamma^2) \ldots\| \cdot \begin{Vmatrix} (\Gamma^1|\Delta^1) & (\Gamma^1|\Delta^2) & \cdots \\ (\Gamma^2|\Delta^1) & (\Gamma^2|\Delta^2) & \cdots \\ (\Gamma^3|\Delta^1) & (\Gamma^3|\Delta^2) & \cdots \\ \cdots & \cdots & \cdots \end{Vmatrix}. \quad (5)$$

The matrix on the right, which we shall write as $\|(\Gamma'|\Delta')\|$, is a unitary matrix in the usual sense: it has the property that its Hermitian conjugate, which is easily seen to be the matrix $\|(\Delta'|\Gamma')\|$ is also its reciprocal, by (1); i.e.

$$\|(\Gamma'|\Delta')\|\cdot\|(\Delta'|\Gamma')\| = 1,$$

where 1 is the unit matrix.

Now if $\|(\Gamma'|\Delta')\|$ represents the transformation from the $\boldsymbol{\Gamma}$ scheme to the $\boldsymbol{\Delta}$ scheme and $\|(\Delta'|E')\|$ the transformation from the $\boldsymbol{\Delta}$ scheme to the $\mathbf{E}$ scheme, the transformation from the $\boldsymbol{\Gamma}$ scheme directly to the $\mathbf{E}$ scheme is given (cf. $2^2 18$) by

$$\|(\Gamma'|E')\| = \|(\Gamma'|\Delta')\|\cdot\|(\Delta'|E')\|. \tag{6}$$

The transformation of the matrix of $\boldsymbol{\alpha}$ from the $\boldsymbol{\Gamma}$ scheme to the $\boldsymbol{\Delta}$ scheme is accomplished (cf. $2^2 19$) by the multiplication of three matrices:

$$\|(\Delta' \mid \alpha \mid \Delta'')\| = \|(\Delta' \mid \Gamma')\|\cdot\|(\Gamma' \mid \alpha \mid \Gamma'')\|\cdot\|(\Gamma'' \mid \Delta'')\|. \qquad (7)^*$$

Suppose now that we are given the matrices of a set $\boldsymbol{\alpha}, \boldsymbol{\beta}, \boldsymbol{\zeta}, \ldots$ of commuting observables in the $\boldsymbol{\Gamma}$ scheme and we wish to find the transformation to the scheme in which these are simultaneously diagonal. We diagonalize $\boldsymbol{\alpha}$ by (3), (4) to obtain the unitary matrix $\|(\Gamma'|\alpha' r')\|$. With this matrix, by (7) we transform the matrix of $\boldsymbol{\beta}$ to the $(\alpha' r')$ scheme. Since $\boldsymbol{\beta}$ commutes with $\boldsymbol{\alpha}$ this matrix will be diagonal with respect to α'; hence we can find a set of states $\boldsymbol{\psi}(\alpha' \beta' s')$ as in § 2^2. Using (6) we find $\|(\Gamma'|\alpha' \beta' s')\|$ by the multiplication of $\|(\Gamma'|\alpha', r')\|$ and $\|(\alpha' r'|\alpha' \beta' s')\|$. This enables us to obtain the matrix of $\boldsymbol{\zeta}$ in the α', β', s' scheme, the diagonalization of which gives states $\boldsymbol{\psi}(\alpha' \beta' \zeta' t')$ etc. See § 4^8 for an example of this procedure carried through in detail.

8. Perturbation theory.

If we know the allowed values and states† of some observable α, we can develop formally a successive approximation method for finding those of an observable which differs but slightly from α, say the observable

$$\beta = \alpha + \epsilon V, \tag{1}$$

where ϵ is supposed to be a small number. Since ϵV may be considered as a perturbation on α, this approximation method is known as the 'perturbation theory.'

Let us denote the mutually distinct proper values of α by $\alpha^1, \alpha^2, \ldots$. If now there is a d_n-fold degeneracy in the value α^n, we shall need another index to distinguish the different states going with this same proper value. We may denote these states by $\psi(\alpha^{nl})$, where $l = 1, 2, \ldots, d_n$. Since the set of ψ's for a given n is determined only to within a unitary transformation, it will be desirable to choose the members of this set in a way significant to the

* This is the equation which is usually written in the form $T = ASA^{-1}$.

† In this discussion the ψ's and α's may be either in the symbolic or Schrödinger scheme.

problem in hand. Let us then choose $\psi(\alpha^{nl})$ in such a way that all matrix components of the type $(\alpha^{nl}|V|\alpha^{nl'})$ are zero unless $l=l'$. This is possible since the transformation which diagonalizes that part of the matrix of V which refers to the level α^n for any given set of $\psi(\alpha^n)$'s will transform this set of ψ's to one satisfying the above requirement.

Now it may be, when we have chosen the states in this way, that the components $(\alpha^{nl}|V|\alpha^{nl})$ with a given n are all different as l runs over the range $1, \ldots, d_n$. It is this special case, practically the only case of importance in applications, which we shall consider now. We shall return to the more general case later.

The observable β, which differs but little from α, will have d_n proper values which coincide with α^n in the limit $\epsilon=0$, and which differ but little from α^n for small ϵ. We shall call these d_n values $\beta^{n1}, \beta^{n2}, \ldots, \beta^{nd_n}$, and shall suppose that each of them may be expressed as a power series in ϵ of the form

$$\beta^{nl}=\beta_0^{nl}+\epsilon\beta_1^{nl}+\epsilon^2\beta_2^{nl}+\ldots, \tag{2}$$

where clearly

$$\beta_0^{nl}=\alpha^n \qquad \text{for all } l. \tag{3}$$

We shall suppose the eigenfunctions which belong to these values to be also expressible as a power series in ϵ, of the form

$$\psi(\beta^{nl})=\psi_0\{\beta^{nl}\}+\epsilon\,\psi_1\{\beta^{nl}\}+\epsilon^2\,\psi_2\{\beta^{nl}\}+\ldots, \tag{4}$$

where $\psi_0\{\beta^{nl}\}$ is a linear combination of the $\psi_0(\alpha^{np})$ for $p=1, \ldots, d_n$. We shall see that the advantage of our particular choice of $\psi(\alpha^{nl})$ lies in the fact that we may take $\psi_0\{\beta^{nl}\}\equiv\psi(\alpha^{nl})$.

Our problem is now to determine formally the coefficients of the various powers of ϵ in (2) and (4) in terms of the matrix components of V and the known properties of the observable α. This we may do by equating to zero the coefficient of each power of ϵ in the equation

$$\beta\,\psi(\beta^{nl})=\beta^{nl}\,\psi(\beta^{nl}) \tag{5a}$$

when this is expanded in the form

$$[\alpha+\epsilon V-\beta_0^{nl}-\epsilon\beta_1^{nl}-\epsilon^2\beta_2^{nl}-\ldots]\,[\psi_0\{\beta^{nl}\}+\epsilon\,\psi_1\{\beta^{nl}\}+\epsilon^2\,\psi_2\{\beta^{nl}\}+\ldots]=0. \tag{5b}$$

This gives us the set of equations:

$$\epsilon^0:\quad [\alpha-\beta_0^{nl}]\,\psi_0\{\beta^{nl}\}=0, \tag{6a}$$

$$\epsilon^1:\quad [\alpha-\beta_0^{nl}]\,\psi_1\{\beta^{nl}\}+[V-\beta_1^{nl}]\,\psi_0\{\beta^{nl}\}=0 \tag{6b}$$

$$\epsilon^2:\quad [\alpha-\beta_0^{nl}]\,\psi_2\{\beta^{nl}\}+[V-\beta_1^{nl}]\,\psi_1\{\beta^{nl}\}-\beta_2^{nl}\,\psi_0\{\beta^{nl}\}=0 \tag{6c}$$

$$\epsilon^3:\quad [\alpha-\beta_0^{nl}]\,\psi_3\{\beta^{nl}\}+[V-\beta_1^{nl}]\,\psi_2\{\beta^{nl}\}-\beta_2^{nl}\,\psi_1\{\beta^{nl}\}-\beta_3^{nl}\,\psi_0\{\beta^{nl}\}=0 \tag{6d}$$

....

If we express $\psi_\kappa\{\beta^{nl}\}$ in terms of the $\psi(\alpha^{n'l'})$, using the notation

$$\psi_\kappa\{\beta^{nl}\}=\sum_{n'l'}\psi(\alpha^{n'l'})\,\{\alpha^{n'l'}|\kappa^{nl}\}, \quad (\kappa=0, 1, 2, \ldots) \tag{7}$$

we can solve these equations in succession. We have introduced the curly braces because the $\psi_\kappa\{\beta^{nl}\}$ do not form a set of orthonormal states, nor are the $\{\alpha^{n'l'}|\kappa^{nl}\}$ coefficients in a unitary transformation. The condition that the $\psi(\beta^{nl})$ form an orthonormal set independent of ϵ gives us the following conditions on these transformation coefficients:

$$\epsilon^0:\quad \sum_{n''l''}\{0^{n'l'}|\alpha^{n''l''}\}\{\alpha^{n''l''}|0^{nl}\}=\delta(n'l',nl) \tag{8a}$$

$$\epsilon^1:\quad \sum_{n''l''}[\{0^{n'l'}|\alpha^{n''l''}\}\{\alpha^{n''l''}|1^{nl}\}+\{1^{n'l'}|\alpha^{n''l''}\}\{\alpha^{n''l''}|0^{nl}\}]=0 \tag{8b}$$

$$\epsilon^2:\quad \sum_{n''l''}[\{0^{n'l'}|\alpha^{n''l''}\}\{\alpha^{n''l''}|2^{nl}\}+\{1^{n'l'}|\alpha^{n''l''}\}\{\alpha^{n''l''}|1^{nl}\}+\{2^{n'l'}|\alpha^{n''l''}\}\{\alpha^{n''l''}|0^{nl}\}]=0$$

$$\ldots\ldots, \tag{8c}$$

where $$\{\kappa^{nl}|\alpha^{n'l'}\}=\overline{\{\alpha^{n'l'}|\kappa^{nl}\}}.$$

The first equation (6a) is just the allowed-values equation for the observable α, and gives us no new information. It tells us that if we take $\beta_0^{nl}=\alpha^n$, $\{\alpha^{n'l'}|0^{nl}\}$ will be zero unless $n'=n$, as we have already observed.

With these facts, the second equation (6b) becomes, when expanded in terms of the $\psi(\alpha^{n''l''})$,

$$\sum_{n''l''}\psi(\alpha^{n''l''})\,[(\alpha^{n''}-\beta_0^{nl})\{\alpha^{n''l''}|1^{nl}\}-\delta(n,n'')\,\beta_1^{nl}\{\alpha^{nl''}|0^{nl}\}$$
$$+\sum_{l'}(\alpha^{n''l''}|V|\alpha^{nl'})\{\alpha^{nl'}|0^{nl}\}]=0. \tag{9}$$

Equating the coefficients of $\psi(\alpha^{nl''})$ to zero in this expression gives

$$[(\alpha^{nl''}|V|\alpha^{nl''})-\beta_1^{nl}]\{\alpha^{nl''}|0^{nl}\}=0. \qquad (l''=1,\ldots,d_n) \tag{10}$$

This is for each value of n and l a set of d_n equations* for the $\{\alpha^{nl''}|0^{nl}\}$. They are satisfied by taking

$$\beta_1^{nl}=(\alpha^{nl}|V|\alpha^{nl});\quad \{\alpha^{nl''}|0^{nl}\}=\delta(l,l''). \tag{11}$$

This makes

$$\psi_0\{\beta^{nl}\}=\psi(\alpha^{nl}). \tag{12}$$

For a given n, if all the $(\alpha^{nl}|V|\alpha^{nl})$ are different, this is to within a phase the only non-vanishing set of solutions of (10) which satisfies the condition (8a).

Equating to zero the coefficient of $\psi(\alpha^{n''l''})$ for $n''\neq n$ in (9) now shows us that

$$\{\alpha^{n''l''}|1^{nl}\}=\frac{(\alpha^{n''l''}|V|\alpha^{nl})}{\alpha^n-\alpha^{n''}}. \qquad (n''\neq n) \tag{13}$$

The third equation (6c) becomes, when expanded in terms of the $\psi(\alpha^{n''l''})$,

$$\sum_{n''l''}\psi(\alpha^{n''l''})[(\alpha^{n''}-\beta_0^{nl})\{\alpha^{n''l''}|2^{nl}\}-\beta_1^{nl}\{\alpha^{n''l''}|1^{nl}\}-\beta_2^{nl}\{\alpha^{n''l''}|0^{nl}\}$$
$$+\sum_{n'l'}(\alpha^{n''l''}|V|\alpha^{n'l'})\{\alpha^{n'l'}|1^{nl}\}]=0. \tag{14}$$

* These equations have the above simple form because of the way we have chosen our $\psi(\alpha^{nl})$. If we had used an arbitrary system of $\psi(\alpha^n)$'s, we would have obtained at this point a set of equations determining the $\psi(\alpha^{nl})$ as we have chosen them. This would lead to a further complication of notation which it seems desirable to avoid.

Equating the coefficient of $\psi(\alpha^{nl})$ to zero gives

$$-\beta_1^{nl}\{\alpha^{nl}|1^{nl}\}-\beta_2^{nl}+\sum_{n'l'}(\alpha^{nl}|V|\alpha^{n'l'})\{\alpha^{n'l'}|1^{nl}\}=0,$$

from which

$$\beta_2^{nl}=\sum_{n'l'}^{n'\neq n}\frac{(\alpha^{nl}|V|\alpha^{n'l'})(\alpha^{n'l'}|V|\alpha^{nl})}{\alpha^n-\alpha^{n'}}. \tag{15}$$

Equating the coefficient of $\psi(\alpha^{nl''})$ $(l''\neq l)$ to zero gives

$$\{\alpha^{nl''}|1^{nl}\}=\sum_{n'l'}^{n'\neq n}\frac{(\alpha^{nl''}|V|\alpha^{n'l'})(\alpha^{n'l'}|V|\alpha^{nl})}{(\alpha^n-\alpha^{n'})(\beta_1^{nl}-\beta_1^{nl''})}. \qquad (l''\neq l) \tag{16}$$

We have now determined $\psi_1\{\beta^{nl}\}$ completely except for the coefficient $\{\alpha^{nl}|1^{nl}\}$, which is not determined by equations (6). Equations (5) and hence (6) determine a set of orthogonal states, but do not take care of normalization. This means that equations (8) will be automatically satisfied for $n'l'\neq nl$ (orthogonality condition), but these equations for $n'l'=nl$ must be expressly considered to secure normalization. (8a) has been satisfied by the choice (11). (8b) tells us that

$$\{1^{nl}|\alpha^{nl}\}+\{\alpha^{nl}|1^{nl}\}=0,$$

or that the real part of $\{\alpha^{nl}|1^{nl}\}$ is zero. We may choose the imaginary part of $\{\alpha^{nl}|1^{nl}\}$ arbitrarily. This is connected with the arbitrariness in phase of $\psi(\beta^{nl})$. $e^{if(\epsilon)}\psi(\beta^{nl})$ is a solution of the same phase for $\epsilon=0$ as $\psi(\beta^{nl})$ if $f(0)=0$. But if $\psi(\beta^{nl})$ is given by (4), we have, if $f(\epsilon)=k_1\epsilon+k_2\epsilon^2+\dots$,

$$\begin{aligned}e^{i(k_1\epsilon+k_2\epsilon^2+\dots)}\psi(\beta^{nl})&=\psi_0+\epsilon(\psi_1+ik_1\psi_0)+\epsilon^2(\psi_2+ik_1\psi_1+[ik_2-k_1^2/2!]\psi_0)+\dots\\&=\psi_0+\epsilon\psi_1'+\epsilon^2\psi_2'+\dots.\end{aligned}$$

This shows that ψ_1 contains an arbitrary imaginary multiple of ψ_0 [i.e. of $\psi(\alpha^{nl})$] which occasions the arbitrariness of $\{\alpha^{nl}|1^{nl}\}$. The most convenient procedure is to set

$$\{\alpha^{nl}|1^{nl}\}=0, \tag{17}$$

since we may obtain any solution from such a particular one by multiplication with a phase factor. With this choice we see that the arbitrariness in ψ_2 is reduced to an imaginary multiple of ψ_0, which we shall also set equal to zero, and so on for the higher approximations.

Finally, setting the coefficient of $\psi(\alpha^{n''l''})$ equal to zero for $n''\neq n$ in (14) gives us the value of $\{\alpha^{n''l''}|2^{nl}\}$. This procedure may be continued as long as one pleases, and it is clear that recursion formulas may be obtained expressing each approximation in terms of the previous one. The value of $\{\alpha^{n''l''}|2^{nl}\}$ is obtained as above. The coefficient of $\psi(\alpha^{nl})$ in the expansion of the fourth equation (6d) gives β_3^{nl}. The coefficient of $\psi(\alpha^{nl''})$ gives $\{\alpha^{nl''}|2^{nl}\}$. The normalization relation (8c) for $n'l'=nl$ determines the real part of $\{\alpha^{nl}|2^{nl}\}$, the imaginary part being arbitrarily set zero, and so on. See the next section for a collection of formulas.

If the $(\alpha^{nl}|V|\alpha^{nl})$ for a given n are not all distinct, (13) is no longer the correct solution. Such a problem should be handled by writing three indices on the states $\psi(\alpha^{nlj})$, where $(\alpha^{nlj}|V|\alpha^{nlj})$ is independent of j. If we are to choose the set of states for a given n and l with significance, we do it so that the expression

$$\sum_{n'l'j'}^{n'\neq n} \frac{(\alpha^{nlj}|V|\alpha^{n'l'j'})(\alpha^{n'l'j'}|V|\alpha^{nlj''})}{\alpha^n - \alpha^{n'}}, \tag{18}$$

which is written down from (15), vanishes unless $j=j''$. In this case we shall find that we have $\psi_0\{\beta^{nlj}\}=\psi(\alpha^{nlj})$, and that β_2^{nlj} will be given by (18) for $j''=j$. The formulas for the coefficients in $\psi(\beta^{nlj})$ will now be much more complicated than before. If (18) for $j=j''$ still has a degeneracy, we should have to use four indices, and so on.

9. Résumé of the perturbation theory.

We know the characteristic functions and values for an observable α and are interested in those of an observable

$$\beta=\alpha+\epsilon V. \tag{1}$$

Denote the distinct eigenvalues of α by α^1, α^2, ... and let α^n be d_n-fold degenerate. Choose the d_n eigenstates $\psi(\alpha^{nl})$ $(l=1, 2, ..., d_n)$ in such a way that $(\alpha^{nl}|V|\alpha^{nl'})$ vanishes unless $l=l'$.* The following applies only to n's such that the d_n values of $(\alpha^{nl}|V|\alpha^{nl})$ are all different. Write

$$\beta^{nl}=\beta_0^{nl}+\epsilon\beta_1^{nl}+\epsilon^2\beta_2^{nl}+\ldots; \tag{2}$$

$$\begin{aligned}\psi(\beta^{nl}) &= \psi_0\{\beta^{nl}\}+\epsilon\,\psi_1\{\beta^{nl}\}+\epsilon^2\psi_2\{\beta^{nl}\}+\ldots\\ &= \sum_{n'l'}\psi(\alpha^{n'l'})\,[\{\alpha^{n'l'}|0^{nl}\}+\epsilon\,\{\alpha^{n'l'}|1^{nl}\}+\epsilon^2\{\alpha^{n'l'}|2^{nl}\}+\ldots].\end{aligned} \tag{3}$$

Then

$$\begin{aligned}
\beta_0^{nl} &= \alpha^n\\
\beta_1^{nl} &= (\alpha^{nl}|V|\alpha^{nl})\\
\beta_2^{nl} &= \sum_{n'l'}^{n'\neq n}\frac{(\alpha^{nl}|V|\alpha^{n'l'})\,(\alpha^{n'l'}|V|\alpha^{nl})}{\alpha^n-\alpha^{n'}}\\
\beta_3^{nl} &= \sum_{n'l'}^{n'\neq n}(\alpha^{nl}|V|\alpha^{n'l'})\,\{\alpha^{n'l'}|2^{nl}\}\\
&\ldots\ldots\ldots\ldots\ldots\ldots\\
\beta_\kappa^{nl} &= \sum_{n'l'}^{n'\neq n}(\alpha^{nl}|V|\alpha^{n'l'})\,\{\alpha^{n'l'}|(\kappa-1)^{nl}\}-\sum_{\lambda=2}^{\kappa-2}\beta_\lambda^{nl}\,\{\alpha^{nl}|(\kappa-\lambda)^{nl}\};
\end{aligned} \tag{4}$$

* One can determine this set of states by diagonalizing that part of the matrix of V which refers to any set of d_n orthogonal states belonging to α^n.

$$
\begin{aligned}
&\{\alpha^{n'l'}|0^{nl}\}=\delta(n'l', nl) \\
&\{\alpha^{n'l'}|1^{nl}\}=\frac{(\alpha^{n'l'}|V|\alpha^{nl})}{\alpha^{n}-\alpha^{n'}} && (n' \neq n) \\
&\{\alpha^{nl'}|1^{nl}\}=\sum_{n''l''}^{n''\neq n}\frac{(\alpha^{nl'}|V|\alpha^{n''l''})(\alpha^{n''l''}|V|\alpha^{nl})}{(\alpha^{n}-\alpha^{n''})(\beta_1^{nl}-\beta_1^{nl'})} && (l' \neq l) \\
&\{\alpha^{nl}|1^{nl}\}=0 \\
&\{\alpha^{n'l'}|2^{nl}\}=-\frac{(\alpha^{n'l'}|V|\alpha^{nl})(\alpha^{nl}|V|\alpha^{nl})}{(\alpha^{n'}-\alpha^{n})^2}-\sum_{n''l''}\frac{(\alpha^{n'l'}|V|\alpha^{n''l''})\{\alpha^{n''l''}|1^{nl}\}}{\alpha^{n'}-\alpha^{n}} && (n' \neq n) \\
&\{\alpha^{nl'}|2^{nl}\}=\frac{\beta_2^{nl}\{\alpha^{nl'}|1^{nl}\}}{\beta_1^{nl'}-\beta_1^{nl}}+\sum_{n''l''}^{n''\neq n}\frac{(\alpha^{nl'}|V|\alpha^{n''l''})\{\alpha^{n''l''}|2^{nl}\}}{\beta_1^{nl}-\beta_1^{nl'}} && (l' \neq l) \qquad (5) \\
&\{\alpha^{nl}|2^{nl}\}=-\tfrac{1}{2}\sum_{n'l'}|\{\alpha^{n'l'}|1^{nl}\}|^2 \\
&\cdots\cdots\cdots\cdots\cdots\cdots \\
&\{\alpha^{n'l'}|\kappa^{nl}\}=\sum_{\lambda=1}^{\kappa-1}\frac{\beta_\lambda^{nl}\{\alpha^{n'l'}|(\kappa-\lambda)^{nl}\}}{\alpha^{n'}-\alpha^{n}}+\sum_{n''l''}\frac{(\alpha^{n'l'}|V|\alpha^{n''l''})\{\alpha^{n''l''}|(\kappa-1)^{nl}\}}{\alpha^{n}-\alpha^{n'}} && (n' \neq n) \\
&\{\alpha^{nl'}|\kappa^{nl}\}=\sum_{\lambda=1}^{\kappa-1}\frac{\beta_{\lambda+1}^{nl}\{\alpha^{nl'}|(\kappa-\lambda)^{nl}\}}{\beta_1^{nl'}-\beta_1^{nl}}+\sum_{n''l''}^{n''\neq n}\frac{(\alpha^{nl'}|V|\alpha^{n''l''})\{\alpha^{n''l''}|\kappa^{nl}\}}{\beta_1^{nl}-\beta_1^{nl'}} && (l' \neq l) \\
&\{\alpha^{nl}|\kappa^{nl}\}=-\tfrac{1}{2}\sum_{\lambda=1}^{\kappa-1}\sum_{n'l'}\overline{\{\alpha^{n'l'}|\lambda^{nl}\}}\{\alpha^{n'l'}|(\kappa-\lambda)^{nl}\}.
\end{aligned}
$$

10. Remarks on the perturbation theory.

It may not at first sight seem clear why we have set down these complicated formulas for the perturbation of a degenerate system when we might by a simple device have reduced the degenerate system to one which is non-degenerate. If we add to the matrix of α the diagonal elements $(\alpha^{nl}|\epsilon V|\alpha^{nl})$ of the perturbation matrix, we have a diagonal matrix of which we know the characteristic values and functions and which is non-degenerate by hypothesis. The rest of the matrix of ϵV may now be considered as the perturbation; and the much simplified formulas to which those of § 9[2] reduce when the system is non-degenerate are applicable. But our perturbation theory is expected to converge rapidly only for small perturbations—a small perturbation being one which causes shifts in the eigenvalues small compared to the original distance between eigenvalues.* Now if we take as the unperturbed levels those for which the degeneracy is removed by the diagonal

* This statement is based on the following rough considerations. Let us represent by V a number of the order of magnitude of an element of the perturbation matrix V, and let us suppose that the difference between two diagonal elements of the type $(\alpha^{nl}|V|\alpha^{nl})$ and $(\alpha^{nl'}|V|\alpha^{nl'})$ (i.e. a difference $\beta_1^{nl'}-\beta_1^{nl}$ such as occurs in the denominators of 9[2]4 and 9[2]5) is also of order V. Let $\alpha_{nn'}$ be a number of the order of magnitude of the difference $\alpha^{n}-\alpha^{n'}$ between two adjacent degenerate levels of α. Then the κth term in 9[2]2 is of the order of magnitude

$$\epsilon^\kappa\beta_\kappa^{nl}\sim\left(\frac{\epsilon V}{\alpha_{nn'}}\right)^{\kappa-1}\epsilon V, \tag{1}$$

while the κth term in $(\alpha^{n'l'}|\beta^{nl})$ is of the order of magnitude

$$\epsilon^\kappa\{\alpha^{n'l'}|\kappa^{nl}\}\sim\left(\frac{\epsilon V}{\alpha_{nn'}}\right)^{\kappa}. \tag{2}$$

Since the displacements caused by the perturbation are of the order of ϵV and the original distance

elements of the perturbation, we have no reason to suppose that the non-diagonal elements will have an effect small compared to the effect of the diagonal elements, i.e. small compared to the distance between unperturbed levels. Hence if we wish rapid convergence we shall probably need to consider the set of states going with an unperturbed level as a unit and the perturbation as a whole, as we have done in the last sections.* It should be remarked that if we do use the device suggested above we obtain for the coefficient of ϵ^2 in β^{nl}

$$\sum_{n'l'}^{n'\neq n} \frac{(\alpha^{nl}|V|\alpha^{n'l'})(\alpha^{n'l'}|V|\alpha^{nl})}{\alpha^n+(\alpha^{nl}|\epsilon V|\alpha^{nl})-\alpha^{n'}-(\alpha^{n'l'}|\epsilon V|\alpha^{n'l'})}, \tag{4}$$

which contains in the denominator the distance between the first-order eigenvalues instead of that between the zero-order eigenvalues. For calculations just to the second order this formula will at times be convenient, and is probably as accurate as the former.

To obtain a slight acquaintance with the perturbation theory, let us consider in detail a simple finite case for which we can obtain an exact solution. Let us take the matrices of α and V to be

α	α^{12}	α^{11}	α^0
α^{12}	10	0	0
α^{11}	0	10	0
α^0	0	0	0

V	α^{12}	α^{11}	α^0
α^{12}	5	0	10
α^{11}	0	0	30
α^0	10	30	0

(5)

[where we have already chosen the states $\psi(\alpha^{11})$ and $\psi(\alpha^{12})$ in such a way that $(\alpha^{11}|V|\alpha^{12})=0$] and consider the characteristic values and states of the observable $\beta=\alpha+\epsilon V$. From the formulas $9^2 4$ and $9^2 5$ we find that to terms in ϵ^4

$$\begin{aligned} \beta^0 &= \quad\quad -100\epsilon^2+5\epsilon^3+997{\cdot}5\epsilon^4, \\ \beta^{11} &= 10 \quad + 90\epsilon^2-180\epsilon^3-3690\epsilon^4, \\ \beta^{12} &= 10+5\epsilon+10\epsilon^2+175\epsilon^3+2692{\cdot}5\epsilon^4; \end{aligned} \tag{6}$$

---

between levels is of the order $\alpha_{nn'}$, we see that we cannot expect the theory to converge rapidly unless these displacements are small compared to this distance.

If the differences $\beta_1^{nl'}-\beta_1^{nl}$ are of smaller order of magnitude than V, (1) and (2) are too small. In particular the magnitude of the elements $\{\alpha^{nl'}|\kappa^{nl}\}$ connecting two states which originally belonged to the same unperturbed level is strongly affected. The above criterion is good provided the differences in first order shifts are of the same size as the shifts themselves. If this is not true, one might take as a guide the requirement that an expression of the type

$$\frac{\epsilon V}{\alpha_{nn'}}\,\frac{V}{\beta_1^{nl'}-\beta_1^{nl}} \tag{3}$$

should be small compared to unity for rapid convergence.

The question of the convergence of the perturbation theory has been considered by WILSON, Proc. Roy. Soc. **A122**, 589; **A124**, 176 (1929). He shows that for finite matrices the theory converges for all values of the parameter ϵ. For infinite matrices no such definite results were obtained. [This means that (1) and (2) can be correct only for small κ.]

* In fact, if two or more unperturbed levels lie very close, it is advisable to consider them as one degenerate level in the perturbation scheme, adding the differences to the diagonal elements of the perturbation.

while to terms in ϵ^3

$$
\begin{aligned}
\psi(\beta^0) &= \psi(\alpha^0)\{1 - 5\epsilon^2 + 0\cdot5\epsilon^3\} \\
&\quad + \psi(\alpha^{11})\{-3\epsilon + 45\epsilon^3\} \\
&\quad + \psi(\alpha^{12})\{-\epsilon + 0\cdot5\epsilon^2 + 14\cdot75\epsilon^3\}, \\
\psi(\beta^{11}) &= \psi(\alpha^0)\{3\epsilon - 6\epsilon^2 - 190\cdot5\epsilon^3\} \\
&\quad + \psi(\alpha^{11})\{1 - 22\cdot5\epsilon^2 - 558\epsilon^3\} \\
&\quad + \psi(\alpha^{12})\{-6\epsilon - 96\epsilon^2 - 1131\epsilon^3\}, \\
\psi(\beta^{12}) &= \psi(\alpha^0)\{\epsilon + 17\cdot5\epsilon^2 + 250\cdot75\epsilon^3\} \\
&\quad + \psi(\alpha^{11})\{6\epsilon + 93\epsilon^2 + 1108\cdot5\epsilon^3\} \\
&\quad + \psi(\alpha^{12})\{1 - 18\cdot5\epsilon^2 - 575\cdot5\epsilon^3\}.
\end{aligned} \tag{7}
$$

The eigenvalues (6) are plotted in Fig. 1^2 for values of ϵ from $-0\cdot1$ to $+0\cdot1$ to terms of the 0, 1, 2, 3, and 4th order in ϵ. The curves marked ∞ give the values obtained by a direct diagonalization of the matrix $\alpha+\epsilon V$.* Fig. 2^2 shows the coefficients in the corresponding states to 1, 2, and 3d order and as given by the exact calculation.

The situation is as follows. In the zeroth approximation there is a doubly degenerate level at 10, and one state at 0. To the first order the level at 10 splits into 2 states of eigenvalues 10 and $10+5\epsilon$. In the second order these states are 'repelled' by that at 0 with strengths $90\epsilon^2$ and $10\epsilon^2$, which soon results in a crossing of the two states for $\epsilon>0$. A crossing of the true levels will be shown in § 11^2 to be impossible. The third and fourth approximations include the third and fourth terms in the power-series expansion of the true eigenvalues (marked ∞). The second-order perturbation theory for this case has little accuracy for a displacement of over one per cent. The crossing of the levels in this approximation contributes somewhat to this inaccuracy, the values for $\epsilon<0$ being better than those for $\epsilon>0$. The fact that the second approximation for $\epsilon<0$ is better than the third or fourth must be regarded as accidental. The eigenfunctions $\psi(\beta^{11})$ and $\psi(\beta^{12})$, which for $\epsilon<0$ are over 92 per cent. $\psi(\alpha^{11})$ and $\psi(\alpha^{12})$, rapidly change character for $\epsilon>0$, $\psi(\beta^{11})$ becoming, at $\epsilon=0\cdot1$, 70 per cent. $\psi(\alpha^{12})$, and *vice versa*. The approximation to these functions is seen to be no better for $\epsilon<0$ than for $\epsilon>0$.

11. Perturbation caused by a single state.†

It is seen that in general the second-order perturbation theory may be interpreted as a repulsion between each pair of levels resulting from the first-order theory, of magnitude equal to the absolute square of the matrix element joining these two levels divided by the unperturbed distance between the levels, or in the form 10^24, divided by the first-order distance between the levels. It is of interest to see what one can say rigorously about this 'repulsive' effect of perturbations.

* Where two curves are not separated in Figs. 1^2 and 2^2, their ordinates differ at most by $0\cdot01$ unit.

† Brillouin, Jour. de Physique, **3**, 379 (1932); MacDonald, Phys. Rev. **43**, 830 (1933).

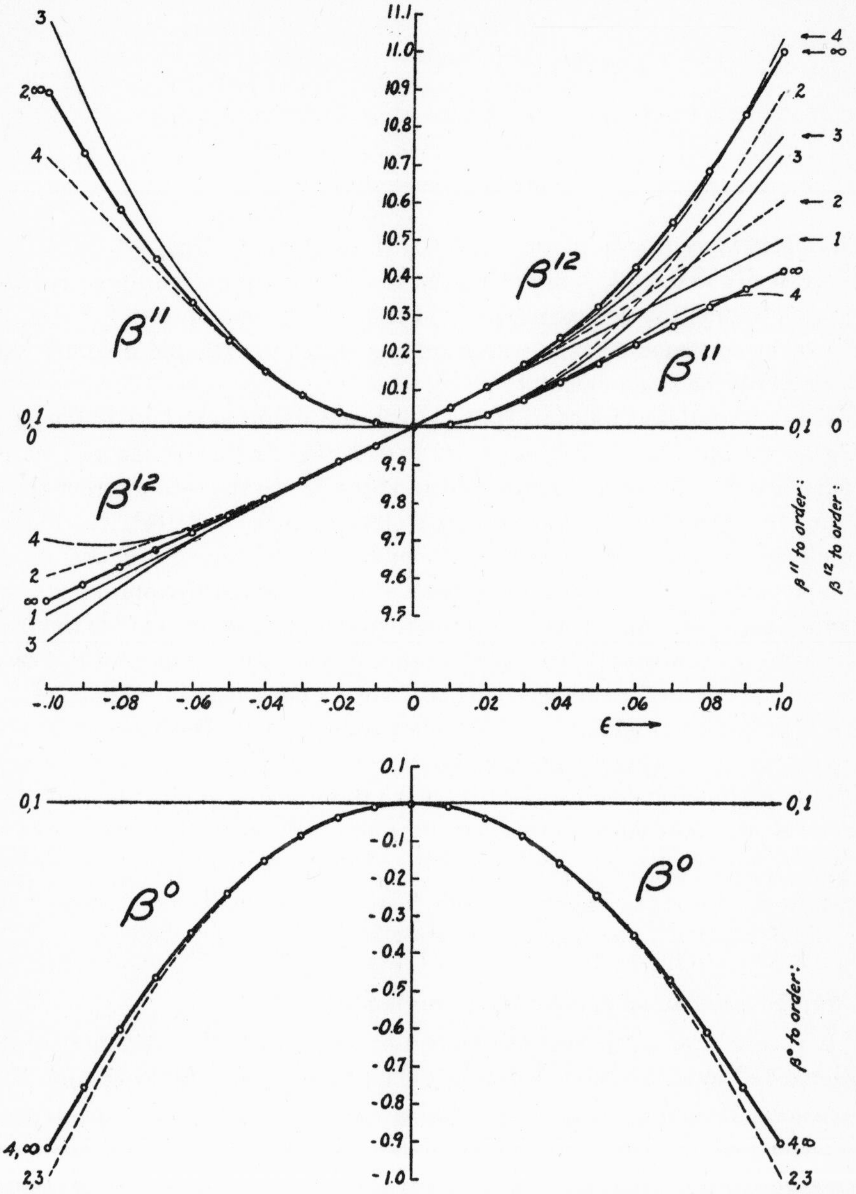

Fig. 1². Successive approximations to the eigenvalues of $\alpha + \epsilon V$ of (5) as functions of ϵ.

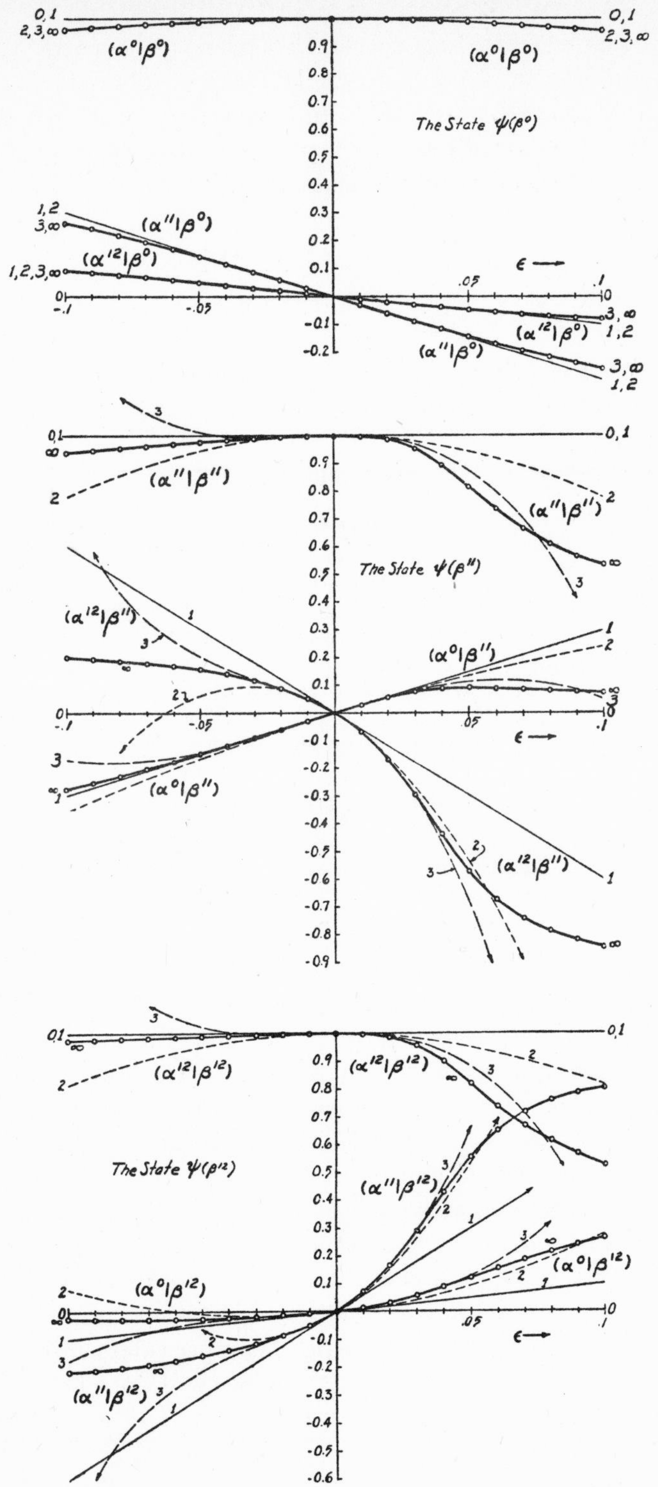

Fig. 2^2. Transformation coefficients in several approximations for the eigenstates of $\alpha + \epsilon V$ of (5) as functions of ϵ.

Suppose that we have diagonalized that portion of a matrix referring to a certain set $\psi(A^1)$, $\psi(A^2)$, ..., $\psi(A^n)$ of states, obtaining eigenvalues $A^1, A^2, \ldots, A^n$, and that we wish to know the effect on this set of levels of the inclusion of another state $\psi(X)$ in our diagonalization. That is, what can we say in general about the eigenvalues of the matrix

$$\left\| \begin{matrix} A^1 & 0 & \ldots & 0 & \alpha_1 \\ 0 & A^2 & \ldots & 0 & \alpha_2 \\ \ldots & \ldots & \ldots & \ldots & \ldots \\ 0 & 0 & \ldots & A^n & \alpha_n \\ \bar{\alpha}_1 & \bar{\alpha}_2 & \ldots & \bar{\alpha}_n & X \end{matrix} \right\|, \tag{1}$$

where we let $(A^i|\quad|X)=\alpha_i$; $(X|\quad|X)=X$? We shall suppose all the α's to be different from zero, since a state with $\alpha=0$ may be removed from the matrix entirely.

If the state going with the eigenvalue λ of this matrix be

$$\psi(\lambda)=\sum_{i=1}^{n}\psi(A^i)\,(A^i|\lambda)+\psi(X)\,(X|\lambda),$$

we have, from (1), the equations (cf. $7^2 2$),

$$(A^j-\lambda)\,(A^j|\lambda)+\alpha_j\,(X|\lambda)=0, \qquad (j=1,2,\ldots,n) \tag{2a}$$

$$\sum_{i=1}^{n}\bar{\alpha}_i\,(A^i|\lambda)+(X-\lambda)\,(X|\lambda)=0. \tag{2b}$$

Substituting the values of $(A^i|\lambda)$ from (2a) in (2b) gives the equation in λ,

$$\sum_{i=1}^{n}\frac{|\alpha_i|^2}{\lambda-A^i}=\lambda-X, \tag{3}$$

which is equivalent to the secular equation obtained by setting the determinant of (2) equal to zero. The roots of this equation give the allowed values of λ.

The left side of (3), which is seen to become infinite at $\lambda=A^i$, is such a function of λ as shown by the heavy curves of Fig. 3^2 for the special case noted. The individual summands which are added to give this curve are indicated by the light rectangular hyperbolas. With these same values of the A's and α's, the value of X may be chosen at will. With the particular choice $X=X^{(1)}=13$, the right side of (3) is represented, as a function of λ, by the leftmost straight line of unit slope. The abscissas of the encircled points of intersection of this line with the heavy curve give the five allowed values of λ. The allowed values for two other positions of the 'perturbing level' X are also shown: one for a position $X=X^{(2)}=23$ within the group of levels A^i, and one for a position $X=X^{(3)}=36$ above the whole group of A's.

It is clear from this picture that the $(n+1)$ proper values of the matrix (1) lie, if we assume that $A^1<A^2<\ldots<A^n$, one below A^1, one between A^1 and A^2, one between A^2 and A^3, ..., one between A^{n-1} and A^n, and one above A^n.

This is independent of the value of X and of the α's, provided none of the α's vanish.*

It is interesting to speak of this in the following way. The perturbation between a level X and a set of levels $A^1, \ldots, A^n$ which lie entirely above it pushes each level of the set up, *but not above the original position of the next higher level*, while X itself is correspondingly pushed down. The perturbation between *any* level X and the set $A^1, \ldots, A^n$ pushes up each level of the set

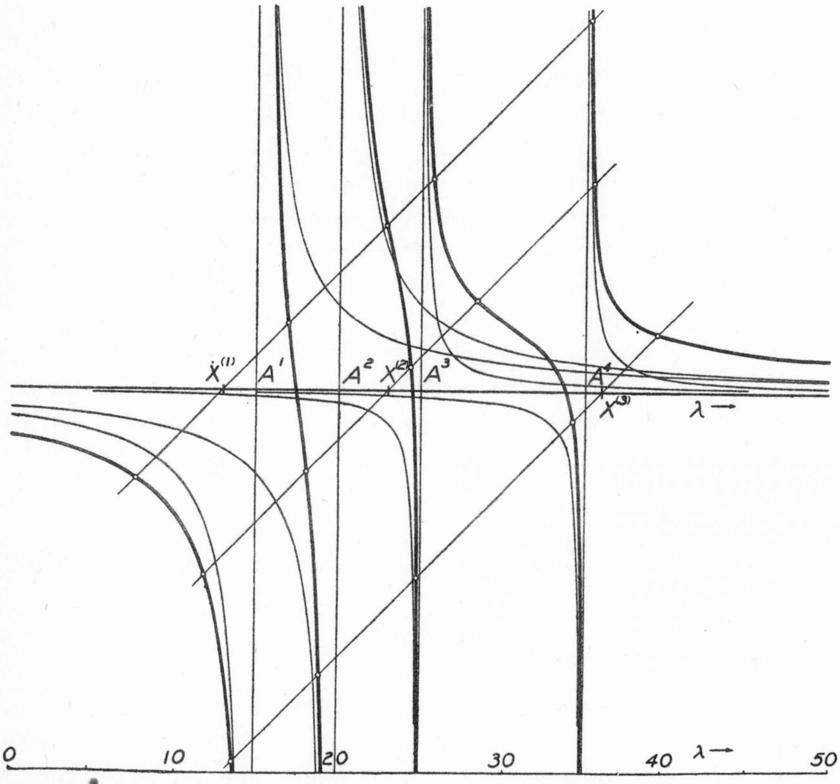

Fig. 3[2]. Graphical determination of the roots of (1). In this example, $A^1 = 15$, $A^2 = 20$, $A^3 = 25$, $A^4 = 35$; $|\alpha_1| = 5$, $|\alpha_2| = 5$, $|\alpha_3| = 2$, $|\alpha_4| = 2$. The roots are indicated for three different values of X: $X^{(1)} = 13$, $X^{(2)} = 23$, and $X^{(3)} = 36$.

which lies above X (but not above the original position of the next higher level); pushes down each level of the set which lies below X (but not below the original position of the next lower level); *while X itself is pushed up or down according to whether the sum*

$$\Sigma \frac{|\alpha_i|^2}{X - A^i}$$

* If two of the A's are equal, say $A^i = A^j$, there is also a root at A^i.

calculated for those levels below X is greater or less than this sum calculated for those levels above X. Hence with regard to the displacement of X one may say that the second-order perturbation theory gives the correct *direction* of motion even when it does not give an accurate value of the amount.

There is, of course, no rigorous meaning to the correlation in the preceding paragraph of the eigenvalues of (1) with the 'unperturbed levels' $A^1, A^2, \ldots, A^n, X$. However, when all the 'perturbations' are small, the states belonging to these eigenvalues will be very closely

$$\psi(A^1),\ \psi(A^2),\ \ldots,\ \psi(A^n),\ \psi(X).$$

The curves of Fig. 3^2 are seen to represent, after a fashion, the potential along the λ-axis of the plot due to infinite lines of doublets perpendicular to the paper at points A^1, A^2, A^3, and A^4 of linear moment $\frac{3}{4}|\alpha_1|^2$, $\frac{3}{4}|\alpha_2|^2$, $\frac{3}{4}|\alpha_3|^2$, and $\frac{3}{4}|\alpha_4|^2$ in the direction of increasing λ. One obtains an eigenvalue whenever this potential equals $\lambda - X$. The dipoles to the left of an eigenvalue raise the potential and hence tend to push the eigenvalue up while those to the right lower the potential and tend to push the eigenvalue down.

At any given eigenvalue the relative amounts contributed to the potential by the different strings of doublets are rather closely related to the relative contributions of the corresponding states to the *eigenstate* in question. These relative contributions to the eigenstate are shown by (2a)—

$$(A^j|\lambda) = -\frac{\alpha_j}{A^j - \lambda}(X|\lambda) \qquad (4)$$

—to be proportional to the quantity $\dfrac{\alpha_j}{A^j - \lambda}$ instead of to the quantity $\dfrac{|\alpha_j|^2}{A^j - \lambda}$ which measures the contribution to the potential.

One sees in Fig. 1^2 an illustration of the 'forces' which keep an eigenvalue from passing the next unperturbed level in the fact that β^{11} could not exceed the first order value of β^{12}, even though β^0 tried very hard to push it past this position.

It should be mentioned that equation (3) is often very easily solved numerically for the values of λ, especially for values which lie close to one of the A's so that one term on the left side of the equation is very sensitive to small changes in λ whereas the other terms are relatively insensitive. In this way the roots β^{11} and β^{12} of the problem in § 10^2 were easily obtained to six significant figures—the root β^0 being then given by the diagonal sum rule. In such a case as that of § 10^2 the roots must be obtained very accurately to give moderate accuracy in the transformation coefficients because of the occurrence of $A^j - \lambda$ in (4).

12. The analysis of non-commuting vectors.

In quantum mechanics, the components of the vectors of physical interest—position, momentum, and angular momentum—become linear operators in place of pure numbers. The manipulation of such 'non-commuting' vectors is facilitated by a set of formulas modified from those of the usual vector analysis.

If $\boldsymbol{A}$ and $\boldsymbol{B}$ are two vector observables ($\boldsymbol{A}=A_x\boldsymbol{i}+A_y\boldsymbol{j}+A_z\boldsymbol{k}$, where A_x, A_y, and A_z are scalar observables),* the commutators connecting the components of $\boldsymbol{A}$ and $\boldsymbol{B}$ transform like the components of a dyadic; hence it is convenient to write

$$[\boldsymbol{A},\boldsymbol{B}]=[A_x,B_x]\boldsymbol{ii}+[A_x,B_y]\boldsymbol{ij}+\dots \tag{1}$$

Two vectors will be said to commute when this dyadic vanishes, i.e. when each component of one commutes with each component of the other. This dyadic is not the dyadic $\boldsymbol{AB}-\boldsymbol{BA}$, but is related to it as follows:

$$[\boldsymbol{A},\boldsymbol{B}]=\boldsymbol{AB}-(\boldsymbol{BA})_C=\boldsymbol{AB}-\boldsymbol{BA}-(\boldsymbol{B}\times\boldsymbol{A})\times\mathfrak{J}, \tag{2}$$

where $\mathfrak{J}$ is the unit dyadic $\boldsymbol{ii}+\boldsymbol{jj}+\boldsymbol{kk}$ and $_C$ indicates the conjugate (transposed) dyadic.† Hence in general $[\boldsymbol{B},\boldsymbol{A}]\neq-[\boldsymbol{A},\boldsymbol{B}]$, but from (2) we see that

$$[\boldsymbol{B},\boldsymbol{A}]=-[\boldsymbol{A},\boldsymbol{B}]_C. \tag{3}$$

With these conventions, for a single particle having position vector $\boldsymbol{r}$ and momentum vector $\boldsymbol{p}$, the set of commutators 5^21 may be written as

$$[\boldsymbol{r},\boldsymbol{r}]=0, \quad [\boldsymbol{p},\boldsymbol{p}]=0, \quad [\boldsymbol{r},\boldsymbol{p}]=-[\boldsymbol{p},\boldsymbol{r}]=i\hbar\mathfrak{J}. \tag{4}$$

In a similar fashion, the commutators of a scalar observable X with the components of a vector observable $\boldsymbol{A}$ form the vector

$$[X,\boldsymbol{A}]=-[\boldsymbol{A},X]=[X,A_x]\boldsymbol{i}+[X,A_y]\boldsymbol{j}+[X,A_z]\boldsymbol{k}. \tag{5}$$

(This is the vector $X\boldsymbol{A}-\boldsymbol{A}X$.) *A scalar will be said to commute with a vector when this vector vanishes.*

The following generalizations of 5^22 will be found very useful

$$[\boldsymbol{A}\cdot\boldsymbol{B},\boldsymbol{C}]=\boldsymbol{A}\cdot[\boldsymbol{B},\boldsymbol{C}]-[\boldsymbol{C},\boldsymbol{A}]\cdot\boldsymbol{B}, \tag{6}$$

$$[\boldsymbol{A}\times\boldsymbol{B},\boldsymbol{C}]=\boldsymbol{A}\times[\boldsymbol{B},\boldsymbol{C}]-([\boldsymbol{C},\boldsymbol{A}]\times\boldsymbol{B})_C, \tag{7a}$$

$$[\boldsymbol{C},\boldsymbol{A}\times\boldsymbol{B}]=[\boldsymbol{C},\boldsymbol{A}]\times\boldsymbol{B}-(\boldsymbol{A}\times[\boldsymbol{B},\boldsymbol{C}])_C, \tag{7b}$$

$$[X,\boldsymbol{A}\times\boldsymbol{B}]=[X,\boldsymbol{A}]\times\boldsymbol{B}+\boldsymbol{A}\times[X,\boldsymbol{B}]. \tag{8}$$

* In this section we shall use light-faced letters for ordinary observables, reserving bold-faced letters for vectors in the Gibbs sense.

† See GIBBS-WILSON, *Vector Analysis*, Chapter v, for a complete discussion of the ordinary theory of dyadics. It should be pointed out that in combining non-commutative vectors, the order of the factors must be preserved; thus

$$\boldsymbol{A}\cdot\boldsymbol{B}=A_xB_x+A_yB_y+A_zB_z, \quad (\boldsymbol{A}\times\boldsymbol{B})_x=A_yB_z-A_zB_y, \text{ etc.}$$

With non-commuting vectors all associative and distributive properties of ordinary vector analysis are preserved; other elementary operations must be performed in accordance with the formulas

$$\boldsymbol{A}\cdot\boldsymbol{B}=\boldsymbol{B}\cdot\boldsymbol{A}+[\boldsymbol{A},\boldsymbol{B}]_S, \tag{9a}$$

$$\boldsymbol{A}\times\boldsymbol{B}=-\boldsymbol{B}\times\boldsymbol{A}+[\boldsymbol{A},\boldsymbol{B}]_\times, \tag{9b}$$

where $_S$ and $_\times$ indicate the scalar and vector obtained by inserting dots and crosses respectively between the two vectors of each dyad in a dyadic—the scalar and vector of the dyadic as defined by Gibbs;

$$\boldsymbol{A}\cdot\boldsymbol{B}\times\boldsymbol{C}=\boldsymbol{A}\times\boldsymbol{B}\cdot\boldsymbol{C}, \tag{10}$$

$$\boldsymbol{A}\times(\boldsymbol{B}\times\boldsymbol{C})=\boldsymbol{B}\boldsymbol{A}\cdot\boldsymbol{C}-\boldsymbol{A}\cdot\boldsymbol{B}\boldsymbol{C}-[\boldsymbol{B},\boldsymbol{A}]\cdot\boldsymbol{C} \tag{11a}$$

$$=\boldsymbol{A}\cdot\boldsymbol{C}\boldsymbol{B}-\boldsymbol{A}\cdot\boldsymbol{B}\boldsymbol{C}-\boldsymbol{A}\cdot[\boldsymbol{C},\boldsymbol{B}], \tag{11b}$$

$$(\boldsymbol{A}\times\boldsymbol{B})\times\boldsymbol{C}=\boldsymbol{B}\boldsymbol{A}\cdot\boldsymbol{C}-\boldsymbol{A}\boldsymbol{B}\cdot\boldsymbol{C}-[\boldsymbol{B},\boldsymbol{A}]\cdot\boldsymbol{C} \tag{12a}$$

$$=\boldsymbol{A}\cdot\boldsymbol{C}\boldsymbol{B}-\boldsymbol{A}\boldsymbol{B}\cdot\boldsymbol{C}-\boldsymbol{A}\cdot[\boldsymbol{C},\boldsymbol{B}]. \tag{12b}$$

The application of these formulas to the ordinary vector operator 'del' (∇) has been discussed by Shortley and Kimball.*

* SHORTLEY and KIMBALL, Proc. Nat. Acad. Sci. **20**, 82 (1934).

CHAPTER III

ANGULAR MOMENTUM

Because of the symmetry of the atomic model, the dynamical variables analogous to the components of angular momentum in classical dynamics play a fundamental rôle in the theory of atomic spectra. The properties of these observables, and of the electron spin, which we shall develop in this chapter, have constant application throughout the theory of atomic spectra.

NOTE. In this chapter and henceforth we shall no longer distinguish by a difference in type face between quantities of a symbolic nature and quantities in the Schrödinger representation. While this explicit distinction has been found very useful in the exposition of the theory, because of the complete symmetry which has been shown to exist between the two schemes no confusion will be entailed by using the same type for each in what follows. In this way, bold-faced type is released for its customary rôle, that of indicating a *vector* quantity in the restricted three-dimensional Gibbs sense.

1. Definition of angular momentum.

In classical mechanics if the position vector of a particle relative to an origin O is $\mathbf{r}$ and its linear momentum is $\mathbf{p}$, its angular momentum about O is defined to be

$$\mathbf{L} = \mathbf{r} \times \mathbf{p}. \tag{1}$$

We tentatively define the angular momentum of a particle in quantum mechanics as a vector given by this same formula in terms of the position and momentum vector observables.

From the basic commutation rules (12^24) we find that the vector $\mathbf{L}$ does not commute with itself, i.e. that the three observables L_x, L_y, L_z do not commute with each other. Instead we find, from a double application of 12^27, that

$$\begin{aligned}[\mathbf{L},\mathbf{L}] = [\mathbf{r}\times\mathbf{p},\, \mathbf{r}\times\mathbf{p}] &= \mathbf{r}\times[\mathbf{p},\mathbf{r}\times\mathbf{p}] - ([\mathbf{r}\times\mathbf{p},\mathbf{r}]\times\mathbf{p})_C \\ &= \mathbf{r}\times[\mathbf{p},\mathbf{r}]\times\mathbf{p} - (\mathbf{r}\times[\mathbf{p},\mathbf{r}]\times\mathbf{p})_C \\ &= -i\hbar\,[\mathbf{r}\times(\mathbf{p}\times\mathfrak{J}) - \{\mathbf{r}\times(\mathbf{p}\times\mathfrak{J})\}_C];\end{aligned}$$

since twice the antisymmetric part of the dyadic $\mathbf{r}\times(\mathbf{p}\times\mathfrak{J})$ is $(\mathbf{r}\times\mathbf{p})\times\mathfrak{J}$, this becomes

$$[\mathbf{L},\mathbf{L}] = -i\hbar\,(\mathbf{r}\times\mathbf{p})\times\mathfrak{J} = -i\hbar\mathbf{L}\times\mathfrak{J}. \tag{2}$$

The question now arises: if we were to regard this commutation rule as the definition of an angular-momentum vector, would this be equivalent to the definition by means of (1)? It turns out that (2) is more general than (1), and that this is just the extra generality that is needed to fit electron spin into the picture. So we shall suppose (1) to hold for a special kind of angular momentum which we shall call *orbital* angular momentum. The general angular momentum defined by

$$[\boldsymbol{J}, \boldsymbol{J}] = -i\hbar \boldsymbol{J} \times \mathfrak{J}, \tag{3}$$

or in terms of components by

$$[J_x, J_y] = i\hbar J_z, \quad [J_y, J_z] = i\hbar J_x, \quad [J_z, J_x] = i\hbar J_y, \tag{3'}$$

will be denoted hereafter by $\boldsymbol{J}$, while the letter $\boldsymbol{L}$ will be reserved for orbital angular momentum.

2. Allowed values of angular momentum.

We proceed now to investigate the properties of the observables J_x, J_y, J_z, which are consequences of $1^3 3$. The squared resultant angular momentum

$$\boldsymbol{J}^2 = J_x^2 + J_y^2 + J_z^2 \tag{1}$$

is readily shown by application of $12^2 6$ to commute with the vector $\boldsymbol{J}$ and hence with all three components J_x, J_y, and J_z. As a consequence there exist simultaneous eigenstates for $\boldsymbol{J}^2$ and any one component, say J_z. We can find the allowed values $\boldsymbol{J}^{2\prime}$ and J'_z by the following procedure. Let $\psi(\gamma \boldsymbol{J}^{2\prime} J'_z)$ be an eigenstate of $\boldsymbol{J}^2$, J_z and a set* Γ of observables which make up together with $\boldsymbol{J}^2$ and J_z a complete set of independent commuting observables for the system in question. We shall require that Γ commute not only with J_z and $\boldsymbol{J}^2$, but also with J_x and J_y, for a reason to be given presently. Then

$$\begin{aligned} (J_x^2 + J_y^2 + J_z^2)\, \psi(\gamma \boldsymbol{J}^{2\prime} J'_z) &= \boldsymbol{J}^{2\prime}\, \psi(\gamma \boldsymbol{J}^{2\prime} J'_z) \\ J_z \psi(\gamma \boldsymbol{J}^{2\prime} J'_z) &= J'_z \psi(\gamma \boldsymbol{J}^{2\prime} J'_z). \end{aligned} \tag{2}$$

From these relations it follows that

$$(J_x^2 + J_y^2)\, \psi(\gamma \boldsymbol{J}^{2\prime} J'_z) = (\boldsymbol{J}^{2\prime} - J'^2_z)\, \psi(\gamma \boldsymbol{J}^{2\prime} J'_z). \tag{3}$$

Since J_x and J_y are real observables, we see from this that $|J'_z| \leqslant \sqrt{\boldsymbol{J}^{2\prime}}$, since the allowed values of $(J_x^2 + J_y^2)$ and of $\boldsymbol{J}^2$ are essentially positive (cf. § 4^2, Problem 3). Now, using the commutation rules $1^3 3$, we find that

$$J_z (J_x \pm iJ_y) = (J_x \pm iJ_y)(J_z \pm \hbar). \tag{4}$$

Operating with the two sides of this equation on $\psi(\gamma \boldsymbol{J}^{2\prime} J'_z)$ gives

$$J_z (J_x \pm iJ_y)\, \psi(\gamma \boldsymbol{J}^{2\prime} J'_z) = (J'_z \pm \hbar)(J_x \pm iJ_y)\, \psi(\gamma \boldsymbol{J}^{2\prime} J'_z). \tag{5}$$

But this equation is of the form of the allowed-values equation and asserts that unless

$$(J_x \pm iJ_y)\, \psi(\gamma \boldsymbol{J}^{2\prime} J'_z) \tag{6}$$

vanishes, it is an eigenstate of J_z belonging to the eigenvalue $J'_z \pm \hbar$. It is also an eigenstate of $\boldsymbol{J}^2$ belonging to the eigenvalue $\boldsymbol{J}^{2\prime}$ and of Γ belonging to the eigenvalues γ since $\boldsymbol{J}^2$ and Γ commute with J_x and J_y. (This is the reason

* In Chapter II we used capital Greek letters to denote complete sets of commuting observables; hereafter we shall use these letters to fill out complete sets—our individual observables will invariably be Roman letters of special significance. For convenience we shall denote a set of eigenvalues of Γ by γ or γ' in place of the previous Γ' or Γ''.

for the requirement that Γ commute with J_x and J_y.) Hence starting with a given pair of allowed simultaneous eigenvalues $\boldsymbol{J}^{2\prime}$, J_z', we in general find a whole series of allowed simultaneous eigenvalues

$$\ldots;\ \boldsymbol{J}^{2\prime},\ J_z'-\hbar;\quad \boldsymbol{J}^{2\prime},\ J_z';\quad \boldsymbol{J}^{2\prime},\ J_z'+\hbar;\quad \ldots. \tag{7}$$

But we have already seen that the values of J_z' which may occur in simultaneous eigenstates with $\boldsymbol{J}^{2\prime}$ are bounded. Hence this series must have a lowest member $\boldsymbol{J}^{2\prime}$, J_z^0, and a highest member $\boldsymbol{J}^{2\prime}$, J_z^1. From (6) we see that the corresponding states must satisfy the relations

$$(J_x-iJ_y)\,\psi(\gamma\boldsymbol{J}^{2\prime}\,J_z^0)=0,$$
$$(J_x+iJ_y)\,\psi(\gamma\boldsymbol{J}^{2\prime}\,J_z^1)=0,$$

for otherwise the left sides of these equations would be eigenstates belonging to $J_z^0-\hbar$ and $J_z^1+\hbar$, contrary to hypothesis. Operation by (J_x+iJ_y) on the first of these equations leads to

$$(J_x^2+J_y^2+\hbar J_z)\,\psi(\gamma\boldsymbol{J}^{2\prime}\,J_z^0)=(\boldsymbol{J}^2-J_z^2+\hbar J_z)\,\psi(\gamma\boldsymbol{J}^{2\prime}\,J_z^0)$$
$$=[\boldsymbol{J}^{2\prime}-(J_z^0)^2+\hbar J_z^0]\,\psi(\gamma\boldsymbol{J}^{2\prime}\,J_z^0)=0;$$

from which, since $\psi(\gamma\boldsymbol{J}^{2\prime}\,J_z^0)\neq 0$ by hypothesis,

$$\boldsymbol{J}^{2\prime}-(J_z^0)^2+\hbar J_z^0=0. \tag{8a}$$

Similarly, operation by (J_x-iJ_y) on the second equation gives

$$\boldsymbol{J}^{2\prime}-(J_z^1)^2-\hbar J_z^1=0. \tag{8b}$$

From these equations (8) we find that

$$(J_z^1+J_z^0)\,(J_z^0-J_z^1-\hbar)=0;$$

since $J_z^1\geqslant J_z^0$ by hypothesis, this shows that $J_z^0=-J_z^1$. Now the difference $J_z^1-J_z^0$ must be a positive integer, or zero, times $\hbar$. This integer we shall, in accord with the traditional notation, write as $2j$, where j is restricted to the values $0, \frac{1}{2}, 1, \frac{3}{2}, 2, \ldots$. Hence

$$J_z^0=-j\hbar,\quad J_z^1=j\hbar, \tag{9}$$

and from (8)

$$\boldsymbol{J}^{2\prime}=j\,(j+1)\,\hbar^2. \tag{10}$$

This restricts the allowed values of $\boldsymbol{J}^2$ to the numbers $j\,(j+1)\,\hbar^2$. With any one of these values the set of allowed values of J_z forming simultaneous eigenstates is determined by (9) as

$$j\hbar,\quad (j-1)\,\hbar,\quad (j-2)\,\hbar,\quad \ldots,\quad -j\hbar.$$

For these allowed values of J_z we shall use the notation $m_j\hbar$, where m_j takes on the values $j, j-1, \ldots, -j$. For convenience we shall often omit the subscript j, writing

$$J_z'=m\hbar. \qquad (m=j,\, j-1, \ldots, -j) \tag{11}$$

Since J_x, J_y, and J_z enter the commutation rules symmetrically, the allowed values of J_x and J_y are the same as those of J_z. This completes the solution of the allowed-values problem.

3. The matrices of angular momentum.

Further development of the method used in the preceding section enables us to find the matrix components of J_x and J_y in the representation in which $\boldsymbol{J}^2$ and J_z are diagonal. It will be convenient to denote the value of $\boldsymbol{J}^2$ by giving the value of j and of J_z by giving m. The first step is the normalization of the ψ's occurring in 2^35. That equation tells us that

$$\psi(\gamma j m\pm 1) = N_\pm (J_x \pm iJ_y)\,\psi(\gamma j m), \tag{1}$$

where $N_\pm$ is the factor necessary to normalize $\psi(\gamma j m\pm 1)$ when $\psi(\gamma j m)$ is normalized in the sense of 1^21. Since $(J_x \pm iJ_y)^\dagger = (J_x \mp iJ_y)$, we find, using 1^22, that

$$\begin{aligned}\bar\psi(\gamma j m\pm 1)\,\psi(\gamma j m\pm 1) &= \overline{N_\pm (J_x \pm iJ_y)\,\psi(\gamma j m)}\,N_\pm (J_x \pm iJ_y)\,\psi(\gamma j m)\\ &= \bar N_\pm N_\pm\,\bar\psi(\gamma j m)\,(J_x \mp iJ_y)\,(J_x \pm iJ_y)\,\psi(\gamma j m)\\ &= \bar N_\pm N_\pm\,\bar\psi(\gamma j m)\,[\boldsymbol{J}^2 - J_z(J_z \pm \hbar)]\,\psi(\gamma j m)\\ &= \bar N_\pm N_\pm\,\hbar^2[j(j+1) - m(m\pm 1)].\end{aligned}$$

If this is to equal unity, we must have, in (1),

$$N_\pm = \frac{e^{i\delta}}{\hbar\sqrt{(j\mp m)(j\pm m+1)}}, \tag{2}$$

where δ is an arbitrary real number. This is the arbitrary phase which occurs in any ψ because two states are not distinct unless they are linearly independent. Such an indeterminacy of phase can have no effect on any of the results of physical significance. For this reason it is permissible for convenience to make an explicit choice of phase. We shall, therefore, mean by $\psi(\gamma j m)$ that set of states connected by the relation

$$(J_x \pm iJ_y)\,\psi(\gamma j m) = \hbar\sqrt{(j\mp m)(j\pm m+1)}\,\psi(\gamma j m\pm 1), \tag{3}$$

which is obtained from (1) and (2) by setting $\delta = 0$.

We may now immediately obtain the matrix components of J_x and J_y. The general element of the matrix of J_x,

$$\begin{aligned}(\gamma j' m'|J_x|\gamma' j m) = \bar\psi(\gamma j' m')\,J_x\,\psi(\gamma' j m) &= \tfrac12\bar\psi(\gamma j' m')\,(J_x + iJ_y)\,\psi(\gamma' j m)\\ &\quad + \tfrac12\bar\psi(\gamma j' m')\,(J_x - iJ_y)\,\psi(\gamma' j m),\end{aligned}$$

is seen from (3) to be zero unless $\gamma = \gamma'$, $j' = j$, $m' = m\pm 1$. The non-vanishing elements have the values

$$\left.\begin{aligned}(\gamma j m{+}1|J_x|\gamma j m) &= \tfrac12\hbar\sqrt{(j-m)(j+m+1)}\\ (\gamma j m{-}1|J_x|\gamma j m) &= \tfrac12\hbar\sqrt{(j+m)(j-m+1)}.\end{aligned}\right\} \tag{4}$$

Similarly, the non-vanishing components of J_y are found to have the values

$$\left.\begin{aligned}(\gamma j m{+}1|J_y|\gamma j m) &= -\tfrac12 i\hbar\sqrt{(j-m)(j+m+1)}\\ (\gamma j m{-}1|J_y|\gamma j m) &= \tfrac12 i\hbar\sqrt{(j+m)(j-m+1)}.\end{aligned}\right\} \tag{5}$$

The relation $$(\gamma j m|J_y|\gamma j m') = e^{i\frac{\pi}{2}(m'-m)}(\gamma j m|J_x|\gamma j m') \tag{6}$$
connecting the components of J_y with those of J_x will be found useful.

Formulas (4) and (5) may by the use of vector notation* be combined into the one formula

$$(\gamma j m\pm 1|\boldsymbol{J}|\gamma j m) = (\boldsymbol{i}\mp i\boldsymbol{j})\tfrac{1}{2}\hbar\sqrt{(j\mp m)(j\pm m+1)}. \tag{7}$$

γ here represents the eigenvalues of a set of observables which commute with $\boldsymbol{J}$. Let us consider in the $\gamma j m$ scheme the matrix of such an observable, say K, which commutes with $\boldsymbol{J}$. This matrix will of course be diagonal with respect to j and m, but by the following argument we can show it to be entirely independent of m. We write

$$K\psi(\gamma' j m) = \sum_\gamma \psi(\gamma j m)(\gamma j m|K|\gamma' j m).$$

Now, from (3)

$$(J_x \pm iJ_y)K\psi(\gamma' j m) = \sum_\gamma \psi(\gamma j m\pm 1)\hbar\sqrt{(j\mp m)(j\pm m+1)}(\gamma j m|K|\gamma' j m).$$

Similarly

$$K(J_x\pm iJ_y)\psi(\gamma' j m)$$
$$= \sum_\gamma \psi(\gamma j m\pm 1)\hbar\sqrt{(j\mp m)(j\pm m+1)}(\gamma j m\pm 1|K|\gamma' j m\pm 1).$$

These two expressions are equal by assumption; on comparing coefficients we have the result

$$(\gamma j m|K|\gamma' j m) = (\gamma j m\pm 1|K|\gamma' j m\pm 1) \quad ([K,\boldsymbol{J}]=0) \tag{8}$$

for all values of m such that the coefficient $(j\mp m)(j\pm m+1)$ does not vanish. This theorem is of importance in cases where the Hamiltonian commutes with various angular momenta (see § 1^7).

We may at this point show, in a similar fashion, that the unitary transformation between two schemes of eigenstates belonging to jm is entirely independent of m. Let

$$\psi(\alpha j m) = \sum_\beta \chi(\beta j m)(\beta j m|\alpha j m). \tag{9}$$

We need not here sum over j' and m' in χ because $\chi(\beta j' m')$ is orthogonal to $\psi(\alpha j m)$ unless $j'=j$, $m'=m$. Then

$$\psi(\alpha j m-1) = \sum_\beta \chi(\beta j m-1)(\beta j m-1|\alpha j m-1). \tag{10}$$

But from (3) we have

$$\psi(\alpha j m-1) = (J_x - iJ_y)\psi(\alpha j m)/\hbar\sqrt{(j+m)(j-m+1)}$$
$$= \sum_\beta \chi(\beta j m-1)(\beta j m|\alpha j m),$$

using the expansion (9). On comparing this with (10), we see that

$$(\beta j m|\alpha j m) = (\beta j m-1|\alpha j m-1). \tag{11}$$

* $(\alpha'|\boldsymbol{J}|\alpha'') = \boldsymbol{i}(\alpha'|J_x|\alpha'') + \boldsymbol{j}(\alpha'|J_y|\alpha'') + \boldsymbol{k}(\alpha'|J_z|\alpha'')$.

Hence this transformation is entirely independent of m. For this reason, in writing the transformation coefficients joining two such systems of states we shall ordinarily omit the m, writing merely $(\beta j|\alpha j)$.

PROBLEMS

1. Show that if $j=\frac{1}{2}$ the matrix representing the component of angular momentum in the direction specified by direction cosines l, m, n is

$$\tfrac{1}{2}\hbar \left\| \begin{matrix} n & l-im \\ l+im & -n \end{matrix} \right\|,$$

where the rows and columns are labelled by $(j=\frac{1}{2}, m=\frac{1}{2})$ and $(j=\frac{1}{2}, m=-\frac{1}{2})$.

2. Prove by direct application of the allowed-values equation (§ 7[2]) that the allowed values of this matrix are $\pm\frac{1}{2}\hbar$.

3. Find the eigen-ψ's for the states in which $j=\frac{1}{2}$ and the component of angular momentum along the direction (l, m, n) has the values $\pm\frac{1}{2}\hbar$. How does the eigen-ψ for the value $+\frac{1}{2}\hbar$ in the direction (l, m, n) compare with that for $-\frac{1}{2}\hbar$ in the direction $(-l, -m, -n)$?

4. Write out the matrices for J_x, J_y, J_z when $j=1$ and verify that $J_xJ_y-J_yJ_x=i\hbar J_z$.

5. Find the matrix for $J_x^2+J_y^2$ and verify that its allowed values are those of $\boldsymbol{J}^2-J_z^2$; i.e. $[j(j+1)-m^2]\hbar^2$.

4. Orbital angular momentum.

Let us now consider orbital angular momentum in the Schrödinger representation and learn some of the properties of the ψ functions which will be useful in later work. Introducing spherical polar coordinates r, θ, φ, we find that

$$L_z=-i\hbar\left(x\frac{\partial}{\partial y}-y\frac{\partial}{\partial x}\right)=-i\hbar\frac{\partial}{\partial\varphi}. \tag{1}$$

The Schrödinger equation for this observable is

$$-i\hbar\frac{\partial}{\partial\varphi}\psi=L_z'\psi, \tag{2}$$

which has the solution

$$\psi(\alpha L_z')=\frac{1}{\sqrt{2\pi}}A(\alpha L_z')\,e^{iL_z'\varphi/\hbar},$$

where $A(\alpha L_z')$ is independent of the coordinate φ. The Schrödinger wave function must be a single-valued function of position, so L_z' must be an integral multiple of $\hbar$:

$$L_z'=m_l\hbar. \qquad (m_l,\text{ an integer}) \tag{3}$$

We shall write

$$\psi(\alpha m_l)=A(\alpha m_l)\Phi(m_l), \tag{4}$$

where

$$\Phi(m_l)=\frac{1}{\sqrt{2\pi}}e^{im_l\varphi} \tag{5}$$

is seen to be normalized in the sense

$$\int_0^{2\pi}\bar{\Phi}(m_l)\Phi(m_l')\,d\varphi=\delta(m_l,m_l'). \tag{6}$$

The requirement (3) rules out fractional values for the z component of orbital angular momentum, and hence also excludes fractional values of the quantum number l, where

$$\boldsymbol{L}^{2\prime} = l(l+1)\hbar^2, \tag{7}$$

since for a given value of l the associated values of m_l are $l, l-1, \ldots, -l$.

Now let us find the dependence on θ of the Schrödinger representative $\psi(\gamma l m_l)$ of a simultaneous eigenstate of Γ, $\boldsymbol{L}^2$ and L_z. In terms of polar coordinates the operators for L_x and L_y are given by

$$\begin{aligned} L_x + iL_y &= \hbar e^{i\varphi}\left(\frac{\partial}{\partial\theta} + i\cot\theta\frac{\partial}{\partial\varphi}\right) \\ L_x - iL_y &= \hbar e^{-i\varphi}\left(-\frac{\partial}{\partial\theta} + i\cot\theta\frac{\partial}{\partial\varphi}\right), \end{aligned} \tag{8}$$

as may readily be calculated by transformation from Cartesian coordinates. We shall write

$$\psi(\gamma l m_l) = B(\gamma l m_l)\,\Theta(l m_l)\,\Phi(m_l), \tag{9}$$

where $B(\gamma l m_l)$ is independent of θ and φ. Since $m_l \not> l$, we have from $3^3 3$

$$(L_x + iL_y)\,\Theta(ll)\,\Phi(l) = 0.$$

Since $\Phi(l)$ is known from (5), this gives us the differential equation

$$\frac{\partial\,\Theta(ll)}{\partial\theta} - l\cot\theta\,\Theta(ll) = 0,$$

from which

$$\Theta(ll) = (-1)^l\sqrt{\frac{(2l+1)!}{2}}\,\frac{1}{2^l l!}\sin^l\theta. \tag{10}$$

The coefficient of $\sin^l\theta$ has here been chosen in such a way that $\Theta(ll)$ be normalized in the sense

$$\int_0^\pi \Theta^2(ll)\sin\theta\,d\theta = 1.$$

The phase has been taken as $(-1)^l$ for convenience in later work.

Now having

$$\psi(\gamma ll) = B(\gamma ll)\,\Theta(ll)\,\Phi(l), \tag{11}$$

we may, by repeated applications of $3^3 3$ and (8), find the $\psi(\gamma l m_l)$ for all m_l. In particular, since (8) act only on the coordinates θ and φ, the factor $B(\gamma ll)$ will recur for each m_l. Hence the factor $B(\gamma l m_l)$ of (9) does not depend on m_l and we may finally write

$$\psi(\gamma l m_l) = B(\gamma l)\,\Theta(l m_l)\,\Phi(m_l). \tag{12}$$

Since (10) has been normalized, these Θ's will be normalized for any m_l. Θ's for the same l but different m_l will not in general be orthogonal, since the Φ's take care of this orthogonality. However, Θ's of different l and the same m_l will be orthogonal since $\psi(\gamma l m_l)$ is orthogonal to $\psi(\gamma l' m_l)$ for $l \neq l'$, and insofar as a state ψ's belonging to l and m_l is concerned, the function $B(\gamma l)$ in (12) is quite arbitrary.

Now to obtain these Θ's explicitly. By repeated applications of $3^3 3$ we find

$$\sqrt{\frac{(l\pm m+k)!}{(l\mp m-k)!}}\,\psi(\gamma l\, m\pm k)=\sqrt{\frac{(l\pm m)!}{(l\mp m)!}}\left(\frac{L_x\pm iL_y}{\hbar}\right)^k\psi(\gamma l m). \tag{13}$$

From (8) we have

$$\hbar^{-1}(L_x\pm iL_y)\,e^{im\varphi}f(\theta)=\mp e^{i(m\pm1)\varphi}\sin^{1\pm m}\theta\frac{d}{d\cos\theta}[\sin^{\mp m}\theta f(\theta)],$$

from which, by iteration,

$$\hbar^{-k}(L_x\pm iL_y)^k e^{im\varphi}f(\theta)=(\mp1)^k e^{i(m\pm k)\varphi}\sin^{k\pm m}\theta\frac{d^k}{(d\cos\theta)^k}[\sin^{\mp m}\theta f(\theta)]. \tag{14}$$

Using these relations and the known form of $\Theta(l\,l)$, we find the general formula

$$\Theta(l\,m)=(-1)^l\sqrt{\frac{2l+1}{2}\frac{(l+m)!}{(l-m)!}}\frac{1}{2^l l!}\frac{1}{\sin^m\theta}\frac{d^{l-m}}{(d\cos\theta)^{l-m}}\sin^{2l}\theta, \tag{15*}$$

and in particular

$$\left.\begin{aligned}\Theta(l\,0)&=\sqrt{\frac{2l+1}{2}}\frac{1}{2^l l!}\frac{d^l}{(d\cos\theta)^l}(\cos^2\theta-1)^l\\ &=\sqrt{\frac{2l+1}{2}}P_l(\cos\theta).\end{aligned}\right\} \tag{16}$$

The polynomial in $\cos\theta$ which follows the irrational factor is the Legendre polynomial $P_l(\cos\theta)$, this particular expression for it being known as Rodrigues' formula.

We can also use (14) to find expressions for $\Theta(l\,m)$ in terms of the derivatives of the Legendre polynomials, the result being

$$(m>0)\left\{\begin{aligned}\Theta(l\,m)&=(-1)^m\sqrt{\frac{2l+1}{2}\frac{(l-m)!}{(l+m)!}}\sin^m\theta\frac{d^m}{(d\cos\theta)^m}P_l(\cos\theta)\\ \Theta(l\,-m)&=+\sqrt{\frac{2l+1}{2}\frac{(l-m)!}{(l+m)!}}\sin^m\theta\frac{d^m}{(d\cos\theta)^m}P_l(\cos\theta).\end{aligned}\right\} \tag{17}$$

Hence $$\Theta(l\,m)=(-1)^m\Theta(l\,-m). \tag{18}$$

The natural choice of phases which we have made here leads to a rather curious occurrence of the factor -1 only for positive odd values of m. If we had approached the problem through the usual form of the theory of

* For $l=0, 1, 2, 3$, the explicit forms of the Θ's are

$\Theta(0\,0)=\sqrt{\tfrac{1}{2}}$

$\Theta(1\,0)=\sqrt{\tfrac{3}{2}}\cos\theta$ $\qquad\Theta(1\,\pm1)=\mp\sqrt{\tfrac{3}{4}}\sin\theta$

$\Theta(2\,0)=\sqrt{\tfrac{5}{8}}(2\cos^2\theta-\sin^2\theta)$ $\qquad\Theta(2\,\pm1)=\mp\sqrt{\tfrac{15}{4}}\cos\theta\sin\theta$ $\qquad\Theta(2\,\pm2)=\sqrt{\tfrac{15}{16}}\sin^2\theta$

$\Theta(3\,0)=\sqrt{\tfrac{7}{8}}(2\cos^3\theta-3\cos\theta\sin^2\theta)$ $\qquad\Theta(3\,\pm1)=\mp\sqrt{\tfrac{21}{32}}(4\cos^2\theta\sin\theta-\sin^3\theta)$ $\qquad\Theta(3\,\pm2)=\sqrt{\tfrac{105}{16}}\cos\theta\sin^2\theta$

$\Theta(3\,\pm3)=\mp\sqrt{\tfrac{35}{32}}\sin^3\theta.$

spherical harmonics the natural tendency would have been to choose the normalizing factors with omission of the $(-1)^m$ in these formulas.*

From the form of the operators (8) together with $3^3 3$ we find the relations

$$\frac{\partial}{\partial\theta}\Theta(l\,m) = \tfrac{1}{2}\sqrt{(l-m)(l+m+1)}\,\Theta(l\,m+1) - \tfrac{1}{2}\sqrt{(l+m)(l-m+1)}\,\Theta(l\,m-1), \qquad (19)$$

$$m\cot\theta\,\Theta(l\,m) = -\tfrac{1}{2}\sqrt{(l-m)(l+m+1)}\,\Theta(l\,m+1) - \tfrac{1}{2}\sqrt{(l+m)(l-m+1)}\,\Theta(l\,m-1), \qquad (20)$$

which are useful in some calculations. We quote here for reference three other relations which may be obtained as special cases of results derived by matrix methods in $\S 9^3$:

$$\left.\begin{aligned}
\cos\theta\,\Theta(l\,m) &= \Theta(l+1\,m)\sqrt{\frac{(l+1-m)(l+1+m)}{(2l+1)(2l+3)}} + \Theta(l-1\,m)\sqrt{\frac{(l-m)(l+m)}{(2l-1)(2l+1)}} \\
\sin\theta\,\Theta(l\,m) &= -\Theta(l+1\,m+1)\sqrt{\frac{(l+m+1)(l+m+2)}{(2l+1)(2l+3)}} + \Theta(l-1\,m+1)\sqrt{\frac{(l-m)(l-m-1)}{(2l-1)(2l+1)}} \\
&= \Theta(l+1\,m-1)\sqrt{\frac{(l-m+1)(l-m+2)}{(2l+1)(2l+3)}} - \Theta(l-1\,m-1)\sqrt{\frac{(l+m)(l+m-1)}{(2l-1)(2l+1)}}.
\end{aligned}\right\} \qquad (21)$$

Of great importance in subsequent calculations is the relation known as the spherical-harmonic addition theorem, which expresses the Legendre polynomials of the angle ω between the directions (θ, φ) and (θ'', φ'') in terms of spherical harmonics of (θ, φ) and (θ'', φ''). The formula is

$$P_l(\cos\omega) = \frac{4\pi}{2l+1}\sum_{m=-l}^{l}\Theta(l\,m)\,\Theta''(l\,m)\,\Phi(m)\,\bar{\Phi}''(m), \qquad (22)$$

where
$$\cos\omega = \cos\theta\cos\theta'' + \sin\theta\sin\theta''\cos(\varphi-\varphi'').$$

This relation may be derived as follows:

$$P_l(\cos\omega) = \sqrt{\frac{4\pi}{2l+1}}\,\Theta'(l\,0)\,\Phi'(0), \qquad (23)$$

where θ' and φ' express the direction of $\boldsymbol{r}$ as referred to a coordinate system with pole along $\boldsymbol{k}'$ (see Fig. 1^3). Since this is an eigen-ψ of $\boldsymbol{L}^2$ belonging to l, it may be expanded in terms of the eigenfunctions referred to $\boldsymbol{k}$ as pole:

$$\Theta'(l\,0)\,\Phi'(0) = \sum_{m=-l}^{l}\Theta(l\,m)\,\Phi(m)\,E_{lm}, \qquad (24)$$

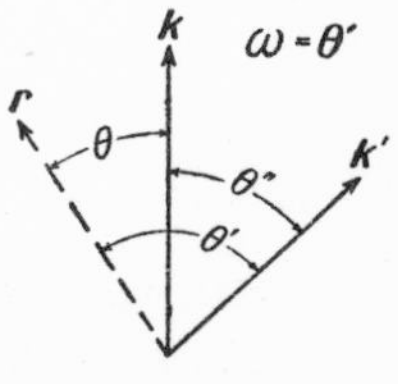

Fig. 1^3.

where the E_{lm} will of course depend parametrically on the angles θ'' and φ'' which relate $\boldsymbol{k}'$ to $\boldsymbol{i}, \boldsymbol{j}, \boldsymbol{k}$. Now $L_{z'}\Theta'(l\,0)\Phi'(0) = 0$, where $L_{z'}$ is the component of angular momentum in the direction $\boldsymbol{k}'$. But from vector analysis

$$L_{z'} = L_z\cos\theta'' + \tfrac{1}{2}(L_x + iL_y)\sin\theta'' e^{-i\varphi''} + \tfrac{1}{2}(L_x - iL_y)\sin\theta'' e^{i\varphi''}.$$

* This question of phase choice has caused some confusion with regard to the relative phases of the matrix components of angular momentum and of electrostatic interaction. See the discussion by UFFORD and SHORTLEY, Phys. Rev. **42**, 167 (1932).

Hence, from (24),

$$\hbar^{-1}L_{z'}\Theta'(l\,0)\Phi'(0)=\sum_m[\Theta(l\,m)\Phi(m)\,m\cos\theta''E_{lm}$$
$$+\Theta(l\,m+1)\Phi(m+1)\,\tfrac{1}{2}\sqrt{(l-m)(l+m+1)}\sin\theta''e^{-i\varphi''}E_{lm}$$
$$+\Theta(l\,m-1)\Phi(m-1)\,\tfrac{1}{2}\sqrt{(l+m)(l-m+1)}\sin\theta''e^{i\varphi''}E_{lm}]=0.$$

Equating the coefficient of $\Theta(l\,m)\Phi(m)$ to zero gives the recurrence relation

$$m\cos\theta''E_{lm}+\tfrac{1}{2}\sqrt{(l-m+1)(l+m)}\sin\theta''e^{-i\varphi''}E_{l\,m-1}$$
$$+\tfrac{1}{2}\sqrt{(l+m+1)(l-m)}\sin\theta''e^{i\varphi''}E_{l\,m+1}=0.$$

On comparing this with (20), we see that if we choose E_{ll} to be

$$\Theta''(l\,l)\,\bar{\Phi}''(l)f(\theta'',\varphi''),$$

where f is arbitrary, then

$$E_{lm}=\Theta''(l\,m)\,\bar{\Phi}''(m)f(\theta'',\varphi'').$$

Now in the special case $\theta''=0$, we see from (24) that $E_{l0}=1$. Hence since $P_l(1)=1$,* (16) shows that $f(0,\varphi'')=\sqrt{\dfrac{4\pi}{2l+1}}$. We have then

$$P_l(\cos\omega)=\frac{4\pi}{2l+1}\sum_m\Theta(l\,m)\Phi(m)\,\Theta''(l\,m)\,\bar{\Phi}''(m)\,g(\theta'',\varphi''),\tag{25}$$

where $g(0,\varphi'')=1$. But from Fig. 1^3 it is clear that $P_l(\cos\omega)$ must be a symmetric function of $\theta\varphi$ and $\theta''\varphi''$. Interchanging $\theta\varphi$ and $\theta''\varphi''$ in (25) gives

$$P_l(\cos\omega)=\frac{4\pi}{2l+1}\sum_m\Theta''(l\,m)\Phi''(m)\,\Theta(l\,m)\,\bar{\Phi}(m)\,g(\theta,\varphi).\tag{26}$$

But (25) and (26) are equal only if $g(\theta'',\varphi'')=g(\theta,\varphi)$ for all pairs of directions. This can be true only if g is constant and hence unity everywhere. This proves the theorem (22).

PROBLEM

Show from (1) and (8) that

$$\boldsymbol{L}^2=-\frac{\hbar^2}{\sin\theta}\frac{\partial}{\partial\theta}\left(\sin\theta\,\frac{\partial}{\partial\theta}\right)-\frac{\hbar^2}{\sin^2\theta}\frac{\partial^2}{\partial\varphi^2},\tag{27}$$

an expression which is essentially the angular part of the Laplace operator.

5. Spin angular momentum.

In 1925 Uhlenbeck and Goudsmit† discovered that great simplifications could be made in the formal description of atomic spectra by assuming that electrons possess an intrinsic angular momentum whose component in any direction is restricted to the values $\pm\frac{1}{2}\hbar$. This is known as the hypothesis of electron spin. An essential part of the hypothesis is that the electron also possesses an intrinsic magnetic moment whose component is $\mp e\hbar/2\mu c$ in a

* This follows immediately upon expanding the lth derivative of $(\cos\theta-1)^l(\cos\theta+1)^l$ in Rodrigues' formula by Leibnitz' theorem for the differentiation of a product, and then setting $\cos\theta=1$.

† Uhlenbeck and Goudsmit, Naturwiss. **13**, 953 (1925); Nature, **117**, 264 (1926).

direction in which the component of angular momentum is $\pm\frac{1}{2}\hbar$. The opposite signs are natural enough on the simple view that the magnetic moment originates in the classical electrodynamic effect of a rotating negative charge distribution. However, the ratio of magnetic moment to angular momentum (the magneto-mechanical ratio) has twice the value which is given by the classical theory and holds for the orbital motion of the electron.

Pauli* has shown that a quantum-mechanical description of the spin that is adequate for most purposes can be obtained by treating the spin as an angular-momentum observable of the type considered in §§ 2^3 and 3^3, but with j restricted to the value $\frac{1}{2}$. Further developments by Dirac† have shown the spin to be intimately associated with relativistic effects, but we shall usually treat spin in the simpler way introduced by Pauli.

We postulate then for the electron a spin angular momentum $\boldsymbol{S}$, independent of the orbital angular momentum $\boldsymbol{L}$. The proper values of the z component, S_z, of the spin are restricted to $\pm\frac{1}{2}\hbar$; hence $\boldsymbol{S}^2$ has always the proper value $\frac{1}{2}(\frac{1}{2}+1)\hbar^2$. If we write the proper value of $\boldsymbol{S}^2$ as $s(s+1)\hbar^2$ and the proper values of S_z as $m_s\hbar$, we always have

$$s=\tfrac{1}{2} \quad m_s=\pm\tfrac{1}{2}.$$

From $3^3 4$ and $3^3 5$ we find for the components of $\boldsymbol{S}$ the *Pauli spin matrices*

$$\|(\alpha\, m_s|S_x|\alpha\, m_s')\| = \begin{array}{c|cc|} & \frac{1}{2} & -\frac{1}{2} \\ \hline \frac{1}{2} & 0 & 1 \\ -\frac{1}{2} & 1 & 0 \\ \hline \end{array}\ \frac{\hbar}{2} \qquad \|(\alpha\, m_s|S_y|\alpha\, m_s')\| = \begin{array}{c|cc|} & \frac{1}{2} & -\frac{1}{2} \\ \hline \frac{1}{2} & 0 & -i \\ -\frac{1}{2} & i & 0 \\ \hline \end{array}\ \frac{\hbar}{2} \qquad \|(\alpha\, m_s|S_z|\alpha\, m_s')\| = \begin{array}{c|cc|} & \frac{1}{2} & -\frac{1}{2} \\ \hline \frac{1}{2} & 1 & 0 \\ -\frac{1}{2} & 0 & -1 \\ \hline \end{array}\ \frac{\hbar}{2}. \qquad (1)$$

In vector notation these matrices are given by

$$\|(\alpha\, m_s|\boldsymbol{S}|\alpha\, m_s')\| = \begin{array}{c|cc|} & \frac{1}{2} & -\frac{1}{2} \\ \hline \frac{1}{2} & \boldsymbol{k} & \boldsymbol{i}-i\boldsymbol{j} \\ -\frac{1}{2} & \boldsymbol{i}+i\boldsymbol{j} & -\boldsymbol{k} \\ \hline \end{array}\ \frac{\hbar}{2} \qquad (1')$$

With the introduction of the spin, the set x, y, z is no longer a complete set of observables for a single electron, but to these must be added a fourth observable, say S_z. (Any component of the spin angular momentum would serve as well.) The Schrödinger ψ then becomes the function $(x'\,y'\,z'\,S_z'|\ \ )$ which we shall write as $\psi_{xyz\sigma}$. Here we have introduced the letter σ for $S_z'/\hbar$ when used in the sense of a Schrödinger coordinate; it is convenient thus to introduce a notation different from the previous m_s for the special case of

* Pauli, Zeits. für Phys. **43**, 601 (1927).

† Dirac, *Quantum Mechanics*, Chapter XII.

Schrödinger expansions. The 'coordinate' σ is, of course, restricted to the values $\pm\frac{1}{2}$.

An eigenstate of S_z with quantum number m_s depends on the coordinate σ through the factor $\delta(\sigma, m_s)$, that is,

$$\psi_{xyz\sigma}(\alpha\, m_s) = \psi_{xyz}(\alpha)\,\delta(\sigma, m_s). \tag{2}$$

The factors $\delta(\sigma, \frac{1}{2})$ and $\delta(\sigma, -\frac{1}{2})$ form an orthonormal set of functions of σ in the sense that

$$\sum_{\sigma=-\frac{1}{2}}^{+\frac{1}{2}} \delta(\sigma, m_s)\,\delta(\sigma, m_s') = \delta(m_s, m_s'); \tag{3}$$

that this set is complete is obvious. We may then write any function ψ as

$$\psi = \psi_{xyz\frac{1}{2}}\,\delta(\sigma, \tfrac{1}{2}) + \psi_{xyz-\frac{1}{2}}\,\delta(\sigma, -\tfrac{1}{2}), \tag{4}$$

and represent this by the matrix

$$\left\| \begin{matrix} \psi_{xyz\frac{1}{2}} \\ \psi_{xyz-\frac{1}{2}} \end{matrix} \right\|. \tag{4'}$$

The result of operation of $\boldsymbol{S}$ on ψ is then given by the multiplication of (1′) by (4′).* When ψ is expressed in this form, $|\psi_{xyz\frac{1}{2}}|^2$ gives the probability density for finding the electron at xyz with spin $+\frac{1}{2}$; $|\psi_{xyz-\frac{1}{2}}|^2$ the probability density at this point with spin $-\frac{1}{2}$.

The direct product $\bar{\phi}\psi$ of two eigenfunctions now implies, in the Schrödinger scheme, summation over σ as well as integration over x, y, z:

$$\bar{\phi}\psi \simeq \sum_{\sigma=-\frac{1}{2}}^{\frac{1}{2}} \int \bar{\phi}_{xyz\sigma}\,\psi_{xyz\sigma}\,dx\,dy\,dz. \tag{5}$$

PROBLEM

From the results of Problem 3, § 3^3, write out in the form (4′) the most general ψ for which the component of spin in the direction whose direction cosines are (l, m, n) has the value $+\frac{1}{2}\hbar$.

6. Vector addition of angular momenta.

Consider a system in which $\boldsymbol{J}_1$ and $\boldsymbol{J}_2$ are two commuting angular-momentum vectors. (Two vectors are said to commute when each component of one commutes with each component of the other, i.e. when the dyadic $[\boldsymbol{J}_1, \boldsymbol{J}_2]$ vanishes.) Then the states of the system can be represented in terms of the scheme

$$\phi(\gamma j_1 j_2 m_1 m_2) \tag{1}$$

in which j_1 and m_1 are the quantum numbers labelling eigenstates of $\boldsymbol{J}_1^2$ and J_{1z}; j_2 and m_2 eigenstates of $\boldsymbol{J}_2^2$ and J_{2z}; and γ eigenstates of a set of observables Γ which commute with $\boldsymbol{J}_1$ and $\boldsymbol{J}_2$. We wish now to investigate the question of the 'addition' of these two vectors to obtain states characterized by proper values of the sum and the z component of the sum. If we can

* By the use of two ψ functions, which are to be correlated with our $\psi_{xyz\frac{1}{2}}$ and $\psi_{xyz-\frac{1}{2}}$, Darwin [Proc. Roy. Soc. **A116**, 227 (1927)] discovered equations which are equivalent to those obtained by use of Pauli matrices; Darwin's formulation however does not bring out the physical significance of the two functions.

learn to add two vectors, we can add any number by repeating the process. These considerations are of fundamental importance for the theory of atomic spectra, for they give us the basis of the rigorous quantum-mechanical transcription of the older vector-coupling characterizations of the energy states of atoms.

We introduce, then, the vector resultant

$$\boldsymbol{J} = \boldsymbol{J}_1 + \boldsymbol{J}_2$$

of our two angular momenta. The states ϕ above are eigenstates of the complete set

$$\Gamma, \boldsymbol{J}_1^2, \boldsymbol{J}_2^2, J_{1z}, J_{2z}$$

of independent commuting observables. But the set

$$\Gamma, \boldsymbol{J}_1^2, \boldsymbol{J}_2^2, \boldsymbol{J}^2, J_z$$

is seen to be also a set of independent commuting observables, and since it contains as many observables as the above, it should also be complete. An eigenstate of this set we shall denote by

$$\psi(\gamma j_1 j_2 j m). \tag{2}$$

Consider now the subspace characterized by the eigenvalues γ, j_1, and j_2. From 2[3]11 we see that this subspace is $(2j_1+1)(2j_2+1)$ dimensional, corresponding to the number of ways of independently assigning m_1 and m_2 to the ϕ's:

$$m_1 = j_1, j_1 - 1, \ldots, -j_1,$$
$$m_2 = j_2, j_2 - 1, \ldots, -j_2.$$

The question which immediately arises is the following: What are the $(2j_1+1)(2j_2+1)$ values of j, m which characterize the ψ's of this subspace? The method of answering this query is apparent when we realize that not only the quantum numbers γ, j_1, and j_2, but also the quantum number m, are common to the two modes of representation. From the fact that $J_z = J_{1z} + J_{2z}$, it follows that $m = m_1 + m_2$. Now since $\boldsymbol{J}$ is an angular momentum satisfying the commutation rules 1[3]3, any j which occurs in this subspace must do so accompanied by the corresponding $2j+1$ values of m. The largest m which occurs in a ϕ is $m = j_1 + j_2$, when $m_1 = j_1$ and $m_2 = j_2$. Hence the largest m occurring in a ψ must be $j_1 + j_2$ and therefore the largest j occurring must also be $j_1 + j_2$. There are two ϕ's with $m = j_1 + j_2 - 1$, namely $m_1 = j_1$, $m_2 = j_2 - 1$ and $m_1 = j_1 - 1$, $m_2 = j_2$. Hence there must be two ψ's with this value of m. One of these ψ's we know to be that going with $j = j_1 + j_2$; the other must belong to a $j = j_1 + j_2 - 1$. In general there will be three ϕ's belonging to $m = j_1 + j_2 - 2$, corresponding to the three values of (m_1, m_2):

$$(j_1, j_2 - 2) \quad (j_1 - 1, j_2 - 1) \quad (j_1 - 2, j_2). \tag{3}$$

This means that we must have in addition to the values $j = j_1 + j_2$ and $j = j_1 + j_2 - 1$ also the value $j = j_1 + j_2 - 2$. However, if either j_1 or j_2 were $\frac{1}{2}$

we should not have had all three of the states (3), and consequently would have obtained no new value of j. In general, the least value j_1+j_2-n of j is obtained when $j_s-n=-j_s$, where j_s is the smaller of j_1 and j_2. This gives $n=2j_s$, and the least value of j as $|j_1-j_2|$. From this point on, the degeneracy of each m must be proper to account for the occurrence of the values

$$j=j_1+j_2,\ j_1+j_2-1,\ \ldots,\ |j_1-j_2|. \tag{4}$$

That with these j's go the proper number $(2j_1+1)(2j_2+1)$ of values of m is easily verified.

To summarize, if we have two commuting angular momenta whose squared magnitudes are $j_1(j_1+1)\hbar^2$ and $j_2(j_2+1)\hbar^2$, then the squared magnitude of the vector sum of the two angular momenta can take on the values $j(j+1)\hbar^2$, where the allowed values of j are given by (4). These values are just those given by the empirical rule used in discussing atomic spectra by the vector-coupling method. One can represent the result pictorially by thinking of vectors of lengths j_1 and j_2 added vectorially, starting with the parallel case which gives the resultant (j_1+j_2) and taking all possible values differing from this by integers down to $|j_1-j_2|$ for the antiparallel case.

Two angular momenta will commute only if they refer to quite independent particles, or to quite independent coordinates of the same particle—for example to two different electrons or to the spin and orbital motion of the same electron. This means, in the last analysis, that we can split any eigenfunction of the type (1) into a sum of the type

$$\phi(\gamma j_1 j_2 m_1 m_2)=\sum_{\alpha\beta}\phi_1(\alpha j_1 m_1)\phi_2(\beta j_2 m_2), \tag{5}$$

where $\boldsymbol{J}_1$ operates only on ϕ_1, $\boldsymbol{J}_2$ only on ϕ_2.

7. The matrix of $(\boldsymbol{J}_1+\boldsymbol{J}_2)^2$.

It will be convenient for future reference to calculate the matrix elements of $\boldsymbol{J}^2$ in the scheme $\phi(\gamma j_1 j_2 m_1 m_2)$.

If we write $\boldsymbol{J}^2$ in the form

$$\boldsymbol{J}^2=(\boldsymbol{J}_1^2+\boldsymbol{J}_2^2+2J_{1z}J_{2z})+(J_{1x}+iJ_{1y})(J_{2x}-iJ_{2y})+(J_{1x}-iJ_{1y})(J_{2x}+iJ_{2y}), \tag{1}$$

it follows immediately from $2^3 2$ and $3^3 3$ that

$$\begin{aligned}\boldsymbol{J}^2\phi(\gamma j_1 j_2 m_1 m_2)&=[j_1(j_1+1)+j_2(j_2+1)+2m_1m_2]\hbar^2\phi(\gamma j_1 j_2 m_1 m_2)\\&+\hbar^2\sqrt{(j_1-m_1)(j_1+m_1+1)(j_2+m_2)(j_2-m_2+1)}\,\phi(\gamma j_1 j_2\, m_1+1\, m_2-1)\\&+\hbar^2\sqrt{(j_1+m_1)(j_1-m_1+1)(j_2-m_2)(j_2+m_2+1)}\,\phi(\gamma j_1 j_2\, m_1-1\, m_2+1).\end{aligned} \tag{2}$$

The matrix components we seek are these coefficients of ϕ according to the definition $2^2 14$. These components are seen to be diagonal with respect to $m=m_1+m_2$, as required by the fact that $\boldsymbol{J}^2$ commutes with J_z.

We shall need particularly the matrix of $\boldsymbol{L}\cdot\boldsymbol{S}=\boldsymbol{S}\cdot\boldsymbol{L}$, where $\boldsymbol{L}^2$ has the value $l(l+1)\hbar^2$, and $\boldsymbol{S}^2$ the value $\frac{1}{2}(\frac{1}{2}+1)\hbar^2$. The components are seen to have the values

$$(\gamma s l m_s m_l|\boldsymbol{L}\cdot\boldsymbol{S}|\gamma' s l' m_s' m_l')=\hbar^2\,\delta(\gamma l m;\gamma' l' m')$$
$$\cdot\{\delta(m_s,m_s')\,m_l m_s+\tfrac{1}{2}\delta(m_s',m_s\pm 1)\sqrt{(l-m+\tfrac{1}{2})(l+m+\tfrac{1}{2})}\},\quad (3)$$

where $m=m_s+m_l$, $m'=m_s'+m_l'$.

8. Matrix of T in the jm scheme. Selection rule on j.

We shall return to the question of the transformation amplitudes for vector addition in § 14[3]. Before discussing that question it will be fruitful to investigate by matrix methods the properties in the jm scheme, where $\boldsymbol{J}$ is any angular momentum, of the matrix of a vector $\boldsymbol{T}$ which satisfies the commutation rule

$$[\boldsymbol{J},\boldsymbol{T}]=[\boldsymbol{T},\boldsymbol{J}]=-i\hbar\boldsymbol{T}\times\mathfrak{J},\quad (1)$$

in the notation of § 12[2]. We shall indicate this relation by saying that $\boldsymbol{T}$ satisfies this commutation rule *with respect to* $\boldsymbol{J}$. This means that the components of $\boldsymbol{T}$ have the following commutators with respect to the components of $\boldsymbol{J}$:

$$\begin{array}{lll} [J_x,T_x]=0 & [J_x,T_y]=i\hbar T_z & [J_x,T_z]=-i\hbar T_y \\ [J_y,T_y]=0 & [J_y,T_z]=i\hbar T_x & [J_y,T_x]=-i\hbar T_z \\ [J_z,T_z]=0 & [J_z,T_x]=i\hbar T_y & [J_z,T_y]=-i\hbar T_x. \end{array}\quad (2)$$

From 12[2]9 we find that

$$\boldsymbol{T}\times\boldsymbol{J}=-\boldsymbol{J}\times\boldsymbol{T}+2i\hbar\boldsymbol{T},\quad \boldsymbol{J}\cdot\boldsymbol{T}=\boldsymbol{T}\cdot\boldsymbol{J}.\quad (3)$$

This commutation rule applies to a large class of vectors:

(*a*) Any angular momentum satisfies it with respect to itself (cf. 1[3]3).

(*b*) If $\boldsymbol{J}=\boldsymbol{J}_1+\boldsymbol{J}_2+\dots$, where each addend commutes with all the other addends, each of the addends satisfies this commutation rule with respect to $\boldsymbol{J}$.

(*c*) Both $\boldsymbol{r}$ and $\boldsymbol{p}$ satisfy this commutation rule with respect to $\boldsymbol{L}$; this follows from a simple calculation of $[\boldsymbol{r}\times\boldsymbol{p},\boldsymbol{r}]$ and $[\boldsymbol{r}\times\boldsymbol{p},\boldsymbol{p}]$ by means of 12[2]7a.

(*d*) Hence if $\boldsymbol{J}=\boldsymbol{L}_1+\boldsymbol{L}_2+\dots+\boldsymbol{S}_1+\boldsymbol{S}_2+\dots$, the coordinate $\boldsymbol{r}$ and the momentum $\boldsymbol{p}$ of any electron satisfy this commutation rule with respect to $\boldsymbol{J}$. ($\boldsymbol{L}_i$ is the orbital moment of the ith electron, $\boldsymbol{S}_i$ the spin moment.)

(*e*) Any linear combination of vectors which satisfy this rule with respect to $\boldsymbol{J}$ will also satisfy it.

(*f*) A calculation similar to that of § 1[3] shows that the cross product of any two vectors (and hence of any number of vectors) which satisfy this rule will also satisfy it. Hence any vector formed by addition and cross multiplication from vectors which satisfy this commutation rule will also satisfy it.

Another important fact which follows immediately from 12^26 is that the scalar product of any two vectors* which satisfy this commutation rule with respect to $\boldsymbol{J}$ will commute with J_x, J_y, J_z, and hence with $\boldsymbol{J}^2$:

$$[\boldsymbol{J}, \boldsymbol{T}_1\cdot\boldsymbol{T}_2]=0, \quad [\boldsymbol{J}^2, \boldsymbol{T}_1\cdot\boldsymbol{T}_2]=0. \tag{4}$$

This is independent of whether $\boldsymbol{T}_1$ commutes with $\boldsymbol{T}_2$ or not.

We shall now consider the problem of obtaining the matrices of T_x, T_y, T_z in a representation in which $\boldsymbol{J}^2$, J_z, and a set A of observables which commute with $\boldsymbol{J}$ are diagonal. We shall first obtain a selection rule on j, i.e. a condition on $j'-j$ necessary for the non-vanishing of a matrix component connecting the states j and j'. This we may do by a method outlined by Dirac (p. 158).

Using the relation 12^26, we find that

$$[\boldsymbol{J}^2, \boldsymbol{T}]=\boldsymbol{J}\cdot[\boldsymbol{J}, \boldsymbol{T}]-[\boldsymbol{T},\boldsymbol{J}]\cdot\boldsymbol{J}= -i\hbar(\boldsymbol{J}\cdot\boldsymbol{T}\times\mathfrak{J}-\boldsymbol{T}\times\mathfrak{J}\cdot\boldsymbol{J})$$
$$= -i\hbar(\boldsymbol{J}\times\boldsymbol{T}-\boldsymbol{T}\times\boldsymbol{J})= -2i\hbar(\boldsymbol{J}\times\boldsymbol{T}-i\hbar\boldsymbol{T}).$$

From this we have

$$[\boldsymbol{J}^2,[\boldsymbol{J}^2,\boldsymbol{T}]]= -2i\hbar[\boldsymbol{J}^2,(\boldsymbol{J}\times\boldsymbol{T}-i\hbar\boldsymbol{T})]= -2i\hbar\{\boldsymbol{J}\times[\boldsymbol{J}^2,\boldsymbol{T}]-i\hbar[\boldsymbol{J}^2,\boldsymbol{T}]\}$$
$$= -2i\hbar\{-2i\hbar\boldsymbol{J}\times(\boldsymbol{J}\times\boldsymbol{T}-i\hbar\boldsymbol{T})-i\hbar(\boldsymbol{J}^2\boldsymbol{T}-\boldsymbol{T}\boldsymbol{J}^2)\}$$
$$=2\hbar^2(\boldsymbol{J}^2\boldsymbol{T}+\boldsymbol{T}\boldsymbol{J}^2)-4\hbar^2\boldsymbol{J}(\boldsymbol{J}\cdot\boldsymbol{T}),$$

using 12^211a to expand $\boldsymbol{J}\times(\boldsymbol{J}\times\boldsymbol{T})$. But

$$[\boldsymbol{J}^2,[\boldsymbol{J}^2,\boldsymbol{T}]]\equiv[\boldsymbol{J}^2,(\boldsymbol{J}^2\boldsymbol{T}-\boldsymbol{T}\boldsymbol{J}^2)]\equiv\boldsymbol{J}^4\boldsymbol{T}-2\boldsymbol{J}^2\boldsymbol{T}\boldsymbol{J}^2+\boldsymbol{T}\boldsymbol{J}^4.$$

Hence

$$\boldsymbol{J}^4\boldsymbol{T}-2\boldsymbol{J}^2\boldsymbol{T}\boldsymbol{J}^2+\boldsymbol{T}\boldsymbol{J}^4=2\hbar^2(\boldsymbol{J}^2\boldsymbol{T}+\boldsymbol{T}\boldsymbol{J}^2)-4\hbar^2\boldsymbol{J}(\boldsymbol{J}\cdot\boldsymbol{T}). \tag{5}$$

Take the matrix component of this equation referring to the states $\alpha\, j\, m$ and $\alpha'\, j'\, m'$, where $j'\neq j$. Since $\boldsymbol{J}\cdot\boldsymbol{T}$ commutes with $\boldsymbol{J}$ [by (4)], this component will vanish for the last term in the equation. From the rest we obtain:

$$\hbar^4[j^2(j+1)^2-2j(j+1)j'(j'+1)+j'^2(j'+1)^2](\alpha j m|\boldsymbol{T}|\alpha' j' m')$$
$$=2\hbar^4[j(j+1)+j'(j'+1)](\alpha j m|\boldsymbol{T}|\alpha' j' m').$$

The bracket on the left is

$$[j(j+1)-j'(j'+1)]^2=(j-j')^2(j+j'+1)^2,$$

while

$$2[j(j+1)+j'(j'+1)]=(j+j'+1)^2+(j-j')^2-1.$$

Hence

$$[(j-j')^2(j+j'+1)^2-(j+j'+1)^2-(j-j')^2+1](\alpha j m|\boldsymbol{T}|\alpha' j' m')=0,$$

or

$$[(j+j'+1)^2-1][(j-j')^2-1](\alpha j m|\boldsymbol{T}|\alpha' j' m')=0. \qquad (j\neq j')$$

In order to obtain a non-vanishing matrix component one of the brackets must vanish. The first cannot since $j'\neq j$ and $j, j'\geqslant 0$. The second vanishes only when $j'-j=\pm 1$. Hence for a non-vanishing matrix component we must have

$$j'-j=0, \pm 1. \tag{6}$$

* An important special instance of this is the square of such a vector.

*The only non-vanishing matrix elements of any component of a vector of the type **T** in the $j\ m$ scheme are those for which* (6) *is satisfied.* This is of course a necessary, not a sufficient, condition for a non-vanishing component.

We may obtain a relation which will be useful later by taking a matrix component of equation (5) diagonal in j, say that joining the states $\alpha\ j\ m$ and $\alpha'\ j\ m'$:

$$j(j+1)\hbar^2(\alpha j m|\boldsymbol{T}|\alpha' j m') = (\alpha j m|\boldsymbol{J}|\alpha j m')(\alpha j m'|\boldsymbol{J}\cdot\boldsymbol{T}|\alpha' j m'), \quad (7)$$

since $\boldsymbol{J}\cdot\boldsymbol{T}$ is diagonal with respect to j and m and since $\boldsymbol{J}$ is diagonal with respect to α by hypothesis.

9. Dependence of the matrix of T on m.*

It will be convenient to consider, in place of T_x and T_y individually, the observable $\mathscr{T} = T_x - iT_y$ since from the matrix of $\mathscr{T}$ we can obtain those of T_x and T_y by using the relations $2T_x = \mathscr{T}^\dagger + \mathscr{T}$ and $2iT_y = \mathscr{T}^\dagger - \mathscr{T}$. These relations obtain because T_x and T_y are assumed to be real observables.

For the matrix of $\mathscr{T}$ we can very easily obtain a selection rule on m. We have

$$[J_z, \mathscr{T}] = [J_z, T_x - iT_y] = [J_z, T_x] - i[J_z, T_y] = i\hbar T_y - \hbar T_x = -\hbar\mathscr{T}.$$

Hence
$$J_z\mathscr{T} - \mathscr{T}J_z = -\hbar\mathscr{T}.$$

Take the $\alpha\ j\ m$; $\alpha'\ j'\ m'$ matrix component:

$$m\hbar(\alpha j m|\mathscr{T}|\alpha' j' m') - (\alpha j m|\mathscr{T}|\alpha' j' m')m'\hbar = -\hbar(\alpha j m|\mathscr{T}|\alpha' j' m')$$

or
$$(m - m' + 1)(\alpha j m|\mathscr{T}|\alpha' j' m') = 0.$$

Then
$$(\alpha j m|\mathscr{T}|\alpha' j' m') = 0 \text{ unless } m' = m + 1,$$

and
$$(\alpha j m|\mathscr{T}^\dagger|\alpha' j' m') = 0 \text{ unless } m' = m - 1.$$

Hence the only non-vanishing matrix component of T_x or T_y are those for which

$$m' = m \pm 1. \quad (1a)$$

Since T_z commutes with J_z, the only non-vanishing components of T_z are those for which

$$m' = m. \quad (1b)$$

We shall now obtain the dependence of the matrix of $\mathscr{T}$ on m. If we denote $J_x - iJ_y$ by $\mathscr{J}$, we have

$$[\mathscr{J}, \mathscr{T}] = [J_x - iJ_y, T_x - iT_y] = [J_x, T_x] - i[J_y, T_x] - i[J_x, T_y] - [J_y, T_y] = 0.$$

The matrix of $\mathscr{J}$ is known. From 3[3]3 the only non-vanishing components are

$$(\alpha j m-1|\mathscr{J}|\alpha j m) = \sqrt{(j+m)(j-m+1)}\,\hbar. \quad (j \geqslant m \geqslant -j+1) \quad (2)‡$$

* The derivation of the selection rule on m is essentially as in DIRAC, p. 158. The method of the rest of the section follows § 29 of BORN and JORDAN, *Elementare Quantenmechanik.*

‡ Although the range $j \geqslant m \geqslant -j+1$ is necessarily all that leads to non-vanishing matrix components, it is important to note that this equation, as well as all the following, is true for the full range $j \geqslant m \geqslant -j$. We shall later wish to take some sums over m and shall sum over this latter range.

Take the most general matrix component which satisfies the selection rules of the equation

$$\mathscr{J}\mathscr{T}=\mathscr{T}\mathscr{J}.$$

This is

$$(\alpha j\,m-1|\mathscr{J}|\alpha j\,m)\,(\alpha j\,m|\mathscr{T}|\alpha' j'\,m+1)$$
$$=(\alpha j\,m-1|\mathscr{T}|\alpha' j'\,m)\,(\alpha' j'\,m|\mathscr{J}|\alpha' j'\,m+1)\qquad (j'-j=0,\ \pm 1)$$

or, from (2),

$$\sqrt{(j+m)(j-m+1)}\,(\alpha j\,m|\mathscr{T}|\alpha' j'\,m+1)$$
$$=(\alpha j\,m-1|\mathscr{T}|\alpha' j'\,m)\sqrt{(j'+m+1)(j'-m)}.\qquad (j'-j=0,\ \pm 1)\quad (3)$$

If $\underline{j'=j}$, we have

$$\frac{(\alpha j\,m|\mathscr{T}|\alpha' j\,m+1)}{\sqrt{(j-m)(j+m+1)}}=\frac{(\alpha j\,m-1|\mathscr{T}|\alpha' j\,m)}{\sqrt{(j-m+1)(j+m)}};$$

and since this holds for any value of m we see from the form of the relation that each ratio must be independent of m. We shall denote this ratio by $(\alpha j\vdots T\vdots\alpha' j)$, this quantity being independent of m. Hence we find, for the dependence on m of the elements of $\mathscr{T}$ diagonal in j:

$$(\alpha j\,m|\mathscr{T}|\alpha' j\,m+1)=(\alpha j\vdots T\vdots\alpha' j)\sqrt{(j-m)(j+m+1)}.\qquad (4)$$

If $\underline{j'=j-1}$, (3) becomes

$$\sqrt{(j+m)(j-m+1)}\,(\alpha j\,m|\mathscr{T}|\alpha' j-1\,m+1)$$
$$=(\alpha j\,m-1|\mathscr{T}|\alpha' j-1\,m)\sqrt{(j+m)(j-m-1)}.$$

Multiply through by $\sqrt{(j-m)/(j+m)}$ and rewrite as

$$\frac{(\alpha j\,m|\mathscr{T}|\alpha' j-1\,m+1)}{\sqrt{(j-m-1)(j-m)}}=\frac{(\alpha j\,m-1|\mathscr{T}|\alpha' j-1\,m)}{\sqrt{(j-m)(j-m+1)}}=(\alpha j\vdots T\vdots\alpha' j-1).$$

Again each ratio is independent of m and hence is set equal to $(\alpha j\vdots T\vdots\alpha' j-1)$. Then

$$(\alpha j\,m|\mathscr{T}|\alpha' j-1\,m+1)=(\alpha j\vdots T\vdots\alpha' j-1)\sqrt{(j-m)(j-m-1)}.\qquad (5)$$

If $\underline{j'=j+1}$, (3) becomes

$$\sqrt{(j+m)(j-m+1)}\,(\alpha j\,m|\mathscr{T}|\alpha' j+1\,m+1)$$
$$=(\alpha j\,m-1|\mathscr{T}|\alpha' j+1\,m)\sqrt{(j+m+2)(j-m+1)}.$$

Multiply by $\sqrt{(j+m+1)/(j-m+1)}$ and rewrite:

$$\frac{(\alpha j\,m|\mathscr{T}|\alpha' j+1\,m+1)}{\sqrt{(j+m+2)(j+m+1)}}=\frac{(\alpha j\,m-1|\mathscr{T}|\alpha' j+1\,m)}{\sqrt{(j+m+1)(j+m)}}=-(\alpha j\vdots T\vdots\alpha' j+1).$$

Hence

$$(\alpha j\,m|\mathscr{T}|\alpha' j+1\,m+1)=-(\alpha j\vdots T\vdots\alpha' j+1)\sqrt{(j+m+1)(j+m+2)}.\quad (6)$$

We shall now determine the dependence of the matrix of T_z on m. We have

$$[\mathscr{J}^\dagger,\mathscr{T}]=[J_x+iJ_y,\mathscr{T}]=[\mathscr{J},\mathscr{T}]+2i[J_y,T_x-iT_y]=2\hbar T_z;$$

that is,

$$2\hbar T_z=\mathscr{J}^\dagger\mathscr{T}-\mathscr{T}\mathscr{J}^\dagger.\qquad (7)$$

Now the only non-vanishing matrix elements of $\mathscr{J}^\dagger$ are

$$(\alpha j\, m|\mathscr{J}^\dagger|\alpha j\, m-1) = \sqrt{(j+m)(j-m+1)}\,\hbar. \quad (j \geqslant m \geqslant -j+1)$$

Since we know the matrices of $\mathscr{J}^\dagger$ and $\mathscr{T}$ we can determine the matrix of T_z directly from (7). We have from (1b) the selection rule $m' = m$.

Then for $j = j'$,

$$\begin{aligned} 2\hbar\,(\alpha j\, m|T_z|\alpha' j\, m) &= (\alpha j\vdots T\vdots \alpha' j)\sqrt{(j+m)(j-m+1)}\sqrt{(j-m+1)(j+m)}\,\hbar \\ &\quad - (\alpha j\vdots T\vdots \alpha' j)\sqrt{(j-m)(j+m+1)}\sqrt{(j+m+1)(j-m)}\,\hbar \\ &= 2m\hbar\,(\alpha j\vdots T\vdots \alpha' j), \end{aligned}$$

or

$$(\alpha j\, m|T_z|\alpha' j\, m) = m\,(\alpha j\vdots T\vdots \alpha' j). \tag{8}$$

Also, since T_z is real, we see that $(\alpha j\vdots T\vdots \alpha' j) = \overline{(\alpha' j\vdots T\vdots \alpha j)}$.

For $j' = j-1$, we have from (7):

$$\begin{aligned} 2\hbar\,(\alpha j\, m|T_z|\alpha' j-1\, m) &= (\alpha j\, m|\mathscr{J}^\dagger|\alpha j\, m-1)\,(\alpha j\, m-1|\mathscr{T}|\alpha' j-1\, m) \\ &\quad - (\alpha j\, m|\mathscr{T}|\alpha' j-1\, m+1)\,(\alpha' j-1\, m+1|\mathscr{J}^\dagger|\alpha' j-1\, m) \\ &= (\alpha j\vdots T\vdots \alpha' j-1)\sqrt{(j-m+1)(j-m)}\sqrt{(j+m)(j-m+1)}\,\hbar \\ &\quad - (\alpha j\vdots T\vdots \alpha' j-1)\sqrt{(j-m)(j-m-1)}\sqrt{(j+m)(j-m-1)}\,\hbar \\ &= 2\sqrt{j^2-m^2}\,(\alpha j\vdots T\vdots \alpha' j-1)\,\hbar \end{aligned}$$

or

$$(\alpha j\, m|T_z|\alpha' j-1\, m) = \sqrt{j^2-m^2}\,(\alpha j\vdots T\vdots \alpha' j-1). \tag{9}$$

For $j' = j+1$ one obtains in an analogous way from (7)

$$(\alpha j\, m|T_z|\alpha' j+1\, m) = \sqrt{(j+1)^2-m^2}\,(\alpha j\vdots T\vdots \alpha' j+1). \tag{10}$$

Since T_z is real we see by comparison of this with (9) that

$$(\alpha j\vdots T\vdots \alpha' j-1) = \overline{(\alpha' j-1\vdots T\vdots \alpha j)}.$$

Hence the matrix $(\alpha j\vdots T\vdots \alpha' j')$ *as we have defined it is Hermitian.* This fact enables us at once to obtain the matrix of $\mathscr{T}^\dagger$. For instance, from (5) we have

$$\begin{aligned} (\alpha' j-1\, m+1|\mathscr{T}^\dagger|\alpha j\, m) &= \overline{(\alpha j\, m|\mathscr{T}|\alpha' j-1\, m+1)} \\ &= (\alpha' j-1\vdots T\vdots \alpha j)\sqrt{(j-m)(j-m-1)}. \end{aligned}$$

The other components of $\mathscr{T}^\dagger$ are found in a similar way.

Collecting all our results, we have the following table of the non-vanishing matrix components of $\boldsymbol{T}$:

$$\begin{aligned} (\alpha j\, m|\boldsymbol{T}|\alpha' j+1\, m\pm 1) &= \mp(\alpha j\vdots T\vdots \alpha' j+1)\,\tfrac{1}{2}\sqrt{(j\pm m+1)(j\pm m+2)}\,(\boldsymbol{i}\pm i\boldsymbol{j}) \\ (\alpha j\, m|\boldsymbol{T}|\alpha' j+1\, m) &= (\alpha j\vdots T\vdots \alpha' j+1)\sqrt{(j+1)^2-m^2}\,\boldsymbol{k} \\ (\alpha j\, m|\boldsymbol{T}|\alpha' j\, m\pm 1) &= (\alpha j\vdots T\vdots \alpha' j)\,\tfrac{1}{2}\sqrt{(j\mp m)(j\pm m+1)}\,(\boldsymbol{i}\pm i\boldsymbol{j}) \\ (\alpha j\, m|\boldsymbol{T}|\alpha' j\, m) &= (\alpha j\vdots T\vdots \alpha' j)\,m\,\boldsymbol{k} \\ (\alpha j\, m|\boldsymbol{T}|\alpha' j-1\, m\pm 1) &= \pm(\alpha j\vdots T\vdots \alpha' j-1)\,\tfrac{1}{2}\sqrt{(j\mp m)(j\mp m-1)}\,(\boldsymbol{i}\pm i\boldsymbol{j}) \\ (\alpha j\, m|\boldsymbol{T}|\alpha' j-1\, m) &= (\alpha j\vdots T\vdots \alpha' j-1)\sqrt{j^2-m^2}\,\boldsymbol{k}. \end{aligned} \tag{11}$$

From its definition, if T_x, T_y and T_z commute with J^2, the matrix $(\alpha j \vdots T \vdots \alpha' j')$ is diagonal in j; if they commute with A it is diagonal in α. These remarks hold in particular for J_x, J_y and J_z, for which

$$(\alpha j \vdots J \vdots \alpha' j') = \hbar \delta_{jj'} \delta_{\alpha\alpha'},$$

so that in this case the formulas (11) agree with $3^3 7$.

From the above formulas and from $3^3 11$, we see that in going from the scheme $\alpha j m$ to a scheme $\beta j m$, the matrix $\|(\alpha j \vdots T \vdots \alpha' j')\|$ is transformed by $\|(\alpha j | \beta j)\|$, just like the matrix of an observable:

$$(\beta j \vdots T \vdots \beta' j') = \sum_{\alpha\alpha'} (\beta j | \alpha j)(\alpha j \vdots T \vdots \alpha' j')(\alpha' j' | \beta' j'). \tag{12}$$

Here we have used the notation $(\beta j | \alpha j) \equiv (\beta j m | \alpha j m)$ justified by the proof in § 3^3 that this element is independent of m. Hence we see that in every respect $\|(\alpha j \vdots T \vdots \alpha' j')\|$ behaves like the matrix of a real observable.

10. The matrices of J_1 and J_2, where $J_1 + J_2 = J$.*

If J is the sum of two commuting angular momenta J_1 and J_2, we can obtain the complete matrices of J_1 and J_2. Consider the representation $\gamma j_1 j_2 j m$, where the observables Γ commute with J_1, J_2, J. (If J is the total angular momentum of an atom; $J_1 = S$, its total spin momentum; and $J_2 = L$, its total orbital momentum, some of the γ's would naturally represent the n's, l's and s's of the individual electrons.) Then the matrices of J_1 and J_2 will be diagonal in γ, j_1, and j_2. Since they will not depend on γ, we shall not carry the label γ. For brevity we shall later drop also the labels j_1 and j_2.

$9^3 11$ gives us the dependence of these matrices on m; hence we need now determine only the $(j_1 j_2 j \vdots J_1 \vdots j_1 j_2 j')$ and $(j_1 j_2 j \vdots J_2 \vdots j_1 j_2 j')$. For the case $j' = j$ these follow easily from $8^3 7$, which becomes for the z component of J_1:

$$j(j+1)\hbar^2 (j_1 j_2 j m | J_{1z} | j_1 j_2 j m)$$
$$= (j_1 j_2 j m | J_1 \cdot J | j_1 j_2 j m)(j_1 j_2 j m | J_z | j_1 j_2 j m);$$

or, from $9^3 11$

$$j(j+1)\hbar (j_1 j_2 j \vdots J_1 \vdots j_1 j_2 j) = (j_1 j_2 j m | J_1 \cdot J | j_1 j_2 j m). \tag{1}$$

But
$$J_2^2 = (J - J_1)^2 = J^2 - 2J_1 \cdot J + J_1^2;$$
or
$$J_1 \cdot J = \tfrac{1}{2}(J_1^2 - J_2^2 + J^2).$$

Using this relation, (1) gives

$$(\gamma j_1 j_2 j \vdots J_1 \vdots \gamma j_1 j_2 j) = \frac{j_1(j_1+1) - j_2(j_2+1) + j(j+1)}{2j(j+1)} \hbar, \tag{2a}$$

and the corresponding element of J_2 is obtained by interchanging subscripts 1 *and* 2.

* §§ 10 and 11 follow Güttinger and Pauli, Zeits. für Phys. **67**, 743 (1931).

The components not diagonal in j are more complicated. We use the relation

$$[J_{1x}-iJ_{1y}, J_{1z}] = -i\hbar J_{1y} - i(i\hbar J_{1x}) = \hbar(J_{1x}-iJ_{1y}),$$

or

$$(J_{1x}-iJ_{1y})J_{1z} - J_{1z}(J_{1x}-iJ_{1y}) = \hbar(J_{1x}-iJ_{1y}).$$

Take the $j_1 j_2 j m; j_1 j_2 j m+1$ component of this equation, using $9^3 4$ to $9^3 10$ (omitting the labels j_1, j_2):

$$\begin{aligned}
&(j{\vdots}J_1{\vdots}j)^2\sqrt{(j-m)(j+m+1)}\,(m+1)\\
&\quad -(j{\vdots}J_1{\vdots}j+1)\sqrt{(j+m+1)(j+m+2)}\,(j+1{\vdots}J_1{\vdots}j)\sqrt{(j+1)^2-(m+1)^2}\\
&\quad +(j{\vdots}J_1{\vdots}j-1)\sqrt{(j-m)(j-m-1)}\,(j-1{\vdots}J_1{\vdots}j)\sqrt{j^2-(m+1)^2}\\
&\quad -(j{\vdots}J_1{\vdots}j)^2 m\sqrt{(j-m)(j+m+1)}\\
&\quad -(j{\vdots}J_1{\vdots}j+1)\sqrt{(j+1)^2-m^2}\,(j+1{\vdots}J_1{\vdots}j)\sqrt{(j-m+1)(j-m)}\\
&\quad +(j{\vdots}J_1{\vdots}j-1)\sqrt{j^2-m^2}\,(j-1{\vdots}J_1{\vdots}j)\sqrt{(j+m)(j+m+1)}\\
&\quad =\hbar(j{\vdots}J_1{\vdots}j)\sqrt{(j-m)(j+m+1)},
\end{aligned}$$

or $$(j{\vdots}J_1{\vdots}j)^2 - |(j{\vdots}J_1{\vdots}j+1)|^2(2j+3) + |(j{\vdots}J_1{\vdots}j-1)|^2(2j-1) = \hbar(j{\vdots}J_1{\vdots}j).$$

Now, from (2a)

$$\begin{aligned}
&\hbar(j{\vdots}J_1{\vdots}j) - (j{\vdots}J_1{\vdots}j)^2\\
&\quad = \frac{[j_1(j_1+1)-j_2(j_2+1)+j(j+1)]\hbar^2}{[2j(j+1)]^2}[j_2(j_2+1)-j_1(j_1+1)+j(j+1)].
\end{aligned}$$

Hence, with the abbreviation

$$p = j_1(j_1+1) - j_2(j_2+1),$$

we obtain the relation

$$|(j{\vdots}J_1{\vdots}j-1)|^2(2j-1) - |(j{\vdots}J_1{\vdots}j+1)|^2(2j+3) = \frac{[j(j+1)]^2-p^2}{[2j(j+1)]^2}\hbar^2. \quad (3)$$

We get another relation in a similar fashion by taking the $j_1 j_2 j m; j_1 j_2 j m$ component of the identity

$$J_1^2 = \tfrac{1}{2}(J_{1x}+iJ_{1y})(J_{1x}-iJ_{1y}) + \tfrac{1}{2}(J_{1x}-iJ_{1y})(J_{1x}+iJ_{1y}) + J_{1z}^2.$$

This becomes

$$\begin{aligned}
&j_1(j_1+1)\hbar^2\\
&\quad = |(j{\vdots}J_1{\vdots}j-1)|^2 j(2j-1) + |(j{\vdots}J_1{\vdots}j+1)|^2(j+1)(2j+3) + (j{\vdots}J_1{\vdots}j)^2 j(j+1),
\end{aligned}$$

or, from (2a), and with the abbreviation

$$q = j_1(j_1+1) + j_2(j_2+1),$$

$$\begin{aligned}
&|(j{\vdots}J_1{\vdots}j-1)|^2 j(2j-1) + |(j{\vdots}J_1{\vdots}j+1)|^2(j+1)(2j+3)\\
&\quad = \frac{-p^2+2qj(j+1)-j^2(j+1)^2}{4j(j+1)}\hbar^2. \quad (4)
\end{aligned}$$

We can solve (3) and (4) for $|(j\vdots J_1\vdots j-1)|^2$ and $|(j\vdots J_1\vdots j+1)|^2$. For the former we obtain

$$|(j\vdots J_1\vdots j-1)|^2[(2j+1)(2j-1)]=\frac{-j^4+(2q+1)j^2-p^2}{4j^2}\hbar^2.$$

Now $\quad 2q+1=(j_1-j_2)^2+(j_1+j_2+1)^2, \quad p^2=(j_1-j_2)^2(j_1+j_2+1)^2;$

hence

$$|(j\vdots J_1\vdots j-1)|^2=\frac{-j^4+(j_1-j_2)^2j^2+(j_1+j_2+1)^2j^2-(j_1-j_2)^2(j_1+j_2+1)^2}{4j^2(4j^2-1)}\hbar^2$$

$$=\frac{[j^2-(j_1-j_2)^2][(j_1+j_2+1)^2-j^2]}{4j^2(4j^2-1)}\hbar^2. \qquad (5)$$

Since $|(j\vdots J_1\vdots j+1)|^2=|(j+1\vdots J_1\vdots j)|^2$, this quantity may be obtained from (5) by substitution of $j+1$ for j.

We have now arrived at the point where we should choose the relative phases of the states of different j. This we do conveniently if we take $(j\vdots J_1\vdots j-1)$ to be real and positive for all j. This means that the non-diagonal elements of J_{1z} are all real and positive. Since the matrix of $\boldsymbol{J}=\boldsymbol{J}_1+\boldsymbol{J}_2$ must be diagonal with respect to j, we have $(j\vdots J_2\vdots j-1)=-(j\vdots J_1\vdots j-1)$. Hence if the non-diagonal elements of J_{1z} are all positive, those of J_{2z} are all negative. An alternative choice of phase would reverse these signs and hence reverse the rôles of $\boldsymbol{J}_1$ and $\boldsymbol{J}_2$. We therefore write

$$\begin{aligned}(\gamma j_1 j_2 j\vdots J_1\vdots \gamma j_1 j_2 j-1)&\\ =(\gamma j_1 j_2 j-1\vdots J_1\vdots \gamma j_1 j_2 j)&\\ =\{\pm\}\hbar\sqrt{\frac{(j-j_1+j_2)(j+j_1-j_2)(j_1+j_2+1+j)(j_1+j_2+1-j)}{4j^2(2j-1)(2j+1)}}&,\\ (\gamma j_1 j_2 j\vdots J_2\vdots \gamma j_1 j_2 j-1)&\\ =(\gamma j_1 j_2 j-1\vdots J_2\vdots \gamma j_1 j_2 j)& \qquad (2b)^*\\ =\{\mp\}\ \textit{the same expression.}&\end{aligned}$$

In the following pages we must be consistent in choosing everywhere either the top or bottom sign wherever the symbols $\{\pm\}$ and $\{\mp\}$ occur. We shall carry both signs as a matter of convenience in the sections immediately following. However, elsewhere *we shall always use* (2b) *with the upper sign. In adding* $\boldsymbol{J}_1$ *and* $\boldsymbol{J}_2$ *we choose the relative phases of states of different* j *so that the non-diagonal elements of* J_{1z} *are positive, of* J_{2z} *negative*

* With the abbreviations $\quad P(j)=(j-j_1+j_2)(j+j_1+j_2+1)$

$$Q(j)=(j_1+j_2-j)(j+j_1-j_2+1),$$

which we shall use in the next section, this formula becomes

$$(\gamma j_1 j_2 j\vdots J_1\vdots \gamma j_1 j_2 j-1)=\{\pm\}\hbar\frac{\sqrt{P(j)\,Q(j-1)}}{2j\sqrt{4j^2-1}}. \qquad (2b')$$

It is to be noted that P is a function only of j_1 and $j+j_2$ while Q is a function only of j_1 and $j-j_2$.

At this point we may also conveniently make a definite choice of the phases of the states $\psi(\gamma j_1 j_2 j m)$ of $6^3 2$ relative to the states $\phi(\gamma j_1 j_2 m_1 m_2)$ of $6^3 1$. These phases are completely specified by the above choice of relative phases and the following specification for the state of largest j and m (namely $j=m=j_1+j_2$):

$$\psi(\gamma j_1 j_2 j_1+j_2 j_1+j_2)=\phi(\gamma j_1 j_2 j_1 j_2). \tag{6}$$

11. Matrix of a vector $\boldsymbol{P}$ which commutes with $\boldsymbol{J}_1$.

We shall now consider the matrix of a vector $\boldsymbol{P}$ which commutes with $\boldsymbol{J}_1$, but which satisfies $8^3 1$ with respect to $\boldsymbol{J}$, where $\boldsymbol{J}=\boldsymbol{J}_1+\boldsymbol{J}_2$. Because of this latter relation we see that $\boldsymbol{P}$ must satisfy $8^3 1$ with respect to $\boldsymbol{J}_2$ also. The dependence of $\boldsymbol{P}$ on m is given by $9^3 11$; we shall proceed to find its dependence on j in the $\gamma j_1 j_2 j m$ scheme. (If $\boldsymbol{J}_1=\boldsymbol{S}$, $\boldsymbol{P}$ may be $e\Sigma \boldsymbol{r}_i$, the total electric moment, or may be $\boldsymbol{L}_i$, the orbital momentum of the ith electron, or $\boldsymbol{r}_i$ or $\boldsymbol{p}_i$, the coordinate or momentum of the ith electron, etc.) The matrix of $\boldsymbol{P}$ will necessarily be diagonal in j_1, so for convenience we may omit the label j_1. We shall also omit for the moment the γ; γ' which will occur in all matrix components of $\boldsymbol{P}$ ($\boldsymbol{J}_1$ and $\boldsymbol{J}_2$ are necessarily diagonal in γ). Because $\boldsymbol{P}$ satisfies $8^3 1$ with respect to $\boldsymbol{J}_2$, we have the *selection rule* $j_2'-j_2=\pm 1$ or zero.

Consider the $\gamma j_1 j_2 j' m; \gamma' j_1 j_2' j'' m+1$ component of the equation

$$(J_{1x}-iJ_{1y})\,P_z=P_z\,(J_{1x}-iJ_{1y}). \tag{1}$$

First for $j'=j-1, j''=j+1$ (from $9^3 6$ and $9^3 10$):

$$-(j_2 j-1\!\vdots\! J_1\!\vdots\! j_2 j)\sqrt{(j+m)(j+m+1)}\,(j_2 j\!\vdots\! P\!\vdots\! j_2' j+1)\sqrt{(j+m+2)(j-m)}$$
$$=-(j_2 j-1\!\vdots\! P\!\vdots\! j_2' j)\sqrt{(j-m)(j+m)}\,(j_2' j\!\vdots\! J_1\!\vdots\! j_2' j+1)\sqrt{(j+m+1)(j+m+2)},$$

or
$$(j_2 j-1\!\vdots\! J_1\!\vdots\! j_2 j)\,(j_2 j\!\vdots\! P\!\vdots\! j_2' j+1)=(j_2 j-1\!\vdots\! P\!\vdots\! j_2' j)\,(j_2' j\!\vdots\! J_1\!\vdots\! j_2' j+1). \tag{2}$$

Since the matrix of J_1 is known, this furnishes a recursion formula for

$$(j_2 j\!\vdots\! P\!\vdots\! j_2' j+1).$$

Now consider the case $j'=j, j''=j+1$. Here we obtain

$$(j_2 j\!\vdots\! J_1\!\vdots\! j_2 j)\,(j_2 j\!\vdots\! P\!\vdots\! j_2' j+1)\,(j-m)$$
$$-(j_2 j\!\vdots\! J_1\!\vdots\! j_2 j+1)\,(j_2 j+1\!\vdots\! P\!\vdots\! j_2' j+1)\,(m+1)+(j_2 j\!\vdots\! P\!\vdots\! j_2' j)\,(j_2' j\!\vdots\! J_1\!\vdots\! j_2' j+1)\,m$$
$$-(j_2 j\!\vdots\! P\!\vdots\! j_2' j+1)\,(j_2' j+1\!\vdots\! J_1\!\vdots\! j_2' j+1)\,(j-m+1)=0.$$

Since this relation is independent of m, the coefficients of m and of unity must each vanish, i.e.

$$[(j_2 j\!\vdots\! J_1\!\vdots\! j_2 j)j-(j_2' j+1\!\vdots\! J_1\!\vdots\! j_2' j+1)\,(j+1)]\,(j_2 j\!\vdots\! P\!\vdots\! j_2' j+1)$$
$$-(j_2 j\!\vdots\! J_1\!\vdots\! j_2 j+1)\,(j_2 j+1\!\vdots\! P\!\vdots\! j_2' j+1)=0,$$

$$[-(j_2 j\!\vdots\! J_1\!\vdots\! j_2 j)+(j_2' j+1\!\vdots\! J_1\!\vdots\! j_2' j+1)]\,(j_2 j\!\vdots\! P\!\vdots\! j_2' j+1)$$
$$-(j_2 j\!\vdots\! J_1\!\vdots\! j_2 j+1)\,(j_2 j+1\!\vdots\! P\!\vdots\! j_2' j+1)+(j_2' j\!\vdots\! J_1\!\vdots\! j_2' j+1)\,(j_2 j\!\vdots\! P\!\vdots\! j_2' j)=0.$$

Subtraction of the second of these from the first gives

$$[(j_2 j\vdots J_1\vdots j_2 j)(j+1)-(j_2' j+1\vdots J_1\vdots j_2' j+1)(j+2)](j_2 j\vdots P\vdots j_2' j+1)$$
$$-(j_2' j\vdots J_1\vdots j_2' j+1)(j_2 j\vdots P\vdots j_2' j)=0. \quad (3)$$

In the same way we obtain from (1) for the case $j'=j$, $j''=j-1$ the two equations

$$[(j_2 j\vdots J_1\vdots j_2 j)(j+1)-(j_2' j-1\vdots J_1\vdots j_2' j-1)j](j_2 j\vdots P\vdots j_2' j-1)$$
$$+(j_2 j\vdots J_1\vdots j_2 j-1)(j_2 j-1\vdots P\vdots j_2' j-1)=0,$$
$$[(j_2 j\vdots J_1\vdots j_2 j)-(j_2' j-1\vdots J_1\vdots j_2' j-1)](j_2 j\vdots P\vdots j_2' j-1)$$
$$+(j_2 j\vdots J_1\vdots j_2 j-1)(j_2 j-1\vdots P\vdots j_2' j-1)-(j_2' j\vdots J_1\vdots j_2' j-1)(j_2 j\vdots P\vdots j_2' j)=0.$$

Subtraction of the first of these from the second gives

$$[-(j_2 j\vdots J_1\vdots j_2 j)j+(j_2' j-1\vdots J_1\vdots j_2' j-1)(j-1)](j_2 j\vdots P\vdots j_2' j-1)$$
$$-(j_2' j\vdots J_1\vdots j_2' j-1)(j_2 j\vdots P\vdots j_2' j)=0. \quad (4)$$

If we can determine $(j_2 j\vdots P\vdots j_2' j+1)$ from (2), (3) will give us $(j_2 j\vdots P\vdots j_2' j)$, and (4), $(j_2 j\vdots P\vdots j_2' j-1)$.

For the case $j_2'=j_2$, (2) gives us

$$\frac{(j_2 j\vdots P\vdots j_2 j+1)}{(j_2 j\vdots J_1\vdots j_2 j+1)}=\frac{(j_2 j-1\vdots P\vdots j_2 j)}{(j_2 j-1\vdots J_1\vdots j_2 j)}=-(j_2\vdots P\vdots j_2).$$

We see that these ratios must be independent of j, as implied by the notation $(j_2\vdots P\vdots j_2)$. Hence

$$(j_2 j\vdots P\vdots j_2 j+1)=-(j_2\vdots P\vdots j_2)(j_2 j\vdots J_1\vdots j_2 j+1), \quad (5)$$

where the last factor is given by $10^3 2$b.

From (3) we obtain, using (5),

$$-[(j_2 j\vdots J_1\vdots j_2 j)(j+1)-(j_2 j+1\vdots J_1\vdots j_2 j+1)(j+2)](j_2\vdots P\vdots j_2)=(j_2 j\vdots P\vdots j_2 j),$$

or, from $10^3 2$a

$$(j_2 j\vdots P\vdots j_2 j)=(j_2\vdots P\vdots j_2)\frac{j_2(j_2+1)-j_1(j_1+1)+j(j+1)}{2j(j+1)}. \quad (6)$$

Similarly, from (4), or by taking the complex conjugate of (5), we have

$$(j_2 j\vdots P\vdots j_2 j-1)=-(j_2\vdots P\vdots j_2)(j_2 j\vdots J_1\vdots j_2 j-1). \quad (7)$$

For the case $j_2'=j_2-1$, (2) becomes, on substitution from $10^3 2$b′,

$$\{\pm\}\frac{\sqrt{P(j)\,Q(j-1)}}{2j\sqrt{4j^2-1}}(j_2 j\vdots P\vdots j_2-1 j+1)$$
$$=\{\pm\}(j_2 j-1\vdots P\vdots j_2-1 j)\frac{\sqrt{P(j)\,Q(j+1)}}{2(j+1)\sqrt{4(j+1)^2-1}}$$

or
$$\frac{2(j+1)\sqrt{4(j+1)^2-1}\,(j_2 j\vdots P\vdots j_2-1 j+1)}{\sqrt{Q(j)\,Q(j+1)}}$$
$$=\frac{2j\sqrt{4j^2-1}\,(j_2 j-1\vdots P\vdots j_2-1 j)}{\sqrt{Q(j-1)\,Q(j)}}=-(j_2\vdots P\vdots j_2-1).$$

This gives us $(j_2\, j \vdots P \vdots j_2-1\, j+1)$. Knowing this, (3) and (4) give us

$$(j_2\, j \vdots P \vdots j_2-1\, j) \quad \text{and} \quad (j_2\, j \vdots P \vdots j_2-1\, j-1).$$

The $(j_2\, j \vdots P \vdots j_2+1\, j')$ are essentially the complex conjugates of these quantities.

Using for conciseness the abbreviations*

$$P(j) = (j-j_1+j_2)(j+j_1+j_2+1),$$
$$Q(j) = (j_1+j_2-j)(j+j_1-j_2+1),$$
$$R(j) = j(j+1) - j_1(j_1+1) + j_2(j_2+1),$$

we find the following complete table of formulas for the dependence on j of the matrix of a vector $\boldsymbol{P}$ which commutes with $\boldsymbol{J}_1$, where $\boldsymbol{J}_1+\boldsymbol{J}_2=\boldsymbol{J}$:

$$
\begin{aligned}
(\gamma j_1 j_2 j \vdots P \vdots \gamma' j_1 j_2{+}1\, j{+}1) &= \phantom{\{\pm\}}(\gamma j_1 j_2 \vdots P \vdots \gamma' j_1 j_2{+}1) \frac{\sqrt{P(j+2)\,P(j+1)}}{2(j+1)\sqrt{(2j+1)(2j+3)}} \\
(\gamma j_1 j_2 j \vdots P \vdots \gamma' j_1 j_2{+}1\, j) &= \{\pm\}(\gamma j_1 j_2 \vdots P \vdots \gamma' j_1 j_2{+}1) \frac{\sqrt{P(j+1)\,Q(j-1)}}{2j(j+1)} \\
(\gamma j_1 j_2 j \vdots P \vdots \gamma' j_1 j_2{+}1\, j{-}1) &= -\,(\gamma j_1 j_2 \vdots P \vdots \gamma' j_1 j_2{+}1) \frac{\sqrt{Q(j-1)\,Q(j-2)}}{2j\sqrt{(2j-1)(2j+1)}} \\
(\gamma j_1 j_2 j \vdots P \vdots \gamma' j_1 j_2\, j{+}1) &= \{\mp\}(\gamma j_1 j_2 \vdots P \vdots \gamma' j_1 j_2) \frac{\sqrt{P(j+1)\,Q(j)}}{2(j+1)\sqrt{(2j+1)(2j+3)}} \\
(\gamma j_1 j_2 j \vdots P \vdots \gamma' j_1 j_2\, j) &= \phantom{\{\pm\}}(\gamma j_1 j_2 \vdots P \vdots \gamma' j_1 j_2) \frac{R(j)}{2j(j+1)} \\
(\gamma j_1 j_2 j \vdots P \vdots \gamma' j_1 j_2\, j{-}1) &= \{\mp\}(\gamma j_1 j_2 \vdots P \vdots \gamma' j_1 j_2) \frac{\sqrt{P(j)\,Q(j-1)}}{2j\sqrt{(2j-1)(2j+1)}} \\
(\gamma j_1 j_2 j \vdots P \vdots \gamma' j_1 j_2{-}1\, j{+}1) &= -\,(\gamma j_1 j_2 \vdots P \vdots \gamma' j_1 j_2{-}1) \frac{\sqrt{Q(j)\,Q(j+1)}}{2(j+1)\sqrt{(2j+1)(2j+3)}} \\
(\gamma j_1 j_2 j \vdots P \vdots \gamma' j_1 j_2{-}1\, j) &= \{\pm\}(\gamma j_1 j_2 \vdots P \vdots \gamma' j_1 j_2{-}1) \frac{\sqrt{P(j)\,Q(j)}}{2j(j+1)} \\
(\gamma j_1 j_2 j \vdots P \vdots \gamma' j_1 j_2{-}1\, j{-}1) &= \phantom{\{\pm\}}(\gamma j_1 j_2 \vdots P \vdots \gamma' j_1 j_2{-}1) \frac{\sqrt{P(j)\,P(j-1)}}{2j\sqrt{(2j-1)(2j+1)}}.
\end{aligned}
\tag{8}
$$

When we seek the corresponding components for a vector $\boldsymbol{Q}$ which commutes with $\boldsymbol{J}_2$, since the only essential distinction between $\boldsymbol{J}_1$ and $\boldsymbol{J}_2$ is the opposite choice of signs in $10^3$2b, we find that *the dependence on j of $\boldsymbol{Q}$ is given by* (8) *with j_1 and j_2 interchanged and the double signs inverted.* Since $\boldsymbol{J}_2$ is itself of the type $\boldsymbol{P}$, and $\boldsymbol{J}_1$ of the type $\boldsymbol{Q}$, the formulas $10^3$2 are special cases of (8).

* We note that $P(j)$ and $Q(j)$ are essentially positive quantities since $|j_1-j_2| \leq j \leq j_1+j_2$. $R(j)$ takes on negative values in certain instances.

The question of obtaining further information about the factors

$$(\gamma j_1 j_2 \vdots P \vdots \gamma' j_1 j_2')$$

now arises. Since $\boldsymbol{P}$ satisfies the commutation rule 8^31 with respect to $\boldsymbol{J}_2$ and commutes with $\boldsymbol{J}_1$, we can express directly from 9^311 the dependence on m_2 of the matrix of $\boldsymbol{P}$ in the $\gamma j_1 j_2 m_1 m_2$ scheme. This matrix will of course be diagonal in j_1 and m_1 and by 3^38 the elements will be entirely independent of m_1. Further, a vector having the properties of $\boldsymbol{P}$ will usually, if not always, be of such a character that when a resolution of the type 6^35 is made of the states in question the vector will operate only on ϕ_2, not on ϕ_1. In this case the matrix elements will be independent also of the value of j_1. This will be true in all cases which we shall experience; hence we shall here treat these elements as independent of j_1.

We shall, for example, have from the first formula of 9^311:

$$(\gamma j_1 j_2 m_1 m_2|\boldsymbol{P}|\gamma' j_1 j_2{+}1\, m_1 m_2{\pm}1)$$
$$= \mp (\gamma j_1 j_2 \vdots P \vdots \gamma' j_1 j_2{+}1)\,(\boldsymbol{i} \pm i\boldsymbol{j})\,\tfrac{1}{2}\sqrt{(j_2 \pm m_2 + 1)(j_2 \pm m_2 + 2)}. \quad (9)$$

There occur here on the right the factors $(\gamma j_1 j_2 \vdots P \vdots \gamma' j_1 j_2') = (\gamma j_2 \vdots P \vdots \gamma' j_2')$ for which in accordance with 9^311 we use exactly the same notation as we have adopted for the factors occurring in (8). That these factors actually are equal is shown very simply by a calculation of $(\gamma j_1 j_2 m_1 m_2|\boldsymbol{P}|\gamma' j_1 j_2' m_1 m_2')$ from 9^311 and of $(\gamma j_1 j_2 j m|\boldsymbol{P}|\gamma' j_1 j_2' j' m')$ from 9^311 and (8) for the case

$$m_1 = j_1, \quad m_2 = j_2, \quad m_2' = j_2', \quad j = m = j_1 + j_2, \quad j' = m' = j_1 + j_2',$$

in which we know from 10^36 that these two matrix components are equal.

Knowing that the factor $(\gamma j_2 \vdots P \vdots \gamma' j_2')$ *of* (8) *has the same properties with respect to the vector* $\boldsymbol{J}_2$ *as the* $(\alpha j \vdots T \vdots \alpha' j')$ *of* 9^311 *has with respect to* $\boldsymbol{J}$ *enables us to repeat in many cases the considerations of* §§ 10^3 *or* 11^3 *with respect to this factor.* In particular if $\boldsymbol{J}_2 = \boldsymbol{P} + \boldsymbol{P}'$, we can determine this factor completely by the formulas 10^32; or if $\boldsymbol{J}_2 = \boldsymbol{J}_2' + \boldsymbol{J}_2''$ and $\boldsymbol{P}$ commutes with $\boldsymbol{J}_2'$, formulas of the type (8) are again applicable to determine the dependence of the matrix of $\boldsymbol{P}$ on j_2, and the present discussion may be repeated. We shall find many applications for such calculations.

12. Matrix of $\boldsymbol{P}\cdot\boldsymbol{Q}$.*

It will be convenient at this point to obtain, in the $\gamma j_1 j_2 j m$ scheme, the matrix of the scalar product of a vector $\boldsymbol{P}$ which commutes with $\boldsymbol{J}_1$ and a vector $\boldsymbol{Q}$ which commutes with $\boldsymbol{J}_2$, both of which satisfy the commutation rules 8^31 with respect to $\boldsymbol{J}$. From 8^34 we see that this product $\boldsymbol{P}\cdot\boldsymbol{Q}$ will be

* Johnson, Phys. Rev. **38**, 1635 (1931).

diagonal with respect to j and m. The elements of $\boldsymbol{P}\cdot\boldsymbol{Q}$ are easily shown from $9^3 11$ to be given by the expression

$$
\begin{aligned}
&(\gamma j_1 j_2 j m|\boldsymbol{P}\cdot\boldsymbol{Q}|\gamma' j_1' j_2' j m) \\
&\quad = \sum_{\gamma'' j' m'} (\gamma j_1 j_2 j m|\boldsymbol{P}|\gamma'' j_1 j_2' j' m')\cdot(\gamma'' j_1 j_2' j' m'|\boldsymbol{Q}|\gamma' j_1' j_2' j m) \\
&\quad = \sum_{\gamma''} (\gamma j_1 j_2 j \vdots P \vdots \gamma'' j_1 j_2' j+1)(\gamma'' j_1 j_2' j+1 \vdots Q \vdots \gamma' j_1' j_2' j)(j+1)(2j+3) \\
&\quad + \sum_{\gamma''} (\gamma j_1 j_2 j \vdots P \vdots \gamma'' j_1 j_2' j)(\gamma'' j_1 j_2' j \vdots Q \vdots \gamma' j_1' j_2' j)\, j(j+1) \\
&\quad + \sum_{\gamma''} (\gamma j_1 j_2 j \vdots P \vdots \gamma'' j_1 j_2' j-1)(\gamma'' j_1 j_2' j-1 \vdots Q \vdots \gamma' j_1' j_2' j)\, j(2j-1). \qquad (1)
\end{aligned}
$$

When the formulas $11^3 8$ giving the dependence on j are substituted here, we obtain

$$
\begin{aligned}
&(\gamma j_1 j_2 j m|\boldsymbol{P}\cdot\boldsymbol{Q}|\gamma' j_1 j_2 j m) \\
&\quad = \tfrac{1}{2}\sum_{\gamma''} (\gamma j_2 \vdots P \vdots \gamma'' j_2)(\gamma'' j_1 \vdots Q \vdots \gamma' j_1)\{j(j+1) - j_1(j_1+1) - j_2(j_2+1)\} \\
&(\gamma j_1 j_2 j m|\boldsymbol{P}\cdot\boldsymbol{Q}|\gamma' j_1 j_2-1\, j m) \\
&\quad = \{\pm\}\tfrac{1}{2}\sum_{\gamma''} (\gamma j_2 \vdots P \vdots \gamma'' j_2-1)(\gamma'' j_1 \vdots Q \vdots \gamma' j_1)\sqrt{(j+j_2-j_1)(j+j_1-j_2+1)(j+j_1+j_2+1)(j_1+j_2-j)} \\
&(\gamma j_1 j_2 j m|\boldsymbol{P}\cdot\boldsymbol{Q}|\gamma' j_1-1\, j_2 j m) \\
&\quad = \{\mp\}\tfrac{1}{2}\sum_{\gamma''} (\gamma j_2 \vdots P \vdots \gamma'' j_2)(\gamma'' j_1 \vdots Q \vdots \gamma' j_1-1)\sqrt{(j+j_1-j_2)(j+j_2-j_1+1)(j+j_1+j_2+1)(j_1+j_2-j)} \\
&(\gamma j_1 j_2 j m|\boldsymbol{P}\cdot\boldsymbol{Q}|\gamma' j_1-1\, j_2-1\, j m) \\
&\quad = -\tfrac{1}{2}\sum_{\gamma''} (\gamma j_2 \vdots P \vdots \gamma'' j_2-1)(\gamma'' j_1 \vdots Q \vdots \gamma' j_1-1)\sqrt{(j_1+j_2+j+1)(j_1+j_2+j)(j_1+j_2-j)(j_1+j_2-j-1)} \\
&(\gamma j_1 j_2 j m|\boldsymbol{P}\cdot\boldsymbol{Q}|\gamma' j_1+1\, j_2-1\, j m) \\
&\quad = \tfrac{1}{2}\sum_{\gamma''} (\gamma j_2 \vdots P \vdots \gamma'' j_2-1)(\gamma'' j_1 \vdots Q \vdots \gamma' j_1+1)\sqrt{(j+j_2-j_1-1)(j+j_2-j_1)(j+j_1-j_2+1)(j+j_1-j_2+2)}. \qquad (2)
\end{aligned}
$$

The only other non-vanishing elements are those obtained by taking the complex conjugates of the last four above. As in the previous formulas we may interchange P and Q, j_1 and j_2 if we invert the double signs. (Compare the second and third elements.)

13. Sum rules.

There are certain sums of scalar products and absolute squares of matrix elements of $\boldsymbol{T}$ and $\boldsymbol{P}$ which will be of interest to us in connection with intensities of spectral radiation.

From $9^3 11$ we easily find (the calculation is similar to that used in obtaining $12^3 1$) that

$$
\begin{aligned}
&\sum_{m''} (\alpha j m|\boldsymbol{T}|\alpha'' j'' m'')\cdot(\alpha'' j'' m''|\boldsymbol{T}|\alpha' j' m') \\
&\qquad = \delta_{jj'}\delta_{mm'}(\alpha j \vdots T \vdots \alpha'' j'')(\alpha'' j'' \vdots T \vdots \alpha' j)\,\Xi(j, j''), \qquad (1)
\end{aligned}
$$

where

$$
\Xi(j, j+1) = (j+1)(2j+3); \quad \Xi(j, j) = j(j+1); \quad \Xi(j, j-1) = j(2j-1). \qquad (2)
$$

In particular, then

$$\sum_{m'} |(\alpha j m|\boldsymbol{T}|\alpha' j' m')|^2 = |(\alpha j \vdots T \vdots \alpha' j')|^2 \,\Xi(j,j'). \qquad (1')^*$$

These sums are independent of the m value of the state $\alpha j m$ occurring on the left. They are not symmetric in j and j', but become symmetric if multiplied by $(2j+1)$ to accomplish the summation over m.

It will be convenient to obtain now a relation which will in § 7[4] be interpreted as indicating that the total radiation from an atom is isotropic and unpolarized. Using the relation

$$\sum_{m=-j}^{+j} m^2 = \tfrac{1}{3} j(j+1)(2j+1),$$

we readily find from 9[3]11 that

$$\sum_{m=-j}^{j} |(\alpha j m|\boldsymbol{T}|\alpha' j' m+1)|^2 = \sum_m |(\alpha j m|\boldsymbol{T}|\alpha' j' m)|^2$$
$$= \sum_m |(\alpha j m|\boldsymbol{T}|\alpha' j' m-1)|^2 = \tfrac{1}{3}(2j+1)|(\alpha j \vdots T \vdots \alpha' j')|^2 \,\Xi(j,j'). \qquad (3)$$

For the case of the vector $\boldsymbol{P}$ of § 11[3], we should like to go on and sum (1) over the j values of the final state, obtaining the values of

$$\sum_{j'' m''} (\gamma j_1 j_2 j m|\boldsymbol{P}|\gamma'' j_1 j_2'' j'' m'') \cdot (\gamma'' j_1 j_2'' j'' m''|\boldsymbol{P}|\gamma' j_1 j_2' j' m')$$
$$= \delta_{jj'}\delta_{mm'} \sum_{j''} (\gamma j_1 j_2 j \vdots P \vdots \gamma'' j_1 j_2'' j'')(\gamma'' j_1 j_2'' j'' \vdots P \vdots \gamma' j_1 j_2' j)\,\Xi(j,j'') \qquad (4)$$

for $j_2'' = j_2+1, j_2$, and j_2-1. The straightforward evaluation of these quantities by substitution from 11[3]8 is algebraically very complicated. This unfriendly complication may be avoided by a transformation to the $\gamma j_1 j_2 m_1 m_2$ scheme as follows:

$$(4) = \delta_{jj'}\delta_{mm'} \sum_{j'' m''} \sum_{m_1 m_2 m_2'} \sum_{m_1'' m_2'' m_2'''} (\gamma j_1 j_2 j m|\gamma j_1 j_2 m_1 m_2)$$
$$(\gamma j_1 j_2 m_1 m_2|\boldsymbol{P}|\gamma'' j_1 j_2'' m_1 m_2')(\gamma'' j_1 j_2'' m_1 m_2'|\gamma'' j_1 j_2'' j'' m'')$$
$$\cdot (\gamma'' j_1 j_2'' j'' m''|\gamma'' j_1 j_2'' m_1'' m_2'')(\gamma'' j_1 j_2'' m_1'' m_2''|\boldsymbol{P}|\gamma' j_1 j_2' m_1'' m_2''')$$
$$(\gamma' j_1 j_2' m_1'' m_2'''|\gamma' j_1 j_2' j m).$$

The sum over $j'' m''$ may be performed at once to give a factor $\delta_{m_1 m_1''}\delta_{m_2' m_2''}$:

$$(4) = \delta_{jj'}\delta_{mm'} \sum_{m_1 m_2 m_2'''} (\gamma j_1 j_2 j m|\gamma j_1 j_2 m_1 m_2)(\gamma j_1 j_2' m_1'' m_2'''|\gamma' j_1 j_2' j m)$$
$$\sum_{m_2'} (\gamma j_1 j_2 m_1 m_2|\boldsymbol{P}|\gamma'' j_1 j_2'' m_1 m_2') \cdot (\gamma'' j_1 j_2'' m_1 m_2'|\boldsymbol{P}|\gamma' j_1 j_2' m_1 m_2''').$$

* $|(\beta|\boldsymbol{T}|\beta')|^2 = (\beta|\boldsymbol{T}|\beta')\cdot(\beta'|\boldsymbol{T}|\beta) = |(\beta|T_x|\beta')|^2 + |(\beta|T_y|\beta')|^2 + |(\beta|T_z|\beta')|^2$.
Cf. footnote, p. 49.

From the discussion at the end of § 11[3], we know that the value of the last sum is given by (1) as

$$\delta_{j_2 j_2'}\,\delta_{m_2 m_2'''}\,(\gamma j_2 \vdots P \vdots \gamma'' j_2'')\,(\gamma'' j_2'' \vdots P \vdots \gamma' j_2)\,\Xi(j_2, j_2'').$$

The factor remaining is then

$$\sum_{m_1 m_2} (\gamma j_1 j_2 j m|\gamma j_1 j_2 m_1 m_2)\,(\gamma' j_1 j_2 m_1 m_2|\gamma' j_1 j_2 j m),$$

which is unity since the values of these transformation coefficients are independent of γ. Finally then, (4) becomes

$$\begin{aligned}\sum_{j'' m''} (\gamma j_1 j_2 j m|\boldsymbol{P}|\gamma'' j_1 j_2'' j'' m'')\cdot(\gamma'' j_1 j_2'' j'' m''|\boldsymbol{P}|\gamma' j_1 j_2' j' m') \\ = \delta_{jj'}\,\delta_{mm'} \sum_{j''} (\gamma j_1 j_2 j \vdots P \vdots \gamma'' j_1 j_2'' j'')\,(\gamma'' j_1 j_2'' j'' \vdots P \vdots \gamma' j_1 j_2' j)\,\Xi(j, j'') \\ = \delta_{jj'}\,\delta_{mm'}\,\delta_{j_2 j_2'}\,(\gamma j_2 \vdots P \vdots \gamma'' j_2'')\,(\gamma'' j_2'' \vdots P \vdots \gamma' j_2)\,\Xi(j_2, j_2''),\end{aligned} \tag{5}$$

and in particular

$$\sum_{j' m'} |(\gamma j_1 j_2 j m|\boldsymbol{P}|\gamma' j_1 j_2' j' m')|^2 = |(\gamma j_2 \vdots P \vdots \gamma' j_2')|^2\,\Xi(j_2, j_2'). \tag{5'}$$

Here j' is summed over the values $j-1$, j and $j+1$. These sums are independent not only of the m value, but of the j value of the state occurring on the left.

14. Transformation amplitudes for vector addition.

To complete the discussion of the addition of two angular-momentum vectors we need to know how to express the eigenstates (6[3]2)

$$\psi(\gamma j_1 j_2 j m) = \psi(j m)$$

in terms of the eigenstates (6[3]1)

$$\phi(\gamma j_1 j_2 m_1 m_2) = \phi(m_1 m_2).$$

As indicated, one does not need in this discussion to write the $\gamma j_1 j_2$ explicitly, since they do not occur as summation indices in the transformation. Moreover, since $\phi(m_1 m_2)$ is an eigenstate of $J_z = J_{1z} + J_{2z}$ with eigenvalue $m_1 + m_2$, it is orthogonal to any eigenstate of J_z with another eigenvalue, e.g. $\psi(j m)$ for $m \neq m_1 + m_2$. Hence, in the relation

$$\psi(j m) = \Sigma\, \phi(m_1 m_2)\,(m_1 m_2|j m), \tag{1}$$

the coefficients $(m_1 m_2|j m)$ will contain a factor $\delta(m_1 + m_2, m)$.

One way to obtain a transformation between these two schemes is to diagonalize the matrix obtained in § 7[3] of $\boldsymbol{J}^2$ in the $m_1 m_2$ scheme. This method does not, however, give us a set of states $\psi(j m)$ with the particular phases we have agreed in §§ 3[3] and 10[3] to use; hence none of the matrix elements we have calculated would be valid in such a scheme. We can, though, by using the formulas of §§ 3[3] and 10[3] themselves, obtain a set of recursion formulas which completely determine the transformation coefficients with the proper phases.

For the state $\psi(j_1+j_2\, j_1+j_2)$ of highest j and m there is only one non-vanishing coefficient, namely (cf. 10^36)

$$(j_1\, j_2|j_1+j_2\, j_1+j_2) = 1. \tag{2}$$

Knowing the state $\psi(j\,m)$, i.e. all the coefficients $(m_1\, m_2|j\,m)$ for the given j and m, 3^33 gives us the state $\psi(j\,m-1)$. Expand 3^33 in the $m_1\, m_2$ scheme:

$$\sum_{m_1 m_2} (J_x - iJ_y)\phi(m_1\, m_2)\,(m_1\, m_2|j\,m)$$
$$= \hbar\sqrt{(j+m)\,(j-m+1)} \sum_{m_1 m_2} \phi(m_1\, m_2)\,(m_1\, m_2|j\,m-1).$$

Since $J_x - iJ_y = (J_{1x} - iJ_{1y}) + (J_{2x} - iJ_{2y})$, we obtain from a reapplication of 3^33:

$$\Sigma\hbar\sqrt{(j_1+m_1)\,(j_1-m_1+1)}\,\phi(m_1-1\, m_2)\,(m_1\, m_2|j\,m)$$
$$+\Sigma\hbar\sqrt{(j_2+m_2)\,(j_2-m_2+1)}\,\phi(m_1\, m_2-1)\,(m_1\, m_2|j\,m)$$
$$= \hbar\sqrt{(j+m)\,(j-m+1)}\,\Sigma\phi(m_1\, m_2)\,(m_1\, m_2|j\,m-1),$$

or, equating coefficients of $\phi(m_1\, m_2)$:

$$\sqrt{(j+m)\,(j-m+1)}\,(m_1\, m_2|j\,m-1)$$
$$= \sqrt{(j_1+m_1+1)\,(j_1-m_1)}\,(m_1+1\, m_2|j\,m)$$
$$+\sqrt{(j_2+m_2+1)\,(j_2-m_2)}\,(m_1\, m_2+1|j\,m). \tag{3}$$

This is the desired recursion relation.

From §§ 9^3 and 10^3 we obtain the recursion formula which steps down the value of j. From 9^311 we find that

$$J_{1z}\psi(j\,m) = m\,(j\!\vdots\! J_1\!\vdots\! j)\,\psi(j\,m) + \sqrt{(j-m)\,(j+m)}\,(j-1\!\vdots\! J_1\!\vdots\! j)\,\psi(j-1\,m)$$
$$+\sqrt{(j-m+1)\,(j+m+1)}\,(j+1\!\vdots\! J_1\!\vdots\! j)\,\psi(j+1\,m).$$

This becomes, on expanding and equating coefficients of $\phi(m_1\, m_2)$ as before,

$$\sqrt{(j-m)\,(j+m)}\,(j-1\!\vdots\! J_1\!\vdots\! j)\,(m_1\, m_2|j-1\,m) = [m_1 - m\,(j\!\vdots\! J_1\!\vdots\! j)]\,(m_1\, m_2|j\,m)$$
$$-\sqrt{(j-m+1)\,(j+m+1)}\,(j+1\!\vdots\! J_1\!\vdots\! j)\,(m_1\, m_2|j+1\,m), \tag{4}$$

where the $(j'\!\vdots\! J_1\!\vdots\! j)$ are given by 10^32. This relation gives us the state $\psi(j-1\,m)$ when we know the states $\psi(j\,m)$ and $\psi(j+1\,m)$.

With the initial condition (2), the two recursion formulas (3) and (4) determine completely all transformation coefficients for the addition of a given j_1 and j_2. A general formula for the transformation coefficients is very difficult to obtain from these relations. The general solution has however been given by Wigner* by the use of group-theoretical methods as the following expression:

$$(j_1 j_2 m_1 m_2|j_1 j_2 j m) = \delta(m, m_1+m_2)$$
$$\cdot\sqrt{\frac{(j+j_1-j_2)!(j-j_1+j_2)!(j_1+j_2-j)!(j+m)!(j-m)!(2j+1)}{(j+j_1+j_2+1)!(j_1-m_1)!(j_1+m_1)!(j_2-m_2)!(j_2+m_2)!}}$$
$$\cdot\sum_\kappa \frac{(-1)^{\kappa+j_2+m_2}(j+j_2+m_1-\kappa)!(j_1-m_1+\kappa)!}{(j-j_1+j_2-\kappa)!(j+m-\kappa)!\kappa!(\kappa+j_1-j_2-m)!}. \quad (5)$$

In this summation κ takes on all integral values consistent with the factorial notation, the factorial of a negative number being meaningless.

This formula is so complex that the calculation of coefficients by its use is as tedious as the direct use of the recursion formulas. But by making tables which show the result of adding any j_1 to $j_2 = 0, \frac{1}{2}, 1, \frac{3}{2}, 2, \ldots$, we obtain a convenient way of evaluating these transformation coefficients. For the trivial case $j_2 = 0$, we have

$$(j_1 0 m_1 0|j_1 0 j m) = \delta(j, j_1)\,\delta(m, m_1),$$

i.e. $$\psi(j_1 0 j_1 m_1) = \phi(j_1 0 m_1 0). \quad (6)$$

In Tables 1[3] to 4[3] are given the values for $j_2 = \frac{1}{2}, 1, \frac{3}{2}$, and 2, which are sufficient for most cases in atomic spectra: one seldom wishes to add two angular momenta the lesser of which is greater than two.†

In the addition of two angular momenta j_a and j_b we shall wish to know the effect of the reversal of the association of j_a and j_b with the j_1 and j_2 of the above formulas. If we associate j_a with j_1, j_b with j_2, we obtain a state $\psi(j_a j_b j m)$ as a certain linear combination of the states $\phi(j_a j_b m_a m_b)$. When we associate j_b with j_1, j_a with j_2, we obtain a state $\psi(j_b j_a j m)$ as a certain linear combination of the states $\phi(j_b j_a m_b m_a)$. We assume the basic states $\phi(j_a j_b m_a m_b)$ and $\phi(j_b j_a m_b m_a)$ to be identical (cf. 6[3]5). The states $\psi(j_a j_b j m)$ and $\psi(j_b j_a j m)$ must be essentially equal; but that they will have different phases is seen from the fact that for the $\psi(j_a j_b j m)$ the non-diagonal elements of the matrix of J_{az} will be positive while for the $\psi(j_b j_a j m)$ these elements will be negative (cf. § 10[3]). Hence, for successive values of j, these states must change their relative phase. However, for $j = j_a + j_b$ the two states are equal since for $m = j_a + j_b$ they both equal $\phi(j_a j_b j_a j_b)$, and

* Wigner, *Gruppentheorie*, p. 206.

† These tables may be calculated from the recursion formulas alone or from (5) alone. Formulas (2) and (3) determine immediately the element in the upper left corner. Repeated application of (3) gives then the top row; from this (4) determines directly the other rows. The bottom row is obtained very simply from (5), the summation in which reduces to one term for this case. For the top row, however, this sum becomes very complicated. The most convenient method of computation combines the use of (5) and of the recursion formulas, especially of (4) for the passage from one row to the next.

TABLE 1[3]. $(j_1\,\tfrac{1}{2}\,m_1\,m_2|j_1\,\tfrac{1}{2}\,j\,m)$

$j=$	$m_2=\frac{1}{2}$	$m_2=-\frac{1}{2}$
$j_1+\frac{1}{2}$	$\sqrt{\frac{j_1+m+\frac{1}{2}}{2j_1+1}}$	$\sqrt{\frac{j_1-m+\frac{1}{2}}{2j_1+1}}$
$j_1-\frac{1}{2}$	$-\sqrt{\frac{j_1-m+\frac{1}{2}}{2j_1+1}}$	$\sqrt{\frac{j_1+m+\frac{1}{2}}{2j_1+1}}$

TABLE 2[3]. $(j_1\,1\,m_1\,m_2|j_1\,1\,j\,m)$*

$j=$	$m_2=1$	$m_2=0$	$m_2=-1$
j_1+1	$\sqrt{\frac{(j_1+m)(j_1+m+1)}{(2j_1+1)(2j_1+2)}}$	$\sqrt{\frac{(j_1-m+1)(j_1+m+1)}{(2j_1+1)(j_1+1)}}$	$\sqrt{\frac{(j_1-m)(j_1-m+1)}{(2j_1+1)(2j_1+2)}}$
j_1	$-\sqrt{\frac{(j_1+m)(j_1-m+1)}{2j_1(j_1+1)}}$	$\frac{m}{\sqrt{j_1(j_1+1)}}$	$\sqrt{\frac{(j_1-m)(j_1+m+1)}{2j_1(j_1+1)}}$
j_1-1	$\sqrt{\frac{(j_1-m)(j_1-m+1)}{2j_1(2j_1+1)}}$	$-\sqrt{\frac{(j_1-m)(j_1+m)}{j_1(2j_1+1)}}$	$\sqrt{\frac{(j_1+m+1)(j_1+m)}{2j_1(2j_1+1)}}$

TABLE 3[3]. $(j_1\,\tfrac{3}{2}\,m_1\,m_2|j_1\,\tfrac{3}{2}\,j\,m)$†

$j=$	$m_2=\frac{3}{2}$	$m_2=\frac{1}{2}$
$j_1+\frac{3}{2}$	$\sqrt{\frac{(j_1+m-\frac{1}{2})(j_1+m+\frac{1}{2})(j_1+m+\frac{3}{2})}{(2j_1+1)(2j_1+2)(2j_1+3)}}$	$\sqrt{\frac{3(j_1+m+\frac{1}{2})(j_1+m+\frac{3}{2})(j_1-m+\frac{3}{2})}{(2j_1+1)(2j_1+2)(2j_1+3)}}$
$j_1+\frac{1}{2}$	$-\sqrt{\frac{3(j_1+m-\frac{1}{2})(j_1+m+\frac{1}{2})(j_1-m+\frac{3}{2})}{2j_1(2j_1+1)(2j_1+3)}}$	$-(j_1-3m+\frac{3}{2})\sqrt{\frac{j_1+m+\frac{1}{2}}{2j_1(2j_1+1)(2j_1+3)}}$
$j_1-\frac{1}{2}$	$\sqrt{\frac{3(j_1+m-\frac{1}{2})(j_1-m+\frac{1}{2})(j_1-m+\frac{3}{2})}{(2j_1-1)(2j_1+1)(2j_1+2)}}$	$-(j_1+3m-\frac{1}{2})\sqrt{\frac{j_1-m+\frac{1}{2}}{(2j_1-1)(2j_1+1)(2j_1+2)}}$
$j_1-\frac{3}{2}$	$-\sqrt{\frac{(j_1-m-\frac{1}{2})(j_1-m+\frac{1}{2})(j_1-m+\frac{3}{2})}{2j_1(2j_1-1)(2j_1+1)}}$	$\sqrt{\frac{3(j_1+m-\frac{1}{2})(j_1-m-\frac{1}{2})(j_1-m+\frac{1}{2})}{2j_1(2j_1-1)(2j_1+1)}}$

$j=$	$m_2=-\frac{1}{2}$	$m_2=-\frac{3}{2}$
$j_1+\frac{3}{2}$	$\sqrt{\frac{3(j_1+m+\frac{3}{2})(j_1-m+\frac{1}{2})(j_1-m+\frac{3}{2})}{(2j_1+1)(2j_1+2)(2j_1+3)}}$	$\sqrt{\frac{(j_1-m-\frac{1}{2})(j_1-m+\frac{1}{2})(j_1-m+\frac{3}{2})}{(2j_1+1)(2j_1+2)(2j_1+3)}}$
$j_1+\frac{1}{2}$	$(j_1+3m+\frac{3}{2})\sqrt{\frac{j_1-m+\frac{1}{2}}{2j_1(2j_1+1)(2j_1+3)}}$	$\sqrt{\frac{3(j_1+m+\frac{3}{2})(j_1-m-\frac{1}{2})(j_1-m+\frac{1}{2})}{2j_1(2j_1+1)(2j_1+3)}}$
$j_1-\frac{1}{2}$	$-(j_1-3m-\frac{1}{2})\sqrt{\frac{j_1+m+\frac{1}{2}}{(2j_1-1)(2j_1+1)(2j_1+2)}}$	$\sqrt{\frac{3(j_1+m+\frac{1}{2})(j_1+m+\frac{3}{2})(j_1-m-\frac{1}{2})}{(2j_1-1)(2j_1+1)(2j_1+2)}}$
$j_1-\frac{3}{2}$	$-\sqrt{\frac{3(j_1+m-\frac{1}{2})(j_1+m+\frac{1}{2})(j_1-m-\frac{1}{2})}{2j_1(2j_1-1)(2j_1+1)}}$	$\sqrt{\frac{(j_1+m-\frac{1}{2})(j_1+m+\frac{1}{2})(j_1+m+\frac{3}{2})}{2j_1(2j_1-1)(2j_1+1)}}$

* From Wigner.

† Calculated by F. Seitz.

TABLE 4³. $(j_1\, 2\, m_1\, m_2 | j_1\, 2\, j\, m)$*

$j=$	$m_2=2$	$m_2=1$	$m_2=0$
j_1+2	$\sqrt{\frac{(j_1+m-1)(j_1+m)(j_1+m+1)(j_1+m+2)}{(2j_1+1)(2j_1+2)(2j_1+3)(2j_1+4)}}$	$\sqrt{\frac{(j_1-m+2)(j_1+m+2)(j_1+m+1)(j_1+m)}{(2j_1+1)(j_1+1)(2j_1+3)(j_1+2)}}$	$\sqrt{\frac{3(j_1-m+2)(j_1-m+1)(j_1+m+2)(j_1+m+1)}{(2j_1+1)(2j_1+2)(2j_1+3)(j_1+2)}}$
j_1+1	$-\sqrt{\frac{(j_1+m-1)(j_1+m)(j_1+m+1)(j_1-m+2)}{2j_1(j_1+1)(j_1+2)(2j_1+1)}}$	$-(j_1-2m+2)\sqrt{\frac{(j_1+m+1)(j_1+m)}{2j_1(2j_1+1)(j_1+1)(j_1+2)}}$	$m\sqrt{\frac{3(j_1-m+1)(j_1+m+1)}{j_1(2j_1+1)(j_1+1)(j_1+2)}}$
j_1	$\sqrt{\frac{3(j_1+m-1)(j_1+m)(j_1-m+1)(j_1-m+2)}{(2j_1-1)2j_1(j_1+1)(2j_1+3)}}$	$(1-2m)\sqrt{\frac{3(j_1-m+1)(j_1+m)}{(2j_1-1)j_1(2j_1+2)(2j_1+3)}}$	$\frac{3m^2-j_1(j_1+1)}{\sqrt{(2j_1-1)j_1(j_1+1)(2j_1+3)}}$
j_1-1	$-\sqrt{\frac{(j_1+m-1)(j_1-m)(j_1-m+1)(j_1-m+2)}{2(j_1-1)j(j_1+1)(2j_1+1)}}$	$(j_1+2m-1)\sqrt{\frac{(j_1-m+1)(j_1-m)}{(j_1-1)j_1(2j_1+1)(2j_1+2)}}$	$-m\sqrt{\frac{3(j_1-m)(j_1+m)}{(j_1-1)j_1(2j_1+1)(j_1+1)}}$
j_1-2	$\sqrt{\frac{(j_1-m-1)(j_1-m)(j_1-m+1)(j_1-m+2)}{(2j_1-2)(2j_1-1)2j_1(2j_1+1)}}$	$-\sqrt{\frac{(j_1-m+1)(j_1-m)(j_1-m-1)(j_1+m-1)}{(j_1-1)(2j_1-1)j_1(2j_1+1)}}$	$\sqrt{\frac{3(j_1-m)(j_1-m-1)(j_1+m)(j_1+m-1)}{(2j_1-2)(2j_1-1)j_1(2j_1+1)}}$

$j=$	$m_2=-1$	$m_2=-2$
j_1+2	$\sqrt{\frac{(j_1-m+2)(j_1-m+1)(j_1-m)(j_1+m+2)}{(2j_1+1)(j_1+1)(2j_1+3)(j_1+2)}}$	$\sqrt{\frac{(j_1-m-1)(j_1-m)(j_1-m+1)(j_1-m+2)}{(2j_1+1)(2j_1+2)(2j_1+3)(2j_1+4)}}$
j_1+1	$(j_1+2m+2)\sqrt{\frac{(j_1-m+1)(j_1-m)}{j_1(2j_1+1)(2j_1+2)(j_1+2)}}$	$\sqrt{\frac{(j_1-m-1)(j_1-m)(j_1-m+1)(j_1+m+2)}{j_1(2j_1+1)(j_1+1)(2j_1+4)}}$
j_1	$(2m+1)\sqrt{\frac{3(j_1-m)(j_1+m+1)}{(2j_1-1)j_1(2j_1+2)(2j_1+3)}}$	$\sqrt{\frac{3(j_1-m-1)(j_1-m)(j_1+m+1)(j_1+m+2)}{(2j_1-1)j_1(2j_1+2)(2j_1+3)}}$
j_1-1	$-(j_1-2m-1)\sqrt{\frac{(j_1+m+1)(j_1+m)}{(j_1-1)j_1(2j_1+1)(2j_1+2)}}$	$\sqrt{\frac{(j_1-m-1)(j_1+m)(j_1+m+1)(j_1+m+2)}{(j_1-1)j_1(2j_1+1)(2j_1+2)}}$
j_1-2	$-\sqrt{\frac{(j_1-m-1)(j_1+m+1)(j_1+m)(j_1+m-1)}{(j_1-1)(2j_1-1)j_1(2j_1+1)}}$	$\sqrt{\frac{(j_1+m-1)(j_1+m)(j_1+m+1)(j_1+m+2)}{(2j_1-2)(2j_1-1)2j_1(2j_1+1)}}$

* Calculated by C. W. Ufford.

the states for other values of m are given directly in terms of this by $3^3 3$. Hence the states in question must satisfy the relation*

$$\psi(j_b\, j_a\, j\, m) = (-1)^{j_a+j_b-j}\,\psi(j_a\, j_b\, j\, m). \tag{7}$$

PROBLEM

Investigate the relation of the preceding formulas to the classical vector-coupling picture.

(*a*) Show that if the spin of the electron is known to have the value $+\frac{1}{2}$ in a direction making the angle θ with the z axis, the probability that the z component of spin have the value $+\frac{1}{2}$ is $\cos^2(\theta/2)$ and that it have the value $-\frac{1}{2}$ is $\sin^2(\theta/2)$. [PAULI, Zeits. für Phys. **43**, 601 (1927).]

(*b*) According to the vector-coupling picture when $j=l+\frac{1}{2}$ the spin is parallel to $\boldsymbol{l}$ as a vector and when $\boldsymbol{j}$ has the component m the angle it makes with the z axis is $\cos\theta = m/(l+\frac{1}{2})$. Compute $\cos^2(\theta/2)$ and $\sin^2(\theta/2)$ for this angle and compare with the square of the coefficients in the first row of Table 1^3.

For the case $j=l-\frac{1}{2}$ one has still to use $\cos\theta = m/(l+\frac{1}{2})$ instead of $m/(l-\frac{1}{2})$ as suggested by the vector-coupling picture. This is an example of the difference between quantum-mechanical formulas and those given by classical methods, although, consistently with the correspondence principle (§ 4^4), the two agree for large values of l.

---

* This agrees with the relation

$$(j_a\, j_b\, m_a\, m_b | j_a\, j_b\, j\, m) = (-1)^{j_a+j_b-j}(j_b\, j_a\, m_b\, m_a | j_b\, j_a\, j\, m)$$

given by Wigner.

CHAPTER IV

THE THEORY OF RADIATION

"The whole subject of electrical radiation seems working itself out splendidly."
OLIVER LODGE, Phil. Mag., August, 1888.

The theory of atomic spectra divides itself rather sharply into two parts: the theory of the energy levels and the corresponding states, and the theory of the radiative process whereby the spectral lines arise through transitions between states. The preceding chapters have provided us with most of the apparatus needed for an attack on the first part of the problem. In this chapter we present the general theory underlying the radiative process.

That radiation is to be treated according to the general scheme of Maxwell's electromagnetic theory is unquestioned by physicists to-day, but it is also recognized that these laws require some kind of quantum-theoretical modification, to be embraced in a general theory of quantum electrodynamics. Such a theory of quantum electrodynamics cannot be said to exist in definitive form to-day. When it is developed it will probably produce alterations in the Hamiltonian which will change the details of the theory of atomic energy levels as we present it. But the general structure of the theory of the energy levels can hardly be affected since it possesses so many points of close contact with the experimental data. Likewise it is possible at present to develop the theory of radiation along general lines with considerable assurance that such alterations as are later brought by a revision of the theory of the electromagnetic field will not affect our main conclusions.

1. Transition probabilities.

Emission of light takes place when there is a transition of an atom from a state of higher to a state of lower energy—a quantum jump. Likewise absorption takes place by an upward transition that is caused by the action of the fields of the radiation on the atom. We start the consideration of these processes by a somewhat phenomenological method due to Einstein.* The argument is independent of special views concerning the electrodynamics of the transition. Let an atom be in an excited level A, that is, in a level of more than the minimum energy. Einstein then ascribes to it a certain probability per unit time, $\mathsf{A}(A, B)$, of making a spontaneous transition with emission of radiation to each level B of lower energy. The light sent out in the process has the wave-number $(E_A - E_B)/hc$ according to Bohr's rule.

* EINSTEIN, Phys. Zeits. **18**, 121 (1917).

Then if the number of atoms in the level A is $N(A)$, due to spontaneous emission processes

$$\frac{dN(A)}{dt} = -\left[\sum_B \mathsf{A}(A,B)\right] N(A), \tag{1}$$

where the summation is over all states of lower energy. The quantity $\tau(A)$ defined by

$$1/\tau(A) = \sum_B \mathsf{A}(A,B) \tag{2}$$

is called the mean life of the level A, for if the only processes occurring are those of spontaneous emission

$$N(A) = N_0(A)\, e^{-t/\tau(A)}, \tag{3}$$

from which it follows that $\tau(A)$ is the mean time which an atom remains in the excited level.

Einstein also ascribes to the atom two probability coefficients representing the effectiveness of the radiation field in causing transitions. The radiation field is supposed to be isotropic and unpolarized and to have spectral energy $\rho(\sigma)\,d\sigma$ in unit volume in the wave-number range $d\sigma$ at σ. If C is an energy level higher than A (where A is now not necessarily an excited level), then the field produces transitions from A to C by absorption at the rate

$$N(A)\,\mathsf{B}(A,C)\,\rho(\sigma),$$

where σ is the wave number corresponding to the transition, by Bohr's rule. Also the radiation stimulates or induces emission processes from C to A at a rate given by

$$N(C)\,\mathsf{B}(C,A)\,\rho(\sigma).$$

The coefficients A and B are fundamental measures of the interaction of the atom with the radiation field and have the same values whether there is thermal equilibrium or not.

By purely statistical considerations we can find relations between the A's and B's, for we know that in thermal equilibrium

(*a*) the relative numbers of atoms in different levels are given by the Maxwell-Boltzmann law,

$$N(A) = g(A)\, e^{-E_A/kT}, \tag{4}$$

where $g(A)$ is the statistical weight of the level A;

(*b*) the radiation density is given by Planck's formula,

$$\rho(\sigma) = hc\sigma \frac{8\pi\sigma^2}{e^{hc\sigma/kT} - 1}. \tag{5}$$

The argument now uses the principle of detailed balancing, according to which in statistical equilibrium the rates of each elementary process and its inverse are equal. The two kinds of emission are regarded together as one elementary process and the one kind of absorption as its inverse. This seemed strange until Dirac's radiation theory* showed that the two kinds of emission were really related to the same process. Hence in equilibrium we have

$$N(A)\,\mathsf{B}(A,C)\,\rho(\sigma)=N(C)\,[\mathsf{A}(C,A)+\mathsf{B}(C,A)\,\rho(\sigma)]. \tag{6}$$

In order that the last three equations be consistent we must have

$$\begin{aligned} g(A)\,\mathsf{B}(A,C) &= g(C)\,\mathsf{B}(C,A)\\ \mathsf{A}(C,A) &= 8\pi hc\sigma^3\,\mathsf{B}(C,A). \end{aligned} \tag{7}$$

Using the latter relation the total rate of the emission processes can be written in the form

$$N(C)\,\mathsf{A}(C,A)\,[n(\sigma)+1], \quad \text{where} \quad n(\sigma)=\rho(\sigma)/8\pi hc\sigma^3. \tag{8}$$

That this is a significant way of writing it is shown by the Dirac radiation theory, which is a quantum-mechanical outgrowth of the method of Rayleigh and Jeans† and Debye‡ for handling the statistical mechanics of the radiation field. The number of degrees of freedom of the field associated with waves in the wave-number range $d\sigma$ at σ is shown to be $8\pi\sigma^2 d\sigma$ per unit volume; the size of quantum associated with each such degree of freedom is $hc\sigma$, therefore $n(\sigma)$ is the mean number of quanta per degree of freedom for the field coordinates belonging to waves of wave-number σ.

In the Dirac theory the electromagnetic field and the emitting or absorbing matter are treated as a single dynamical system. Because of the rather weak coupling between the two parts, field and matter, it is possible to make fairly precise statements concerning the amount of energy in the field and the amount of energy in the matter. (If we regard two separate systems as one system, the two Hamiltonians commute with one another and so each commutes with their sum, therefore eigenvalues of each are constants of the motion. But if a small interaction is present, this is only approximately true.) The emission process consists of a change in the system whereby energy leaves the matter and raises the quantum number of one of the field's degrees of freedom by one. The absorption process is a change in which one of the field's degrees of freedom has its quantum number diminished by one, the energy going into the matter.

Without going into details we can say that the probability of these internal changes in unit time is proportional to the square of the matrix component of the interaction energy between the field and the matter. The matrix component is taken in a representation in which the states of the two parts are labelled by quantum numbers; the component to use is the one whose initial indices specify the initial state and whose final indices specify the final state. The whole transition probability will be very small unless the

* Dirac, *Quantum Mechanics*, Chapter xi; see also Fermi, Rev. Mod. Phys. **4**, 87 (1932).
† Jeans, *Dynamical Theory of Gases*, Chapter xvi; Phil. Mag. **20**, 953 (1910).
‡ Debye, Ann. der Phys. **33**, 1427 (1910).

initial and final states of the whole system have the same (or almost the same) energy.

Each of the coordinates for a wave of wave-number σ is dynamically equivalent to a harmonic oscillator of the frequency $\nu = c\sigma$. Hence the possible field energies associated with each wave are $h\nu\,(n+\frac{1}{2})$, where n is an integer.

The interaction of the field with the matter is a linear function of the wave amplitudes, so the matrix components, so far as the change in any field-wave's quantum number is concerned, contain as a factor $(n|q|n')$, where n and n' are the initial and final quantum numbers of the degree of freedom in question and q is the numerical value of its amplitude. Since the field coordinates are dynamically like harmonic oscillators, we may take for the matrix components of q the well-known values

$$(n|q|n') = \begin{cases} \sqrt{n+1} & (n' = n+1) \\ \sqrt{n}. & (n' = n-1) \end{cases} \tag{9}$$

All others are zero. As increase in n corresponds to increase in field energy, or emission by the matter, and decrease in n corresponds to absorption, the lack of symmetry is just what we need to get the spontaneous emission as well as the induced: $|(n|q|n+1)|^2$ is proportional to $n+1$, so emission by passage of energy to this degree of freedom can occur even in the absence of radiation $(n=0)$ in it. But $|(n|q|n-1)|^2$ is proportional to n, so absorption by passage of energy from this wave to the matter is simply proportional to the amount of energy already present in that wave in excess of the zero point energy, $\frac{1}{2}h\nu$. The actual emission in the statistical ensemble will therefore be proportional to $[n(\sigma)+1]$ as in (8), where $n(\sigma)$ is the mean number of quanta in all of the degrees of freedom of the right frequency.

This is just a sketch of the theory, given in the hope that it will make clear the way in which the two apparently different kinds of emission are really two parts of a single process. We conclude this outline with the remark that it is not quite true that the only transitions occurring are those in which the atom's change in energy is just opposite to that of the field. This gives rise to the natural breadth of spectral lines, the quantum-mechanical analogue of the fact in classical theory that a radiating oscillator has its amplitude damped out by the loss of energy by radiation. Thus the wave-train emitted in a single process is finite and not monochromatic when expressed as a sum of monochromatic waves with a Fourier integral. This point has been studied in detail by Weisskopf and Wigner.* They find that the relative probability of emission in the range $d\sigma$ at σ, where σ is now measured from the wave-number given by Bohr's rule, is

$$\frac{1}{\pi}\frac{d\sigma/\sigma_0}{1+(\sigma/\sigma_0)^2}, \quad \text{where} \quad \sigma_0 = \frac{1}{4\pi c}\left(\frac{1}{\tau(A)}+\frac{1}{\tau(C)}\right). \tag{10}$$

* Weisskopf and Wigner, Zeits. für Phys. **63**, 54 (1930).

This tells the distribution of energy over the line and relates the half-width, σ_0, to the harmonic mean of the mean lives of the initial and final levels. This line width has no practical importance for ordinary spectroscopy as it is smaller than the width of image produced in spectroscopic instruments owing to their finite resolving power.

2. Classical electromagnetic theory.

We regard radiation as a spatial flow of energy governed essentially by Maxwell's equations. In these we have a scalar field ρ which is the electric charge density in electrostatic units and a vector field $\boldsymbol{I}$ which is the electric current density in electromagnetic units. These fields are connected by an equation of continuity,

$$\operatorname{div}\boldsymbol{I} + \frac{1}{c}\frac{\partial \rho}{\partial t} = 0, \tag{1}$$

which expresses the conservation of electric substance.

Related to ρ and $\boldsymbol{I}$ are two vector fields, the electric field $\mathscr{E}$ and the magnetic field $\mathscr{H}$, which are characterized by Maxwell's equations

$$\operatorname{div}\mathscr{H} = 0, \tag{2a}$$

$$\operatorname{curl}\mathscr{E} + \frac{1}{c}\frac{\partial \mathscr{H}}{\partial t} = 0, \tag{2b}$$

$$\operatorname{div}\mathscr{E} = 4\pi\rho, \tag{2c}$$

$$\operatorname{curl}\mathscr{H} - \frac{1}{c}\frac{\partial \mathscr{E}}{\partial t} = 4\pi \boldsymbol{I}. \tag{2d}$$

These equations express respectively the fundamental laws:

(*a*) Non-existence of a magnetic stuff analogous to electric charge.

(*b*) Faraday's law of electromagnetic induction.

(*c*) The Coulomb law in electrostatics.

(*d*) The Ampere law for the magnetic field due to electric currents, together with Maxwell's displacement-current hypothesis.

There are various auxiliary fields which may be conveniently employed in dealing with these equations, the most important of which are the scalar and vector potentials φ and $\boldsymbol{A}$. These have the property that if they are made to satisfy the equations

$$\begin{aligned} -\Delta\varphi + \frac{1}{c^2}\frac{\partial^2 \varphi}{\partial t^2} &= 4\pi\rho, \\ -\Delta \boldsymbol{A} + \frac{1}{c^2}\frac{\partial^2 \boldsymbol{A}}{\partial t^2} &= 4\pi \boldsymbol{I}, \\ \operatorname{div}\boldsymbol{A} + \frac{1}{c}\frac{\partial \varphi}{\partial t} &= 0; \end{aligned} \tag{3}$$

$\mathscr{E}$ and $\mathscr{H}$, derived from them by the formulas

$$\mathscr{E} = -\text{grad}\varphi - \frac{1}{c}\frac{\partial \boldsymbol{A}}{\partial t}, \qquad \mathscr{H} = \text{curl}\boldsymbol{A}, \tag{4}$$

will satisfy Maxwell's equations. It is to be noticed that if $\rho \equiv 0$ one may set $\varphi = 0$ and derive $\mathscr{E}$ and $\mathscr{H}$ simply from $\boldsymbol{A}$. This is a usual procedure in dealing with electromagnetic waves at points far from their source.

In the theory of the differential equations (3), it is shown that the solution for the scalar potential can be written as

$$\varphi(x' y' z' t') = \int \frac{\rho(x y z t)}{R} dx\, dy\, dz, \tag{5}$$

in which $t = t' - R/c,$

and R is the distance between the volume element $dx\,dy\,dz$ and the point $x'\, y'\, z'$. This is known as the retarded-potential solution because each volume element contributes to the potential according to the value of the charge density there at a time earlier than the instant t' by the time it takes light to travel from the volume element to the point $x'\ y'\ z'$. An exactly similar formula gives $\boldsymbol{A}$ in terms of $\boldsymbol{I}$.

To develop the classical theory of radiation in a form suitable for use in the quantum theory we need to expand this retarded potential formula under the assumptions that ρ and $\boldsymbol{I}$ vary harmonically with the time, i.e. we take $\rho(x y z t)$ as the real part of the expression $\rho(x y z)\, e^{2\pi i \nu t}$. Here $\rho(x y z)$ may be complex, as in the case of a rotating unsymmetrical charge distribution. Hence we write

$$\rho(t) = \mathscr{R}\{\rho e^{2\pi i \nu t}\}, \qquad \boldsymbol{I}(t) = \mathscr{R}\{\boldsymbol{I} e^{2\pi i \nu t}\}, \tag{6}$$

where $\boldsymbol{I}$ is in general a bivector [a vector of the form $(\boldsymbol{I}_r + i\boldsymbol{I}_i)$]. In terms of ρ and $\boldsymbol{I}$ the equation of continuity becomes

$$\text{div}\boldsymbol{I} + ik\rho = 0, \tag{7}$$

where $k = 2\pi\nu/c = 2\pi\sigma.$

3. Expansion of the retarded potential.

In the theory of atomic spectra ρ and $\boldsymbol{I}$ are essentially zero outside a sphere of radius small compared to k^{-1}. This suggests an appropriate development of the retarded-potential formula according to powers of k. In the expression

$$\varphi(x' y' z' t') = e^{2\pi i \nu t'} \int \rho \frac{e^{-ikR}}{R} dv \tag{1}$$

we may use the expansion*

$$\frac{e^{-ikR}}{R} = \sum_{\lambda=0}^{\infty} (2\lambda + 1) \frac{\zeta_\lambda(kr')}{ir'} \frac{\psi_\lambda(kr)}{kr} P_\lambda(\cos\omega), \tag{2}$$

* Bateman, *Partial Differential Equations*, p. 388, Example 4.

where r and r' are the radii to the points $x\ y\ z$ and $x'\ y'\ z'$, ω is the angle between them, and ζ_λ and ψ_λ are Bessel functions:

$$\begin{aligned}\psi_\lambda(kr) &= \sqrt{\tfrac{1}{2}\pi kr}\,J_{\lambda+\frac{1}{2}}(kr)\\ \zeta_\lambda(kr') &= \sqrt{\tfrac{1}{2}\pi kr'}\,[J_{\lambda+\frac{1}{2}}(kr')+(-1)^\lambda iJ_{-\lambda-\frac{1}{2}}(kr')].\end{aligned} \tag{3}$$

This shows us that the potential given by (1) corresponds to a system of diverging spherical waves, since

$$\frac{\zeta_\lambda(kr')}{ir'} = i^\lambda \frac{e^{-ikr'}}{r'}\left(1-\frac{i\lambda(\lambda+1)}{2kr'}+\dots\right). \tag{4}$$

We shall be interested only in the two leading terms in $\zeta_\lambda(kr')$. The first measures the transport of energy, the second that of angular momentum; the higher terms correspond to local fields near the oscillating charge distribution.

Substituting (2) in (1), we find for the amplitude of the λth diverging wave

$$(2\lambda+1)\int \rho \frac{\psi_\lambda(kr)}{kr} P_\lambda(\cos\omega)\,dv.$$

At this point we introduce the classification of terms by powers of kr, instead of by the integer λ. We shall be interested only in the first three terms. Since

$$\frac{\psi_\lambda(kr)}{kr} = \frac{(kr)^\lambda}{1.3.5\dots(2\lambda+1)}\left(1-\frac{k^2r^2}{2(2\lambda+3)}+\dots\right), \tag{5}$$

we find for the leading terms in the scalar potential

$$\begin{aligned}\varphi = \frac{e^{2\pi i\nu t'-ikr'}}{r'}\Big\{&\int \rho P_0\,dv + ik\left(1-\frac{i}{kr'}\right)\int \rho r P_1\,dv\\ &-\frac{k^2}{3}\left[\left(1-\frac{3i}{kr'}\right)\int \rho r^2 P_2\,dv + \tfrac{1}{2}\int \rho r^2 P_0\,dv\right]+\dots\Big\}.\end{aligned} \tag{6}$$

A corresponding expression holds for $\boldsymbol{A}$ on replacing ρ by $\boldsymbol{I}$. Here the terms are expressed in terms of the successive moments of the charge distribution. The first term vanishes, the second gives what is called (electric-) dipole radiation, the third gives (electric-) quadrupole radiation. It has not been necessary thus far in spectroscopy to consider higher terms.

We define the dipole and quadrupole moments of ρ by the equations

$$\boldsymbol{P} = \int \rho \boldsymbol{r}\,dv, \quad \mathfrak{R} = \int \rho \boldsymbol{r}\boldsymbol{r}\,dv. \tag{7}$$

Using the values of the Legendre polynomials, we readily find that

$$\int \rho r P_1\,dv = \boldsymbol{r}_0'\cdot\boldsymbol{P}, \quad \int \rho r^2 P_2\,dv = \tfrac{3}{2}\boldsymbol{r}_0'\cdot\mathfrak{R}\cdot\boldsymbol{r}_0' - \tfrac{1}{2}\mathfrak{R}_S,$$

$$\int \rho r^2 P_0\,dv = \mathfrak{R}_S,$$

where $\mathfrak{R}_S$ is written for the invariant sum of the diagonal terms in $\mathfrak{R}$, and $\boldsymbol{r}'_0$ is a unit vector in the direction r'.

Let us make the corresponding reductions for the vector potential. First we note that from the equation of continuity and the fact that ρ and $\boldsymbol{I}$ vanish everywhere outside a finite sphere we may write

$$ik\int \rho g\, dv = \int \boldsymbol{I}\cdot\nabla g\, dv,$$

where g is any scalar, vector or dyadic function of position that is finite everywhere that ρ and $\boldsymbol{I}$ are not zero. Putting $g=1$, we find $\int \rho\, dv = 0$, as already asserted. Putting $g=\boldsymbol{r}$ and $\boldsymbol{rr}$, we find the relations

$$\int \boldsymbol{I}\, dv = ik\boldsymbol{P}, \quad \int (\boldsymbol{Ir}+\boldsymbol{rI})\, dv = ik\mathfrak{R},$$

which enable us to express the vector potential in terms of the moments of the charge distribution. Because of the occurrence of k in these formulas we need only the first two terms in $\boldsymbol{A}$ to have terms to the order k^2 in the result:

$$\boldsymbol{A} = \frac{e^{2\pi i\nu t' - ikr'}}{r'}\left\{\int \boldsymbol{I}\, dv + ik\left(1-\frac{i}{kr'}\right)\int \boldsymbol{I}r\cos\omega\, dv + \ldots\right\}.$$

The first term gives the part of the vector potential associated with the dipole radiation. The second consists of two parts:

$$\boldsymbol{r}'_0\cdot\int \boldsymbol{rI}\, dv = \tfrac{1}{2}\boldsymbol{r}'_0\cdot\int (\boldsymbol{rI}+\boldsymbol{Ir})\, dv + \tfrac{1}{2}\boldsymbol{r}'_0\cdot\int (\boldsymbol{rI}-\boldsymbol{Ir})\, dv.$$

The first term here is proportional to the quadrupole moment; the second can be written as

$$-\boldsymbol{r}'_0\times\boldsymbol{M}, \quad \text{where} \quad \boldsymbol{M} = \tfrac{1}{2}\int \boldsymbol{r}\times\boldsymbol{I}\, dv, \tag{8}$$

the magnetic moment of the electric current distribution. The second term in $\boldsymbol{A}$ therefore represents part of the electric-quadrupole radiation field and the whole of the magnetic-dipole radiation field.

To summarize, the scalar and vector potentials up to the second power of k in the moments of the charge distribution are

$$\left.\begin{aligned}\varphi &= \frac{e^{2\pi i\nu t - ikr}}{r}\left\{ik\left(1-\frac{i}{kr}\right)\boldsymbol{r}_0\cdot\boldsymbol{P} - \tfrac{1}{2}k^2\left[\boldsymbol{r}_0\cdot\mathfrak{R}\cdot\boldsymbol{r}_0 - \frac{i}{kr}(3\boldsymbol{r}_0\cdot\mathfrak{R}\cdot\boldsymbol{r}_0 - \mathfrak{R}_S)\right]\right\}\\ \boldsymbol{A} &= \frac{e^{2\pi i\nu t - ikr}}{r}\left\{ik\boldsymbol{P} - ik\left(1-\frac{i}{kr}\right)\boldsymbol{r}_0\times\boldsymbol{M} - \tfrac{1}{2}k^2\left(1-\frac{i}{kr}\right)\boldsymbol{r}_0\cdot\mathfrak{R}\right\}.\end{aligned}\right\} \tag{9}$$

Since the coordinates of the volume elements of the charge and current distribution no longer appear, we have here, for convenience in later work, used unprimed letters to denote the position and time for which the potentials are computed.

4. The correspondence principle for emission.

Before calculating the radiation field associated with the potentials of the preceding section let us investigate their connection with the problem of emission in the quantum theory. The most fruitful idea in passing from the earlier atomic theory to the quantum mechanics was the correspondence principle of Bohr. From the principles of quantization developed for conditionally periodic dynamical systems it followed that in the asymptotic limit of transitions between states of large quantum numbers the spectroscopic frequencies of Bohr approached equality with actual frequencies of the mechanical motion. In classical mechanics the energy E of such a system is expressible in terms of a set of constants of the motion, $J_1, J_2, \ldots, J_n$, known as action variables. The quantum conditions restrict these J's to values that are integral multiples of h,

$$J_k = n_k h. \quad (k = 1, 2, \ldots, n;\ n_k, \text{an integer}) \quad (1)$$

The classical frequencies of the motion are given by the formula

$$\nu_k = \frac{\partial E(J_1, J_2, \ldots, J_n)}{\partial J_k}, \quad (2)$$

so that the combination frequencies which occur in a Fourier expansion of the motion are of the form

$$\sum_k \tau_k \nu_k = \sum_k \tau_k \frac{\partial E}{\partial J_k}. \qquad (\tau_k, \text{ integers}) \quad (3)$$

On the other hand, the quantum frequency associated with the transition from the set of quantum numbers

$$(n_1, n_2, \ldots, n_n) \to (n_1 + \tau_1, n_2 + \tau_2, \ldots, n_n + \tau_n)$$

by Bohr's frequency condition is

$$\frac{1}{h}[E(J_1 + \tau_1 h, J_2 + \tau_2 h, \ldots) - E(J_1, J_2, \ldots)]. \quad (4)$$

Developing this in a power series in the τ's, we have

$$\sum_k \tau_k \frac{\partial E}{\partial J_k} + \frac{h}{2} \sum_{kl} \tau_k \tau_l \frac{\partial^2 E}{\partial J_k \partial J_l} + \ldots. \quad (5)$$

The first term is just the classical frequency; the other terms represent quantum corrections. In the special case in which E is a linear function of the J's, which is that of a system of harmonic oscillators, the higher terms vanish so that the correspondence is exact.

Bohr's great idea was that this approximate connection probably holds not only for the frequencies, but for other features of the spectrum as well.

According to the frequency relation we have a definite term in the Fourier analysis of the motion, namely, the combination term with multiples $\tau_1, \tau_2, \ldots, \tau_n$, correlated with a definite quantum transition, namely $(n_1+\tau_1, \ldots) \to (n_1, \ldots)$. Although the radiation process in quantum theory occurs in jumps, whereas the classical radiation process is essentially continuous, Bohr postulated that statistically the transition probabilities will be such that the expected rate of radiation for a particular transition will have a similar asymptotic correlation with the corresponding Fourier component of the motion. Likewise the polarization of the radiation is postulated to be that of the classical radiation due to the corresponding Fourier component.

Thus for the electric-dipole radiation the procedure is to develop the components of $\boldsymbol{P}$ for motion in the state specified by $n_1, n_2, \ldots, n_n$ in a Fourier series, for example,

$$P_x(t) = \sum_\tau [n_1, \ldots | P_x | n_1+\tau_1, \ldots] \exp\left(2\pi i t \sum_k \tau_k \frac{\partial E}{\partial J_k}\right), \tag{6}$$

where each τ runs over all positive and negative integers. The notation for the coefficient is chosen to suggest its analogy with the corresponding matrix component in quantum mechanics. Exploitation of this analogy corresponds to the first step in the formulation of quantum mechanics as it was done by Heisenberg.* Since $\boldsymbol{P}(t)$ is real,

$$[n_1, \ldots | P_x | n_1-\tau_1, \ldots] = \overline{[n_1, \ldots | P_x | n_1+\tau_1, \ldots]}, \tag{7}$$

which is the analogue of the Hermitian condition for matrix components. By the correspondence principle, therefore, the rate of radiation of energy by quantum jumps in the transition $(n_1+\tau_1, \ldots) \to (n_1, \ldots)$ is related to the classical rate of radiation by the Fourier components of $\boldsymbol{P}$ whose frequency corresponds to the quantum frequency in the sense of (5). That is, there is an asymptotic connection between the rate of radiation of this frequency and the classical rate of radiation due to the dipole moment,

$$[n_1, \ldots | \boldsymbol{P} | n_1+\tau_1, \ldots] e^{2\pi i \nu t} + [n_1, \ldots | \boldsymbol{P} | n_1-\tau_1, \ldots] e^{-2\pi i \nu t} = 2\mathscr{R}\{[n_1, \ldots | \boldsymbol{P} | n_1+\tau_1, \ldots] e^{2\pi i \nu t}\}, \tag{8}$$

where $\nu = \sum \tau_k \partial E / \partial J_k$ and $\mathscr{R}$ indicates the real part of the expression in braces.

The expected rate of radiation in quantum theory is

$$h\nu\, \mathsf{A}(n_1+\tau_1, \ldots; n_1, \ldots),$$

where A is the spontaneous-transition probability for the transition in question. It is this which is postulated to be asymptotically equal to the classical rate of radiation.

* HEISENBERG, Zeits. für Phys. 33, 879 (1925).

For large quantum numbers and small τ's the value of this classical rate of radiation does not depend much on whether we use the Fourier components of the motion of the initial state or the final state. But for smaller quantum numbers these two sets of Fourier amplitudes may be greatly different and the correspondence principle does not say which should be used. In addition to these two possibilities one might even consider basing the Fourier analysis on some intermediate motion, say one with quantum numbers $(n_1+\frac{1}{2}\tau_1, n_2+\frac{1}{2}\tau_2, \ldots)$ when considering this jump. All such schemes are equally good asymptotically.

Bohr never offered the correspondence principle as more than an intuitive view that it is in this direction that we should seek for a more complete quantum theory—it was in this direction that Heisenberg found it. It is not our aim here to develop quantum mechanics inductively from the correspondence principle. Rather we wish merely to sketch its place in the theory of atomic spectra. Its importance in leading to the discovery of quantum mechanics cannot be over-emphasized; naturally, after that discovery, it no longer occupies a prominent place in the details of our subject. It is really surprising how many developments were made with the correspondence principle in its original form. The selection rules were obtained by arguments concerning the absence of certain Fourier components. If certain of these were absent in the initial and final state, and in a certain class of intermediate states as well, one felt very confident in the prediction that the associated quantum jumps had zero probability. For a complete account of detailed accomplishments of this kind see Van Vleck, *Quantum Principles and Line Spectra*.*

The lack of symmetry between initial and final states which characterizes the above discussion was remedied, and the 'verschärfung' of the correspondence principle accomplished, by Heisenberg's replacement of the Fourier components by the corresponding matrix components. We shall use the principle in this form.

Therefore we postulate that the radiation field accompanying a spontaneous transition from a state a of higher energy to a state b of lower energy is, as to angular intensity distribution and polarization, the same as that given by the classical theory for a charge distribution whose moments are

$$2\mathscr{R}\{(a|\quad|b)\,e^{2\pi i\nu t}\},$$

where $(a|\quad|b)$ is the matrix component of the kind of electric or magnetic moment under consideration. Moreover, the spontaneous-transition probability is to be taken as $\frac{1}{h\nu}$ times the classical rate of radiation of this charge distribution.

* National Research Council Bulletin No. 54, Washington, 1926.

For an atom the various moments which we shall need are expressed in terms of the coordinates of the electrons by the formulas

$$\boldsymbol{P} = -e\sum_i \boldsymbol{r}_i$$
$$\mathfrak{R} = -e\sum_i \boldsymbol{r}_i\boldsymbol{r}_i \tag{9}$$
$$\boldsymbol{M} = -\frac{e}{2\mu c}\sum_i (\boldsymbol{L}_i + 2\boldsymbol{S}_i).$$

It is the matrix components of these quantities which are to be used in calculating the radiation.

We have preferred to put the theory of emission on the correspondence-principle basis merely to avoid the lengthy calculations needed to derive the results by quantum-mechanical methods. The postulates are justified in Dirac's radiation theory.

5. The dipole-radiation field.

Let us investigate the field associated with the terms in 3[4]9 which depend on the dipole moment, $\boldsymbol{P}$. Using 2[4]4 we find, on retaining only the r^{-1} and r^{-2} terms,

$$\mathscr{E} = \frac{e^{2\pi i(\nu t - \sigma r)}}{r}\left\{\left(k^2 - \frac{ik}{r}\right)(\mathfrak{J} - \boldsymbol{r}_0\boldsymbol{r}_0)\cdot\boldsymbol{P} + \frac{2ik}{r}\boldsymbol{r}_0\boldsymbol{r}_0\cdot\boldsymbol{P}\right\}$$
$$\mathscr{H} = \frac{e^{2\pi i(\nu t - \sigma r)}}{r}\left(k^2 - \frac{ik}{r}\right)\boldsymbol{r}_0 \times \boldsymbol{P}, \tag{1}$$

where $\mathfrak{J}$ is the unit dyadic $\boldsymbol{ii} + \boldsymbol{jj} + \boldsymbol{kk}$.

The flow of energy is given by the Poynting vector,

$$\boldsymbol{S} = \frac{c}{4\pi}\mathscr{E} \times \mathscr{H}. \tag{2}$$

Since $\mathscr{E}$ and $\mathscr{H}$ are given by the real parts of (1), the time average radiation is given by

$$\boldsymbol{S}_{\text{av}} = \tfrac{1}{4}\frac{c}{4\pi}(\mathscr{E} \times \overline{\mathscr{H}} + \overline{\mathscr{E}} \times \mathscr{H}), \tag{3}$$

where $\mathscr{E}$ and $\mathscr{H}$ are as in (1). Therefore at large distances the flow of energy is

$$\boldsymbol{S}_{\text{av}} = \frac{c}{8\pi}\frac{k^4}{r^2}|(\mathfrak{J} - \boldsymbol{r}_0\boldsymbol{r}_0)\cdot\boldsymbol{P}|^2\,\boldsymbol{r}_0. \tag{4}$$

In the theory of atomic spectra $\boldsymbol{P}$ is nearly always given by formulas like 9[3]11. For transitions in which $\Delta m = 0$, the dipole moment is of the form $P\boldsymbol{k}$, directed along the z-axis, and the radiation field is that of a simple linear oscillator. In this case, in any direction of observation, the radiation is linearly polarized with the electric vector in the plane determined by $\boldsymbol{r}_0$ and $\boldsymbol{P}$. The intensity at an angle θ with the z-axis is

$$\boldsymbol{S}_{\text{av}} = \frac{c}{8\pi}\frac{k^4}{r^2}P^2\sin^2\theta\,\boldsymbol{r}_0, \tag{5}$$

so the entire rate of radiation through a large sphere in all directions is

$$\frac{c}{4}k^4P^2\int_0^\pi \sin^3\theta\,d\theta = \frac{16\pi^4\sigma^4 c}{3}P^2. \tag{6}$$

For transitions in which $\Delta m = \pm 1$, the vector $\boldsymbol{P}$ is of the form $\sqrt{\tfrac{1}{2}}P\,(\boldsymbol{i} \pm i\boldsymbol{j})$, so

$$\mathscr{R}\{\boldsymbol{P}e^{2\pi i\nu t}\} = \sqrt{\tfrac{1}{2}}P\,(\boldsymbol{i}\cos 2\pi\nu t \mp \boldsymbol{j}\sin 2\pi\nu t).$$

The vector rotates in the xy plane, in the clockwise direction for $\Delta m = +1$ when viewed from the positive z-direction, and in the opposite sense for $\Delta m = -1$. In either case the intensity in a direction making an angle θ with the z-axis is

$$\boldsymbol{S}_{\text{av}} = \frac{c}{8\pi}\frac{k^4}{r^2}\tfrac{1}{2}(1+\cos^2\theta)\,P^2\boldsymbol{r}_0, \tag{7}$$

and the whole radiation in all directions is the same as (6).

We may now write down the transition probabilities. In view of the postulate of § 4[4], they are

$$\mathsf{A}(a,b) = \frac{64\pi^4\sigma^3}{3h}|(a|\boldsymbol{P}|b)|^2, \tag{8}$$

for the transition from the upper state a to the lower state b. *For either value of* Δm, *we may replace* P^2 *by* $4|(a|\boldsymbol{P}|b)|^2$ *in equations* (5) *and* (7) *to find the angular intensity distribution.*

Before calculating the polarization in the case $\Delta m = \pm 1$, let us consider the general question of the type of wave represented by $\mathscr{R}\{\mathscr{E}e^{2\pi i\nu t}\}$, where $\mathscr{E} = \mathscr{E}_r + i\mathscr{E}_i$, a constant bivector. The variation with time,

$$\mathscr{R}\{\mathscr{E}e^{2\pi i\nu t}\} = \mathscr{E}_r\cos 2\pi\nu t - \mathscr{E}_i\sin 2\pi\nu t, \tag{9}$$

is such that the real $\mathscr{E}$'s end-point describes an ellipse turning from $\mathscr{E}_i$ into $\mathscr{E}_r$. If this direction is counter-clockwise on looking at the plane normal to $\boldsymbol{r}_0$ in the direction *opposite* to $\boldsymbol{r}_0$ (that is, as you would be faced if the light were to enter your eyes), the polarization is called left-elliptic. In general, $\mathscr{E}_r$ and $\mathscr{E}_i$ may be oblique, in which case they form a pair of conjugate semi-diameters of the ellipse. It is usually more convenient to shift the phase so that they are perpendicular and so coincide with the principal axes of the ellipse. We may write $\mathscr{E}e^{2\pi i\nu t} = \mathscr{E}e^{i\delta}e^{2\pi i\nu t - i\delta}$ and choose δ so that the real and imaginary parts of $\mathscr{E}e^{i\delta}$ are perpendicular. This readily leads to the following equation for δ,

$$\tan 2\delta = \frac{2\mathscr{E}_r\cdot\mathscr{E}_i}{\mathscr{E}_i^2 - \mathscr{E}_r^2}.$$

Usually the form of the calculations is such that the location of the principal axes is evident from symmetry. Thus in atomic radiation problems, the axes are along the meridians and latitude circles of the polar coordinates. We shall denote by $\boldsymbol{\theta}_0$ and $\boldsymbol{\varphi}_0$ the unit vectors at each point in the direction of increasing θ and φ respectively so that $\boldsymbol{r}_0$, $\boldsymbol{\theta}_0$, $\boldsymbol{\varphi}_0$ are the unit vectors of a right-handed system.

Then, if $\boldsymbol{P}=\sqrt{\tfrac{1}{2}}P\,(\boldsymbol{i}\pm i\boldsymbol{j})$,

$$\mathscr{E}\propto(\boldsymbol{\theta}_0\boldsymbol{\theta}_0+\boldsymbol{\varphi}_0\boldsymbol{\varphi}_0)\cdot(\boldsymbol{i}\pm i\boldsymbol{j})=(\boldsymbol{\theta}_0\cos\theta\pm i\boldsymbol{\varphi}_0)e^{\pm i\varphi}. \tag{10}$$

The radiation is, therefore, for the upper sign, right circularly polarized in the direction $\theta=0$, right elliptically polarized in directions $0<\theta<\pi/2$, linearly polarized in the equatorial plane, and left elliptically polarized in the 'southern hemisphere.' For this case the electric moment's variation in time is represented by a point rotating in a circle in the xy-plane in the same sense as the rotation of the electric vector in the light wave. The factor $e^{\pm i\varphi}$ occurring in $\mathscr{E}$ attends to the obvious detail that the phase in the light wave sent in the direction φ has a constant relation to the phase of $\boldsymbol{P}$ relative to the direction of observation.

Next we consider the transport of angular momentum in the radiation field. The volume density of electromagnetic momentum in the field is given by $\boldsymbol{S}/c^2$, where $\boldsymbol{S}$ is the Poynting vector. The corresponding density of angular momentum relative to the origin is therefore $\boldsymbol{r}\times\boldsymbol{S}/c^2$, and as this is moving outward with velocity c, the transport of angular momentum through unit area normal to $\boldsymbol{r}_0$ is $\boldsymbol{r}\times\boldsymbol{S}/c$.

Another way to find the angular-momentum transport is to consider the equation

$$\frac{\partial}{\partial t}(\boldsymbol{S}/c^2)+\operatorname{div}(-\mathfrak{T})=0, \tag{11}$$

in which $\mathfrak{T}$ is Maxwell's stress tensor for the electromagnetic field—

$$\mathfrak{T}=\frac{1}{4\pi}[\mathscr{E}\mathscr{E}+\mathscr{H}\mathscr{H}-\tfrac{1}{2}(\mathscr{E}^2+\mathscr{H}^2)\mathfrak{J}].$$

This shows that the negative of the stress tensor gives the transport of electromagnetic momentum, in the sense that $-\mathfrak{T}\cdot\boldsymbol{n}$ gives the amount of momentum crossing unit area normal to $\boldsymbol{n}$ toward positive $\boldsymbol{n}$ in unit time. Hence the momentum transported radially outward is $-\mathfrak{T}\cdot\boldsymbol{r}_0$ and the angular-momentum transport corresponding to this is

$$-\boldsymbol{r}\times(\mathfrak{T}\cdot\boldsymbol{r}_0)=-\frac{1}{4\pi}(\boldsymbol{r}\times\mathscr{E}\mathscr{E}\cdot\boldsymbol{r}_0+\boldsymbol{r}\times\mathscr{H}\mathscr{H}\cdot\boldsymbol{r}_0).$$

Since the leading term in $\boldsymbol{S}$ in the dipole-radiation field is perpendicular to $\boldsymbol{r}$, we need to consider the next term to get the angular-momentum transport. Using (1), we find the time mean angular-momentum transport at large distances to be

$$\boldsymbol{T}_{\mathrm{av}}=i\frac{k^3}{8\pi}[\boldsymbol{\varphi}_0(\boldsymbol{\theta}_0\cdot\boldsymbol{P}\boldsymbol{r}_0\cdot\bar{\boldsymbol{P}}-\boldsymbol{\theta}_0\cdot\bar{\boldsymbol{P}}\boldsymbol{r}_0\cdot\boldsymbol{P})+\boldsymbol{\theta}_0(\boldsymbol{\varphi}_0\cdot\boldsymbol{P}\boldsymbol{r}_0\cdot\bar{\boldsymbol{P}}-\boldsymbol{\varphi}_0\cdot\bar{\boldsymbol{P}}\boldsymbol{r}_0\cdot\boldsymbol{P})]$$

per unit solid angle. If $\boldsymbol{P}$ is real except for a phase factor $e^{i\delta}$, as when $\Delta m = 0$, the radiation of angular momentum vanishes. For $\Delta m = \pm 1$, where $\boldsymbol{P}$ is of the form $\sqrt{\frac{1}{2}}P\,(\boldsymbol{i} \pm i\boldsymbol{j})$, the value of $\boldsymbol{T}_{\text{av}}$ is

$$\boldsymbol{T}_{\text{av}} = \pm \frac{k^3}{8\pi} P^2 \sin\theta \, \boldsymbol{\theta}_0. \tag{12}$$

From the axial symmetry we see that the result of integration over all directions must be a vector along the z-axis. Its value is

$$\mp \boldsymbol{k} \frac{k^3}{4\pi} P^2 \, 2\pi \int_0^\pi \sin^3\theta \, d\theta = \mp \tfrac{1}{3} k^3 P^2 \boldsymbol{k}.$$

The radiation for $\Delta m = +1$ is thus in the direction $-\boldsymbol{k}$ which is in the right direction to compensate for the increase in angular momentum of the atom. Moreover the rate of radiation of angular momentum bears the correct ratio to the rate of radiation of energy to insure that the angular momentum $\hbar$ is radiated in the same time as the energy $h\nu$. This is very satisfactory for the correspondence principle, and was formerly used as a foundation for the selection rule in the angular-momentum quantum numbers.*

It may be objected that we have proved too much, since we found no radiation of angular momentum for $\Delta m = 0$. To be sure there is no change in the z-component of the atom's angular momentum in this case, but there may be a change in the resultant angular momentum, j, according to 9[3]11. The answer is that the classical analogy breaks down here, or better that the classical analogue of the quantum-mechanical state for an assigned j and m must be considered to involve an average over all orientations of the component of $\boldsymbol{J}$ in the xy-plane. This time average of J_x or J_y is zero by symmetry, and also since the diagonal matrix components of J_x and J_y are zero. From this standpoint the perpendicular components of $\boldsymbol{J}$ have an average value of zero before and after radiation, so there is no change in their value, and this agrees with the vanishing value of the angular-momentum transport in this case.

PROBLEM

Show that the magnetic-dipole radiation field is obtained by writing $\mathscr{H}$ for $\mathscr{E}$ and $-\mathscr{E}$ for $\mathscr{H}$ and replacing $\boldsymbol{P}$ by $\boldsymbol{M}$ in (1). Hence show that for magnetic-dipole radiation

$$\mathsf{A}(a, b) = \frac{64\pi^4\sigma^3}{3h} |(a|\boldsymbol{M}|b)|^2. \tag{13}$$

6. The quadrupole-radiation field.

The quadrupole-radiation field may be handled in an analogous fashion. For brevity we omit the discussion of the transport of angular momentum

* RUBINOWICZ, Phys. Zeits. **19**, 440 (1918);
BOHR, *Quantentheorie der Linienspektren*, Part I, p. 48.

in this case. The terms responsible for the energy flow are, from the potentials 3^49,

$$\mathscr{E}=\frac{ik^3}{2}\frac{e^{2\pi i(\nu t-\sigma r)}}{r}(\mathfrak{J}-\boldsymbol{r}_0\boldsymbol{r}_0)\cdot(\mathfrak{N}\cdot\boldsymbol{r}_0)$$

$$\mathscr{H}=\frac{ik^3}{2}\frac{e^{2\pi i(\nu t-\sigma r)}}{r}\boldsymbol{r}_0\times(\mathfrak{N}\cdot\boldsymbol{r}_0), \tag{1}$$

so that the Poynting vector is, from 5^43,

$$\boldsymbol{S}_{\text{av}}=\frac{ck^6}{32\pi r^2}|(\mathfrak{J}-\boldsymbol{r}_0\boldsymbol{r}_0)\cdot(\mathfrak{N}\cdot\boldsymbol{r}_0)|^2\boldsymbol{r}_0. \tag{2}$$

We are not ready-armed from Chapter III with the formulas for the dependence of the matrix components of $\mathfrak{N}$ on m, so we must stop to work them out. $\mathfrak{N}$ is a dyadic formed of the sum of dyads $-e\boldsymbol{r}_i\boldsymbol{r}_i$ for each electron (4^49), and so is a sum of dyads formed from vectors of type $\boldsymbol{T}$ considered in §§ 8^3 *et seq.* We can find the dependence on m of any such dyad by calculating its matrix components from the matrix components of $\boldsymbol{T}$ given by 9^311. Thus,

$$(\alpha j m|\boldsymbol{T}_1\boldsymbol{T}_2|\alpha' j' m')=\sum_{\alpha'' j'' m''}(\alpha j m|\boldsymbol{T}_1|\alpha'' j'' m'')(\alpha'' j'' m''|\boldsymbol{T}_2|\alpha' j' m'). \tag{3}$$

This enables us to see that the selection rules on j and m for $\boldsymbol{T}_1\boldsymbol{T}_2$ are Δj or $\Delta m=\pm 2, \pm 1$, or 0. In this way we may prepare a table of all non-vanishing matrix components for any dyadic made up of a linear combination of dyads of vectors of type $\boldsymbol{T}$. If this be done for a general dyadic, the formulas can be expressed as the sum of the matrix components of the symmetric part and the antisymmetric part. The latter is equivalent to a vector whose matrix components show the same dependence on m as a vector of type $\boldsymbol{T}$. Since any symmetric dyadic is the sum of two symmetric dyads, in making the calculations we may as well put $\boldsymbol{T}_1=\boldsymbol{T}_2=\boldsymbol{T}$ to get a symmetric dyad.

Calculations of this type were first made by Rubinowicz.* It turns out that each matrix component can be expressed as a numerical coefficient multiplying basic dyadics, $\mathfrak{R}(\Delta m)$, which depend on Δm only, except that $(\alpha j m|\boldsymbol{TT}|\alpha' j m)$ also includes an extra term which is a numerical factor times the identical dyadic, $\mathfrak{J}=\boldsymbol{ii}+\boldsymbol{jj}+\boldsymbol{kk}$. These dyadics, normalized so that $\mathfrak{R}(\Delta m):\overline{\mathfrak{R}}(\Delta m)=1$, are

$$\begin{aligned}\mathfrak{R}(\pm 2)&=\tfrac{1}{2}[\boldsymbol{ii}-\boldsymbol{jj}\pm i(\boldsymbol{ij}+\boldsymbol{ji})]\\ \mathfrak{R}(\pm 1)&=\tfrac{1}{2}[\boldsymbol{ki}+\boldsymbol{ik}\pm i(\boldsymbol{kj}+\boldsymbol{jk})]\\ \mathfrak{R}(0)&=\sqrt{\tfrac{2}{3}}[\boldsymbol{kk}-\tfrac{1}{2}\boldsymbol{ii}-\tfrac{1}{2}\boldsymbol{jj}].\end{aligned} \tag{4}$$

They determine the angular intensity distribution and polarization in its

* RUBINOWICZ, Zeits. für Phys. **61**, 338 (1930).

dependence on Δm. We may readily calculate the values of the vectorial factors occurring in (1) for these basic dyadics:

$$\begin{aligned}(\mathfrak{J}-\boldsymbol{r}_0\boldsymbol{r}_0)\cdot\mathfrak{K}(\pm 2)\cdot\boldsymbol{r}_0 &= \tfrac{1}{2}\sin\theta\, e^{\pm 2i\varphi}(\boldsymbol{\theta}_0\cos\theta \pm i\boldsymbol{\varphi}_0)\\ (\mathfrak{J}-\boldsymbol{r}_0\boldsymbol{r}_0)\cdot\mathfrak{K}(\pm 1)\cdot\boldsymbol{r}_0 &= \tfrac{1}{2}e^{\pm i\varphi}(\boldsymbol{\theta}_0\cos 2\theta \pm i\boldsymbol{\varphi}_0\cos\theta)\\ (\mathfrak{J}-\boldsymbol{r}_0\boldsymbol{r}_0)\cdot\mathfrak{K}(0)\cdot\boldsymbol{r}_0 &= -\sqrt{\tfrac{3}{2}}\sin\theta\cos\theta\,\boldsymbol{\theta}_0\\ (\mathfrak{J}-\boldsymbol{r}_0\boldsymbol{r}_0)\cdot\mathfrak{J}\cdot\boldsymbol{r}_0 &= 0;\end{aligned} \tag{5}$$

so that the spherically symmetric part of $(\alpha j m|\boldsymbol{TT}|\alpha' j m)$ does not contribute to the radiation. With these factors it is easy to see that the polarization of the light depends on the direction of observation θ and on Δm according to the scheme

θ	$\Delta m = +2$	$\Delta m = +1$	$\Delta m = 0$
0	—	Right circular	—
0 to $\pi/4$	Right elliptic	Right elliptic	π
$\pi/4$	Right elliptic	σ	π
$\pi/4$ to $\pi/2$	Right elliptic	Left elliptic	π
$\pi/2$	σ	π	—

Here the senses right and left are for the positive values of Δm. They are opposite for negative values of Δm and opposite at an angle $\pi-\theta$ in the 'southern hemisphere.' σ and π mean linear polarization with $\mathscr{E}$ along $\boldsymbol{\varphi}_0$ and $\boldsymbol{\theta}_0$ respectively. For $\Delta m = \pm 1$ the elliptic polarization becomes circular at $\pi/3$ and $2\pi/3$.

First we consider $\Delta j = 0$ for the dyadic $\boldsymbol{r}_i\boldsymbol{r}_i$. Here occur three sums,

$$D_\nu = \sum_{\alpha''}(\alpha j \vdots r_i \vdots \alpha'' j+\nu)(\alpha'' j+\nu \vdots r_i \vdots \alpha' j),$$

for $\nu = 1, 0, -1$. By requiring that $\boldsymbol{r}_i \times \boldsymbol{r}_i = 0$ (cf. 12[2]4 and 12[2]9b) have vanishing matrix components, we find a relation between them,

$$(2j+3)D_1 - D_0 - (2j-1)D_{-1} = 0.$$

If we write

$$D = (j+1)D_1 - jD_{-1},$$

the matrix components in question become

$$\begin{aligned}(\alpha j m|\boldsymbol{r}_i\boldsymbol{r}_i|\alpha' j m\pm 2) &= D\sqrt{(j\mp m)(j\mp m-1)(j\pm m+1)(j\pm m+2)}\,\mathfrak{K}(\pm 2)\\ (\alpha j m|\boldsymbol{r}_i\boldsymbol{r}_i|\alpha' j m\pm 1) &= D\,(2m\pm 1)\sqrt{(j\mp m)(j\pm m+1)}\,\mathfrak{K}(\pm 1)\\ (\alpha j m|\boldsymbol{r}_i\boldsymbol{r}_i|\alpha' j m) &= D\sqrt{\tfrac{2}{3}}\,[3m^2 - j(j+1)]\,\mathfrak{K}(0)\\ &\quad + \tfrac{1}{3}[(j+1)^2(2j+3)\,D_1 - j^2(2j-1)\,D_{-1}]\,\mathfrak{J}.\end{aligned} \tag{6a}$$

Similarly for $\Delta j = -1$ there occur two sums,

$$E_\nu = \sum_{\alpha''}(\alpha j \vdots r_i \vdots \alpha'' j+\nu)(\alpha'' j+\nu \vdots r_i \vdots \alpha' j-1),$$

for $\nu = 0, -1$. The vanishing of the matrix of $\boldsymbol{r}_i \times \boldsymbol{r}_i$ here shows that

$$(j+1)\,E_0 - (j-1)\,E_{-1} = 0.$$

If we let

$$E = E_0 + E_{-1},$$

these matrix components are

$$(\alpha j\, m|\boldsymbol{r}_i\boldsymbol{r}_i|\alpha' j-1\, m\pm 2) = \pm\tfrac{1}{2}E\sqrt{(j\mp m)(j\mp m-1)(j\mp m-2)(j\pm m+1)}\,\mathfrak{K}(\pm 2),$$

$$(\alpha j\, m|\boldsymbol{r}_i\boldsymbol{r}_i|\alpha' j-1\, m\pm 1) = \tfrac{1}{2}E\,(j\pm 2m+1)\sqrt{(j\mp m)(j\mp m-1)}\,\mathfrak{K}(\pm 1),$$

$$(\alpha j\, m|\boldsymbol{r}_i\boldsymbol{r}_i|\alpha' j-1\, m) = \sqrt{\tfrac{3}{2}}Em\sqrt{j^2-m^2}\,\mathfrak{K}(0). \quad (6b)$$

Finally for $\Delta j = -2$ the matrix components are multiples of the single sum

$$F = \sum_{\alpha''}(\alpha j\vdots r_i\vdots\alpha'' j-1)\,(\alpha'' j-1\vdots r_i\vdots\alpha' j-2),$$

the complete expressions being

$$(\alpha j\, m|\boldsymbol{r}_i\boldsymbol{r}_i|\alpha' j-2\, m\pm 2) = \tfrac{1}{2}F\sqrt{(j\mp m)(j\mp m-1)(j\mp m-2)(j\mp m-3)}\,\mathfrak{K}(\pm 2),$$

$$(\alpha j\, m|\boldsymbol{r}_i\boldsymbol{r}_i|\alpha' j-2\, m\pm 1) = \pm F\sqrt{(j^2-m^2)(j\mp m-1)(j\mp m-2)}\,\mathfrak{K}(\pm 1),$$

$$(\alpha j\, m|\boldsymbol{r}_i\boldsymbol{r}_i|\alpha' j-2\, m) = \sqrt{\tfrac{3}{2}}F\sqrt{(j^2-m^2)[(j-1)^2-m^2]}\,\mathfrak{K}(0). \quad (6c)$$

The matrix components for $\Delta j = +1$ or $+2$ can be written down from these since the matrices are Hermitian. The dyadic $\mathfrak{R}$ is of the form $-e\sum_i \boldsymbol{r}_i\boldsymbol{r}_i$, so that its matrix components are formally like those given above with new D, E, and F which are $-e\sum_i$ of the D, E, and F for the individual electrons.

To calculate the total rate of radiation we need to know that the integral of the product of each of the first three expressions in (5) by its complex conjugate over all directions is $4\pi/5$. Hence using the postulate of § 4^4, the total rate of radiation associated with the transition $(\alpha j\, m \to \alpha' j'\, m')$ is

$$\frac{ck^6}{10}|(\alpha j\, m|\mathfrak{R}|\alpha' j'\, m')|^2,$$

where the square of the dyadic indicates the double-dot product of the dyadic with its complex conjugate.

This gives for the corresponding probability of spontaneous emission

$$\mathsf{A}(\alpha j\, m, \alpha' j'\, m') = \frac{32\pi^6\sigma^5}{5h}|(\alpha j\, m|\mathfrak{R}|\alpha' j'\, m')|^2. \quad (7)$$

Examining the formulas for the matrix components we see that transitions between the following values of j: $0\to 0$, $\tfrac{1}{2}\to\tfrac{1}{2}$, and $1 \rightleftarrows 0$, have vanishing intensity. Likewise in the dipole-radiation formulas the transitions $j=0 \to j=0$ have vanishing intensity. These are special cases of a general rule. The 2^n-pole radiation is connected with a change in the resultant angular momentum of the electromagnetic field by n units. Since there is conservation of the total angular momentum, the vector sum of that of the atom plus that of the field remains constant. We must then have

$$|j'-n| \leqslant j \leqslant j'+n,$$

where j is the resultant angular momentum of the atom before radiation and j' its value after the transition.

Brinkman* has developed a group-theoretical method of obtaining the quadrupole-moment matrix components. The next term, or octopole radiation, has been considered briefly by Huff and Houston.†

7. Spectral lines in natural excitation.

When an atom is unperturbed by external fields, its Hamiltonian commutes with the resultant angular momentum $\boldsymbol{J}$. As a consequence ($3^3 8$) an energy level corresponding to a given value of j is $(2j+1)$-fold degenerate, each of the different states being characterized by a different value of m. Such a set of $(2j+1)$ *states* we shall call a *level*. When the atom is perturbed and the different states do not all have the same energy it is still convenient to refer to such a set of states by a generic name, so we shall also use level in this wider sense. Likewise we define the word *line* to mean the radiation associated with all possible transition between the states belonging to two levels. The radiation resulting from a transition between a particular pair of states we call a *component* of the line. All components have the same wave number unless the atom is perturbed by an external field.

The actual emitted intensity of the component from state a to state b in ergs/second is equal to

$$I(a,b) = N(a)\, h\nu\, \mathsf{A}(a,b), \tag{1}$$

where $N(a)$ is the number of atoms in state a and $\mathsf{A}(a,b)$ is the spontaneous emission probability for that transition. This equation holds only when the radiation density present is so small that the induced emission is negligible. In actual practice it is very difficult to know $N(a)$ and hence to make absolute comparisons between theory and experiment for emission intensity. Most work is therefore on relative intensities of related groups of lines.

In most light sources the conditions of excitation of the atoms are sufficiently isotropic that it is safe to assume that the numbers of atoms in the different states of the same level are equal. This is called *natural excitation.* If the excitation occurs in some definitely non-isotropic way, as by absorption from a unidirectional beam of light or by impacts from a unidirectional beam of electrons, large departures from natural excitation may be produced. The study of such effects‡ raises a whole complex of problems somewhat detached from the main body of spectroscopy. We shall always assume natural excitation unless the contrary is stated.

* Brinkman, Dissertation Utrecht, 1932.

† Huff and Houston, Phys. Rev. **38**, 842 (1930).

‡ For a review of this field see Mitchell and Zemansky, *Resonance Radiation and Excited Atoms*, Cambridge University Press, 1934.

The total *intensity* of a line is the sum of the intensities of its components. Hence in natural excitation for the line from level A to level B,

$$I(A,B)=N(a)\,h\nu\,\frac{64\pi^4\sigma^3}{3h}\sum_{ab}|(a|\boldsymbol{P}|b)|^2, \tag{2}$$

in the case of dipole radiation, by $5^4 8$. The sum of the squared matrix components occurring here we shall define as the *strength* of the line, $\mathsf{S}(A, B)$. We shall also find it convenient to call $|(a|\boldsymbol{P}|b)|^2$ the strength of the component a, b and to introduce the partial sums over a and b alone, with the notation

$$\mathsf{S}(a, B)=\sum_b|(a|\boldsymbol{P}|b)|^2,\quad \mathsf{S}(A, b)=\sum_a|(a|\boldsymbol{P}|b)|^2,$$

$$\mathsf{S}(A, B)=\sum_a \mathsf{S}(a, B)=\sum_b \mathsf{S}(A, b)=\sum_{ab}\mathsf{S}(a, b).$$

The intensity of a particular line is therefore proportional to the number of atoms in any one of the initial states, to σ^4 and to $\mathsf{S}(A, B)$. This represents a slight departure from the traditional procedure, where the intensity is written as proportional to the total number of atoms in the initial level, $N(A)$:

$$I(A, B)=N(A)\,h\nu\,\mathsf{A}(A, B).$$

As $N(A)=(2j_A+1)\,N(a)$, this gives

$$\mathsf{A}(A, B)=\frac{1}{2j_A+1}\,\frac{64\pi^4\sigma^3}{3h}\,\mathsf{S}(A, B). \tag{3}$$

Here A is the spontaneous-transition probability of Einstein. It has the disadvantage that it is not symmetrical in the initial and final levels owing to the factor $(2j_A+1)^{-1}$ which relates it to the symmetrical $\mathsf{S}(A, B)$. In later work we shall find formulas for the strengths of certain classes of lines where one level belongs to one group and the other to another group (e.g. multiplets). For the strength of a line it will not matter in such formulas whether A or B is the initial level, whereas if $j_A \neq j_B$, the Einstein A's are different in the two cases. Therefore we shall regard the strengths as the more convenient theoretical measure of line intensities.

Before calculating $\mathsf{S}(A, B)$ let us consider what properties we expect it to have. First, since the excitation is isotropic and there is no external field we expect the radiation to be of equal intensity in all directions, and unpolarized. Second, if the initial conditions are isotropic we expect the physical situation to remain so during the radiation process. This means that the mean life of each state of level A must be the same. The mean life of a state a is the reciprocal of the total transition probability from a to all lower states b, c, d, ... belonging to all lower levels B, C, D, ... That is, it is the reciprocal of

$$\frac{64\pi^4\sigma^3}{3h}[\mathsf{S}(a, B)+\mathsf{S}(a, C)+\ldots].$$

This should have the same value for each state a, otherwise if the excitation is cut off the numbers of atoms in the different states of level A will be unequal after a time, and a non-isotropic situation will have grown out of an isotropic one. Since there is no very simple general relation between the levels $B, C, \ldots$ for all atoms, we expect this to be true for each summand $\mathsf{S}(a, B)$. Likewise because of the symmetry of $|(a|\boldsymbol{P}|b)|^2$ with regard to a and b, it follows that if $\mathsf{S}(a, B)$ is independent of a, then $\mathsf{S}(A, b)$ is independent of b.

We shall now show that the dipole radiation in the transition $\alpha j \to \alpha' j'$ is isotropic and unpolarized. By $5^4 5$ the intensity in the direction θ of radiation plane-polarized in the direction $\boldsymbol{\theta}_0$ is proportional to

$$\sum_{m=-j}^{j} |(\alpha j\, m|\boldsymbol{P}|\alpha' j'\, m)|^2 \sin^2\theta. \qquad (\Delta m = 0) \quad (4a)$$

There will also be right- and left-elliptically polarized radiations whose intensities ($5^4 7$) are respectively proportional to

$$\sum_m \tfrac{1}{2} |(\alpha j\, m|\boldsymbol{P}|\alpha' j'\, m+1)|^2 (1+\cos^2\theta), \quad (\Delta m = +1) \quad (4b)$$

$$\sum_m \tfrac{1}{2} |(\alpha j\, m|\boldsymbol{P}|\alpha' j'\, m-1)|^2 (1+\cos^2\theta). \quad (\Delta m = -1) \quad (4c)$$

That the sum of these three quantities is independent of θ follows from $13^3 3$; hence the intensity of radiation is the same in all directions. The sum of the terms for $\Delta m = +1$ and $\Delta m = -1$ is no longer elliptically polarized because the right-handed radiation from $m \to m+1$ is of the same intensity as the left-handed from $-m \to -m-1$; it is, however, partially plane polarized with relative intensities $\sum |(\alpha j\, m|\boldsymbol{P}|\alpha' j'\, m+1)|^2$ along $\boldsymbol{\varphi}_0$ and $\sum |(\alpha j\, m|\boldsymbol{P}|\alpha' j'\, m+1)|^2 \cos^2\theta$ along $\boldsymbol{\theta}_0$ (cf. $5^4 10$ and $5^4 9$). When the $\boldsymbol{\theta}_0$ component is added to that for $\Delta m = 0$, it follows from $13^3 3$ that the resultant radiation is unpolarized.

The summations over the three values of m_b which go with $m_a = m$ have already been evaluated in $13^3 1'$, where their independence of the value of m was noted. Because of this, summation over m merely introduces the factor $(2j+1)$, so for the whole strength of the line $\alpha\, j$ to $\alpha'\, j'$ we have

$$\mathsf{S}(\alpha j, \alpha' j') = (2j+1)\, |(\alpha j \vdots P \vdots \alpha' j')|^2\, \Xi(j, j'), \qquad (5)$$

where

$$\Xi(j, j+1) = (j+1)(2j+3); \quad \Xi(j, j) = j(j+1); \quad \Xi(j, j-1) = j(2j-1).$$

Everything we have said here about electric-dipole radiation is true for magnetic-dipole, if we replace $\boldsymbol{P}$ by $\boldsymbol{M}$, since the dependence on m of $\boldsymbol{M}$ ($4^4 9$) is the same as that of $\boldsymbol{P}$.

For quadrupole radiation the analogue of (2) is

$$I(A, B) = N(a)\, h\nu \frac{32\pi^6\sigma^5}{5h} \sum_{ab} |(a|\mathfrak{R}|b)|^2. \qquad (6)$$

The strengths for quadrupole lines are defined exactly as for dipole with $\mathfrak{R}$ in place of $\boldsymbol{P}$. Expressions for the strengths of the various quadrupole lines are easily calculated from the values of the matrix components in the previous section. The results are

$$\begin{aligned} \mathsf{S}(\alpha j, \alpha' j) &= \tfrac{2}{3} j(j+1)(2j+1)(2j-1)(2j+3) D^2 \\ \mathsf{S}(\alpha j, \alpha' j-1) &= \tfrac{1}{2} j(j+1)(j-1)(2j+1)(2j-1) E^2 \\ \mathsf{S}(\alpha j, \alpha' j-2) &= j(j-1)(2j+1)(2j-1)(2j-3) F^2. \end{aligned} \tag{7}$$

The partial sums here also have the property of giving unpolarized light of equal intensity in all directions. The spontaneous transition probability analogous to (3) is

$$\mathsf{A}(A, B) = \frac{32\pi^6\sigma^5}{5h} \frac{\mathsf{S}(A, B)}{2j_A + 1}. \tag{8}$$

8. Induced emission and absorption.

In view of the relation found in § 1^4 between the probability coefficients for spontaneous and induced emission and absorption, the formula for $\mathsf{B}(A, B)$ in terms of the matrix components of dipole or quadrupole moment is given by combining $1^4 7$ with $7^4 3$ or $7^4 8$. That gives the result for the interaction of an atom with an isotropic unpolarized field.

There is a further point on which we wish information, namely the probability for absorption of light from a unidirectional polarized beam. From the following argument, we expect this to be simply related to the $\mathsf{A}(a, b)$ for a line component. The emission in the transition $a \to b$ is a spherical wave whose intensity and polarization in different directions we have calculated for the important special cases. Let the angular distribution factor be $f(\theta, \varphi)\, d\omega$, where $\int f\, d\omega = 1$ over all directions, and let $\boldsymbol{l}(\theta, \varphi)$ be the unit bivector ($\boldsymbol{l}\cdot\bar{\boldsymbol{l}} = 1$) normal to the direction of propagation, $\boldsymbol{n}$, which gives the polarization of the wave emitted in the direction θ, φ. We suppose for convenience that $\boldsymbol{l}$ is normalized so that its real and imaginary parts are perpendicular, as discussed in § 5^4.

If $\mathsf{A}(a, b)$ is the whole transition probability ($5^4 8$ or $6^4 7$), then we can say statistically that $\mathsf{A}(a, b) f(\theta, \varphi)\, d\omega$ is the transition probability for processes in which the light is sent out in the solid angle $d\omega$ at θ, φ. This emission takes place with the polarization described by $\boldsymbol{l}$. Hence for the probability of spontaneous transition from a to b with emission in the solid angle $d\omega$ at θ, φ, and with the polarization $\boldsymbol{l}$ we may write

$$\mathsf{A}(a, b, \theta, \varphi, \boldsymbol{l})\, d\omega = \mathsf{A}(a, b) f(\theta, \varphi)\, d\omega. \tag{1}$$

There will also be stimulated emission. This and the spontaneous emission constitute the inverse process to absorption from a beam whose range of directions is $d\omega$ at θ, φ. Of the $8\pi\sigma^2 d\sigma$ waves per unit volume (we say waves as a short expression for degrees of freedom of the field) there are $2\sigma^2 d\sigma\, d\omega$ associated with waves which travel in directions contained in $d\omega$, the factor two arising from the two independent states of polarization associated with each direction of propagation. We are at liberty to choose the independent states of polarization in any convenient way. The natural way in discussing the transitions $a \rightleftarrows b$ of the atom, is to choose $\boldsymbol{l}$ as one of the canonical states of polarization and to introduce $\boldsymbol{m}$, orthogonal to $\boldsymbol{l}$ in the sense $\boldsymbol{l}\cdot\bar{\boldsymbol{m}}=0$, to describe the other.

Let the intensity of the beam be $\rho(\sigma,\boldsymbol{l})\,d\sigma\,d\omega$. This is the volume density of energy in wave-number range $d\sigma$ of waves whose directions are in the cone $d\omega$ and with their polarization described by $\boldsymbol{l}$. We define the probability of absorption (or induced emission) of this radiation so as to cause a transition of the atom from a to b as

$$\mathsf{B}(a,b,\theta,\varphi,\boldsymbol{l})\,\rho(\sigma,\boldsymbol{l})\,d\omega.$$

Then, as in 1^46, if a is the upper state, the balance of emission and absorption between these degrees of freedom and the atom when in thermal equilibrium gives

$$N(a)[\mathsf{A}(a,b,\theta,\varphi,\boldsymbol{l})+\rho(\sigma,\boldsymbol{l})\,\mathsf{B}(a,b,\theta,\varphi,\boldsymbol{l})]=N(b)\,\mathsf{B}(b,a,\theta,\varphi,\boldsymbol{l})\,\rho(\sigma,\boldsymbol{l}).$$

Using the Maxwell-Boltzmann law 1^44 and the Planck law 1^45 divided by 8π to get $\rho(\sigma,\boldsymbol{l})$ and realizing that here the statistical weights are each unity, we have

$$\begin{aligned}\mathsf{B}(a,b,\theta,\varphi,\boldsymbol{l})&=\mathsf{B}(b,a,\theta,\varphi,\boldsymbol{l})\\ \mathsf{A}(a,b,\theta,\varphi,\boldsymbol{l})&=hc\sigma^3\,\mathsf{B}(a,b,\theta,\varphi,\boldsymbol{l}).\end{aligned}\qquad(2)$$

The same equations hold for the other state of polarization, $\boldsymbol{m}$, but as we have chosen $\boldsymbol{l}$ in accordance with the actual polarization of the wave sent out in the emission process, $\mathsf{A}(a,b,\theta,\varphi,\boldsymbol{m})=0$, and so the corresponding absorption coefficient vanishes. Analogous to 1^48 we may write the rate of emission processes in the form

$$N(a)\,\mathsf{A}(a,b,\theta,\varphi,\boldsymbol{l})\,[n(\sigma)+1]\,d\omega,\qquad(3)$$

where $n(\sigma)=\rho(\sigma,\boldsymbol{l})/hc\sigma^3$ is the mean number of quanta per degree of freedom in the beam in the direction and polarization state under consideration.

Let us consider the description of the polarization a little more closely. The bivector $\boldsymbol{l}$ may be written in the form

$$\boldsymbol{l}=\boldsymbol{\theta}_0\cos\tfrac{1}{2}\delta+i\boldsymbol{\varphi}_0\sin\tfrac{1}{2}\delta,\qquad(4)$$

since we have found in § 5^4 and § 6^4 that the elliptic polarization always has $\boldsymbol{\theta}_0$ and $\boldsymbol{\varphi}_0$ for its principal axes when the transitions are between states labelled by m. The range $0\leqslant\delta\leqslant 2\pi$ covers all possible types of polarization according to the scheme

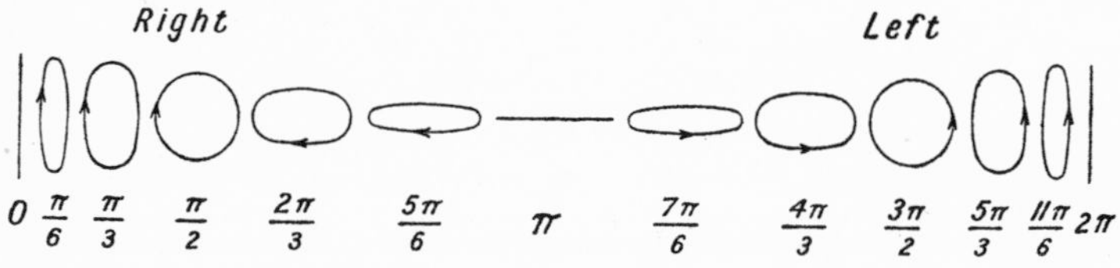

The bivector orthogonal to $\boldsymbol{l}$ is

$$\boldsymbol{m} = \boldsymbol{\theta}_0 \sin\tfrac{1}{2}\delta - i\boldsymbol{\varphi}_0 \cos\tfrac{1}{2}\delta,$$

which represents a wave of the same ellipticity as that for $\boldsymbol{l}$ but with the major and minor axes interchanged and the sense of turning, right or left, reversed.

If the incident light is polarized in some other manner than with regard to one of the canonically chosen forms, $\boldsymbol{l}$ and $\boldsymbol{m}$, say with its electric vector a scalar multiple of the unit bivector $\boldsymbol{e}$, then it may be expressed as the resultant of two components polarized as in $\boldsymbol{l}$ and $\boldsymbol{m}$ by the formula

$$\boldsymbol{e} = \boldsymbol{l}(\bar{\boldsymbol{l}}\cdot\boldsymbol{e}) + \boldsymbol{m}(\bar{\boldsymbol{m}}\cdot\boldsymbol{e}). \tag{5}$$

In this case the absorption is simply $|\bar{\boldsymbol{l}}\cdot\boldsymbol{e}|^2$ times the value it has when the polarization is along $\boldsymbol{l}$. If the incident light is unpolarized, the intensity is distributed equally between each of any two orthogonal states of polarization, so the absorption is half what it would be if the polarization were along $\boldsymbol{l}$.

Finally let us consider the absorption by a group of atoms in a state of natural excitation. The total number of transitions per second from the states b of level B to the states a of level A caused by absorption of light whose polarization is described by $\boldsymbol{e}$ is

$$\rho(\sigma,\boldsymbol{e})\,d\omega \sum_{b,a} \mathsf{B}(b,a,\theta,\varphi,\boldsymbol{l})\,|\bar{\boldsymbol{l}}\cdot\boldsymbol{e}|^2 N(b) = \frac{\rho(\sigma,\boldsymbol{e})\,d\omega}{hc\sigma^3} \sum_{b,a} \mathsf{A}(a,b,\theta,\varphi,\boldsymbol{l})\,|\bar{\boldsymbol{l}}\cdot\boldsymbol{e}|^2 N(b).$$

Using (1) and the expression 5[4]8 for $\mathsf{A}(a,b)$ we find, for example in the case of dipole radiation, that this equals

$$\frac{64\pi^4}{3h^2c}\,\rho(\sigma,\boldsymbol{e})\,d\omega\,N(b) \sum_{b,a} |(a|\boldsymbol{P}|b)|^2\,|\bar{\boldsymbol{l}}\cdot\boldsymbol{e}|^2 f_{a,b}(\theta,\varphi).$$

We have seen in § 7[4] that the total emission of all components of a line in natural excitation is unpolarized and of equal intensity in all directions. The sum here is the total strength of emission per unit solid angle with polarization $\boldsymbol{e}$; this is accordingly $1/8\pi$ times the total strength of the line A, B. Therefore we can express the rate of absorption for the whole line in terms of the strength $\mathsf{S}(A,B)$ as defined in § 7[4] as

$$\frac{8\pi^3}{3h^2c}\,\mathsf{S}(A,B)\,N(b)\,\rho(\sigma,\boldsymbol{e})\,d\omega. \tag{6}$$

This gives the number of quanta per second absorbed from a unidirectional beam of energy density $\rho(\sigma,\boldsymbol{e})\,d\sigma\,d\omega$. As we expect, it depends only on the energy density and is independent of direction and polarization of the light. The Einstein B for isotropic radiation obtained from this relation,

$$\mathsf{B}(B,A) = \frac{8\pi^3}{3h^2c}\,\frac{\mathsf{S}(A,B)}{2j_B+1}, \tag{7}$$

bears the proper relation to $\mathsf{A}(A,B)$ (cf. 1[4]7 and 7[4]3).

The results we have reached by the phenomenological method follow from the Dirac radiation theory. Another derivation which gives correct results for the stimulated processes but does not give the spontaneous emission is to treat the electromagnetic field of the light wave as a perturbation, using the form of the perturbation theory in which the perturbation is regarded as causing transitions.*

9. Dispersion theory. Scattering. Raman effect.

In the preceding section we have considered the absorption of light by matter. This occurs when the light frequency corresponds to a transition frequency to within a small range given by the natural line breadth ($1^4 10$). We wish now to consider the interaction of light and matter when there is no such exact agreement. There are essentially three phenomena:

(*a*) The macroscopic light wave is propagated with a phase velocity c/n instead of c, where n is the index of refraction.

(*b*) There is a scattering of light to all sides without change in the wave-number. This is called Rayleigh scattering.

(*c*) There is a scattering to all sides of light in which the frequency is shifted by an amount corresponding to one of the atomic transitions. This is the Raman scattering. It is unimportant in monatomic gases but we mention it to complete the picture.

As subheads under (*a*) there are special effects like the double refraction produced when the gas is in an external electric or magnetic field (the Kerr effect and the Faraday effect respectively).

All these phenomena belong together in the theory, for they are all connected with the oscillating moments induced in the atoms by the perturbing action of the fields of the incident light wave. When these are calculated by quantum mechanics it is found that they include some terms which oscillate with the same frequency as the light wave. In addition there are terms which oscillate with frequencies $\nu \pm \nu(A, B)$, where $\nu(A, B)$ is the Bohr frequency associated with transition from level A to level B. The latter are responsible for Raman scattering; they are of different frequency and so cannot be coherent with the incident radiation.

The forced oscillations of the same frequency as the incident light are to a certain extent coherent with it (that is, to a certain extent they have a definite phase relation to the phase of the incident light), so it is necessary to consider the phases in summing the scattered waves due to different atoms to find the resultant field. This dynamical theory of refraction goes back to Larmor and Lorentz, but the best modern treatment is by

* DIRAC, *Quantum Mechanics*, p. 173.

Darwin.* The net result of such calculations can be simply stated. The induced dipole moment $\boldsymbol{P}$ is related to the electric vector $\mathscr{E}$ by

$$\boldsymbol{P}=\mathfrak{a}\cdot\mathscr{E}, \tag{1}$$

where $\mathfrak{a}$ is the polarizability tensor. Then if N is the number of molecules per unit volume, the total dipole moment per unit volume is $N\mathfrak{a}\cdot\mathscr{E}$. Assuming that $N\mathfrak{a}\ll\mathfrak{J}$, i.e. that the principal values of $N\mathfrak{a}$ are each small compared to unity, we may neglect the difference between the effective $\mathscr{E}$ on an atom and the macroscopic $\mathscr{E}$; hence the dielectric constant, a tensor, becomes

$$\mathfrak{e}=\mathfrak{J}+4\pi N\mathfrak{a}. \tag{2}$$

Using the form of Maxwell's equations for a material medium devoid of free charge and of unit magnetic susceptibility:

$$\begin{aligned}\mathscr{D}=\mathfrak{e}\cdot\mathscr{E},\qquad \operatorname{div}\mathscr{D}&=0,\qquad \operatorname{div}\mathscr{H}=0,\\ \operatorname{curl}\mathscr{E}=-\frac{1}{c}\frac{\partial\mathscr{H}}{\partial t},&\qquad \operatorname{curl}\mathscr{H}=\frac{1}{c}\frac{\partial\mathscr{D}}{\partial t},\end{aligned} \tag{3}$$

we may find plane-wave solutions in which each vector is a constant multiple of $e^{2\pi i\nu(t-n\boldsymbol{n}\cdot\boldsymbol{r}/c)}$, where n is the index of refraction and $\boldsymbol{n}$ a unit vector in the direction of propagation. The constant amplitudes must satisfy the equations

$$\begin{aligned}\boldsymbol{n}\cdot\mathscr{D}=0,&\quad n\boldsymbol{n}\times\mathscr{E}=\mathscr{H},\\ \boldsymbol{n}\cdot\mathscr{H}=0,&\quad n\boldsymbol{n}\times\mathscr{H}=-\mathscr{D},\end{aligned} \tag{4}$$

so $\mathscr{D}$ and $\mathscr{H}$ are transverse to the direction of propagation. Since the principal values of the dielectric constant differ from unity by small quantities, we may write $n=1+\delta$. Eliminating $\mathscr{H}$ between the right-hand pair of (4), we find that the part of $\mathscr{D}$ which is perpendicular to $\boldsymbol{n}$ satisfies

$$\mathfrak{a}\cdot\mathscr{D}=\frac{\delta}{2\pi N}\mathscr{D}. \tag{5}$$

If $\mathfrak{a}$ is simply a scalar α, any value of $\mathscr{D}$ normal to $\boldsymbol{n}$ is admissible and the index of refraction is

$$n=1+2\pi N\alpha; \tag{6}$$

but if $\mathfrak{a}$ is not a scalar, then (5) shows that $\delta/2\pi N$ is equal to an eigenvalue of the tensor $\mathfrak{a}$ regarded as a two-dimensional tensor in the plane normal to $\boldsymbol{n}$. The two eigenvalues lead to two values of the index of refraction, that is, double refraction for each direction of propagation, $\boldsymbol{n}$. The corresponding eigenvectors give the two types of polarization associated with the waves propagated with the velocities corresponding to the two indices of refraction.

We shall see when we calculate $\mathfrak{a}$ quantum mechanically that it is always Hermitian (with regard to the three axes of space), so the two eigenvalues

* Darwin, Trans. Cambr. Phil. Soc. **23**, 137 (1924);
see also Esmarch, Ann. der Phys. **42**, 1257 (1913);
Oseen, *ibid.* **48**, 1 (1915);
Bothe, *ibid.* **64**, 693 (1921);
Lundblad, Univ. Arskrift, Upsala, 1920.

are real and correspond to propagation without absorption. Choosing two real unit vectors $\boldsymbol{l}$ and $\boldsymbol{m}$ so that $\boldsymbol{l}$, $\boldsymbol{m}$, and $\boldsymbol{n}$ form a right-handed coordinate system, $\mathfrak{a}$ will appear in the form,

$$\mathfrak{a} = \mathfrak{a}_{11}\boldsymbol{ll} + \mathfrak{a}_{12}\boldsymbol{lm} + \bar{\mathfrak{a}}_{12}\boldsymbol{ml} + \mathfrak{a}_{22}\boldsymbol{mm}.$$

There are two cases to distinguish: if $\mathfrak{a}_{12} = \bar{\mathfrak{a}}_{12}$ so the components are all real, then the eigenvectors are real and we have ordinary double refraction, whereas if $\mathfrak{a}_{12} \neq \bar{\mathfrak{a}}_{12}$ the eigenvectors are bivectors and the two fundamental states of polarization are elliptic, or in special cases, circular. It is easy to show that the two eigenvectors $\mathscr{D}_1$ and $\mathscr{D}_2$ are orthogonal: $\mathscr{D}_1 \cdot \overline{\mathscr{D}}_2 = 0$. Therefore the two states of polarization are orthogonal as discussed in § 8^4, so they have the same ellipticity but one is right and the other left, and the major axis of one ellipse is the minor axis of the other.

The foregoing discussion applies to the effect of the waves scattered by the molecules if the medium is absolutely homogeneous, in which case N is the number of molecules or atoms in unit volume. Owing to the thermal motions of the molecules in a gas the medium is not homogeneous; the number of molecules in any volume v at a given instant fluctuates about the mean value Nv, where N is calculated by dividing the total number of molecules by the total volume. The waves scattered by the number of atoms in each volume element which is in excess of the mean do not interfere in regular fashion; hence they give rise to scattered light. This, together with some non-coherent parts of the induced moment of unshifted frequency, gives rise to the Rayleigh scattering. We shall not have occasion to consider in detail the scattered radiation.

The dispersion and the related double refractions produced by external fields are of importance in the theory of spectra in that they provide other methods for measuring line strength than those based directly on emission or absorption. For their theory we need to calculate the polarizability tensor $\mathfrak{a}$ introduced in (1). This is obtained by calculating the perturbation of the states of the atom by the field of the light wave. Then we calculate the matrix components $(\alpha j m|\boldsymbol{P}|\alpha' j' m')_{\mathrm{p}}$ of the electric moment of the atom with regard to these perturbed states. Of these the component $(\alpha j m|\boldsymbol{P}|\alpha j m)_{\mathrm{p}}$ gives rise to an electric moment which is coherent with the incident light wave and which is responsible for the various dispersion phenomena. When the atom is unperturbed by other external fields the energy is independent of m, and the matrix components $(\alpha j m|\boldsymbol{P}|\alpha j m \pm 1)_{\mathrm{p}}$ also contain terms which are of the same frequency as the incident light, but these are not coherent with it and so they contribute an additional amount to the Rayleigh scattering without contributing to the dispersion.*

* See Breit, Rev. Mod. Phys. 5, 106 (1933), for an adequate discussion of this point.

The calculation is based on a slight modification of the perturbation theory of § 8[2], required by the fact that the perturbation depends on the time. To a first approximation the perturbation is

$$-\mathscr{R}\{\boldsymbol{P}\cdot\mathscr{E}e^{2\pi i\nu t}\}, \tag{7}$$

where $\mathscr{R}\{\mathscr{E}e^{2\pi i\nu t}\}$ is the electric vector of the perturbing light wave. This neglects effects connected with quadrupole transitions. The state $(\alpha j m)$ whose unperturbed eigenfunction is $\psi(\alpha j m)\, e^{-iWt/\hbar}$ will to the first order become

$$\psi(\alpha j m)\, e^{-iWt/\hbar} + \psi_1(\alpha j m),$$

where $\psi_1(\alpha j m)$ is chosen so that this expression satisfies ($5^2 3$)

$$\left(H - i\hbar\frac{\partial}{\partial t}\right)\psi_1 = \tfrac{1}{2}\left[\boldsymbol{P}\cdot\mathscr{E}e^{iEt/\hbar} + \boldsymbol{P}\cdot\overline{\mathscr{E}}e^{-iEt/\hbar}\right]\psi(\alpha j m)\, e^{-iWt/\hbar}.$$

Here we have written $E = h\nu$ to represent the frequency of the light, and H for the unperturbed Hamiltonian. Assuming the expansion

$$\psi_1(\alpha j m) = \tfrac{1}{2}\sum_{\alpha' j' m'} \psi(\alpha' j' m')\left[(\alpha' j' m'|\alpha j m)_+ e^{iD_1 t/\hbar} + (\alpha' j' m'|\alpha j m)_- e^{-iD_2 t/\hbar}\right], \tag{8}$$

we find that this is a solution if

$$\begin{aligned} D_1 &= E - W, \quad (\alpha' j' m'|\alpha j m)_+ = \frac{(\alpha' j' m'|\boldsymbol{P}|\alpha j m)\cdot\mathscr{E}}{W' - W + E} \\ D_2 &= E + W, \quad (\alpha' j' m'|\alpha j m)_- = \frac{(\alpha' j' m'|\boldsymbol{P}|\alpha j m)\cdot\overline{\mathscr{E}}}{W' - W - E}. \end{aligned} \tag{9}$$

The matrix component of $\boldsymbol{P}$ with respect to these perturbed states is, to the first order,

$$\begin{aligned}(\alpha j m|\boldsymbol{P}|\alpha j m)_{\mathrm{p}} = (\alpha j m|\boldsymbol{P}|\alpha j m) &+ \bar{\psi}_1(\alpha j m)\, \boldsymbol{P}\, \psi(\alpha j m)\, e^{-iWt/\hbar} \\ &+ \bar{\psi}(\alpha j m)\, \boldsymbol{P}\, \psi_1(\alpha j m)\, e^{iWt/\hbar}.\end{aligned}$$

Using the results just found for $\bar{\psi}_1(\alpha j m)$ and $\psi_1(\alpha j m)$ this can be written as

$$(\alpha j m|\boldsymbol{P}|\alpha j m) + \mathscr{R}\{\mathfrak{a}(\alpha j m)\cdot\mathscr{E}e^{2\pi i\nu t}\}, \tag{10}$$

where $\mathfrak{a}(\alpha j m)$ is the polarizability tensor of the state $\alpha j m$: therefore

$$\mathfrak{a}(\alpha j m) = \sum_{\alpha' j' m'}\left[\frac{(\alpha j m|\boldsymbol{P}|\alpha' j' m')(\alpha' j' m'|\boldsymbol{P}|\alpha j m)}{W' - W + E} + \frac{\overline{(\alpha j m|\boldsymbol{P}|\alpha' j' m')(\alpha' j' m'|\boldsymbol{P}|\alpha j m)}}{W' - W - E}\right]. \tag{11}$$

The dependence of $\mathfrak{a}$ on m can be found by using $9^3 11$ for the matrix components of $\boldsymbol{P}$. When this is done it is found that the tensor can be expressed conveniently as the sum of three parts: an isotropic part which is a multiple of $\mathfrak{J}$, a doubly refracting part which is a multiple of $\mathfrak{K}(0)$ of $6^4 4$, and a gyrotropic part which is a multiple of $i(\boldsymbol{ji} - \boldsymbol{ij})$. When there is natural excitation and no external fields, the last two parts vanish when the sum is taken over the different m values, leaving only the isotropic part. This shows that in natural excitation, in the absence of external fields, the gas is optically isotropic. The matrix components occurring in the isotropic part are proportional to the strengths $\mathsf{S}(\alpha j, \alpha' j')$ of the lines, so that finally the isotropic polarizability α for natural excitation may be written as

$$\alpha(\alpha j) = \frac{2}{3h} \sum_{\alpha' j'} \frac{\nu(\alpha' j', \alpha j)\, \mathsf{S}(\alpha j, \alpha' j')}{(2j+1)[\nu^2(\alpha' j', \alpha j) - \nu^2]}, \tag{12}$$

where $\nu(\alpha' j', \alpha j) = [W(\alpha' j') - W(\alpha j)]/h$. This is the *average*, not the *sum*, of the polarizabilities for the different values of m. The frequency $\nu(\alpha' j', \alpha j)$ is positive for those energy levels $W(\alpha' j')$ which are higher than $W(\alpha j)$, and negative for those lower.

The Kramers-Heisenberg dispersion formula is obtained at once when we use this value of α in (6) to obtain the index of refraction, $n = 1 + \delta$. The term (12) gives the average contribution of atoms in the energy level $W(\alpha j)$ so that the effective average value of α is obtained from the distribution $N(\alpha j)$ of atoms:

$$\alpha = \frac{1}{N} \sum_{\alpha j} \alpha(\alpha j)\, N(\alpha j). \tag{13}$$

In case all the atoms are in the lowest energy level we need here only one term like (12). In this all the $\nu(\alpha' j', \alpha j)$ are positive. The dispersion formula for this case was discovered before quantum mechanics by Ladenburg* from a correspondence principle argument. The existence of the negative dispersion terms for atoms in excited states was recognized by Kramers† and later Kramers and Heisenberg‡ reached the complete formula including Raman scattering from the correspondence principle. The experimental reality of negative dispersion has been very thoroughly demonstrated by Ladenburg and his associates.§

In the discussion of the dispersion formula (12) it is convenient to introduce a new set of quantities called the oscillator strengths, as distinguished from line strengths. This is based on a comparison between any one term of (12) with the term given by classical theory for dispersion due to an isotropic harmonic oscillator of charge e, mass μ, and natural frequency $\nu(\alpha' \alpha)$. The equation of motion of such an oscillator when perturbed by the electric vector is

$$\ddot{\boldsymbol{r}} + [2\pi\, \nu(\alpha' \alpha)]^2 \boldsymbol{r} = \frac{e}{\mu} \mathscr{E} e^{2\pi i \nu t},$$

where the actual motion is $\mathscr{R}\{\boldsymbol{r}\}$. The solution of this equation represents

* Ladenburg, Zeits. für Phys. **4**, 451 (1921).
† Kramers, Nature, **113**, 673 (1924).
‡ Kramers and Heisenberg, Zeits. für Phys. **31**, 681 (1925).
§ Ladenburg, Zeits. für Phys. **48**, 15 (1928);
Kopfermann and Ladenburg, *ibid.* **48**, 26, 51 (1928);
Carst and Ladenburg, *ibid.* **48**, 192 (1928).

the sum of free and forced oscillations. For the latter we find that $\boldsymbol{r} = \boldsymbol{a}e^{2\pi i\nu t}$, where

$$e\boldsymbol{a} = \frac{e^2}{4\pi^2\mu}\frac{1}{\nu^2(\alpha'\alpha) - \nu^2}\mathfrak{J}\cdot\mathscr{E},$$

so the classical value of the polarizability of the oscillator is

$$\alpha_c(\alpha'\alpha) = \frac{e^2}{4\pi^2\mu}\frac{1}{\nu^2(\alpha'\alpha) - \nu^2}.$$

Comparing with (12) we see that the quantum theoretical formula for $\alpha(\alpha j)$ can be written as

$$\alpha(\alpha j) = \sum_{\alpha' j'} \mathsf{f}(\alpha' j', \alpha j)\,\alpha_c(\alpha' j', \alpha j), \tag{14}$$

where

$$\mathsf{f}(\alpha' j', \alpha j) = \frac{8\pi^2\mu}{3e^2 h}\frac{\nu(\alpha' j', \alpha j)\,\mathsf{S}(\alpha j, \alpha' j')}{2j+1}.$$

In other words, each spectral line contributes to the polarizability a term whose form is that of the contribution from a classical oscillator, multiplied by a dimensionless quantity $\mathsf{f}(\alpha' j', \alpha j)$ called the *oscillator strength*. Like the Einstein A's and B's, the f's are not symmetric in the initial and final states owing to the occurrence of the factor $(2j+1)^{-1}$. From the symmetry of S and the antisymmetry of $\nu(\alpha' j', \alpha j)$ we find that

$$(2j+1)\,\mathsf{f}(\alpha' j', \alpha j) = -(2j'+1)\,\mathsf{f}(\alpha j, \alpha' j').$$

Using (13) and (14) it is possible to express the effective polarizability in terms of the f's as the sum of contributions associated with each spectral line:

$$\alpha = \frac{1}{N}\sum \mathsf{f}(\alpha' j', \alpha j)\,\alpha_c(\alpha' j', \alpha j)\,N(\alpha j)\left[1 - \frac{N(\alpha' j')/(2j'+1)}{N(\alpha j)/(2j+1)}\right], \tag{15}$$

in which αj refers to the lower energy level of each line and $\alpha' j'$ to the higher, and the sum is over all the lines in the spectrum. The negative term in the brackets is the correction due to the negative dispersion arising from atoms in excited states.

The oscillator strengths $\mathsf{f}(\alpha' j', \alpha j)$ satisfy an interesting sum rule which was discovered independently by Thomas and by Kuhn* from consideration of the dispersion formula from the standpoint of the correspondence principle. It is a simple consequence of the law of commutation of $\boldsymbol{p}$ and $\boldsymbol{r}$ and is important historically in that it was used by Heisenberg† to find this commutation law in his first paper on matrix mechanics. The rule states that for a system containing a single electron the sum $\sum_{\alpha' j'} \mathsf{f}(\alpha' j', \alpha j)$ is equal to unity. The proof is very simple. From the commutation law 5[2]1 we have

$$\boldsymbol{p}\cdot\boldsymbol{r} - \boldsymbol{r}\cdot\boldsymbol{p} = -3i\hbar.$$

* Thomas, Naturwiss. **13**, 627 (1925);
Kuhn, Zeits. für Phys. **33**, 408 (1925).
† Heisenberg, Zeits. für Phys. **33**, 879 (1925).

For a single electron in an arbitrary potential field the Hamiltonian is

$$H = \frac{1}{2\mu}\boldsymbol{p}^2 + V(\boldsymbol{r}),$$

from which by the commutation law we find that

$$\boldsymbol{p} = -i(\mu/\hbar)(\boldsymbol{r}H - H\boldsymbol{r}),$$

so that the matrix components of $\boldsymbol{p}$ in terms of those of $\boldsymbol{r}$ are

$$(\alpha j m|\boldsymbol{p}|\alpha' j' m') = -2\pi i\mu\, \nu(\alpha'\alpha)\, (\alpha j m|\boldsymbol{r}|\alpha' j' m').$$

Hence a diagonal matrix component of $(\boldsymbol{p}\cdot\boldsymbol{r} - \boldsymbol{r}\cdot\boldsymbol{p})$ is

$$4\pi\mu \sum_{\alpha' j' m'} \nu(\alpha'\alpha)\, |(\alpha j m|\boldsymbol{r}|\alpha' j' m')|^2 = 3\hbar.\text{*}$$

This is independent of m and can therefore be summed over m by multiplying by $2j+1$. Substituting the oscillator strengths $\mathsf{f}(\alpha' j', \alpha j)$ from their definition in (14) this becomes

$$\sum_{\alpha' j'} \mathsf{f}(\alpha' j', \alpha j) = 1,$$

which is the required sum rule.

10. Natural shape of absorption lines.

Let us now consider the actual shape of an absorption line in a gas, taking account of the finite width of spectral lines as given by 1[4]10. The finite width arises in the coupling of the atom and the radiation field and so applies to absorption as well as emission. When we take account of 1[4]10 we find that the absorption probability varies with σ over the natural width of the line. Measuring σ from the centre of the line we may write

$$\mathsf{b}(B, A, \sigma) = \frac{\mathsf{b}(B, A, 0)}{1 + (\sigma/\sigma_0)^2} \tag{1}$$

for the probability that a quantum between σ and $\sigma + d\sigma$ produces a transition in unit time. Then, if $\rho(\sigma)$ is essentially constant in the neighbourhood of the line, the total number of transitions produced by radiation in the neighbourhood of the central frequency is

$$\rho(\sigma)\, N(B) \int_{-\infty}^{\infty} \frac{\mathsf{b}(B, A, 0)\, d\sigma}{1 + (\sigma/\sigma_0)^2} = \rho(\sigma)\, N(B)\, \pi\sigma_0\, \mathsf{b}(B, A, 0).$$

In the ordinary treatment in which we neglect the natural width this total number of transitions is written $\rho(\sigma)\, N(B)\, \mathsf{B}(B, A)$, hence we have

$$\mathsf{b}(B, A, 0) = \mathsf{B}(B, A)/\pi\sigma_0. \tag{2}$$

The intensity of a light beam is measured by the energy $I(\sigma)$ crossing unit area normal to the direction of propagation in unit time in the wave-number range $d\sigma$ at σ. Since $\rho(\sigma)$ is the volume density of energy, evidently

$$I(\sigma) = c\, \rho(\sigma).$$

* Sums of this type, but with ν^2, ν^3, and ν^4 written in place of $\nu(\alpha'\alpha)$, have been evaluated by VINTI, Phys. Rev. **41**, 432 (1932).

The beam goes through thickness Δx in time $\Delta x/c$, so that its diminution in intensity can be found by multiplying the number of transitions it produces in this time by the average size of quantum, $h\nu$, to give the relation

$$-\Delta I(\sigma) = I(\sigma)\, N(B)\frac{h\nu}{c}\frac{\mathsf{b}(B,A,0)}{1+(\sigma/\sigma_0)^2}\Delta x.$$

Hence the intensity after going through a finite thickness x of absorbing material is

$$I(\sigma) = I_0 e^{-\frac{\zeta}{1+(\sigma/\sigma_0)^2}}, \tag{3}$$

where

$$\zeta = \frac{h\nu}{c}\frac{\mathsf{B}(B,A)}{\pi\sigma_0} N(B)\, x.$$

The coefficient of $N(B)\,x$ in the expression for ζ is an area that is characteristic of the line in question. We shall call it the effective area for absorption of that line. Then ζ is the number of atoms in the initial level of the absorption process contained in a cylinder whose base has this area and whose height is the thickness of the absorbing matter.

The amount of energy removed from the incident beam in $d\sigma$ at σ is

$$I_0\left[1 - e^{-\frac{\zeta}{1+(\sigma/\sigma_0)^2}}\right].$$

If we are dealing with a small amount of absorbing matter, we may calculate the total absorption by expanding the exponential function, saving only the first two terms and integrating with respect to σ. This recovers the result $\pi\sigma_0 I_0 \zeta$ or $h\nu\, \mathsf{B}(B,A)\, N(B)\, x/c$ which applies when we ignore the natural line width.

But if there is a larger amount of absorbing matter, so that $\zeta \sim 1$ or $\zeta \gg 1$, we have to calculate

$$I_0 \sigma_0 \int_{-\infty}^{+\infty} \left[1 - e^{-\frac{\zeta}{1+(\sigma/\sigma_0)^2}}\right] d\sigma/\sigma_0 = I_0 \sigma_0 f(\zeta) \tag{4}$$

to get the total absorption. The integrand gives the shape of the absorption line. In Fig. 1[4] it is plotted for several values of ζ. For large ζ the absorption becomes essentially complete in the centre of the band and the curve becomes much broader than the natural breadth. Although the absorption coefficient becomes very small as we go away from the centre of the line, the large amount of matter present gives rise to appreciable absorption at fairly large distances from the centre. The absorption is 50 per cent. for

$$(\sigma/\sigma_0) = \sqrt{(\zeta/\log_e 2) - 1}.$$

For large values of ζ we can estimate the integral by supposing the integrand equal to unity out to this point and zero beyond. This gives for the total absorption

$$I_0 \sigma_0 2\sqrt{\zeta/\log_e 2} = 2{\cdot}39 I_0 \sigma_0 \sqrt{\zeta}.$$

This is remarkably close to the true asymptotic value.

Considerations of this kind were first developed by Ladenburg and Reiche* on the basis of the classical electron theory. They evaluated the integral in (4) and found

$$f(\zeta) = \pi\zeta e^{-\zeta/2}[J_0(i\zeta/2) - iJ_1(i\zeta/2)], \tag{5}$$

where the J's are Bessel functions. Using the known asymptotic expansions we have, for $\zeta \gg 1$,

$$f(\zeta) = 2\sqrt{\pi\zeta} = 3{\cdot}544\sqrt{\zeta}.$$

Adopting any convenient definition for the 'width' of the absorption line, such as the wave-number interval between points where the absorption amounts to some fraction of the total intensity, we see that the width varies as the square root of the active number of atoms in the line of sight. Out on the 'wings' of the line, where the absorption is small, we may develop the

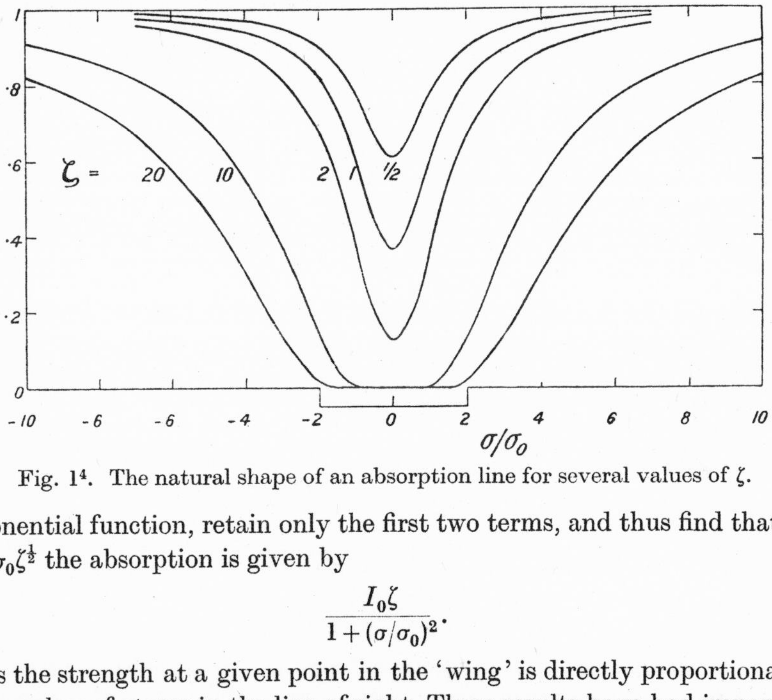

Fig. 1⁴. The natural shape of an absorption line for several values of ζ.

exponential function, retain only the first two terms, and thus find that for $\sigma \gg \sigma_0\zeta^{\frac{1}{2}}$ the absorption is given by

$$\frac{I_0\zeta}{1+(\sigma/\sigma_0)^2}.$$

Thus the strength at a given point in the 'wing' is directly proportional to the number of atoms in the line of sight. These results have had important astrophysical application† to the theory of the Fraunhofer lines in the solar spectrum, but we cannot go into this interesting topic here.

* Ladenburg and Reiche, Ann. der Phys. **42**, 181 (1913).

† Stewart, Astrophys. J. **59**, 30 (1924);
Schütz, Zeits. für Astrophys. **1**, 300 (1930);
Minnaert and Mulders, *ibid.* **2**, 165 (1931);
Unsöld, *ibid.* **2**, 199 (1931).

CHAPTER V

ONE-ELECTRON SPECTRA

The simplest atomic model and the only one for which all details of the spectrum can be accurately calculated is that of hydrogen and the hydrogen-like ions, He II, Li III, etc. This model consists of a heavy nucleus of charge Ze and a single electron, interacting principally according to the Coulomb law. We shall see in later chapters that the spectra of the alkali metals may be treated by regarding one electron as moving in a central non-Coulomb field determined by the nucleus and the average effect of the other electrons. In this chapter we treat the hydrogen-like spectra and the alkali spectra, anticipating for the latter the justification of the assumption that these may to a good approximation be regarded as one-electron spectra.

1. Central-force problem.

We consider the motion of an electron of mass μ in a central field in which its potential energy is given by $U(r)$, where r is its distance from the centre of force. Neglecting electron spin and relativity effects the Hamiltonian for the electron is

$$H=\frac{1}{2\mu}(p_x^2+p_y^2+p_z^2)+U(r). \tag{1}$$

From 8^34 we see that H commutes with L_x, L_y, L_z and $\boldsymbol{L}^2$, and therefore $\boldsymbol{L}^2$ and any component, e.g. L_z, may be taken as constants of the motion. We look therefore for simultaneous eigenstates of H, $\boldsymbol{L}^2$ and L_z. The Schrödinger equation (6^22) for this Hamiltonian is

$$-\frac{\hbar^2}{2\mu}\Delta\phi+U\phi=W\phi.$$

If we express the Laplace operator in polar coordinates and use 4^327, this becomes

$$-\frac{\hbar^2}{2\mu}\frac{1}{r^2}\frac{\partial}{\partial r}\left(r^2\frac{\partial\phi}{\partial r}\right)+\frac{\boldsymbol{L}^2}{2\mu r^2}\phi+U\phi=W\phi. \tag{2}$$

The requirement that ϕ be an eigenstate of $\boldsymbol{L}^2$ and L_z determines its dependence on θ and φ according to 4^312, so we may write

$$\phi=\frac{R_r}{r}\Theta(l\,m_l)\,\Phi(m_l). \tag{3}$$

In this case the equation for the radial factor of the eigenfunction becomes

$$\frac{\hbar^2}{2\mu}\frac{d^2R}{dr^2}+\left[W-U(r)-\frac{\hbar^2}{2\mu}\frac{l\,(l+1)}{r^2}\right]R=0. \tag{4}$$

The allowed values of the energy W are the values for which this equation possesses solutions which vanish at the origin and remain finite at infinity. It will be noted that the value of L_z does not occur in the equation for W, hence each value of W associated with the quantum number l will be $(2l+1)$-fold degenerate. (This result follows also from 3^38 since H commutes with $\boldsymbol{L}$.)

The equation (4) for the radial eigenfunction is formally that of a one-dimensional motion under the effective potential energy

$$U(r)+\frac{\hbar^2}{2\mu}\frac{l(l+1)}{r^2},$$

the second term corresponding to the effective potential energy of the centrifugal force. Starting with the lowest value of W determined by this equation for a particular value of l, it is customary to distinguish the states by means of a 'total' quantum number n which has the value $l+1$ for the lowest state and increases by unity for each higher state, so that $n-l-1$ is the number of nodes of the radial eigenfunction, not counting the origin as a node.

Also in accordance with custom among spectroscopists we introduce a letter code for the values of l, according to the scheme

Value of l:	0	1	2	3	4	5	6	7	8	9	10	11	12	13	14	15	...
Designation:	*s*	*p*	*d*	*f*	*g*	*h*	*i*	*k*	*l*	*m*	*n*	*o*	*q*	*r*	*t*	*u*	...

The energy levels will be labelled by the quantum numbers n and l according to the scheme $W(1s)$, $W(5d)$, etc. This scheme arose in connection with the interpretation of various series of lines in the alkali spectra, known as the sharp, principal, diffuse, and fundamental series. The quantum numbers nl will be said to specify the *configuration* of the electron.

The preceding discussion has been for a fixed centre of force. Let us now consider the two-body problem in which the electron moves under the influence of the same field due to a nucleus of mass M. Let X, Y, Z be the coordinates of the centre of mass of the system and x, y, z the coordinates of the electron relative to the nucleus. Then, as in classical mechanics, the kinetic energy of the whole system becomes

$$T=\tfrac{1}{2}(\mu+M)(\dot{X}^2+\dot{Y}^2+\dot{Z}^2)+\tfrac{1}{2}\mu'(\dot{x}^2+\dot{y}^2+\dot{z}^2), \qquad \left(\mu'=\frac{\mu M}{\mu+M}\right)$$

where μ' is the so-called reduced mass. Expressing this in terms of momenta as in classical dynamics and substituting the Schrödinger operators for the momenta, the Schrödinger equation of the two-body problem becomes

$$-\frac{\hbar^2}{2(\mu+M)}\Delta_X\psi-\frac{\hbar^2}{2\mu'}\Delta_x\psi+U(r)\psi=E\psi,$$

where Δ_x is the Laplacian with respect to the internal coordinates and Δ_X is that with respect to the centroid's coordinates. The variables may be separated by writing ψ as a product of a function of the internal coordinates by another function of the centroid's coordinates. The separate Schrödinger equation for the internal coordinates is then found to be just (2) with the reduced mass in place of the electronic mass, while that for the external coordinates is simply the Schrödinger equation for a free particle of mass $\mu + M$. The total energy E breaks up into the sum of the internal energy W and the energy of translation. The energy of translational motion does not need to be considered further as it does not change appreciably during a radiation process and thus does not affect the line frequency. Hence we may allow for the finite mass of the nucleus simply by using the reduced mass in place of the electronic mass in equation (2).

2. Radial functions for hydrogen.

We shall now consider the radial equation $1^5 4$ for the Coulomb potential function $U(r) = -Ze^2/r$:

$$\frac{\hbar^2}{2\mu}\frac{d^2R}{dr^2} + \left(W + \frac{Ze^2}{r} - \frac{\hbar^2}{2\mu}\frac{l(l+1)}{r^2}\right)R = 0. \tag{1}$$

By making the substitutions

$$W = -\frac{\mu e^4 Z^2}{2\hbar^2 n^2}, \quad r = \frac{\frac{1}{2}\rho n \mathsf{a}}{Z}, \quad \mathsf{a} = \frac{\hbar^2}{\mu e^2}, \tag{2}$$

this equation takes on the more concise form

$$\frac{d^2R}{d\rho^2} + \left(-\frac{1}{4} + \frac{n}{\rho} - \frac{l(l+1)}{\rho^2}\right)R = 0, \tag{3}$$

which is recognized as the equation of the confluent hypergeometric function.* From the form assumed by the equation near $\rho = \infty$ we see that the solution which remains finite there contains the factor $e^{-\rho/2}$. An attempt to solve the equation with a power series in ρ shows that the solution which is finite at the origin starts off with ρ^{l+1}. This suggests the substitution

$$R = \rho^{l+1} e^{-\rho/2} f(\rho),$$

which leads to the following equation for $f(\rho)$:

$$f'' + \left(\frac{2(l+1)}{\rho} - 1\right)f' + \frac{n-l-1}{\rho} f = 0. \tag{4}$$

If this equation be solved in terms of a power series in ρ, it will be found that the series breaks off as a polynomial of degree $n-l-1$ if n is an integer greater than l. The polynomials obtained in this way belong to the set usually known as associated Laguerre polynomials. If n is not an integer the power series defines a solution which becomes infinite like $e^{+\rho}$ as $\rho \to \infty$, so the finite solutions are defined by the requirement that n be integral. In this case, (2) gives just the well-known Bohr levels for hydrogen-like atoms:

$$\sigma = \frac{W}{hc} = -\mathsf{R}\frac{Z^2}{n^2}; \quad \mathsf{R} = \frac{\mu e^4}{4\pi\hbar^3 c}. \tag{5}$$

* Whittaker and Watson, *A Course of Modern Analysis*, Chapter xvi.

The ordinary Laguerre polynomials may be defined by the equation

$$L_n(x) = e^x D^n(x^n e^{-x}), \tag{6}$$

where D is the symbol for differentiation with regard to x. An equivalent form is

$$L_n(x) = (D-1)^n(x^n). \tag{7}$$

If the indicated differentiations are actually performed, this becomes

$$L_n(x) = (-1)^n \left[x^n - \frac{n^2}{1!} x^{n-1} + \frac{n^2(n-1)^2}{2!} x^{n-2} - \ldots + (-1)^n n! \right]. \tag{8}$$

The associated Laguerre polynomial $L_n^m(x)$ is defined by

$$L_n^m(x) = D^m L_n(x). \tag{9}$$

It is the polynomials so defined that satisfy (4), the solution being

$$f(\rho) = L_{n+l}^{2l+1}(\rho), \tag{10}$$

a polynomial of degree $n-l-1$. An equivalent definition for the associated Laguerre polynomial is

$$L_n^m(\rho) = \frac{1}{n!} e^\rho \rho^{-m} D^n(e^{-\rho} \rho^{m+n}). \tag{11}$$

Applying either of the definitions we find that the polynomials are given by the general formula:

$$L_{n+l}^{2l+1}(\rho) = -[(n+l)!]^2 \sum_{\lambda=0}^{n-l-1} \frac{(-\rho)^\lambda}{\lambda!(n-l-1-\lambda)!(2l+1+\lambda)!}. \tag{12}$$

From this form they are easily seen to have the following relation to the hypergeometric function:

$$L_{n+l}^{2l+1}(\rho) = -\frac{[(n+l)!]^2}{(n-l-1)!(2l+1)!} \lim_{\beta\to\infty} F(-n+l+1, \beta, 2l+2, \rho/\beta). \tag{13}$$

Another form of the solution for the radial functions which is due to Eckart* is useful in some calculations. He showed that to within a normalizing factor the radial wave function $R(nl)/r$ is

$$\chi(nl) = \frac{r^l}{(n-l-1)!} \left(\frac{\partial}{\partial z}\right)^{n-l-1} \left\{ e^{zr} \left(z - \frac{Z}{na}\right)^{n+l} \right\}, \tag{14}$$

in which z is set equal to $-Z/na$ after the differentiation is performed. Carrying out the indicated differentiations, it is easily seen that

$$\chi(nl) = \left(\frac{2Z}{na}\right)^{l+1} \rho^l e^{-\rho/2} \frac{1}{(n+l)!} L_{n+l}^{2l+1}(\rho). \tag{15}$$

The Laguerre polynomials may also be studied with the aid of their generating function, according to the following expansion

$$\sum_{\gamma=0}^{\infty} L_{\beta+\gamma}^{\alpha}(x) \frac{t^\gamma}{(\beta+\gamma)!} = (-1)^\alpha \frac{e^{-\frac{xt}{(1-t)}}}{(1-t)^{\alpha+1}}. \tag{16}$$

* Eckart, Phys. Rev. **28**, 927 (1926).

This is the relation used by Schrödinger* for calculating certain integrals involving Laguerre polynomials.

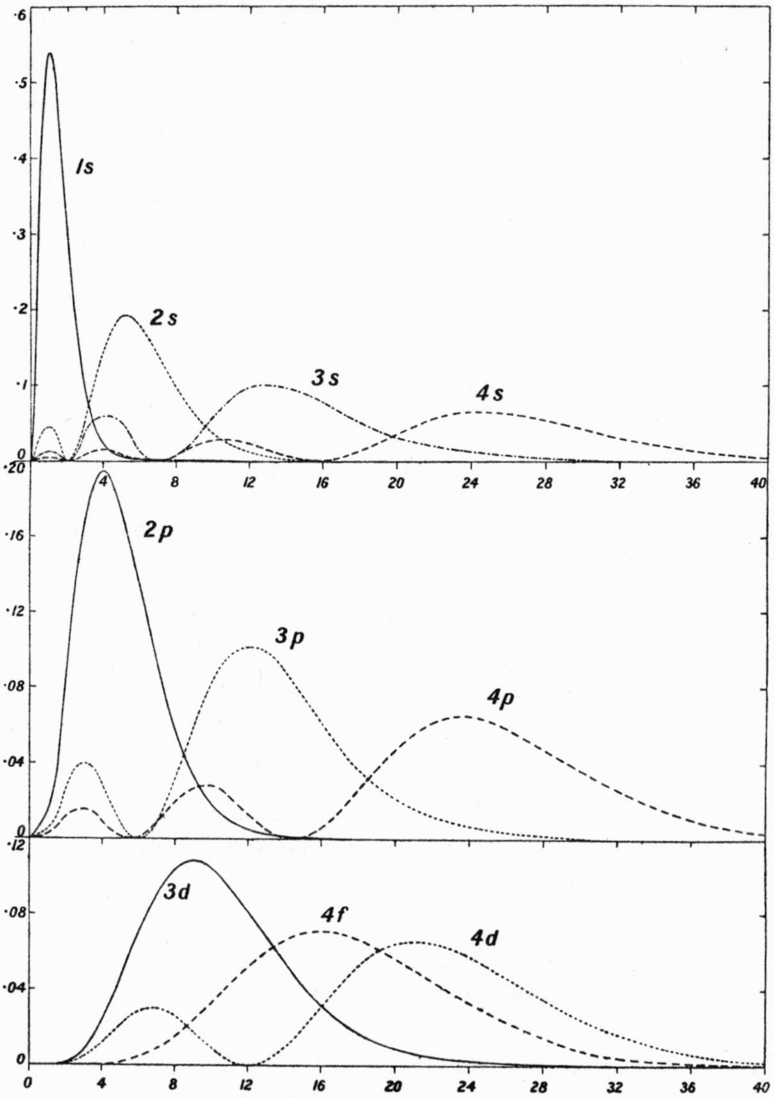

Fig. 1[5]. Radial probability distribution a $R^2(nl)$ for several of the lowest levels in hydrogen. (Abscissa is the radius in atomic units.)

In order to normalize the radial functions we need the result

$$\int_0^\infty \rho^{2l+2}e^{-\rho}[L_{n+l}^{2l+1}(\rho)]^2 d\rho = \frac{2n[(n+l)!]^3}{(n-l-1)!}, \tag{17}$$

which is readily obtained from the generating function or from Eckart's

* SCHRÖDINGER, Ann. der Phys. **80**, 485 (1926).

formula. Therefore the final formula for the normalized radial function $R(nl)$ is

$$R(nl)=\sqrt{\frac{Z\,(n-l-1)!}{n^2\mathrm{a}[(n+l)!]^3}}\,e^{-\rho/2}\rho^{l+1}\,L_{n+l}^{2l+1}(\rho),\qquad \rho=\frac{2Zr}{n\mathrm{a}} \tag{18}$$

which is normalized in the sense $\int_0^\infty R^2(nl)\,dr=1$. Several of these functions are given explicitly in Table 1^5. The probability of finding the electron in dr at r is $R^2(nl)\,dr$; this distribution function is plotted in Fig. 1^5 for some of the lowest states.

TABLE 1^5. *Normalized radial eigenfunctions for $Z=1$.**

$R(1s)=-2re^{-r}$	$R(3p)=-\frac{8}{27\sqrt{6}}r^2e^{-\frac{1}{3}r}(1-\frac{1}{6}r)$
$R(2s)=-\frac{1}{\sqrt{2}}re^{-\frac{1}{2}r}(1-\frac{1}{2}r)$	$R(4p)=-\frac{1}{16}\sqrt{\frac{5}{3}}r^2e^{-\frac{1}{4}r}(1-\frac{1}{4}r+\frac{1}{80}r^2)$
$R(3s)=-\frac{2}{3\sqrt{3}}re^{-\frac{1}{3}r}(1-\frac{2}{3}r+\frac{2}{27}r^2)$	$R(3d)=-\frac{4}{81\sqrt{30}}r^3e^{-\frac{1}{3}r}$
$R(4s)=-\frac{1}{4}re^{-\frac{1}{4}r}(1-\frac{3}{4}r+\frac{1}{8}r^2-\frac{1}{192}r^3)$	$R(4d)=-\frac{1}{64\sqrt{5}}r^3e^{-\frac{1}{4}r}(1-\frac{1}{12}r)$
$R(2p)=-\frac{1}{2\sqrt{6}}r^2e^{-\frac{1}{2}r}$	$R(4f)=-\frac{1}{768\sqrt{35}}r^4e^{-\frac{1}{4}r}$

The average values of various powers of r for the hydrogenic wave functions are given in Table 2^5.†

TABLE 2^5.

k	$\mathrm{a}^{-k}\int_0^\infty r^k R^2(nl)\,dr$
1	$\frac{1}{2Z}[3n^2-l(l+1)]$
2	$\frac{n^2}{2Z^2}[5n^2+1-3l(l+1)]$
3	$\frac{n^2}{8Z^3}[35n^2(n^2-1)-30n^2(l+2)(l-1)+3(l+2)(l+1)l(l-1)]$
4	$\frac{n^4}{8Z^4}[63n^4-35n^2(2l^2+2l-3)+5l(l+1)(3l^2+3l-10)+12]$
-1	$\frac{Z}{n^2}$
-2	$\frac{Z^2}{n^3(l+\frac{1}{2})}$
-3	$\frac{Z^3}{n^3(l+1)(l+\frac{1}{2})l}$
-4	$\frac{Z^4\frac{1}{2}[3n^2-l(l+1)]}{n^5(l+\frac{3}{2})(l+1)(l+\frac{1}{2})l(l-\frac{1}{2})}$

* In this table r is measured in atomic units. The general eigenfunctions for any Z and arbitrary length unit are obtained by multiplying the functions of this table by $\sqrt{Z/\mathrm{a}}$ and replacing r by Zr/a.

† The average values of r^{-5} and r^{-6} may be found in VAN VLECK, Proc. Roy. Soc. **A143**, 679 (1934).

3. The relativity correction.*

We shall now investigate the effect of the relativistic variation of mass with velocity in the central-force problem, particularly in the case of hydrogen.

To obtain the wave equation we take the relativistic expressions for the Hamiltonian (total energy)

$$H=\mu c^2\left[\frac{1}{\sqrt{1-\beta^2}}-1\right]+U(r), \qquad (\beta=v/c) \quad (1)$$

and for the momenta

$$p_x=\frac{\mu\dot{x}}{\sqrt{1-\beta^2}}, \quad p_y=\frac{\mu\dot{y}}{\sqrt{1-\beta^2}}, \quad p_z=\frac{\mu\dot{z}}{\sqrt{1-\beta^2}}, \qquad (2)$$

where μ is the rest mass and v the velocity of the electron. Then

$$\boldsymbol{p}^2=\frac{\mu^2c^2\beta^2}{1-\beta^2}=\mu^2c^2\left(\frac{1}{1-\beta^2}-1\right),$$

so that

$$\frac{1}{\sqrt{1-\beta^2}}=\sqrt{1+\frac{1}{\mu^2c^2}\boldsymbol{p}^2}.$$

Hence

$$H=\mu c^2\left[\sqrt{1+\frac{1}{\mu^2c^2}\boldsymbol{p}^2}-1\right]+U(r)$$

$$=U(r)+\frac{1}{2\mu}\boldsymbol{p}^2-\frac{1}{8\mu^3c^2}\boldsymbol{p}^4+\ldots. \qquad (3)$$

The third and succeeding terms of this expression are very small (in atomic units, $\boldsymbol{p}\sim 1$, $c=137$, $\mu=1$, cf. Appendix) compared to the second and hence may safely be regarded as a perturbation. We shall keep only the third term.

We shall thus solve the equation

$$\left\{U(r)+\frac{1}{2\mu}\boldsymbol{p}^2-\frac{1}{8\mu^3c^2}\boldsymbol{p}^4\right\}\psi=E\psi, \qquad (4)$$

under the assumption that the ψ's and energies do not differ much from those of the non-relativistic equation

$$\left\{U(r)+\frac{1}{2\mu}\boldsymbol{p}^2\right\}\psi(W)=W\psi(W) \qquad (5)$$

which we have considered hitherto. Under this assumption we may set

$$-\frac{1}{8\mu^3c^2}\boldsymbol{p}^4\psi=-\frac{1}{8\mu^3c^2}\boldsymbol{p}^4\psi(W).$$

Now from (5)

$$\frac{1}{2\mu}\boldsymbol{p}^2\psi(W)=(W-U)\psi(W);$$

* See the sketch in § 5 of Dirac's more rigorous treatment.

since $\boldsymbol{p}^2$ commutes with $(W-U)$ we may apply $\boldsymbol{p}^2$ again to obtain

$$-\frac{1}{8\mu^3c^2}\boldsymbol{p}^4\psi(W)=-\frac{1}{2\mu c^2}(W-U)^2\psi(W),$$

or to our approximation

$$-\frac{1}{8\mu^3c^2}\boldsymbol{p}^4\psi=-\frac{1}{2\mu c^2}(E-U)^2\psi. \tag{6}$$

Hence the relativistic central-force equation becomes

$$-\frac{\hbar^2}{2\mu}\Delta\psi+\left\{(U-E)-\frac{1}{2\mu c^2}(U-E)^2\right\}\psi=0. \tag{7}$$

Just as in § 1⁵ we may take this ψ to be an eigenstate of angular momentum with the quantum numbers m_l and l, and obtain the radial equation

$$\frac{\hbar^2}{2\mu}\frac{d^2R}{dr^2}+\left[(E-U)+\frac{1}{2\mu c^2}(E-U)^2-\frac{\hbar^2}{2\mu}\frac{l(l+1)}{r^2}\right]R=0. \tag{8}$$

In the case of hydrogen and hydrogen-like ions we may solve this rigorously. We set $U=-Ze^2/r$ to obtain

$$\frac{d^2R}{dr^2}+\left\{\frac{E^2}{\hbar^2c^2}+\frac{2\mu E}{\hbar^2}+\frac{2\mu Ze^2}{\hbar^2 r}\left(1+\frac{E}{\mu c^2}\right)+\frac{1}{r^2}[\alpha^2Z^2-l(l+1)]\right\}R=0, \tag{9}$$

where α denotes the fine structure constant $e^2/\hbar c=1/137{\cdot}3$. We see that for $c\rightarrow\infty$ this equation reduces to the equation 2⁵1, and that it may be discussed in the same way. For simplicity we rewrite the above equation in the form

$$\frac{d^2R}{dr^2}+\left\{A+\frac{B}{r}+\frac{C}{r^2}\right\}R=0. \tag{10}$$

Near $r=\infty$, the solutions of this become asymptotically

$$e^{\pm\sqrt{-A}\,r},$$

of which the solution with the minus sign remains finite for the energies in which we are interested. At the origin if we attempt to find a power series solution starting as r^γ, we find the indicial equation

$$\gamma(\gamma-1)=-C=l(l+1)-\alpha^2Z^2,$$

whence

$$\gamma_\pm=\pm\sqrt{(l+\tfrac{1}{2})^2-\alpha^2Z^2}+\tfrac{1}{2}.$$

Of these it is γ_+ which gives our previous solutions in the limit $c\rightarrow\infty$. Hence we proceed to find a solution of the form

$$R=e^{-\sqrt{-A}\,r}r^{\gamma_+}\sum_{k=0}^{\infty}b_kr^k. \tag{11}$$

This leads to the following recursion formula for the coefficients b_k:

$$b_{k+1} = \frac{2\sqrt{-A}(\gamma_+ + k) - B}{(\gamma_+ + k + 1)(\gamma_+ + k) + C} b_k. \tag{12}$$

The series $\Sigma b_k r^k$ determined in this way converges as $e^{2\sqrt{-A}\,r}$ except for the particular values of E which cause it to terminate as a finite polynomial. Setting the coefficient $b_{s+1} = 0$ gives the value of E which gives a polynomial of degree s in place of the infinite series. This value of E is found to satisfy

$$\frac{E}{\mu c^2} = \left(1 + \frac{\alpha^2 Z^2}{[s + \sqrt{(l+\frac{1}{2})^2 - \alpha^2 Z^2} + \frac{1}{2}]^2}\right)^{-\frac{1}{2}} - 1. \tag{13}$$

If we take the negative sign for the square root in this expression we get a set of energy levels just greater than $-2\mu c^2$, and converging to a limit at that point, with a continuous spectrum below $E = -2\mu c^2$. Such states of *negative kinetic energy* occur of necessity in any relativistic theory, but here we may just ignore them by falling back on our treatment of the relativistic variation of mass as a mere perturbation on the ordinary energy levels to which (13) reduce if we take the positive square root and let $c \to \infty$.

If we expand (13) in powers of α^2, we obtain, to terms in α^4:

$$E = -\frac{\mu c^2}{2} \frac{\alpha^2 Z^2}{n^2} \left\{1 + \frac{\alpha^2 Z^2}{n^2}\left(\frac{n}{l+\frac{1}{2}} - \frac{3}{4}\right) + \ldots\right\} \tag{14}$$

($n = s + l + 1$) or, in wave numbers

$$\sigma = \frac{E}{hc} = -\frac{\mathsf{R}Z^2}{n^2}\left\{1 + \frac{\alpha^2 Z^2}{n^2}\left(\frac{n}{l+\frac{1}{2}} - \frac{3}{4}\right) + \ldots\right\}. \tag{15}$$

4. Spin-orbit interaction.

We shall now complete the picture of the fine structure in the absence of external fields by considering the effect of the electron spin. The effect of the spin in one-electron spectra arises from the interaction of the magnetic moment of the electron with the effective magnetic field set up by its motion around the nucleus. Here, as in all considerations involving electron spin, we must in the last analysis choose the term in the Hamiltonian which represents this interaction to agree with experiment. From the picture of an electron as a spinning top Thomas* and Frenkel† have obtained a formula which gives experimental agreement, and is of the sort occurring in Dirac's theory of the electron, § 5^5. Their result for the spin-orbit interaction energy of an electron in a central field with potential $U(r)$ is

$$H^I = \frac{1}{2\mu^2 c^2}\left(\frac{1}{r}\frac{\partial U(r)}{\partial r}\right)\boldsymbol{L}\cdot\boldsymbol{S} = \xi(r)\,\boldsymbol{L}\cdot\boldsymbol{S}, \tag{1}$$

where $\boldsymbol{L}$ is the orbital and $\boldsymbol{S}$ the spin angular momentum.

* Thomas, Nature, **117**, 514 (1926).

† Frenkel, Zeits. für Phys. **37**, 243 (1926).

We must then consider the energy levels and eigenfunctions for the Hamiltonian H^0+H^I, where H^0 is the (relativistic or non-relativistic) Hamiltonian without spin interaction. We shall use a representation in which H^0, $\boldsymbol{L}^2$, $\boldsymbol{S}^2$, L_z, and S_z are diagonal. The eigenvalues of H^0 are independent of m_s, the z-component of spin (this follows from 3^38 since H^0 commutes with $\boldsymbol{S}$), so we may form an eigenfunction characterized by $H^{0\prime}=W_{nl}$, s, l, m_s, and m_l merely by multiplying 1^53 by $\delta(\sigma, m_s)$ as in 5^32:

$$\phi(n\,l\,m_s\,m_l)=\frac{R(nl)}{r}\,\Theta(l\,m_l)\,\Phi(m_l)\,\delta(\sigma,m_s). \tag{2}$$

The matrix components of H^I in this scheme are

$$(n\,l\,m_s\,m_l|H^I|n'\,l'\,m_s'\,m_l')$$
$$=\sum_\sigma\iiint\bar{\phi}(n\,l\,m_s\,m_l)\,[\xi(r)\boldsymbol{L}\cdot\boldsymbol{S}]\,\phi(n'\,l'\,m_s'\,m_l')\,r^2\sin\theta\,dr\,d\theta\,d\varphi$$
$$=\int_0^\infty R(nl)\,\xi(r)\,R(n'l')\,dr$$
$$\cdot\left\{\sum_\sigma\int_0^{2\pi}\int_0^\pi\Theta(l\,m_l)\,\bar{\Phi}(m_l)\,\delta(\sigma,m_s)\,[\boldsymbol{L}\cdot\boldsymbol{S}]\,\Theta(l'\,m_l')\,\Phi(m_l')\,\delta(\sigma,m_s')\,\sin\theta\,d\theta\,d\varphi\right\}.$$

The expression in braces is just the matrix component of $\boldsymbol{L}\cdot\boldsymbol{S}$ which was evaluated in 7^33. This matrix component contains a factor $\delta(l,l')$. Hence we can write*

$$(n\,l\,m_s\,m_l|H^I|n'\,l'\,m_s'\,m_l')=\delta(l,l')\,(l\,m_s\,m_l|\boldsymbol{L}\cdot\boldsymbol{S}|l'\,m_s'\,m_l')\int_0^\infty R(nl)\,\xi(r)\,R(n'l')\,dr. \tag{3}$$

This shows that our matrix is rigorously diagonal with respect to l. Although it is not diagonal with regard to n, the perturbation theory shows that the non-diagonal elements in which n's differ have a negligible effect on the energies since states of the same l but different n have large energy differences. This shows that the first approximation of the perturbation theory in which these elements are neglected will be a good approximation. On the other hand if it had turned out that there were matrix components

* This method of calculation is given to show the relation between the Schrödinger and the matrix type of calculation. We should ordinarily say

$$(n\,l\,m_s\,m_l|\xi(r)\,\boldsymbol{L}\cdot\boldsymbol{S}|n'\,l'\,m_s'\,m_l')=\sum_{n''\,l''\,m_s''\,m_l''}(n\,l\,m_s\,m_l|\xi(r)|n''\,l''\,m_s''\,m_l'')(n''\,l''\,m_s''\,m_l''|\boldsymbol{L}\cdot\boldsymbol{S}|n'\,l'\,m_s'\,m_l').$$

Since $\boldsymbol{r}^2$ commutes with $\boldsymbol{L}$ and $\boldsymbol{S}$, the first factor in the summand contains $\delta(l\,m_s m_l;\,l''\,m_s''\,m_l'')$; by 7^33 the second factor contains $\delta(n''l'', n'l')$. Hence the whole sum reduces to

$$\delta(l,l')\,(n\,l\,m_s\,m_l|\xi(r)|n'\,l\,m_s\,m_l)(n'\,l\,m_s\,m_l|\boldsymbol{L}\cdot\boldsymbol{S}|n'\,l\,m_s'\,m_l'),$$

which is seen to equal (3).

joining states of different l and the same n this approximation would have been very poor, at least for hydrogen, where these states lie very close together.

We thus consider the diagonalization of that portion of the matrix of H which corresponds to a given configuration nl. With the notation

$$\zeta_{nl} = \hbar^2 \int_0^\infty R^2(nl)\, \xi(r)\, dr, \tag{4}$$

this portion of the matrix of H may be written as

$$W_{nl} + \hbar^{-2}\zeta_{nl} \| (n\, l\, m_s\, m_l | \boldsymbol{L}\cdot\boldsymbol{S} | n\, l\, m_s'\, m_l') \|. \tag{5}*$$

To our approximation, then, the eigenstates of H are not only eigenstates of $\boldsymbol{L}^2$ and $\boldsymbol{S}^2$ with the values $l(l+1)\hbar^2$ and $\frac{3}{4}\hbar^2$, but also eigenstates of $\boldsymbol{L}\cdot\boldsymbol{S}$; hence they are eigenstates of $\boldsymbol{J}^2 = (\boldsymbol{S}+\boldsymbol{L})^2$.† From $6^3 4$ we see that the quantum number $j\,[\boldsymbol{J}^{2\prime} = j(j+1)\hbar^2]$ may have either of the values $l+\frac{1}{2}$, $l-\frac{1}{2}$, except for $l=0$, when we obtain only $j=\frac{1}{2}$. Since

$$2(\boldsymbol{L}\cdot\boldsymbol{S})' = (\boldsymbol{S}+\boldsymbol{L})^{2\prime} - \boldsymbol{L}^{2\prime} - \boldsymbol{S}^{2\prime} = \{j(j+1) - \tfrac{3}{4} - l(l+1)\}\hbar^2, \tag{6}$$

$$H' = W_{nl} + \hbar^{-2}\zeta_{nl}(\boldsymbol{L}\cdot\boldsymbol{S})' = \begin{cases} W_{nl} + \frac{1}{2}l\zeta_{nl} & \text{when } j = l+\frac{1}{2} \\ W_{nl} - \frac{1}{2}(l+1)\zeta_{nl} & \phantom{\text{when }} j = l-\frac{1}{2}. \end{cases} \tag{7}$$

Hence all except s configurations split into two levels, corresponding to $j=l+\frac{1}{2}$ and $j=l-\frac{1}{2}$. Each of these levels is still $(2j+1)$-fold degenerate, corresponding to the $2j+1$ values of $m\,[J_z' = m\hbar]$. Since all reasonable potential functions are increasing functions of r, *the parameters* ζ_{nl} *are essentially positive*. Therefore the level of the higher j lies above that of the lower; the centre of gravity of the two levels remains at W_{nl} if in determining the centre of gravity the levels are weighted according to their degeneracy.

The (one or) two *levels* into which each configuration is split are together said to constitute a *doublet term*. The *terms* are designated as ^2S ('doublet S'), ^2P, ^2D, ... according to the l value of the configuration from which they arise. The separate levels are designated by adding the value of j as a subscript, thus $^2S_{\frac{1}{2}}$; $^2P_{\frac{1}{2}}$, $^2P_{\frac{3}{2}}$; $^2D_{\frac{3}{2}}$, $^2D_{\frac{5}{2}}$; etc. The quantum number n or the configuration label nl is frequently prefixed to these symbols, and to specify an individual *state* the value of m is given as a superscript, thus $2\,{}^2P_{\frac{3}{2}}^{\frac{3}{2}}$ or $2p\,{}^2P_{\frac{1}{2}}^{-\frac{1}{2}}$.

* W_{nl} represents the matrix whose diagonal elements have the constant value W_{nl} and whose non-diagonal elements vanish, i.e. W_{nl} times the unit matrix.

† Not only to this approximation, but in general, are the eigenstates of H eigenstates of $\boldsymbol{L}^2$, $\boldsymbol{S}^2$, and $\boldsymbol{J}^2$, since H commutes with these three observables. The only non-rigorous quantum number is the quantum number n.

The eigen-ψ's going with these states are given by 14[3]6 and Table 1[3]. For $l=0$

$$\begin{aligned}\psi(ns\,{}^2S_{\frac{1}{2}}^{\frac{1}{2}}) &= \phi(ns\,\tfrac{1}{2}\,0)\\ \psi(ns\,{}^2S_{\frac{1}{2}}^{-\frac{1}{2}}) &= \phi(ns\,-\tfrac{1}{2}\,0),\end{aligned} \tag{8a}$$

and for $l>0$

$$\begin{aligned}\psi(nl\,{}^2L_{l+\frac{1}{2}}^m) &= \sqrt{\frac{l+m+\frac{1}{2}}{2l+1}}\,\phi(nl\,\tfrac{1}{2}\,m-\tfrac{1}{2}) + \sqrt{\frac{l-m+\frac{1}{2}}{2l+1}}\,\phi(nl\,-\tfrac{1}{2}\,m+\tfrac{1}{2})\\ \psi(nl\,{}^2L_{l-\frac{1}{2}}^m) &= \sqrt{\frac{l-m+\frac{1}{2}}{2l+1}}\,\phi(nl\,\tfrac{1}{2}\,m-\tfrac{1}{2}) - \sqrt{\frac{l+m+\frac{1}{2}}{2l+1}}\,\phi(nl\,-\tfrac{1}{2}\,m+\tfrac{1}{2}),\end{aligned} \tag{8b}$$

where the ϕ's on the right are those of (2), labelled by the quantum numbers $n\,l\,m_s\,m_l$. This transformation is of importance in many-electron spectra, where we shall speak of it as the transformation from the $n\,l\,m_s\,m_l$ scheme to the $n\,l\,j\,m$ scheme. In obtaining (8) we have correlated s with j_1, l with j_2, altering the phases in Table 1[3] according to 14[3]7; it is easily verified that the non-diagonal elements $(nl\,{}^2L_{l+\frac{1}{2}}^m|S_z|nl\,{}^2L_{l-\frac{1}{2}}^m)$ are positive as in § 10[3]. *This correlation of* $\boldsymbol{S}$, $\boldsymbol{L}$, $\boldsymbol{J}$ *to* $\boldsymbol{J}_1$, $\boldsymbol{J}_2$, $\boldsymbol{J}$ *respectively we shall adopt as standard.*

For S levels one must not conclude from (7) that the displacement due to spin is zero since $l=0$, for in this case the integral ζ diverges at the origin. This follows from the fact that near the origin the screening of the orbital electrons must become zero, so that $U(r)$ must represent the full potential due to a charge Ze, and hence have the form $-(Ze^2/r)+\text{const}$. With such a potential R starts as r^{l+1}, just as in hydrogen, so that the integrand of (4) becomes essentially $1/r$ for small r when $l=0$. Hence the spin displacement of an S level is essentially indeterminate in this calculation. We shall return to this point in the next section.

In the case of the Coulomb potential, $U(r)=-Ze^2/r$, the value of ζ_{nl} is given at once by Table 2[5]:

$$\zeta_{nl}=\frac{Ze^2\hbar^2}{2\mu^2c^2\mathrm{a}^3}\int_0^\infty r^{-3}\,R^2(nl)\,dr=\frac{e^2\hbar^2}{2\mu^2c^2\mathrm{a}^3}\,\frac{Z^4}{n^3l(l+\frac{1}{2})(l+1)}. \tag{9}$$

With this value of ζ_{nl} the two energy levels given by the spin-orbit interaction are

$$\left.\begin{aligned}H' &= W_{nl}+\frac{\mu c^2}{2}\frac{\alpha^2Z^2}{n^2}\frac{\alpha^2Z^2}{n^2}\frac{n}{(2l+1)(l+1)}, &(j=l+\tfrac{1}{2})\\ H' &= W_{nl}-\frac{\mu c^2}{2}\frac{\alpha^2Z^2}{n^2}\frac{\alpha^2Z^2}{n^2}\frac{n}{(2l+1)l}. &(j=l-\tfrac{1}{2})\end{aligned}\right\} \tag{10}$$

It will be noticed that the spin-orbit interaction gives a correction of the same order of magnitude as the relativity correction 3[5]15; using the value of W_{nl} there found we obtain for the fine-structure levels of the hydrogen-like atoms the formulas

$$
\begin{aligned}
({}^2L_{l+\frac{1}{2}}):\ \sigma &= -\frac{\mathsf{R}Z^2}{n^2}\left[1+\frac{\alpha^2Z^2}{n^2}\left(\frac{n}{l+1}-\frac{3}{4}\right)\right]\\
({}^2L_{l-\frac{1}{2}}):\ \sigma &= -\frac{\mathsf{R}Z^2}{n^2}\left[1+\frac{\alpha^2Z^2}{n^2}\left(\frac{n}{l}-\frac{3}{4}\right)\right].
\end{aligned}
\tag{11}
$$

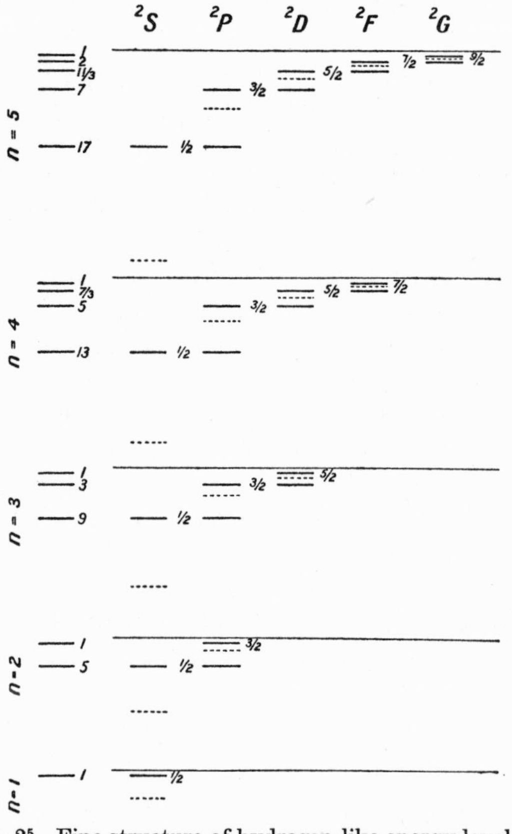

Fig. 2[5]. Fine structure of hydrogen-like energy levels.

These formulas show that the two levels having the same j value but different values of l have the same wave number

$$\sigma = -\frac{\mathsf{R}Z^2}{n^2}\left[1+\frac{\alpha^2Z^2}{n^2}\left(\frac{n}{j+\frac{1}{2}}-\frac{3}{4}\right)\right]. \tag{12}$$

This system of levels, which we have obtained by perturbation theory, is the same to this order of accuracy as that given by a rigorous solution of Dirac's relativistic equations (§ 5[5]) and agrees with experiment (§ 7[5]). The relativistic-spin displacements

$$-\frac{\mathsf{R}\alpha^2Z^4}{4n^4}\left\{\frac{4n}{j+\frac{1}{2}}-3\right\}$$

of these levels are shown in Fig. 2^5 for $n = 1, \ldots, 5$, using $R\alpha^2 Z^4/4n^4$ as a unit in each case. The relativistic displacements given by $3^5 15$ are indicated by the broken lines; each of these splits to form a spin doublet as in (10). Because of the occurrence of n^{-4} in the unit, the actual splitting decreases very rapidly as n increases.

5. Sketch of the relativistic theory.

Dirac* has developed a theory of the electron in an electromagnetic field which satisfies the relativistic requirement of invariance under a Lorentz transformation. The characteristic feature of it is that the Hamiltonian operator is made linear in the momenta in order that the operators $\partial/\partial x$ be on an equal footing with the $\partial/\partial t$ which occurs linearly in the fundamental equation $5^2 3$.

In order to factorize the non-relativistic quadratic Hamiltonian into two linear factors it is found necessary to introduce some new non-commuting observables. It turns out that this introduces changes in the theory which are akin to the electron spin. In other words, the spin is not introduced *ad hoc* but is a consequence of the relativity requirements. If the energy levels W are reckoned to include the rest energy μc^2 of the electron, besides the usual energy levels in the neighbourhood of $+\mu c^2$ the theory also gives negative levels in the neighbourhood of $-\mu c^2$. This was for some time regarded as a serious difficulty for the theory, but now it appears likely that these are connected with the positive electron or 'positron' recently discovered by Anderson.†

In this section we shall give a brief account of the one-electron atom problem as it appears in Dirac's theory. The wave equation for an electron of charge $-e$ in an electromagnetic field whose potentials are φ, $\boldsymbol{A}$ is

$$\left\{-\alpha\mu c^2 - c\boldsymbol{\beta}\cdot\left(\boldsymbol{p}+\frac{e}{c}\boldsymbol{A}\right)-e\varphi\right\}\psi = W\psi, \tag{1}$$

which in the case of a central field, with no magnetic field, becomes

$$\{-\alpha\mu c^2 - c\boldsymbol{\beta}\cdot\boldsymbol{p} + U(r)\}\psi = W\psi. \tag{2}$$

In these equations α and the three components of the vector $\boldsymbol{\beta}$ are real observables whose squares are equal to unity and which anticommute with each other‡ and commute with the positional coordinates and momenta.

* DIRAC, *Quantum Mechanics*, Chapter XII. See also a report by RUMER, Phys. Zeits. **32**, 601 (1931).

† ANDERSON, Phys. Rev. **43**, 491 (1933);
BLACKETT and OCHIALINI, Proc. Roy. Soc. **A139**, 699 (1933).

‡ That is, $\alpha\beta_x + \beta_x\alpha = 0$, $\beta_x\beta_y + \beta_y\beta_x = 0$, etc.

Each quantity has the allowed values ± 1. The quantity α we shall call the *aspect* of the electron. In problems in which relativistic effects are small the first term in (2) is the largest; hence the allowed energy levels are near to the allowed values of $-\alpha\mu c^2$, that is, near to $\pm\mu c^2$ as stated before. The vector $\boldsymbol{\beta}$ is the quantum analogue of the ratio of the velocity of the electron to the velocity of light, as Breit* has shown.

The states ψ for this problem are required, in order that we may find four anticommuting matrices, to have four component functions of position. (This is analogous to our use in § 5^3 of ψ's with two component functions of position to allow for the spin degeneracy.) Hence α and $\boldsymbol{\beta}$ are represented by fourth-order matrices acting on the coordinates which distinguish the four components of ψ, and these we write as a matrix of one column and four rows. We may work with a representation in which α is diagonal and in which α and $\boldsymbol{\beta}$ have the forms

$$\alpha=\begin{Vmatrix}1&0&0&0\\0&1&0&0\\0&0&-1&0\\0&0&0&-1\end{Vmatrix}\quad \beta_x=\begin{Vmatrix}0&0&0&1\\0&0&1&0\\0&1&0&0\\1&0&0&0\end{Vmatrix}\quad \beta_y=\begin{Vmatrix}0&0&0&-i\\0&0&i&0\\0&-i&0&0\\i&0&0&0\end{Vmatrix}\quad \beta_z=\begin{Vmatrix}0&0&1&0\\0&0&0&-1\\1&0&0&0\\0&-1&0&0\end{Vmatrix}. \tag{3}$$

It follows directly from the properties of the components of $\boldsymbol{\beta}$ that the vector $\boldsymbol{\beta}\times\boldsymbol{\beta}$ satisfies the commutation rule

$$[\boldsymbol{\beta}\times\boldsymbol{\beta},\boldsymbol{\beta}\times\boldsymbol{\beta}]=4(\boldsymbol{\beta}\times\boldsymbol{\beta})\times\mathfrak{J}.$$

By comparison of this with $1^3 3$ we see that $\boldsymbol{\beta}\times\boldsymbol{\beta}$ is a constant times an angular momentum. In fact, the components of this vector are proportional to the components of the spin vector of § 5^3 written twice over:

$$\boldsymbol{S}=-\tfrac{1}{4}i\hbar\boldsymbol{\beta}\times\boldsymbol{\beta}, \tag{4}$$

that is,

$$S_x=\begin{Vmatrix}0&\frac12&0&0\\\frac12&0&0&0\\0&0&0&\frac12\\0&0&\frac12&0\end{Vmatrix}\hbar \quad S_y=\begin{Vmatrix}0&-\frac12 i&0&0\\\frac12 i&0&0&0\\0&0&0&-\frac12 i\\0&0&\frac12 i&0\end{Vmatrix}\hbar \quad S_z=\begin{Vmatrix}\frac12&0&0&0\\0&-\frac12&0&0\\0&0&\frac12&0\\0&0&0&-\frac12\end{Vmatrix}\hbar$$

$$=-\tfrac12 i\hbar\beta_y\beta_z \qquad =-\tfrac12 i\hbar\beta_z\beta_x \qquad =-\tfrac12 i\hbar\beta_x\beta_y.$$

It may be shown (Dirac, § 72) that because of the form of the Hamiltonian (1) the electron must be considered as having a spin angular momentum $\boldsymbol{S}$ and a magnetic moment $-\dfrac{e}{\mu c}\boldsymbol{S}$, exactly the values postulated by Uhlenbeck and Goudsmit. S_z is diagonal in the scheme we are using. Hence we may say that the first component of ψ refers to $\alpha'=1$, $S_z'=m_s\hbar=+\frac12\hbar$; the second component to $\alpha'=1$, $m_s=-\frac12$; the third to $\alpha'=-1$, $m_s=+\frac12$; and the fourth to $\alpha'=-1$, $m_s=-\frac12$.

* Breit, Proc. Nat. Acad. Sci. **14**, 553 (1928).

Dirac's Hamiltonian (2) commutes with $\boldsymbol{J}=\boldsymbol{L}+\boldsymbol{S}$, so we may introduce a set of states labelled by j and m. For a definite value of j and m there are two possible values of α' and two possible values of l, namely $j\pm\frac{1}{2}$. The labelling by $(j\,m\,l\,\alpha')$ determines the state except for a function of the radius. We denote such a state by $\chi(j\,m\,l\,\alpha')$. The forms of these states are given by $4^5 8$. For $\alpha'=1$ the third and fourth components of χ vanish and for $\alpha'=-1$ the first and second components vanish. The non-vanishing components of $\chi(j\,m\,l\,\alpha')$ are

$$\left\|\begin{matrix}\sqrt{\dfrac{j+m}{2j}}\,\phi(j-\frac{1}{2}\;m-\frac{1}{2})\\ \sqrt{\dfrac{j-m}{2j}}\,\phi(j-\frac{1}{2}\;m+\frac{1}{2})\end{matrix}\right\| \quad \text{for } l=j-\tfrac{1}{2}, \qquad \left(\frac{2}{\hbar^2}\boldsymbol{L}\cdot\boldsymbol{S}=j-\tfrac{1}{2}\right)$$

$$\left\|\begin{matrix}\sqrt{\dfrac{j-m+1}{2j+2}}\,\phi(j+\frac{1}{2}\;m-\frac{1}{2})\\ -\sqrt{\dfrac{j+m+1}{2j+2}}\,\phi(j+\frac{1}{2}\;m+\frac{1}{2})\end{matrix}\right\| \quad \text{for } l=j+\tfrac{1}{2}, \quad \left(\frac{2}{\hbar^2}\boldsymbol{L}\cdot\boldsymbol{S}=-j-\tfrac{3}{2}\right) \tag{5}$$

where $\phi(l\,m_l)$ is written for the spherical harmonic $\Theta(l\,m_l)\,\Phi(m_l)$. These states may be regarded as labelled by l or by $\frac{2}{\hbar^2}\boldsymbol{L}\cdot\boldsymbol{S}$. They are not suitable in this form to be eigen-ψ's of energy since $\boldsymbol{L}^2$ does not commute with the Hamiltonian. The quantity

$$K=\alpha\left(\frac{2}{\hbar^2}\boldsymbol{L}\cdot\boldsymbol{S}+1\right), \tag{6}$$

whose eigenvalues we denote by k, does commute with the Hamiltonian and so its eigenvalues may serve as quantum numbers. The allowed values of k are $\pm(j+\frac{1}{2})$ and the states labelled by $(j\,m\,k\,\alpha')$ which we shall denote by $\psi(j\,m\,k\,\alpha')$, are related to the χ's by

$$\begin{aligned}\psi(j\,m\,\pm j\pm\tfrac{1}{2}\,+1)&=\chi(j\,m\,j\mp\tfrac{1}{2}\,+1)\\ \psi(j\,m\,\pm j\pm\tfrac{1}{2}\,-1)&=\chi(j\,m\,j\pm\tfrac{1}{2}\,-1).\end{aligned} \tag{7}$$

It is convenient now to transform (2) by introducing the operator $\beta_r=\boldsymbol{r}_0\cdot\boldsymbol{\beta}$, where $\boldsymbol{r}_0=\boldsymbol{r}/r$. Since $\beta_r^2=1$ the second term becomes

$$-c\boldsymbol{\beta}\cdot\boldsymbol{p}=-c\beta_r\boldsymbol{r}_0\cdot\boldsymbol{\beta}\boldsymbol{\beta}\cdot\boldsymbol{p}=-c\beta_r\left(\boldsymbol{r}_0\cdot\boldsymbol{p}+\frac{2i}{\hbar}\frac{\boldsymbol{L}\cdot\boldsymbol{S}}{r}\right).$$

The last form here follows from the fact that the dyad $\boldsymbol{\beta\beta}=\mathfrak{J}-\frac{2i}{\hbar}\boldsymbol{S}\times\mathfrak{J}$. On introducing the Schrödinger operator $(-i\hbar\,\text{grad})$ for $\boldsymbol{p}$, this becomes

$$-c\boldsymbol{\beta}\cdot\boldsymbol{p}=ic\hbar\beta_r\left(\frac{\partial}{\partial r}-\frac{2}{\hbar^2}\frac{\boldsymbol{L}\cdot\boldsymbol{S}}{r}\right)=ic\hbar\beta_r\left(\frac{\partial}{\partial r}-\frac{\alpha K-1}{r}\right). \tag{8}$$

Let us now consider the action of β_r on $\psi(j\,m\,k\,\alpha')$. In the initial representation β_r has the form

$$\beta_r = \left\| \begin{matrix} 0 & 0 & \cos\theta & \sin\theta\, e^{-i\varphi} \\ 0 & 0 & \sin\theta\, e^{i\varphi} & -\cos\theta \\ \cos\theta & \sin\theta\, e^{-i\varphi} & 0 & 0 \\ \sin\theta\, e^{i\varphi} & -\cos\theta & 0 & 0 \end{matrix} \right\|;$$

with the aid of 4[3]21 we readily find that

$$\beta_r \psi(j\,m\,k\,\alpha') = \psi(j\,m\,k\,-\alpha'). \tag{9}$$

The eigen-ψ of the Hamiltonian going with definite values of k, j, m will be a linear combination of $\psi(j\,m\,k+1)$ and $\psi(j\,m\,k-1)$ which we write in the form

$$\Psi = \frac{F(r)}{r}\psi(j\,m\,k+1) - \frac{iG(r)}{r}\psi(j\,m\,k-1). \tag{10}$$

Substituting this in (2) and using (8) and (9) we find the equations which determine $F(r)$ and $G(r)$:

$$-\mu c^2 F + c\hbar\left(\frac{dG}{dr} + \frac{k}{r}G\right) + U(r)\,F = WF,$$

$$+\mu c^2 G - c\hbar\left(\frac{dF}{dr} - \frac{k}{r}F\right) + U(r)\,G = WG.$$

These equations may be discussed in the usual way for $U(r) = -Ze^2/r$ by writing

$$F(r) = e^{-r/a}f(r), \quad G(r) = e^{-r/a}g(r), \quad a = \hbar/\sqrt{m^2c^2 - W^2/c^2}$$

and seeking the conditions for polynomial solutions for $f(r)$ and $g(r)$. The details are given in Dirac (§ 74).*

The allowed discrete energy values found in this way are

$$W = +\mu c^2\left(1 + \frac{\alpha^2 Z^2}{(s + \sqrt{k^2 - \alpha^2 Z^2})^2}\right)^{-\frac{1}{2}}, \tag{11}$$

where $\alpha = e^2/\hbar c$, the fine structure constant introduced in 3[5]9. In this equation the allowed values of s are

$$\begin{aligned} s = 0, 1, 2, 3, \ldots \quad &\text{for} \quad k < 1 \\ 1, 2, 3, \ldots \quad &\text{for} \quad k \geqslant 1. \end{aligned} \tag{12}$$

The relation of the quantum number k to the ordinary doublet notation is given by the scheme:

$$\begin{matrix} & {}^2S_{\frac{1}{2}} & {}^2P_{\frac{1}{2}} & {}^2P_{\frac{3}{2}} & {}^2D_{\frac{3}{2}} & {}^2D_{\frac{5}{2}} & \ldots \\ k = & -1 & 1 & -2 & 2 & -3 & \ldots \end{matrix} \tag{13}$$

The levels which lie close together are those for a definite value of the total quantum number

$$n = s + |k|. \tag{14}$$

* See also Gordon, Zeits. für Phys. **48**, 11 (1928); Darwin, Proc. Roy. Soc. **A118**, 654 (1928).

The lack of symmetry of (12) with regard to the sign of k is just such that for a given value of n we have the same doublets occurring as in the non-relativistic theory, e.g. for $n=3$ we may have

$$\begin{aligned} s=2,\ k&=\pm 1, \text{ giving } {}^2S_{\frac{1}{2}},\ {}^2P_{\frac{1}{2}}, \\ 1,\ k&=\pm 2, \text{ giving } {}^2P_{\frac{3}{2}},\ {}^2D_{\frac{3}{2}}, \\ 0,\ k&=-3, \text{ giving } {}^2D_{\frac{5}{2}}, \end{aligned}$$

so that the lack of the allowed value $s=0$ with positive k is just what is needed to prevent the appearance of half of the ${}^2F$ term and so preserve the structural similarity with the non-relativistic theory. (11) gives, to terms in α^4, the same system of levels as we found in $4^5 12$; (13) and (14) show that levels of the same n and j have rigorously the same energy.

For small values of Z the energy levels (11) are close to $+\mu c^2$. In the ψ's belonging to these levels, the third and fourth components, which refer to $\alpha' = -1$, are large compared to the other two, which refer to states in which $\alpha' = +1$. In other words, the aspect is an almost exact quantum number. Besides the discrete levels there is an allowed continuous spectrum for $W > +\mu c^2$ and for $W < -\mu c^2$. We shall not have any use for the exact radial functions and so do not work out the detailed expressions.

It is instructive to discuss the Dirac equation for the central-field problem (2) in a way which exhibits the connection with the relativistic and spin-orbit effects as discussed in §§ 3^5 and 4^5. Replacing W by $E+\mu c^2$, (2) becomes

$$\left\{-\begin{Vmatrix} 1 & 0 \\ 0 & -1 \end{Vmatrix}\mu c^2 - c\boldsymbol{p}\cdot\begin{Vmatrix} 0 & \boldsymbol{\sigma} \\ \boldsymbol{\sigma} & 0 \end{Vmatrix} + U(r)\right\}\begin{Vmatrix} \psi_+ \\ \psi_- \end{Vmatrix} = (E+\mu c^2)\begin{Vmatrix} \psi_+ \\ \psi_- \end{Vmatrix}$$

in which the dependence of α and $\boldsymbol{\beta}$ on the values of α' is shown explicitly in terms of a vector $\boldsymbol{\sigma}$ whose components are $(2/\hbar)$ times the spin matrices of $5^3 1$. The first equation of the pair implies

$$\psi_+ = \frac{-c}{E+2\mu c^2 - U(r)}\boldsymbol{p}\cdot\boldsymbol{\sigma}\psi_-,$$

which may be used to eliminate ψ_+ from the other equation of the pair to obtain as the equation for ψ_-

$$\frac{1}{2\mu}\boldsymbol{p}\cdot\boldsymbol{\sigma} f(r)\,\boldsymbol{p}\cdot\boldsymbol{\sigma}\psi_- + U(r)\psi_- = E\psi_- \tag{15}$$

in which
$$f(r) = \left[1 + \frac{E-U}{2\mu c^2}\right]^{-1}.$$

In actual atoms $(E-U) \ll 2\mu c^2$ everywhere except very close to the nucleus, so $f(r)\sim 1$. If we simply treat $f(r)$ as exactly equal to 1, ψ_- satisfies the Schrödinger equation for the non-relativistic problem without spin-orbit interaction, since $(\boldsymbol{p}\cdot\boldsymbol{\sigma})^2 = \boldsymbol{p}^2$. However, if we save the first term in the development of $f(r)$ we get both the spin-orbit interaction and the first approximation to the relativity correction.

We write $\boldsymbol{p}\cdot\boldsymbol{\sigma} f(r)\boldsymbol{p}\cdot\boldsymbol{\sigma}\psi = f(r)(\boldsymbol{p}\cdot\boldsymbol{\sigma})^2\psi + [\boldsymbol{p}\cdot\boldsymbol{\sigma}, f(r)]\boldsymbol{p}\cdot\boldsymbol{\sigma}\psi.$

The first term of this becomes simply $f(r)\,\boldsymbol{p}^2\psi$. By a simple application of $5^2 5$, we find that

$$[\boldsymbol{p}\cdot\boldsymbol{\sigma}, f(r)] = \frac{\hbar}{i}\frac{f'(r)}{r}\boldsymbol{r}\cdot\boldsymbol{\sigma},$$

so the second term is

$$\frac{\hbar}{i}\frac{f'(r)}{r}(\boldsymbol{r}\cdot\boldsymbol{\sigma})(\boldsymbol{p}\cdot\boldsymbol{\sigma})\psi = f'(r)\left(-\hbar^2\frac{\partial}{\partial r}+\frac{\hbar\boldsymbol{L}\cdot\boldsymbol{\sigma}}{r}\right)\psi.$$

Now to the first approximation

$$f(r) = 1-\frac{E-U}{2\mu c^2} = 1-\frac{\boldsymbol{p}^2}{4\mu^2c^2},$$

$$f'(r) = \frac{U'(r)}{2\mu c^2}.$$

Hence the equation for ψ_- is

$$\frac{1}{2\mu}\boldsymbol{p}^2\psi_- + U(r)\psi_- - \frac{1}{8\mu^3c^2}\boldsymbol{p}^4\psi_- + \frac{1}{2\mu^2c^2}\frac{1}{r}\frac{\partial U}{\partial r}\boldsymbol{L}\cdot\tfrac{1}{2}\hbar\boldsymbol{\sigma}\psi_- - \frac{\hbar^2U'(r)}{4\mu^2c^2}\frac{\partial\psi_-}{\partial r} = E\psi_-. \tag{16}$$

The first two terms constitute the simple non-spin non-relativistic Hamiltonian, the third is the first approximation to the effect of variable mass as considered in § 3^5, the fourth is the spin-orbit term as considered in § 4^5. The last term is peculiar to the Dirac theory and does not have a simple classical interpretation. This last term, in the case of a Coulomb field, $U = -Ze^2/r$ gives rise to the following change in the energy by a first-order perturbation calculation:

$$-\frac{\hbar^2e^2Z}{4\mu^2c^2}\int_0^\infty \frac{1}{r^2}\mathrm{R}\frac{d\mathrm{R}}{dr}r^2\,dr = \frac{\hbar^2e^2Z}{8\mu^2c^2}\mathrm{R}^2(0),$$

since $\mathrm{R}(\infty)=0$. (Here $\mathrm{R}(r)$ is the whole radial factor $R(nl)/r$ of the wave function.) Hence this term contributes nothing except to S terms, for they are the only ones for which $\mathrm{R}(0)$ is not zero. Using the value of the radial factor from $2^5 18$, the amount contributed to the ns energy by this term is

$$\frac{e^2\hbar^2}{8\mu^2c^2}\frac{4Z^4}{n^3a^3} = \frac{\mu c^2}{2}\frac{\alpha^4Z^4}{n^3},$$

which is the value of the spin-orbit correction of S terms as given in $4^5 10$ if we suppose those formulas to hold without change for S terms. On the other hand, as already remarked, the term in $\boldsymbol{L}\cdot\boldsymbol{S}$ becomes indeterminate for S terms since the integral $\int \frac{1}{r}U'(r)\psi^2 dv$ diverges and $\boldsymbol{L}\cdot\boldsymbol{S}$ vanishes. We now see that that is simply a consequence of the approximation procedure which implies $(E-U) \ll 2\mu c^2$. Near the nucleus we can have $(E-U) \gg 2\mu c^2$ and for S states this is important because of the relatively large probability of the electron's being close to the nucleus in these states. If we avoid using the expansion of $f(r)$ we see that

$$f'(r)=\left[1+\frac{E-U}{2\mu c^2}\right]^{-2}\frac{U'(r)}{2\mu c^2}$$

so near the nucleus the quantity $f'(r)$ approaches a finite limit, hence the radial integral $\int \frac{1}{r} f'(r)\, R^2 dr$ is really finite, so this term vanishes on account of the vanishing of $\boldsymbol{L}\cdot\boldsymbol{S}$. Hence we have shown that the approximation procedure for Dirac's equations when properly applied leads to 4[5]11 even for the energy of the S terms. It is of some interest to note that Darwin had discovered (16) prior to the work of Dirac but without the term in $\partial\psi/\partial r$; as a result his equations gave the hydrogen levels correctly except in regard to the S levels.

6. Intensities in hydrogen.

In order to calculate the transition probabilities or the line strengths in the hydrogen spectrum we need the matrix components of the electric moment, $\boldsymbol{P}=-e\boldsymbol{r}$. Since the results of § 9[3] and § 11[3] are applicable to $\boldsymbol{P}$, all of the hydrogenic intensities are expressible in terms of the integrals

$$\int_0^\infty r\, R(n\,l)\, R(n'\,l-1)\, dr.$$

The evaluation of these integrals is rather difficult. Schrödinger* calculated them by use of the generating function for the Laguerre polynomials. Other simpler calculations have been made by Epstein, Eckart, and Gordon†. Bateman‡ has generalized Epstein's result. According to Gordon the integral above is equal to

$$\frac{\mathrm{a}(-1)^{n'-l}}{4(2l-1)!}\sqrt{\frac{(n+l)!\,(n'+l-1)!}{(n-l-1)!\,(n'-l)!}}\,\frac{(4nn')^{l+1}\,(n-n')^{n+n'-2l-2}}{(n+n')^{n+n'}}$$
$$\cdot\left[F\left(-n_r,-n'_r,2l,-\frac{4nn'}{(n-n')^2}\right)-\left(\frac{n-n'}{n+n'}\right)^2 F\left(-n_r-2,-n'_r,2l,-\frac{4nn'}{(n-n')^2}\right)\right], \tag{1}$$

in which the F's are hypergeometric functions and n_r and n'_r are the radial quantum numbers $n-l-1$, and $n'-l$ respectively.

Special cases of the formula have been written out and simplified, and numerical values of the integrals have been calculated by several persons,

* Schrödinger, Ann. der Phys. **79**, 361 (1926).
† Epstein, Proc. Nat. Acad. Sci. **12**, 629 (1926);
Eckart, Phys. Rev. **28**, 927 (1926);
Gordon, Ann. der Phys. **2**, 1031 (1929).
‡ Bateman, *Partial Differential Equations*, Cambridge University Press, p. 453.

most fully by Kupper.* In Table 3^5 we give some of the most important formulas for special series and in Table 4^5 numerical values for the most important values of the quantum numbers, the latter being based on Kupper's work as revised by Bethe.† The expression (1) is not applicable for $n'=n$. In this case the integral is easily found to have the value

$$\int_0^\infty r\,R(n\,l)\,R(n\,l-1)\,dr = \tfrac{3}{2}an\sqrt{n^2-l^2}. \tag{2}$$

This integral, of course, does not occur in the intensity problem in hydrogen, but is important in the Stark effect (Chapter XVII).

Let us now consider the strengths of the lines in the fine structure of hydrogen.‡ We could go at this directly by calculating the matrix components from the eigenfunctions ($4^5$8): this would provide ample exercise in calculation of matrix components. But it is more instructive to work them out as an application of the general results of Chapter III. In applying § 11^3 we identify $\boldsymbol{J}_1$ with the electron spin, so $j_1=\frac{1}{2}$, and $\boldsymbol{J}_2$ with the orbital angular momentum, so $j_2=l$. The strengths of the lines $n\,l\,{}^2L_j \rightarrow n'\,l-1\,{}^2(L-1)_{j'}$ are given by $7^4$5 in terms of the quantities $(n\,l\,{}^2L_j\vdots P\vdots n'\,l-1\,{}^2(L-1)_{j'})$. The dependence of these on j' and j is given in $11^3$8, where they are expressed in terms of the $(n\,l\vdots P\vdots n'\,l-1)$. According to the discussion just following $11^3$8, this quantity is really independent of the spin and so, as in $11^3$9, we may express it in terms of the nlm_sm_l scheme of states in which the spin and orbital angular momenta are not coupled. Since $P_z=-er\cos\theta$ we have

$$(n\,l\,m_s\,m_l|P_z|n'\,l-1\,m_s\,m_l)$$
$$= -e\int_0^\infty r\,R(n\,l)\,R(n'\,l-1)\,dr\int_0^\pi \cos\theta\,\Theta(l\,m_l)\,\Theta(l-1\,m_l)\sin\theta\,d\theta.$$

Using $4^3$21, we integrate over θ to obtain

$$(n\,l\,m_s\,m_l|P_z|n'\,l-1\,m_s\,m_l) = -\frac{e}{\sqrt{4l^2-1}}\int_0^\infty r\,R(n\,l)\,R(n'\,l-1)\,dr\cdot\sqrt{l^2-m_l^2}.$$

Similarly $$(n\,l\,m_s\,m_l|P_z|n'\,l\,m_s\,m_l)=0,$$

so for a non-vanishing component l must change by one unit. Using $9^3$11, we have

$$(n\,l\,m_s\,m_l|P_z|n'\,l-1\,m_s\,m_l) = (n\,l\vdots P\vdots n'\,l-1)\sqrt{l^2-m_l^2};$$

hence, finally,

$$(n\,l\vdots P\vdots n'\,l-1) = -\frac{e}{\sqrt{4l^2-1}}\int_0^\infty r\,R(n\,l)\,R(n'l-1)\,dr = \mathsf{s}(n\,l,n'\,l-1). \tag{3}$$

* Kupper, Ann. der Phys. **86**, 511 (1928);
Sugiura, Jour. de Phys. **8**, 113 (1927);
Slack, Phys. Rev. **31**, 527 (1928);
Maxwell, Phys. Rev. **38**, 1664 (1931).

† Bethe, Handbuch der Physik **24/1**, 2ᵈ ed., 442 (1933).

‡ Sommerfeld and Unsöld, Zeits. für Phys. **36**, 259; **38**, 237 (1926);
Bechert, Ann. der Phys. **6**, 700 (1930);
Saha and Banerji, Zeits. für Phys. **68**, 704 (1931).

The quantities occurring in $7^4 5$ are thus completely expressed in terms of the integrals over the radial eigenfunctions.

The calculations may be exemplified by a detailed consideration of the line strengths in the fine structure of H_α, the ensemble of the $n=3 \rightarrow n=2$

TABLE 3^5. *Values of* $\left[\int_0^\infty r\, R(n\,l)\, R(n'\,l-1)\, dr\right]^2$ *in atomic units for* $n' \neq n$.

$n\,l$	$n'\,l-1$	
np	$1s$	$2^8 n^7 (n-1)^{2n-5} (n+1)^{-2n-5}$
	$2s$	$2^{17} n^7 (n^2-1)(n-2)^{2n-6}(n+2)^{-2n-6}$
	$3s$	$2^8 3^7 n^7 (n^2-1)(n-3)^{2n-8}(7n^2-27)^2(n+3)^{-2n-8}$
	$4s$	$2^{26} 3^{-2} n^7 (n^2-1)(n-4)^{2n-10}(23n^4-288n^2+768)^2(n+4)^{-2n-10}$
	$5s$	$2^8 3^{-2} 5^9 n^7 (n^2-1)(n-5)^{2n-12}(91n^6-2545n^4+20625n^2-46875)^2(n+5)^{-2n-12}$
nd	$2p$	$2^{19} 3^{-1} n^9 (n^2-1)(n-2)^{2n-7}(n+2)^{-2n-7}$
	$3p$	$2^{11} 3^9 n^9 (n^2-1)(n^2-4)(n-3)^{2n-8}(n+3)^{-2n-8}$
	$4p$	$2^{30} 3^{-1} 5^{-1} n^9 (n^2-1)(n^2-4)(n-4)^{2n-10}(9n^2-80)^2(n+4)^{-2n-10}$
	$5p$	$2^{11} 3^{-3} 5^9 n^9 (n^2-1)(n^2-4)(n-5)^{2n-12}(67n^4-1650n^2+9375)^2(n+5)^{-2n-12}$
nf	$3d$	$2^{13} 3^9 5^{-1} n^{11} (n^2-1)(n^2-4)(n-3)^{2n-9}(n+3)^{-2n-9}$
	$4d$	$2^{38} 3^{-2} 5^{-1} n^{11} (n^2-1)(n^2-4)(n^2-9)(n-4)^{2n-10}(n+4)^{-2n-10}$
	$5d$	$2^{13} 3^{-2} 5^{11} 7^{-1} n^{11} (n^2-1)(n^2-4)(n^2-9)(n-5)^{2n-12}(11n^2-175)^2(n+5)^{-2n-12}$

TABLE 4^5. *Values of* $\left[\int_0^\infty r\, R(n\,l)\, R(n'\,l-1)\, dr\right]^2$ *in atomic units.*

	$2p$	$3p$	$4p$	$5p$	$6p$	$7p$	$8p$
$1s$	1·66	0·267	0·093	0·044	0·024	0·015	0·010
$2s$	27·0	9·4	1·64	0·60	0·29	0·17	0·10
$3s$	0·9	162	29·9	5·1	1·9	0·9	0·5
$4s$	0·15	6·0	540	72·6	11·9	5·7	2·1
$5s$	0·052	0·9	21·2	1125	134	41·4	21·8
$6s$	0·025	0·33	2·9		2835		
$7s$	0·014	0·16	1·4			5292	
$8s$	0·009	0·09	0·8				9072

	$3d$	$4d$	$5d$	$6d$	$7d$	$8d$
$2p$	22·52	2·92	0·95	0·44	0·242	0·149
$3p$	101·2	57·2	8·8	3·0	1·44	0·82
$4p$	1·7	432	121·9	19·3	7·7	3·2
$5p$	0·23	9·1	1181·25	203	36	12·3
$6p$	0·08	1·3		2592		
$7p$	0·03	0·5			4961·25	
$8p$	0·02	0·2				8640

	$4f$	$5f$	$6f$	$7f$	$8f$
$3d$	104·6	11·0	3·2	1·4	0·8
$4d$	252·0	197·8	26·9	8·6	3·9
$5d$	2·75	900			
$6d$	0·32		2187		
$7d$	0·08			4410	
$8d$	0·04				7920

transitions. This is made up of three *doublets*, $3\,^2P \to 2\,^2S$, $3\,^2D \to 2\,^2P$ and $3\,^2S \to 2\,^2P$. Using 11^38 we have

$$(3\,^2D_{\frac{5}{2}}\vdots P\vdots 2\,^2P_{\frac{3}{2}}) = \sqrt{\tfrac{3}{5}}(3d\vdots P\vdots 2p),$$
$$(3\,^2D_{\frac{3}{2}}\vdots P\vdots 2\,^2P_{\frac{3}{2}}) = \sqrt{\tfrac{4}{15}}(3d\vdots P\vdots 2p),$$
$$(3\,^2D_{\frac{3}{2}}\vdots P\vdots 2\,^2P_{\frac{1}{2}}) = \sqrt{\tfrac{5}{3}}(3d\vdots P\vdots 2p).$$

From Table 3^5 and (3), $\quad (3d\vdots P\vdots 2p)^2 = \dfrac{e^2\mathsf{a}^2}{15}\dfrac{2^{22}3^8}{5^{13}};$

hence from 7^45 we find for the strengths of the lines,

$$\begin{array}{llll}
 & \text{(a)} & \mathsf{S}(3\,^2D_{\frac{5}{2}}\to 2\,^2P_{\frac{3}{2}}) = 9.2^{24}3^75^{-14}e^2\mathsf{a}^2 & (6912) \\
 & \text{(b)} & \mathsf{S}(3\,^2D_{\frac{3}{2}}\to 2\,^2P_{\frac{3}{2}}) = 1. \quad ,, & (\ 768) \\
 & \text{(c)} & \mathsf{S}(3\,^2D_{\frac{3}{2}}\to 2\,^2P_{\frac{1}{2}}) = 5. \quad ,, & (3840) \\
\text{In the same way,} & \text{(d)} & \mathsf{S}(3\,^2P_{\frac{3}{2}}\to 2\,^2S_{\frac{1}{2}}) = 2.2^{21}3^65^{-12}e^2\mathsf{a}^2 & (1600) \quad (4) \\
 & \text{(e)} & \mathsf{S}(3\,^2P_{\frac{1}{2}}\to 2\,^2S_{\frac{1}{2}}) = 1. \quad ,, & (\ 800) \\
 & \text{(f)} & \mathsf{S}(3\,^2S_{\frac{1}{2}}\to 2\,^2P_{\frac{1}{2}}) = 1.2^{16}3^75^{-12}e^2\mathsf{a}^2 & (\ \ 75) \\
 & \text{(g)} & \mathsf{S}(3\,^2S_{\frac{1}{2}}\to 2\,^2P_{\frac{3}{2}}) = 2. \quad ,, & (\ 150)
\end{array}$$

It will be observed that the relative strengths of the lines in the *same* doublet bear simple integral ratios, 9:1:5 and 2:1. This is a special case of a general result for multiplets discussed fully in § 2^9. The relative strengths of the different doublets are not so simple since these involve different radial integrals. The figures in parentheses at the right in this table give the relative strength of each line, the common factor being $2^{16}3^65^{-14}e^2\mathsf{a}^2$.

We next consider the theoretical values of the Einstein **A** transition probabilities. For a particular line these are expressed in terms of the strengths by 7^43. It is convenient to express σ with the Rydberg constant as unit, and the strengths in atomic units, $e^2\mathsf{a}^2$. Then the numerical coefficient in 7^43 becomes

$$\frac{64\pi^4\mathsf{R}^3}{3h}e^2\mathsf{a}^2 = \tfrac{1}{6}\alpha^3\tau^{-1} = 2{\cdot}662\times 10^9\,\text{sec}^{-1}, \tag{5}$$

where α is the fine-structure constant and τ the atomic time unit (see Appendix). The absolute values of the transition probabilities involved in the first Balmer line are therefore

$$\begin{array}{ll}
\text{(a)} & \mathsf{A}(3\,^2D_{\frac{5}{2}}\to 2\,^2P_{\frac{3}{2}}) = 0{\cdot}643\times 10^8\,\text{sec}^{-1} \\
\text{(b)} & \mathsf{A}(3\,^2D_{\frac{3}{2}}\to 2\,^2P_{\frac{3}{2}}) = 0{\cdot}107 \\
\text{(c)} & \mathsf{A}(3\,^2D_{\frac{3}{2}}\to 2\,^2P_{\frac{1}{2}}) = 0{\cdot}536 \quad \text{(sum} = 0{\cdot}643) \\
\text{(d)} & \mathsf{A}(3\,^2P_{\frac{3}{2}}\to 2\,^2S_{\frac{1}{2}}) = 0{\cdot}223 \qquad (6) \\
\text{(e)} & \mathsf{A}(3\,^2P_{\frac{1}{2}}\to 2\,^2S_{\frac{1}{2}}) = 0{\cdot}223 \\
\text{(f)} & \mathsf{A}(3\,^2S_{\frac{1}{2}}\to 2\,^2P_{\frac{1}{2}}) = 0{\cdot}021 \\
\text{(g)} & \mathsf{A}(3\,^2S_{\frac{1}{2}}\to 2\,^2P_{\frac{3}{2}}) = 0{\cdot}042.
\end{array}$$

The values illustrate several important properties of the **A**'s. The sum of the two transition probabilities from $^2D_{\frac{3}{2}}$ is equal to the single transition probability from $^2D_{\frac{5}{2}}$. This is a general result true in more complex spectra (§ 2[9]). The two transition probabilities in $^2P \rightarrow {}^2S$ are equal, but the intensity ratio of the two lines is 2:1 because the **A**'s have to be multiplied by the statistical weights of the initial levels to give intensities. On the other hand, for $^2S \rightarrow {}^2P$ the ratio 2:1 appears in the transition probabilities themselves, so that in both cases the intensity ratio is 2:1 although in the first case it is 'due to' a factor in the statistical weights while in the second it is 'due to' a factor in the transition probabilities. The quantities $(2J+1)\mathbf{A}$, which are proportional to the strengths, are thus symmetrical with regard to initial and final states, while the **A**'s are not.

Suppose we have an experimental arrangement whereby at $t=0$ equal numbers of atoms are put into each of the 18 states belonging to $n=3$ and that the atoms are not subject to any external disturbance. Then the relative intensity of the different doublets would depend on the time after the excitation at which the observation is made. The number of atoms in any level A after time t would be reduced by the factor $e^{-k_A t}$, where $k_A = \sum_B \mathbf{A}(A, B)$, the sum being taken over all of the levels B of energy lower than A. In the particular case of the first Balmer line, the $3s$ and $3d$ states can only go to $2p$, but $3p$ can go to $1s$ as well as $2s$. We may calculate that

$$\mathbf{A}(3\,^2P \rightarrow 1\,^2S_{\frac{1}{2}}) = 1{\cdot}64 \times 10^8 \text{ sec}^{-1}$$

for either $3\,^2P_{\frac{3}{2}}$ or $3\,^2P_{\frac{1}{2}}$, and so the mean lives [the reciprocals of the total transition probabilities $k_A = k(nl)$] of each level of the three initial configurations have the values

Configuration	Mean life
$3s$	$16{\cdot}0 \times 10^{-8}$ sec
$3p$	$0{\cdot}54 \times 10^{-8}$ sec
$3d$	$1{\cdot}56 \times 10^{-8}$ sec

Since the mean lives of the two component levels of ^2P and ^2D are equal respectively, the relative intensity of the lines originating from them will not change with time. But the relative intensity of the lines originating in levels of different L value does change with time. The $3\,^2S \rightarrow 2\,^2P$ will be emitted for a much longer time than the others on account of the greater lives of their initial states.

Such an experiment has never been performed. The actual studies of the relative intensity of the lines are made with an electrical discharge where the excitation and emission reach a steady state. The atoms are subject to disturbing electric fields and to collisions which tend to produce a mobile equilibrium between the states of differing l and the same n. The result is that when the disturbance is sufficiently great all of the states of the same n have the same mean life, corresponding to a total transition probability $\bar{k}(n)$ which is a weighted average of the individual total transition probabilities:

$$\bar{k}(n) = \frac{1}{n^2} \sum_{l=0}^{n-1} (2l+1)\, k(nl).$$

TABLE 5[5]. *Transition probabilities and mean lives for hydrogen.*

Initial config.	$k(nl)$ 10^8 sec^{-1}	Life 10^{-8} sec	Total transition probability (10^8 sec^{-1}) to final configuration indicated:*				
			$2p$	$3p$	$4p$	$5p$	
$2s$	0						
$3s$	0·063	16	0·063				
$4s$	0·043	23	0·025	0·018			
$5s$	0·027$_7$	36	0·012$_7$	0·008$_5$	0·006$_5$		
$6s$	0·0176	57	0·007$_3$	0·0051	0·0035	0·0017	
			$1s$	$2s$	$3s$	$4s$	$5s$
					$3d$	$4d$	$5d$
$2p$	6·25	0·16	6·25				
$3p$	1·86	0·54	1·64	0·22			
$4p$	0·81	1·24	0·68	0·095	0·030		
					0·003		
$5p$	0·415	2·40	0·34	0·049	0·016	0·007$_5$	
					0·001$_5$	0·002	
$6p$	0·243	4·1	0·195	0·029	0·0096	0·0045	0·0021
					0·0007	0·0009	0·0010
			$2p$	$3p$	$4p$	$5p$	
					$4f$	$5f$	
$3d$	0·64	1·56	0·64				
$4d$	0·274	3·65	0·204	0·070			
$5d$	0·142	7·0	0·094	0·034	0·014		
					0·000$_5$		
$6d$	0·080	12·6	0·048	0·0187	0·0086	0·0040	
					0·0002	0·0004	
			$3d$	$4d$	$5d$		
					$5g$		
$4f$	0·137	7·3	0·137				
$5f$	0·071	14·0	0·045	0·026			
$6f$	0·0412	24·3	0·0210	0·0129	0·0072		
					0·0001		
			$4f$	$5f$			
$5g$	0·042$_5$	23·5	0·042$_5$				
$6g$	0·0247	40·5	0·0137	0·0110			
			$5g$				
$6h$	0·0164	61	0·0164				

* This is $\sum_{j'=l'-\frac{1}{2}}^{l'+\frac{1}{2}} \mathsf{A}(nlj \to n'l'j')$, a quantity which is independent of the value of j.

In Table 5[5] some of the important values of the transition probabilities for hydrogen are given together with the mean lives of the individual states, and in Table 6[5] are given the average mean lives for several values of n. The discussion of the experimental results of intensity studies in hydrogen is taken up in the next section.

TABLE 6[5]. *Average mean life* $1/\bar{k}(n)$.

n	(10^{-8} sec)
2	0·21
3	1·02
4	3·35
5	8·8
6	19·6

7. Experimental results for hydrogenic spectra.

Let us consider briefly the relation of the foregoing theoretical results to the experimental data. We have found in § 2[5] that the energy levels in hydrogenic atoms are given by

$$W = -\frac{\mu e^4 Z^2}{2\hbar^2 n^2},$$

where μ is the reduced mass (§ 1[5]). In spectroscopy we are concerned directly with the wave-numbers of the lines which are obtained as the differences of the term values

$$\sigma = -\frac{\mathsf{R}_{\mathrm{H}} Z^2}{n^2}, \quad \mathsf{R}_{\mathrm{H}} = \frac{\mu_{\mathrm{H}} e^4}{4\pi\hbar^3 c}. \tag{1}$$

The universal constant R_{H} occurring in this expression is known as the Rydberg constant for hydrogen. The quantity obtained by using the reduced mass of the electron with respect to any other nucleus is known as the Rydberg constant for that atom and we write R_∞ (or R) for the Rydberg constant obtained by using the electronic mass in this formula.

The spectrum of hydrogen divides itself naturally into a number of series of 'lines', each series being the set of lines having a common final n value:

		Wave-lengths (Ångströms)	
Final n	Name	First member	Series limit
1	Lyman	1216	912
2	Balmer	6562	3648
3	Paschen	18571	8208
4	Brackett	40500	14600

The Lyman series lies deep in the ultra-violet, the Balmer series is the prominent feature of the visible hydrogen spectrum, the other two are in the infra-red.

The Rydberg constant is one of the best known atomic constants. A critical study of all the data made by Birge* led him to adopt the values

$$\mathsf{R}_{\mathrm{H}} = 109677{\cdot}759 \pm 0{\cdot}05\ \mathrm{cm}^{-1}$$
$$\mathsf{R}_{\mathrm{He}} = 109722{\cdot}403 \pm 0{\cdot}05\ \ ,,$$
$$\mathsf{R}_{\infty} = 109737{\cdot}42\ \pm 0{\cdot}06\ \ ,, .$$

The smallness of the difference between R_{H} and R_{He} means that some He II lines are very close to certain H I lines. Thus for the first Balmer line, H_α,

$$\sigma_{\mathrm{H}} = \mathsf{R}_{\mathrm{H}}\left(\frac{1}{2^2} - \frac{1}{3^2}\right) = \tfrac{5}{36}\mathsf{R}_{\mathrm{H}},$$

and close to it, in the He II spectrum, is

$$\sigma_{\mathrm{He}} = 4\mathsf{R}_{\mathrm{He}}\left(\frac{1}{4^2} - \frac{1}{6^2}\right) = \tfrac{5}{36}\mathsf{R}_{\mathrm{He}}.$$

By making an accurate measure of the separation of these lines and taking the ratio of the nuclear masses as known from the chemical atomic-weight determinations, it is possible to measure the mass of the proton in atomic units, that is, the ratio of its mass to that of the electron. The most accurate study of this kind has been made by Houston† who finds

$$M_{\mathrm{H}}/\mu = 1838{\cdot}2 \pm 1{\cdot}8.$$

Taking this in combination with the Faraday constant in electrolysis, the value of e/μ for the electron becomes

$$e/\mu = (1{\cdot}7602 \pm 0{\cdot}0018) \times 10^7\ \textsc{emu}.$$

This is in good agreement with the recent precision measurements on free electrons in deflecting fields by Perry and Chaffee, and Kirchner.‡ Thus the theoretical effect of the finite mass of the nucleus is fully verified. This effect was the means used in the discovery of the hydrogen isotope, deuterium, by Urey.§

The relativity-spin structure for hydrogen and He II is very fine, the distance of the extreme components of the nth energy level being

$$\Delta\sigma = \alpha^2 \mathsf{R}\, Z^4 (n-1)/n^4$$

according to $4^5 12$. For the final states of the Balmer series in hydrogen

$$\Delta\sigma = 0{\cdot}365\ \mathrm{cm}^{-1},$$

while the states of $n = 4$ in ionized helium have a splitting three times as great. Since the higher levels show a much smaller splitting, the observed pattern in a Balmer line is that of two close groups of lines with this separation.

* Birge, Rev. Mod. Phys. **1**, 1 (1929).

† Houston, Phys. Rev. **30**, 608 (1927).

‡ Perry and Chaffee, Phys. Rev. **36**, 904 (1930); Kirchner, Ann. der Phys. **8**, 975 (1931); **12**, 503 (1932).

§ Urey, Brickwedde and Murphy, Phys. Rev. **40**, 1, 464 (1932); J. Chem. Phys. **1**, 512 (1933).

Sommerfeld and Unsöld* have discussed the modern interpretation of the fine structure for several lines. We confine our attention to H_α. The theoretical pattern for this line is shown in Fig. 3^5, where the ordinates are proportional to the relative intensities of the lines as calculated in the preceding section.

The broadening of the lines by the Doppler effect due to the thermal velocities of the atoms in the source is considerable, so the experiments are performed with the discharge tube immersed in liquid air. According to the

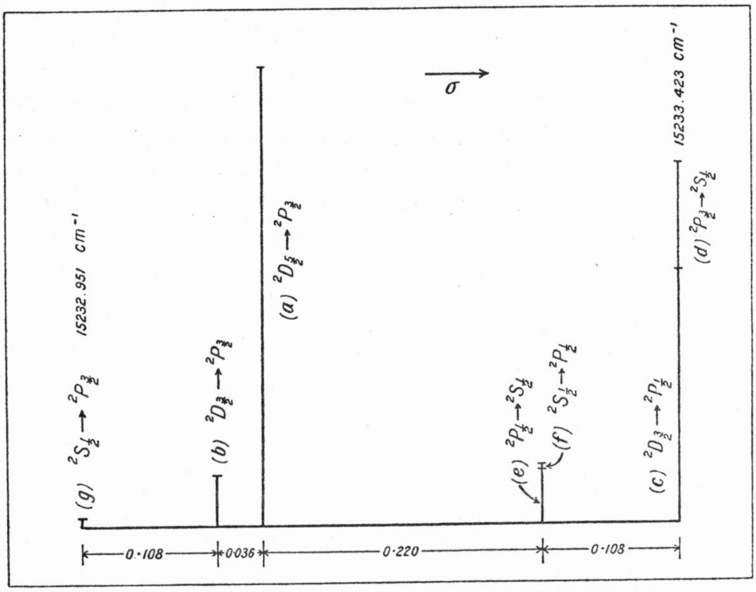

Fig. 3^5. Theoretical fine structure of H_α.

theory of the Doppler effect, an atom emitting wave-number σ_0 if moving toward the observer with speed u will appear to emit radiation of wave-number σ where

$$\sigma = \sigma_0(1 + u/c).$$

If the distribution of velocities is according to Maxwell's law for absolute temperature T, the fraction of all atoms whose speed toward any direction is between u and $u+du$ is given by

$$dn = \sqrt{\frac{m}{2\pi kT}}\, e^{-mu^2/2kT} du;$$

hence the intensity of radiation between σ and $\sigma + d\sigma$ is proportional to

$$\sqrt{\frac{mc^2}{2\pi kT\sigma_0^2}}\, e^{-\frac{(\sigma-\sigma_0)^2}{2a^2}} d\sigma. \qquad \left(a^2 = \frac{kT\sigma_0^2}{mc^2}\right)$$

* SOMMERFELD and UNSÖLD, Zeits. für Phys. **36**, 259; **38**, 237 (1926); see also SLATER, Proc. Nat. Acad. Sci. **11**, 732 (1925).

Numerically, if T is expressed in thousands of degrees and m is expressed in atomic weight units,

$$a/\sigma_0 = 0{\cdot}958 \times 10^{-5}\sqrt{T/m}.$$

The liquid-air bath does not prevent some warming of the gas during the discharge and the observed widths correspond to a temperature of about 200° K., that is, $a \sim 0{\cdot}07$ cm^{-1}, so the width of the lines is comparable with some of the intervals in the pattern. With this in mind we expect that (g) will be too weak to show, that (a) and (b) will be lumped together and that (e)+(f) will produce an asymmetry on the curve produced by (c)+(d).

This is exactly what was found experimentally.* Especially important is the occurrence of (e)+(f), since on the original fine-structure theory of Sommerfeld this component is forbidden. Its presence therefore lends support to the present interpretation of the fine-structure levels. The relative intensities are found to depend somewhat on the discharge conditions.†

The relative intensities of the absorption of the first three Balmer lines has been studied by Snoek.‡ He found the relative values to be, for the two fine-structure groups arising from the two initial levels,

$$H_\alpha : H_\beta : H_\gamma = 100:18{\cdot}8:7{\cdot}4 \qquad (\text{from } j=\tfrac{3}{2})$$

$$100:20{\cdot}2:8{\cdot}5 \qquad (\text{from } j=\tfrac{1}{2}).$$

The theoretical ratios for absorption from the $j=\frac{3}{2}$ or from $2p_{\frac{1}{2}}$ are

$$100:17{\cdot}6:6{\cdot}3,$$

while those for absorption from the $2s_{\frac{1}{2}}$ are

$$100:24:9{\cdot}9.$$

The experimental values from $j=\frac{3}{2}$ are in good agreement with the theory and those from $j=\frac{1}{2}$ correspond quite closely to a mean between the theoretical values for $2s_{\frac{1}{2}}$ and $2p_{\frac{1}{2}}$.

The same question was studied by Carst and Ladenburg§ by measuring the amount of the anomalous dispersion around H_α and H_β. The uncertain element here is the amount of correction for negative dispersion due to atoms in the $n=3$ and $n=4$ states. They conclude that the oscillator strength ratio is between 4·66 and 5·91 for $H_\alpha : H_\beta$. Since the oscillator strengths are proportional to the absorption, we see from the foregoing figures that the

* The principal papers are:
HANSEN, Ann. der Phys. **78**, 558 (1925);
KENT, TAYLOR and PEARSON, Phys. Rev. **30**, 266 (1927);
HOUSTON and HSIEH, Phys. Rev. **45**, 263 (1934);
WILLIAMS and GIBBS, Phys. Rev. **45**, 475 (1934);
SPEDDING, SHANE and GRACE, Phys. Rev. **47**, 38 (1935).

† See BETHE, Handbuch der Physik, **24/1**, 2d ed., 452 (1933).

‡ SNOEK, Diss. Utrecht (1929); Archives Neerlandaises, **12**, 164 (1929); Zeits. für Phys. **50**, 600 (1928); **52**, 654 (1928).

§ CARST and LADENBURG, Zeits. für Phys. **48**, 192 (1928).

theory gives 4·16 if all atoms are in the $2s$ states and 5·68 if all are in the $2p$ states, while the average according to statistical weights is 5·37.

Ornstein and Burger* studied the relative intensity in emission of Balmer and Paschen lines having the same initial states so as to eliminate the uncertainty as to the number of atoms in the initial states. The results were

		Obs.	Theory
H_β/P_α	$(4\to2):(4\to3)$	2·6	3·55
H_γ/P_β	$(5\to2):(5\to3)$	2·5	3·4
H_δ/P_γ	$(6\to2):(6\to3)$	2·0	3·2

The theoretical values are based on assumption that each initial state has the same number of atoms in it. The departures indicate that this is not the case and that there must be an extra supply of atoms in f states since these contribute to the Paschen lines but not to the Balmer lines.

An interesting absolute measurement of the life time of the excited states of He^+ was made by Maxwell.† He excited the helium by electron impact occurring in a narrow electron beam. The excited ions produced were drawn out by a transverse electric field which did not sensibly affect the original electron beam because of the presence of a controlling longitudinal magnetic field. The ions move various distances in the transverse field before radiating, corresponding to the probability distribution of life times. By studying the spatial distribution of intensity of light he was able to infer the mean lives of the excited atoms. For the $n=6$ states of He^+ he found a mean average life of $(1\cdot1\pm0\cdot2)\times10^{-8}$ sec, while the theoretical value is $1\cdot17\times10^{-8}$ sec. The mean life is $\frac{1}{16}$ that of hydrogen since the frequency cubed varies as Z^6 and the radius squared varies as Z^{-2} in the expression for the transition probability.

8. General structure of the alkali spectra.

In the next chapter we shall formulate the problem of the motion of N electrons in the field of a nucleus and shall see that there is a limit to the number of electrons of each nl value that can occur in the atom. When for a given value of nl the maximum number is present, we speak of these as forming a closed shell (§ 5^6). If we have an atom in which all the electrons but one are in closed shells, the mutual interaction of the electrons is greatly simplified and the energy-level scheme is, to a good approximation, just that of a single electron moving in a central field (§ 10^6). This effective central field for the extra electron outside closed shells is the resultant field of the nucleus and of the other electrons, so for neutral atoms it is of the form $-e^2/r$ at large distances from the nucleus and is equal to $-(Ze^2/r)+C$ at small distances, where C is the constant potential at the origin due to the electrons in closed shells.

* ORNSTEIN and BURGER, Zeits. für Phys. **62**, 636 (1930).
† MAXWELL, Phys. Rev. **38**, 1664 (1931).

Since this is a central-field problem, we may apply § 1[5] and classify the states by quantum numbers n and l. The angular factors will be those corresponding to precise values of angular momentum. Since the potential energy at each distance is less than in the simple Coulomb law, the energy corresponding to each nl is lower than for the corresponding hydrogen state. Moreover, since the difference between the potential energy and the Coulomb law is greater at small values of r, the difference is greater for smaller values of l.

It is convenient to discuss the empirical data by introduction of the effective quantum number $n^\star$ defined by

$$\sigma(nl) = -\mathsf{R}/n^{\star 2}, \tag{1}$$

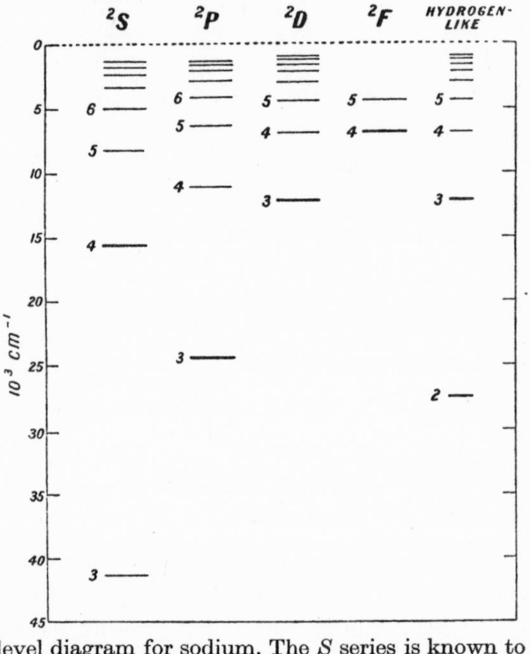

Fig. 4[5]. Energy level diagram for sodium. The S series is known to $n = 14$, the P to $n = 59$, the D to $n = 15$, and the F to $n = 5$. The only doublet terms which have been resolved are the first seven of the P series.

where R is the Rydberg constant for the atom in question. The dependence of $n^\star$ on n is almost linear, so that the difference $(n - n^\star) = \Delta$, which is called the quantum defect, is almost constant in a particular series (constant l, varying n) of terms. These facts are illustrated in Fig. 5[5] where the empirical values of the quantum defects are plotted for the various series of the alkali spectra. Each plotted point on this scale is big enough to cover the variation of Δ with n along a series. This shows clearly that the f terms even in Cs are approximately hydrogen-like although all others in Cs show large displacements from the hydrogenic values.

The much smaller dependence of Δ on n in a given series is shown in Fig. 6[5] for several typical series. This shows the difference, $\Delta(nl)$ minus the Δ for the first member of the series, as a function of n. It will be noticed that Δ increases with n in the d series and decreases in the s and p series.

The various empirical formulas for the terms in a spectral series which are in use among spectroscopists may all be regarded as formulas for the dependence of $n^\star$ or Δ on n. The simplest is the Rydberg formula which simply regards Δ as a constant. Other special formulas which are in use are

$$\Delta = a + b\,\sigma(nl) \qquad \text{(Ritz)},$$
$$\Delta = a + b/n^2 \qquad \text{(Ritz)},$$
$$\Delta = a + b/n \qquad \text{(Hicks)}.$$

In the first of these the deviation of Δ from constancy is proportional to the term itself which means that the quantity represented is contained implicitly in the formula which represents it.

In considering the variation of the one-electron spectra along an iso-electronic sequence, it is desirable to take account of the large dependence on the degree of ionization by writing

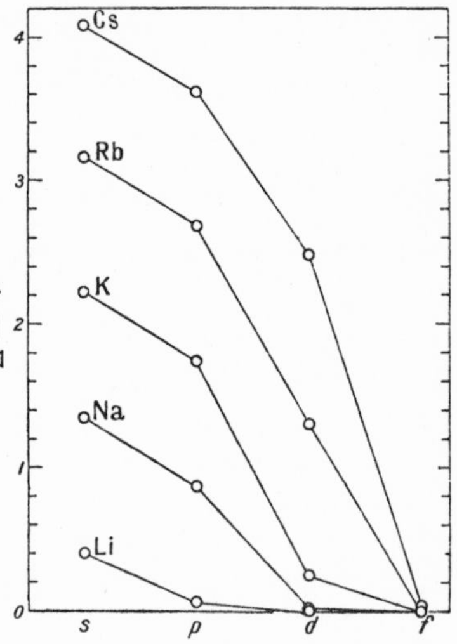

Fig. 5[5]. Variation with l and Z of quantum defect Δ in alkali spectra.

$$\sigma(nl) = -\mathsf{R}Z_0^2/n^{\star 2} \qquad (2)$$

as the equation defining $n^\star$, where Z_0 is the net charge of the ion around which the single electron moves (one for neutral atoms, two for singly-charged ions, etc.). For higher values of the nuclear charge in an iso-electronic sequence the departure from the Coulomb law caused by the inner group of electrons is smaller in proportion to the whole potential energy. Therefore we expect the terms to become more hydrogen-like as Z_0 increases in such a sequence. This means that the $\Delta(nl)$ tend to zero as Z_0 increases. This is illustrated in Fig. 7[5] where Δ for the s and p terms of the eleven-electron sequence, Na I, Mg II, Al III, and Si IV is plotted against Z_0.

In the orbital form of the atomic theory there was a sharp distinction between penetrating and non-penetrating orbits, the former being those which extend into small enough values of r that the field has a non-Coulomb character. This distinction is not so sharp in quantum mechanics as all of the eigenfunctions have non-vanishing values near the origin. Nevertheless the states of the higher l values penetrate less, as in the old theory. For the non-penetrating orbits the old theory gave an explanation of the slight departures from hydrogenic values by assuming that the core of the atom, composed of the inner groups of electrons, is polarized by the field of the outer electron, so that the electron's motion was perturbed by an interaction term, $-\alpha e^2/2r^4$, where α is the polarizability of the core. Using values of α inferred from ion refractivity data, this gives a fairly satisfactory account of the deviations.

The alkali spectra have, of course, a doublet structure caused by the spin-orbit interaction discussed in § 4^5. Since ζ_{nl} is essentially positive, ${}^2L_{l+\frac{1}{2}} > {}^2L_{l-\frac{1}{2}}$, i.e. the level with higher j value is higher in energy. This

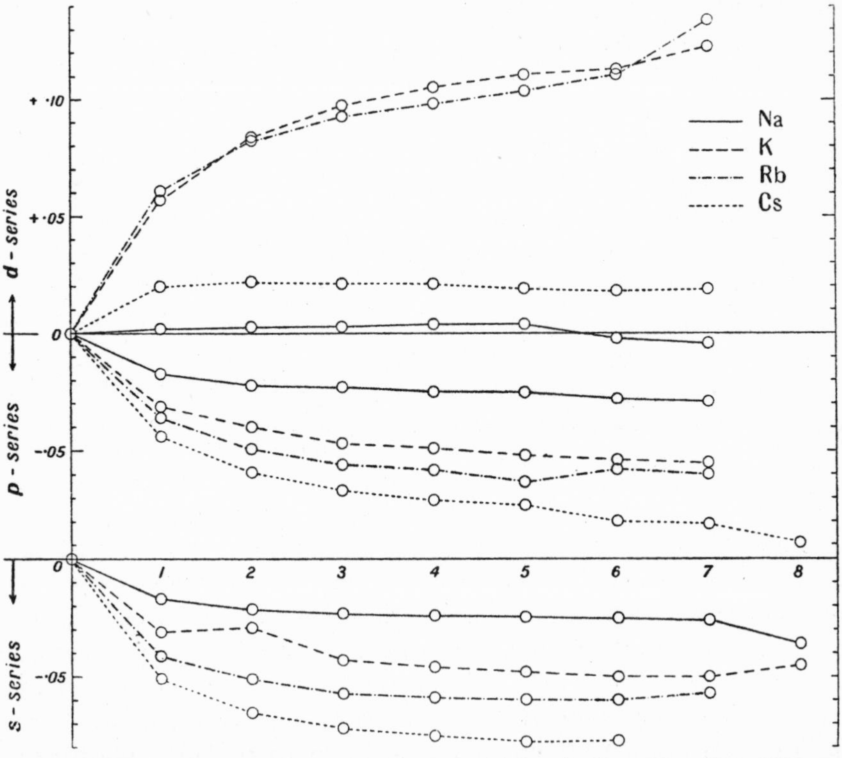

Fig. 6^5. Variation with n of the quantum defect Δ in alkali spectra. The difference between the quantum defect of each term and that of the lowest term of the series to which it belongs is plotted against the difference between the total quantum numbers of these terms.

order is called *normal*. For fields not deviating too greatly from Coulomb fields we may expect the value of the doublet interval to be given roughly by $4^5 11$, which can be written in the form

$$ {}^2L_{l+\frac{1}{2}} - {}^2L_{l-\frac{1}{2}} = \alpha^2 \mathsf{R} \frac{Z^4}{n^3 l(l+1)} = 5{\cdot}822 \frac{Z^4}{n^3 l(l+1)} \text{ cm}^{-1}. \tag{3} $$

If this form has any kind of approximate validity for the alkalis, we expect that the doublet interval will vary about as $1/n^3$. The absolute value of the intervals we do not expect to be given at all by the formula. The approximate variation as $1/n^3$ is shown in Fig. 8[5], in which is plotted, on logarithmic scales, the doublet interval as ordinate and n as abscissa for all the known doublets of the alkalis. From this it is clear that in the main the variation is like n^{-3} or n^{-4}, but the diagram also shows some striking exceptions, thus far unexplained. For example the D series of both Rb and Cs show a remarkable discontinuity in their values so that in the case of Rb for $n=11$ the interval takes a jump from an expected value of about 1 cm^{-1} up to 6 cm^{-1}, which is twice as big as the interval for $n=5$, the first member of the series. In the case of Cs the intervals for $n=13$ and 14 take a similar sudden jump upward and become nearly equal to the P intervals for the same n's.

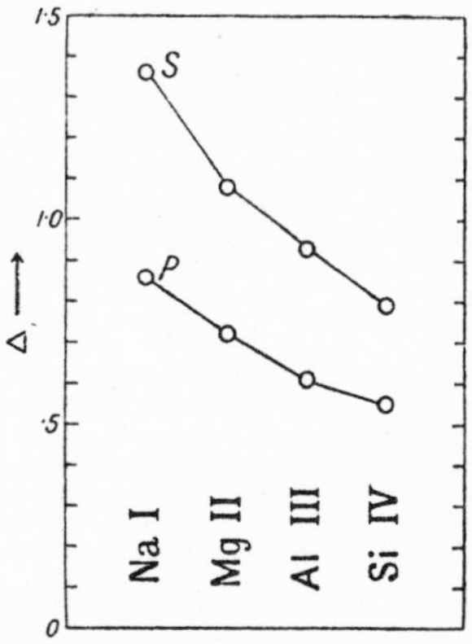

Fig. 7[5]. Variation of Δ in the Na-like iso-electronic sequence.

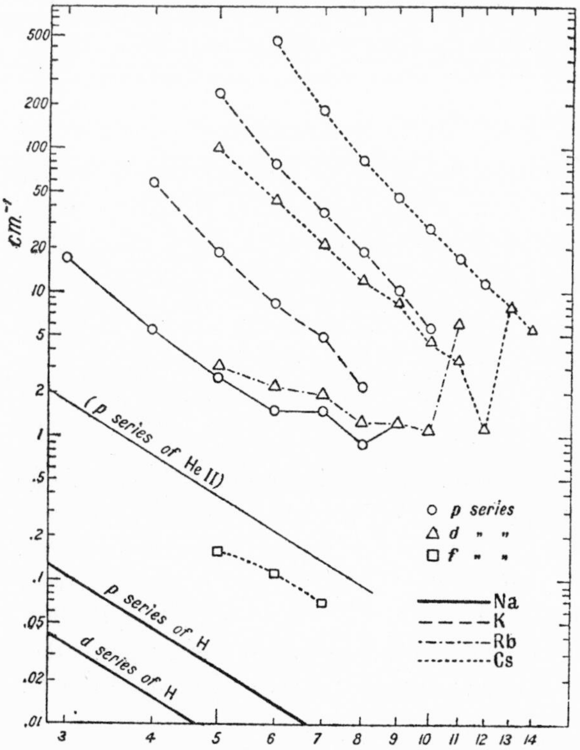

Fig. 8[5]. Variation of doublet interval with n in alkali spectra.

Data on doublet intervals for F series are scarce, indicating that the interval is usually small. But in the case of Cs it is definitely known that the doublet F terms are inverted, that is, the level with the higher j value has the lower energy. The separations, however, are very small:

$^2F_{\frac{5}{2}} - {}^2F_{\frac{3}{2}}$ in Cs.

n	4	5	6	7	8	9	10
$\Delta\sigma$	−0·2	−0·2	−0·1	−0·0	−0·1	−0·1	−0·0

The first order perturbation theory is quite inadequate for explanation of this inversion since ζ is essentially positive.

We turn now to the relative intervals of corresponding terms in D and P series. According to the hydrogen formula these intervals should be as 1:3. We expect them to be less than this in the alkalis, since the p eigenfunctions have appreciable values for smaller values of r than the d eigenfunctions. Since the deviation from the Coulomb law is in the sense of stronger fields than Coulomb at smaller distances, this tends to make ζ for a p state greater than for a d state in a larger ratio than for hydrogen. That those expectations are borne out is shown in a table for Cs:

n	6	7	8	9	10	11	12	13	14
$\dfrac{^2D_{\frac{5}{2}} - {}^2D_{\frac{3}{2}}}{^2P_{\frac{3}{2}} - {}^2P_{\frac{1}{2}}}$	0·078	0·115	0·135	0·182	0·164	0·192	0·096	0·964	1·02

The anomaly of the last two terms, already mentioned, is here striking.

In the case of the iso-electronic sequence Li I, Be II, etc. the interval for $2p$ is known and agrees quite well in absolute value with the value given by the theory for hydrogen:

	Li I	Be II	B III	C IV	N V	O VI
$2p$ interval is	0·34	6·6	34·0	107·4	259·1	533·8
Value by (3)	0·364	5·82	29·5	93·2	227·4	471·6
S	2·019	1·937	1·884	1·858	1·838	1·816

Various somewhat empirical modifications of the doublet interval formula (3) have been proposed. Landé† from an analysis of the classical penetrating-orbit picture proposed replacing Z^4 by $Z_i^2 Z_0^2$, where Z_0 is the net ion charge as in (2) and Z_i is an effective nuclear charge in the inner region to which the orbit penetrates, and replacing n^3 by n^{*3}. Another modification that has been used a great deal by Millikan and Bowen‡ regards the entire orbit as determined by a Coulomb field of effective nuclear charge $(Z-S)$, where S

† Landé, Zeits. für Phys. **25**, 46 (1924).
‡ Millikan and Bowen, Phys. Rev. **23**, 764 (1924); **24**, 209, 233 (1924).

represents the amount by which the full nuclear charge is diminished by screening of the field by the inner electrons. If one chooses values of S to fit the observed $2p$ intervals in the Li I to O VI sequence, they are found (third row in the above table) to be nearly equal to two, which is a reasonable value. In the older theory the formula (3) was derived as a relativity effect and therefore should have applied to the interval between s and p terms rather than to the ^2P interval. This puzzle was solved by the interpretation of the doublet formula as due to the spin-orbit interaction.

9. Intensities in alkali spectra.

After we have found a central field which gives a satisfactory representation of the energy levels of a one-electron spectrum, we may use the eigenfunctions of this field to calculate the radial integrals

$$\int_0^\infty r\,R(n\,l)\,R(n'\,l-1)\,dr$$

which are needed for the intensity problem. The theory goes exactly as for hydrogen (§ 6^5) with these integrals replacing the ones based on hydrogenic eigenfunctions.

The first point to be noticed is the 2:1 strength ratio in the principal series $n\,^2P \rightarrow {}^2S$, which was already calculated in our discussion of hydrogen. This value* is observed for the first doublet in Na and K. For Rb the experimental ratio for the second doublet is 2·5 while in Cs the ratios for the second and third doublets are 4·0 and 4·7, although it is 2·0 for the first doublet. The explanation of these departures, due to Fermi, is considered in § 5^{15}.

Next we consider the relative strengths of the doublets in the principal series. Most of the experimental data is expressed in terms of effective f values for the unresolved doublets, defined in a way exactly analogous to the definition in $9^4$14 of the f values for lines. For the doublet of wave number σ connecting the low-energy configuration nl and the higher configuration $n'l'$,

$$\mathsf{f}(n'l', nl) = \frac{1}{6(2l+1)}\left(\frac{\sigma}{\mathsf{R}}\right)\frac{\mathsf{S}(n'l', nl)}{e^2\mathsf{a}^2}, \tag{1}$$

where S is the total strength of the doublet. Calculations of the f values for the principal series $2s-np$ of Li have been made by Trumpy† using radial eigenfunctions based on a Hartree field (§ 8^{14}). The ratios of the f values of successive doublets were measured by Filipov‡ by the anomalous-dispersion method. The striking thing is the very large ratio, 136, of the first to the second doublet. In Table 7^5 the experimental values of Filipov

* The experimental data are summarized by KORFF and BREIT, Rev. Mod. Phys. **4**, 471 (1932).
† TRUMPY, Zeits. für Phys. **61**, 54 (1930); **66**, 720 (1930).
‡ FILIPOV, Zeits. für Phys. **69**, 526 (1931).

are given as calculated from the measured ratios assuming the first one to have the theoretical value calculated by Trumpy. To show how much the values are affected by the departure of the Li eigenfunctions from the H functions, we have given in the last column the values obtained by using hydrogenic radial integrals together with the Li frequencies in the theoretical formula.

TABLE 7[5]. f *values in the Li* $2s-np$ *series.*

n	Filipov exptl.	Trumpy theory	Li cm^{-1} with H integrals
2	75000×10^{-5}	75000×10^{-5}	123000×10^{-5}
3	549	551	86000
4	478	471	18000
5	314	253	7000
6	192		
7	128		
8	91		
9	67·6		
10	52·0		
11	40·6		
12	32·7		
13	26·6		

Similar calculations for the $3s-np$ series in Na have been made independently by Prokofjew, Trumpy and Sugiura.* Prokofjew's method of obtaining the central field is discussed in § 3[14]; Sugiura's method is similar. Trumpy used a Hartree field. The experimental ratios were measured by Filipov and Prokofjew† and as in the table for Li have been reduced to absolute values by assuming agreement with Trumpy's calculations for the first line. These are given in Table 8[5]. The small inset table gives the values as calculated by Prokofjew for several other doublets.

Values for the principal series of Cs (for which no theoretical calculations have been made) and of the ratios of the first two doublets of the principal series in K and Rb are given in the report by Korff and Breit (*loc. cit.*). Zwaan‡ has calculated the spontaneous transition probabilities

$$\mathsf{A}(4p\to3d)=0{\cdot}13\times10^{8}\,\text{sec}^{-1}\quad\text{and}\quad\mathsf{A}(4p\to4s)=1{\cdot}55\times10^{8}\,\text{sec}^{-1}$$

for Ca II using eigenfunctions of a field determined by the method of § 3[14]. The values are of interest in connection with Milne's theory of the support of calcium in the solar chromosphere by radiation pressure.§

* PROKOFJEW, Zeits. für Phys. **58**, 255 (1929);
TRUMPY, *ibid.* **61**, 54 (1929);
SUGIURA, Phil. Mag. **4**, 495 (1927).
† FILIPOV and PROKOFJEW, Zeits. für Phys. **56**, 458 (1929).
‡ ZWAAN, Diss. Utrecht, 1929.
§ MILNE, Monthly Notices R.A.S. **84**, 354 (1924); **85**, 111 (1924); **86**, 8 (1925).

TABLE 8^5. f *values in the Na* $3s-np$ *series.*

n	Exptl.	Trumpy theory	Prokofjew theory	Sugiura theory
3	97550×10^{-5}	97550×10^{-5}	97960×10^{-5}	97280×10^{-5}
4	1403	1440	1426	1440
5	205	241	221	560
6	63·1	98	73	280
7	25·6			
8	13·4			
9	8·11			
10	5·37			
11	3·84			
12	2·84			
13	2·17			
14	1·73			
15	1·40			
16	1·16			
17	0·92			
18	0·75			

Transition	Prokofjew
$3p-4s$	0·163
$3p-3d$	0·832
$3p-4d$	0·108
$4s-4p$	1·35

10. Zeeman effect.*

When an atom with one valence electron is placed in a magnetic field, the energy levels are split into several components, giving characteristic Zeeman patterns. The interaction energy which produces these displacements consists of two parts, that arising from the spin of the electron and that arising from its orbital motion.

On the spin hypothesis (§ 5^3) an electron has a component of magnetic moment of magnitude $\mp e\hbar/2\mu c$ in the direction in which the component of spin angular momentum is $\pm\frac{1}{2}\hbar$. Since the energy of a particle of magnetic moment $\boldsymbol{M}$ in a field $\mathscr{H}$ is $-\boldsymbol{M}\cdot\mathscr{H}$, the interaction energy of the magnetic electron may be written

$$\frac{e}{\mu c}\boldsymbol{S}\cdot\mathscr{H}=\left(\frac{e}{2\mu c}\right)\mathscr{H}\cdot 2\boldsymbol{S}. \qquad (1)$$

The effect of the magnetic field on the orbital motion of the electron is best obtained by using the vector potential $\boldsymbol{A}$, where $\mathscr{H}=\text{curl}\boldsymbol{A}$. The Hamiltonian for a particle of charge $-e$ in the field of potential $\boldsymbol{A}$ is obtained by writing $\boldsymbol{p}+(e/c)\boldsymbol{A}$ in place of the momentum $\boldsymbol{p}$ in the Hamiltonian for the case without the magnetic field.† Then, in place of the term $\boldsymbol{p}^2/2\mu$, we have

$$\frac{1}{2\mu}\left(\boldsymbol{p}+\frac{e}{c}\boldsymbol{A}\right)^2=\frac{1}{2\mu}\boldsymbol{p}^2+\frac{e}{2\mu c}(\boldsymbol{p}\cdot\boldsymbol{A}+\boldsymbol{A}\cdot\boldsymbol{p})+\frac{e^2}{2\mu c^2}\boldsymbol{A}^2.$$

Now since $\boldsymbol{p}=-i\hbar\,\text{grad}$, we obtain, using a formula of vector analysis,

$$\boldsymbol{p}\cdot\boldsymbol{A}\psi=-i\hbar\,\text{div}(\boldsymbol{A}\psi)=-i\hbar\psi\,\text{div}\boldsymbol{A}-i\hbar\boldsymbol{A}\cdot\text{grad}\psi=\boldsymbol{A}\cdot\boldsymbol{p}\psi$$

* HEISENBERG and JORDAN, Zeits. für Phys. **37**, 263 (1926);
C. G. DARWIN, Proc. Roy. Soc. **A115**, 1 (1927);
K. DARWIN, *ibid.* **A118**, 264 (1928).

† See CONDON and MORSE, *Quantum Mechanics*, pp. 26–27. (Note that their development is for a particle of charge $+e$.)

if we choose $\boldsymbol{A}$ as a solenoidal vector (i.e. so that $\operatorname{div}\boldsymbol{A}=0$: this may always be done). Hence the energy due to the magnetic field may be written as

$$\frac{e}{\mu c}\boldsymbol{A}\cdot\boldsymbol{p}+\frac{e^2}{2\mu c^2}\boldsymbol{A}^2. \tag{2}$$

For a uniform magnetic field $\mathscr{H}$, we may take $\boldsymbol{A}=\frac{1}{2}(\mathscr{H}\times\boldsymbol{r})$.*

We shall now make a rough calculation of the order of magnitude of the terms in (2) which will show that the second is entirely negligible for all magnetic fields thus far used in the laboratory. In atomic units (see Appendix), $e=1$, $\mu=1$, $c=137$, r and p for an electron are of the order of 1. The usual magnitude of $\mathscr{H}$ in Zeeman-effect measurements is from 10,000 to 30,000 gauss, i.e. from 0·6 to $1{\cdot}8\times10^{-3}$ atomic units. The larger of these values gives $\boldsymbol{A}\sim0{\cdot}9\times10^{-3}$ atomic units, and the first term of (2) $\sim0{\cdot}65\times10^{-5}$ atomic units of energy $\simeq1{\cdot}5$ cm^{-1}. This is an energy which may easily be measured spectroscopically, since the limit of spectroscopic accuracy in the visible is of the order of 0·001 cm^{-1}. On the other hand, even if we take $\mathscr{H}$ as 200,000 gauss (which is about the highest field anyone has ever succeeded in using), the second term represents an energy of only 0·0002 cm^{-1}. This magnitude is certainly at present undetectable.

Hence we shall consider only the first term of (2). When we set $\boldsymbol{A}=\frac{1}{2}(\mathscr{H}\times\boldsymbol{r})$, this interaction becomes

$$\frac{e}{\mu c}\boldsymbol{A}\cdot\boldsymbol{p}=\frac{e}{2\mu c}\mathscr{H}\times\boldsymbol{r}\cdot\boldsymbol{p}=\frac{e}{2\mu c}\mathscr{H}\cdot\boldsymbol{r}\times\boldsymbol{p}=\frac{e}{2\mu c}\mathscr{H}\cdot\boldsymbol{L}. \tag{3}$$

When this is added to (1) we obtain as the whole effect of the magnetic field the energy

$$H^M=\frac{e}{2\mu c}\mathscr{H}\cdot(\boldsymbol{L}+2\boldsymbol{S}). \tag{4}$$

In particular if we take the z axis in the direction of $\mathscr{H}$ this interaction energy becomes

$$H^M=\frac{e\mathscr{H}}{2\mu c}(L_z+2S_z)=\frac{e\mathscr{H}}{2\mu c}(J_z+S_z)=o\,(J_z+S_z), \tag{5}$$

where o is a coefficient proportional to the magnetic field strength

$$o=\frac{e\mathscr{H}}{2\mu c}. \tag{6}\dagger$$

Thus the energy (5) arising from the interaction of the atom and the magnetic field is diagonal in a representation in which L_z and S_z are diagonal, i.e. in the representation $4^5 2$.

* To show that this represents a uniform magnetic field $\mathscr{H}$, we have

$$\operatorname{curl}\boldsymbol{A}=\tfrac{1}{2}\operatorname{curl}(\mathscr{H}\times\boldsymbol{r})=\tfrac{1}{2}\{\boldsymbol{r}\cdot\operatorname{grad}\mathscr{H}-\boldsymbol{r}\operatorname{div}\mathscr{H}-\mathscr{H}\cdot\operatorname{grad}\boldsymbol{r}+\mathscr{H}\operatorname{div}\boldsymbol{r}\}.$$

Since $\mathscr{H}$ is constant, the first two terms are zero. The third term is $-\mathscr{H}$ and the fourth $3\mathscr{H}$, hence $\operatorname{curl}\boldsymbol{A}=\mathscr{H}$. Similarly $\operatorname{div}\boldsymbol{A}=0$ as required.

† Note that this is 2π times the o as usually defined (the classical Larmor frequency).

Weak-field case.

In the case in which the Zeeman splitting is small compared to the spin doubling we can consider (5) as a perturbation on each of the two levels obtained by the spin interaction separately. We must thus calculate the matrix components $(nl\,^2L_j^m|H^M|nl\,^2L_j^{m'})$ for the different degenerate states of the level $nl\,^2L_j$. These are easily obtained from $4^5 8$:

$$(nl\,^2L_{l\pm\frac{1}{2}}^m|o\,(J_z+S_z)|nl\,^2L_{l\pm\frac{1}{2}}^{m'})$$

$$=\left\{\sqrt{\frac{l\pm m+\frac{1}{2}}{2l+1}}\,\bar{\phi}(nl\,\tfrac{1}{2}\,m-\tfrac{1}{2})\pm\sqrt{\frac{l\mp m+\frac{1}{2}}{2l+1}}\,\bar{\phi}(nl\,-\tfrac{1}{2}\,m+\tfrac{1}{2})\right\}o\hbar$$

$$\cdot\left\{\sqrt{\frac{l\pm m'+\frac{1}{2}}{2l+1}}(m'+\tfrac{1}{2})\,\phi(nl\,\tfrac{1}{2}\,m'-\tfrac{1}{2})\pm\sqrt{\frac{l\mp m'+\frac{1}{2}}{2l+1}}(m'-\tfrac{1}{2})\,\phi(nl\,-\tfrac{1}{2}\,m'+\tfrac{1}{2})\right\}$$

$$=\hbar o\,\delta(m,m')\left\{(m+\tfrac{1}{2})\frac{l\pm m+\frac{1}{2}}{2l+1}+(m-\tfrac{1}{2})\frac{l\mp m+\frac{1}{2}}{2l+1}\right\}$$

$$=\hbar om\frac{2l+1\pm 1}{2l+1}\,\delta(m,m'). \tag{7}$$

This matrix is already diagonal so that the displacements of the different Zeeman components of the level $^2L_{l\pm\frac{1}{2}}$ are given by

$$(^2L_{l\pm\frac{1}{2}}^m)=\hbar om\frac{2l+1\pm 1}{2l+1}=\hbar omg. \tag{8}$$

The factor $\dfrac{2l+1\pm 1}{2l+1}$ is known as the *Landé g-factor*. This derivation holds for s levels as well as for those with $l>0$.

Thus, for weak fields, to the accuracy with which $4^5 8$ represents the eigen-ψ's of the components of the spin doublets, each level is split symmetrically into $2j+1$ equally spaced states, the splitting being proportional to the magnetic field, and being independent of the value of the total quantum number n and of the atom in question. The interval between the states after splitting is determined completely by g, which depends only on l and j. For a given l, the level of higher j is more widely split than the level of lower j. A table of g-values is given below:

g-values for doublet spectra.

	^2S	^2P	^2D	^2F	^2G	^2H
$j=l+\frac{1}{2}$	2	$\frac{4}{3}$	$\frac{6}{5}$	$\frac{8}{7}$	$\frac{10}{9}$	$\frac{12}{11}$
$j=l-\frac{1}{2}$	—	$\frac{2}{3}$	$\frac{4}{5}$	$\frac{6}{7}$	$\frac{8}{9}$	$\frac{10}{11}$

In terms of wave-numbers the displacement (8) is given by

$$\sigma=4{\cdot}670\times 10^{-5}\mathscr{H}mg, \tag{9}$$

where $\mathscr{H}$ is in gauss.* The accuracy with which this formula holds for all

* Note the accuracy of our estimate of $1{\cdot}5$ cm^{-1} as the magnetic perturbation caused by a field of 30,000 gauss.

doublets of all one-electron atoms and ions was early discovered experimentally, and is expressed by *Preston's Rule.** Indeed, one of the methods of obtaining e/μ is by the accurate measurement of this Zeeman effect.

The strong-field case—Paschen-Back effect.

We have been considering the case where the Zeeman effect is small compared to the spin interaction. We shall now consider the other extreme, where the magnetic field is so strong, or the spin-orbit interaction so small, that the spin splitting is negligible compared to the magnetic splitting. In this case we start with the degenerate configuration nl and the system of eigenfunctions $\phi(nlm_s m_l)$ in which H^M is diagonal. The magnetic energy is then given by

$$(nlm_s m_l|o\,(J_z+S_z)|nlm_s m_l)=\hbar o\,(m+m_s). \tag{10}$$

Hence each configuration is split symmetrically into $2l+3$ equally spaced components corresponding to the $2l+3$ possible values, $l+1,\ l,\ \ldots,\ -(l+1)$, of $m+m_s\,(=m_l+2m_s)$. Of these the two highest and the two lowest are non-degenerate, while the others are doubly degenerate corresponding to the two ways of obtaining a given value of $(m_l)+(2m_s)$: $(m_l)+(1)$, $(m_l+2)+(-1)$.

If we now superpose a *small* spin-orbit interaction we get a displacement for the non-degenerate components represented by $\hbar^{-2}\zeta(m_s m_l|\boldsymbol{L}\cdot\boldsymbol{S}|m_s m_l)=\zeta m_l m_s$ (cf. $7^3 3$). For the degenerate components we note that because of the $\delta(m,m')$ in $7^3 3$ there are no matrix components of $\boldsymbol{L}\cdot\boldsymbol{S}$ joining the two states, so that each state is displaced by the amount $\zeta m_l m_s$, which may or may not split the component.

This state of affairs is known as the Paschen-Back effect, the eigenfunctions in this case being those labelled by $nlm_s m_l$.

The transition case.

We have been able to determine very simply the Zeeman splitting in two extreme cases, that in which the magnetic interaction is small compared to the spin-orbit, in which case the eigenfunctions are labelled by j and m; and that in which the magnetic interaction is large compared to the spin-orbit, in which case the eigenfunctions are labelled by m_s and m_l. In order to obtain the transition between these two extreme cases we must apply the spin-orbit and magnetic perturbations simultaneously, treating

$$H^I+H^M=\xi(r)\,\boldsymbol{L}\cdot\boldsymbol{S}+o\,(J_z+S_z) \tag{11}$$

as a perturbation.

We use the fundamental system of eigenfunctions labelled by $nlm_s m_l$, in which the matrix of H^M is diagonal and the matrix of H^I is given by $4^5 3$. We shall again consider just the matrix for a given nl. From $7^3 3$ we see that this matrix will split up according to m. The values $m=l+\frac{1}{2}$ and $-(l+\frac{1}{2})$ are

* This rule applies also to atoms with more than one electron outside of closed shells. See § 2^{16}.

realizable in just one way, namely with $m_l = \pm l$, $m_s = \pm \frac{1}{2}$ respectively. Hence the energies corresponding to these states are obtained immediately from $7^3 3$:

$$\begin{aligned} m &= l+\tfrac{1}{2} & \epsilon &= \tfrac{1}{2}l\zeta + \hbar o\,(l+1) \\ m &= -(l+\tfrac{1}{2}) & \epsilon &= \tfrac{1}{2}l\zeta - \hbar o\,(l+1). \end{aligned} \tag{12a}$$

For the other values of m we obtain the second-order secular equation:

$$\begin{matrix} (m_s=-\frac{1}{2}) \\ (m_s=+\frac{1}{2}) \end{matrix} \begin{vmatrix} -\frac{1}{2}\zeta\,(m+\frac{1}{2}) + \hbar o\,(m-\frac{1}{2}) - \epsilon & \frac{1}{2}\zeta\sqrt{(l+m+\frac{1}{2})(l-m+\frac{1}{2})} \\ \frac{1}{2}\zeta\sqrt{(l+m+\frac{1}{2})(l-m+\frac{1}{2})} & \frac{1}{2}\zeta\,(m-\frac{1}{2}) + \hbar o\,(m+\frac{1}{2}) - \epsilon \end{vmatrix} = 0,$$

which has the two solutions

$$4\epsilon_{\pm} = 4\hbar o m - \zeta \pm \sqrt{(4\hbar o m - \zeta)^2 - 4\{\hbar^2 o^2 (4m^2-1) - 4\hbar\zeta o m - \zeta^2 l\,(l+1)\}}. \tag{12b}$$

Here ζ represents the magnitude of the spin perturbation while $\hbar o$ represents the magnitude of the magnetic perturbation. If we neglect $(\hbar o/\zeta)^2$ compared to 1, we obtain

$$\begin{aligned} \epsilon_+ &= \hbar o m \left(\frac{2l+2}{2l+1}\right) + \tfrac{1}{2}l\zeta \\ \epsilon_- &= \hbar o m \left(\frac{2l}{2l+1}\right) - \tfrac{1}{2}(l+1)\,\zeta, \end{aligned} \tag{13}$$

which agrees with the weak-field case as given by (8) and $4^5 7$. If we neglect $(\zeta/\hbar o)^2$ we obtain

$$\begin{aligned} \epsilon_+ &= \hbar o\,(m+\tfrac{1}{2}) + \tfrac{1}{2}(m-\tfrac{1}{2})\,\zeta \\ \epsilon_- &= \hbar o\,(m-\tfrac{1}{2}) - \tfrac{1}{2}(m+\tfrac{1}{2})\,\zeta, \end{aligned} \tag{14}$$

which agrees with the results obtained in the strong-field case.

The complete transition (12) is plotted in Fig. 9^5 for the case of a ^2P term. The splitting is plotted in units of ζ as a function of $\hbar o/\zeta$, the ratio of magnetic to spin interaction. The broken lines at the left are given by (13) while those at the right are given by (14).

The best experimental data on the complete Zeeman effect are furnished by the work of Kent* on the Li doublet λ6708, which arises from the transition $2p\,^2P \rightarrow 2s\,^2S$. For most doublet spectra the doublet interval is so large, i.e. ζ is so large, that the highest utilizable magnetic fields will correspond to only small values of $\hbar o/\zeta$; hence a good Paschen-Back effect is not obtainable. But the doublet splitting of lithium is very small; the $2p\,^2P$ has a splitting of only $0{\cdot}338\,\text{cm}^{-1}$ $(=\frac{3}{2}\zeta)$. Kent used a maximum field of 44,200 gauss, and since $\hbar o = 4{\cdot}670 \times 10^{-5}\mathscr{H}\,\text{cm}^{-1}$, this corresponds to $\hbar o/\zeta = 9{\cdot}16$, a value even beyond the limits of Fig. 9^5.

The best way to compare our curves with Kent's data, which is not of sufficient accuracy to be fairly reduced to term values, is by plotting the line pattern which should result from a transition between the ^2P of Fig. 9^5 and

* Kent, Astrophys. J. **40**, 337 (1914).

a ^2S. From (12a) we see that a ^2S is split uniformly into two states with energies $\hbar o$ and $-\hbar o$; hence to the scale of Fig. 9^5 a ^2S would be represented by two lines through the origin of slopes $+1$ and -1 corresponding to $m = \pm\frac{1}{2}$.

In Fig. 10^5 we have plotted the line pattern for transitions between a ^2P and a ^2S, using full lines for perpendicular (σ) components ($\Delta m = \pm 1$) and broken lines for parallel (π) components ($\Delta m = 0$). Kent's observed patterns are plotted on the same figure, using circles for σ components and crosses for π components. The abscissas are fixed by the relation $\hbar o/\zeta = 20{\cdot}8 \times 10^{-5}\mathscr{H}$

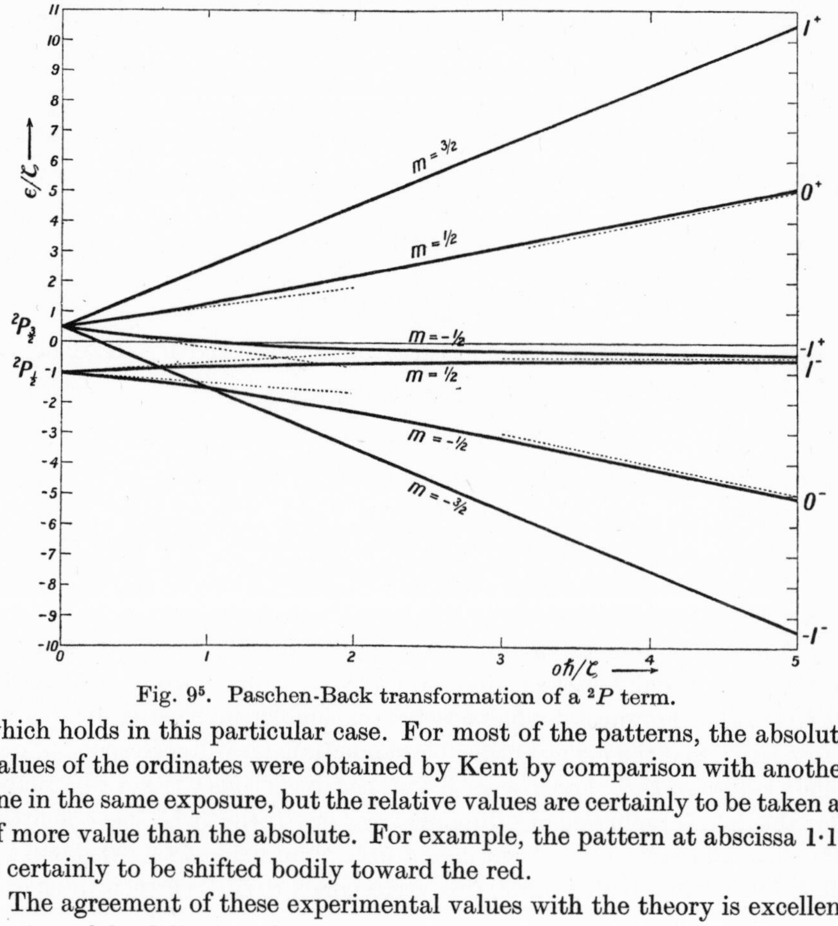

Fig. 9^5. Paschen-Back transformation of a ^2P term.

which holds in this particular case. For most of the patterns, the absolute values of the ordinates were obtained by Kent by comparison with another line in the same exposure, but the relative values are certainly to be taken as of more value than the absolute. For example, the pattern at abscissa 1·12 is certainly to be shifted bodily toward the red.

The agreement of these experimental values with the theory is excellent in view of the difficulty of accurate measurement of such fine patterns. It is pleasing that the theory of the Zeeman effect gives, purely from symmetry considerations, the absolute values of the perturbations, so that the pattern of Fig. 10^5 is completely determined in absolute value if one knows the doublet splitting.

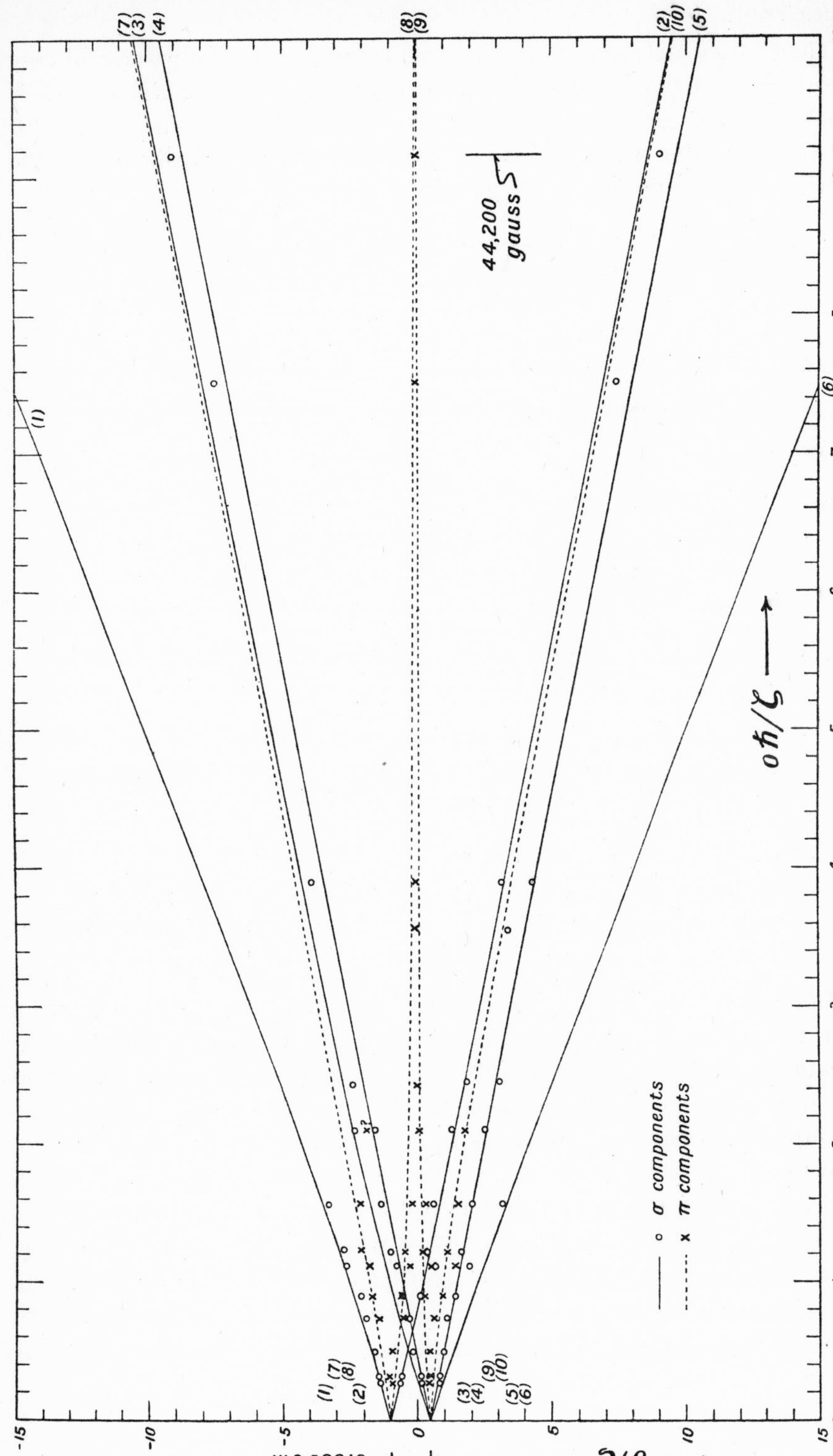

Fig. 10^5. Kent's data on the Zeeman effect of the Li multiplet $2\,^2P \rightarrow 2\,^2S$.

Intensities.

The relative strengths of the allowed transitions in the weak-field case may be obtained by use of the matrix components calculated in §§ 9^3 and 11^3. The details are given in § 4^{16}. Fig. 11^5 shows the expected line pattern for $^2P \rightarrow {}^2S$, with the relative strengths for observation perpendicular to the magnetic field indicated by the lengths of the lines.

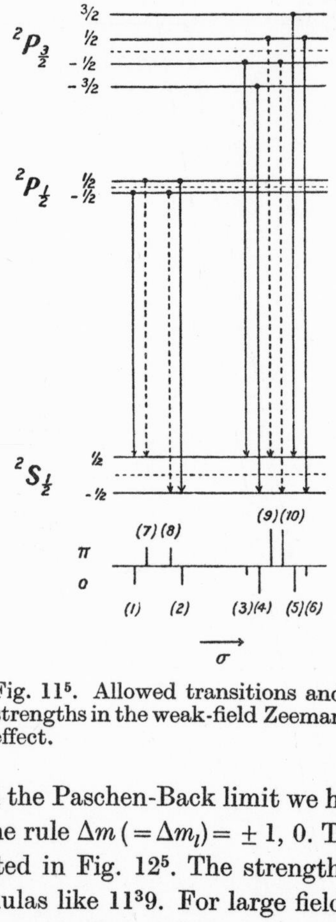

Fig. 11^5. Allowed transitions and strengths in the weak-field Zeeman effect.

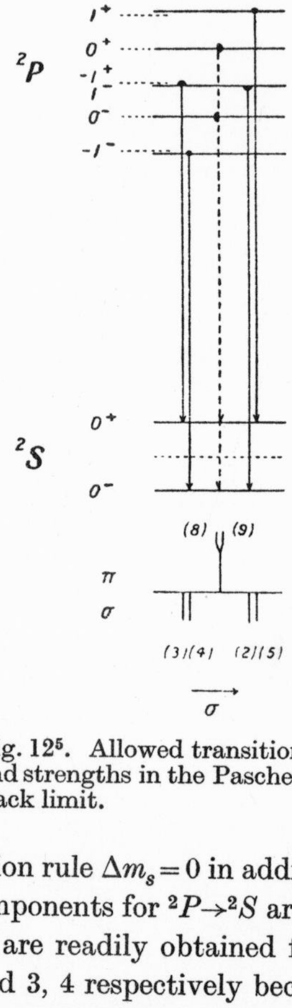

Fig. 12^5. Allowed transitions and strengths in the Paschen-Back limit.

In the Paschen-Back limit we have the selection rule $\Delta m_s = 0$ in addition to the rule $\Delta m\,(=\Delta m_l) = \pm 1, 0$. The allowed components for $^2P \rightarrow {}^2S$ are indicated in Fig. 12^5. The strengths in this case are readily obtained from formulas like $11^3 9$. For large fields lines 2, 5 and 3, 4 respectively become asymptotically parallel to and equidistant on either side from the outer components of a normal Lorentz triplet of interval $\hbar o$. Hence Kent's pattern in which these lines are unresolved forms almost exactly a Lorentz triplet.

For intermediate fields, we may at once obtain the strengths from the eigenfunctions. Since $\psi(np\,^2P^{\frac32}_{\frac32}) = \phi(np\,\tfrac12\,1)$ and $\psi(n's\,^2S^{\frac12}_{\frac12}) = \phi(n's\,\tfrac12\,0)$, the eigenfunctions for these states are independent of field; hence the strength of component (5) is independent of field. Similarly the strength of (4) is independent of field. For $m = \pm\frac{1}{2}$, we may find from a set of equations like $7^2 2$ the coefficients in the expansion

$$\chi(np\, m\, \epsilon_+) = \alpha\,\phi(np\, {-\tfrac{1}{2}}\, m{+\tfrac{1}{2}}) + \beta\,\phi(np\, \tfrac{1}{2}\, m{-\tfrac{1}{2}}),$$
$$\chi(np\, m\, \epsilon_-) = \beta\,\phi(np\, {-\tfrac{1}{2}}\, m{+\tfrac{1}{2}}) - \alpha\,\phi(np\, \tfrac{1}{2}\, m{-\tfrac{1}{2}}).$$

From these we find

$$|(np\, m\, \epsilon_+|\boldsymbol{P}|n's\, {-\tfrac{1}{2}}\, 0)|^2 = \alpha^2 |(np\, {-\tfrac{1}{2}}\, m{+\tfrac{1}{2}}|\boldsymbol{P}|n's\, {-\tfrac{1}{2}}\, 0)|^2,$$
$$|(np\, m\, \epsilon_-|\boldsymbol{P}|n's\, {-\tfrac{1}{2}}\, 0)|^2 = \beta^2 |(np\, {-\tfrac{1}{2}}\, m{+\tfrac{1}{2}}|\boldsymbol{P}|n's\, {-\tfrac{1}{2}}\, 0)|^2.$$

So the strength of the component from one of the states to ${}^2S_{\frac{1}{2}}^{-\frac{1}{2}}$ is α^2, from the other, β^2 times the strength from $np\, {-\frac{1}{2}}\, m{+\frac{1}{2}}$ in the Paschen-Back limit. Since $\alpha^2 + \beta^2 = 1$, the sum of these strengths is a constant.

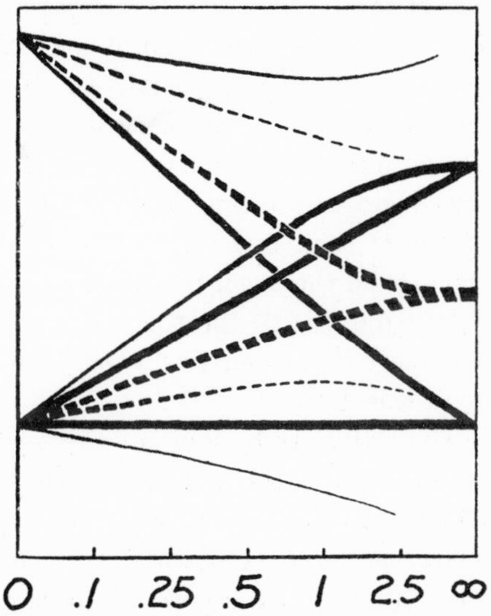

Fig. 13^5. The variation in strength of the components of ${}^2P \rightarrow {}^2S$ in a Paschen-Back transition. (The numbers are values of $\hbar o/\zeta$.)

In Fig. 13^5, which is taken from the paper by K. Darwin, are indicated roughly by the widths of the lines the intensities of the transitions of Fig. 10^5. Referring to the numbering of Fig. 10^5, lines (4) and (5) are of constant strength; the sums of the strengths of (6) and (2) and of (3) and (1), for the σ components, of (7) and (9), and of (10) and (8), for the π components, are independent of $\mathscr{H}$. While no intensity measurements have been made with sufficient accuracy to check quantitatively these changes in intensity, we may note that Kent observes none of the lines which are forbidden in the Paschen-Back limit at values of $\hbar o/\zeta$ greater than 2·1.

CHAPTER VI

THE CENTRAL-FIELD APPROXIMATION*

We shall now proceed to lay the foundation for the theory of spectra of atoms containing more than one electron. The mode of describing spectra that is used by working spectroscopists everywhere is based on the idea that the atoms can be treated to a fairly good approximation by regarding the electrons as moving in a central field and not interacting with each other. This is made the starting-point for a calculation in which the interactions actually occurring are treated as perturbations. We shall see that the relative importance of different kinds of interactions varies a great deal from element to element, and that this is in a sense the origin of the different spectroscopic and chemical behaviour of different elements. In all atoms there will be terms in the Hamiltonian which represent the magnetic interactions of the electronic orbits and spins and terms which represent the Coulomb repulsion of the several electrons. At present it is not known how to give a satisfactory theory of atoms with more than one electron which takes into account relativity effects: in fact, even the exact relativistic treatment of hydrogen which allows for the finite mass of the proton is not known. But relativity effects are usually small so that one can give a fairly satisfactory treatment of the subject in spite of this defect.

In this chapter we shall also study some of the features of the atomic problem which have their origin in the fact that all electrons are believed to be dynamically equivalent. This fact provides a natural place in the theory for an empirical rule known as *Pauli's exclusion principle* and makes possible a fairly complete understanding of the periodic table of the elements.

1. The Hamiltonian for many-electron atoms.

According to the nuclear model, we regard an atom as made up of a central massive positively charged nucleus, surrounded by a number of electrons. This dynamical model is described in the theory by a Hamiltonian function whose proper values give the allowed energy levels and whose proper functions are of use in calculating various properties of the atom. Since for all atoms the nuclear mass is more than 1800 times that of the electrons, we may approximately regard the nucleus as a fixed centre of force instead of treating its coordinates as dynamical variables. This amounts to treating the nucleus as of infinite mass: the correction to finite nuclear mass is treated in § 1^{18}. The principal interactions between the particles are due to the Coulomb

* The method used in this and the next chapter is due to SLATER, Phys. Rev. **34**, 1293 (1929).

electrostatic forces. For most purposes we can neglect the relativistic variation of mass with velocity; thus for a system of N electrons moving about a nucleus of charge Ze we are led to the approximate Hamiltonian

$$H = \sum_{i=1}^{N}\left(\frac{1}{2\mu}\boldsymbol{p}_i^2 - \frac{Ze^2}{r_i}\right) + \sum_{i>j=1}^{N}\frac{e^2}{r_{ij}}. \tag{1}$$

Here subscripts 1, 2, ..., N serve to distinguish the different electrons, r_i is the distance of the ith electron from the nucleus and r_{ij} is the mutual distance of the ith and j^{th} electrons.

In addition to the terms written down it is essential to treat the magnetic interactions of the electronic orbits and spins. The exact way of doing this along the lines of a generalization of Dirac's relativistic theory of the electron to a system of several electrons is not known; some work in this direction and its relation to the helium spectrum we shall report in § 7^7. An approximate allowance for the spin-orbit interaction may be made by including for each electron a term of the form $\xi(r_i)\,\boldsymbol{L}_i\cdot\boldsymbol{S}_i$ such as we have introduced in § 4^5 for the doublet structure of one-electron spectra. We shall adopt this as our working hypothesis and to a large extent base the theory of atomic spectra on the quantum-mechanical properties of the approximate Hamiltonian

$$H = \sum_{i=1}^{N}\left(\frac{1}{2\mu}\boldsymbol{p}_i^2 - \frac{Ze^2}{r_i} + \xi(r_i)\,\boldsymbol{L}_i\cdot\boldsymbol{S}_i\right) + \sum_{i>j=1}^{N}\frac{e^2}{r_{ij}}. \tag{2}$$

In this Hamiltonian, the terms expressing the mutual repulsion of the electrons prevent a separation of variables. These terms cannot very well be neglected as 'small' and treated later by perturbation theory, for although if Z is fairly great any one of them is small compared to a Ze^2/r_i term, there are so many of them that their total effect is comparable with the interaction between the electrons and the nucleus. The procedure generally used is based on the idea of screening, according to which the greater part of the mutual repulsion terms is taken into account in the approximate solution on which a perturbation theory treatment is based.

The mutual repulsion terms, being all positive, tend to cancel the negative terms which represent the attraction of the nucleus. If r_i is large compared to all the other r_j, then $r_{ij}\sim r_i$ and $e^2/r_{ij}\sim e^2/r_i$. Since there are $N-1$ values of j associated with a particular value of i, one sees that at large distances the electron moves in a field that approximates to $(Z-N+1)e^2/r_i$. We say that the other $(N-1)$ electrons have screened off the force field of the nucleus. Now as r_i diminishes to a value comparable with the values of the other r's, this compensation becomes less exact. From potential theory we know that if the other electrons have a distribution with

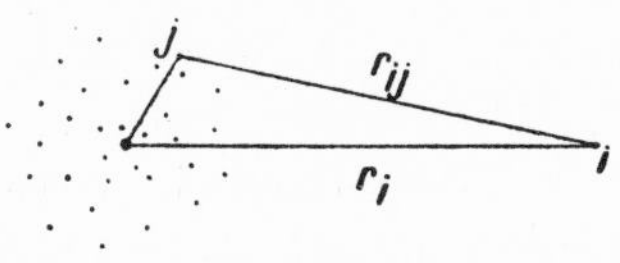

Fig. 1^6.

spherical symmetry, then inside a shell of radius a and total charge $-e'$ the potential due to the shell is constant and equal to $-e'/a$; hence the potential energy of an electron inside such a shell is ee'/a. For an electron near the nucleus, r_i is small compared to the other r_j, so its potential energy as a function of r_i will approach the form

$$-\frac{Ze^2}{r_i}+\frac{(N-1)e^2}{a}$$

in which a is the harmonic mean of the radii of the shells in which the other electrons are distributed. From these general remarks we see that a large part of the effect of the mutual repulsion of the electrons can be allowed for by starting with a Hamiltonian in which the potential energy of each electron at distance r from the centre is such a function $U(r)$ that

$$U(r)\approx -\frac{Ze^2}{r}+C \qquad \text{for } r \text{ small,}$$

$$\approx -\frac{(Z-N+1)e^2}{r} \qquad \text{for } r \text{ large.}$$

Later we shall see that in this characterization of $U(r)$, r small and r large mean respectively $r \ll \mathsf{a}/Z$ and $r \gg \mathsf{a}/(Z-N+1)$, where a is the Bohr hydrogen radius of $2^5 2$.

Our programme will be to build up a systematic theory of the spectra of atoms based on the use of such a screened potential-energy function as the starting point of the perturbation theory. We shall see that a large amount of general information about spectra can be obtained with hardly any more detailed assumption about $U(r)$ than this. In Chapter XIV the theory will be extended by detailed consideration of the question of the best assumptions for $U(r)$ and the actual evaluation of quantities that depend on the choice made.

The approximate Hamiltonian E which we make the starting point of our calculation takes the form

$$E=\sum_i\left[\frac{1}{2\mu}\mathbf{p}_i^2+U(r_i)\right], \tag{3}$$

while the perturbation potential will consist of the difference

$$H-E=V=\sum_{i=1}^{N}\left[\xi(r_i)\,\mathbf{L}_i\cdot\mathbf{S}_i-\frac{Ze^2}{r_i}-U(r_i)\right]+\sum_{i>j=1}^{N}\frac{e^2}{r_{ij}}. \tag{4}$$

2. Equivalence degeneracy.

The Schrödinger equation for the allowed values of E ($1^6 3$) takes the form

$$\sum_{i=1}^{N}\left[-\frac{\hbar^2}{2\mu}\Delta_i+U(r_i)\right]\phi=E'\phi. \tag{1}$$

Since the left side of this is the sum of a number of parts each of which refers to the coordinates of only one electron, it is evident that the variables can be separated if we write

$$\left.\begin{aligned} \phi &= \prod_{i=1}^{N} u_i(a^i) = u_1(a^1)\,u_2(a^2)\,u_3(a^3)\ldots \\ E' &= \sum_{i=1}^{N} E'(a^i). \end{aligned}\right\} \tag{2}$$

Since the Hamiltonian E is a symmetric function of the coordinates of the different electrons, the N equations for the N factors of the wave function are all of the same form so that it is not necessary to carry a subscript to show to which electron's coordinates we refer; each u is the same function of its coordinates for the same set of quantum numbers a^i. Physically this means simply that in the absence of interaction each electron moves in the central field $U(r)$ exactly as it would move if the other electrons were not present. The equation for each u is

$$\left[-\frac{\hbar^2}{2\mu}\Delta + U(r)\right]u(a^i) = E'(a^i)\,u(a^i). \tag{3}$$

For convenience, we denote by a^i an *individual set* of four quantum numbers which specifies the state of motion of a *single* electron in a central field. Thus a^i represents a set of values $(n\,l\,m_s\,m_l)$ or $(n\,l\,j\,m)$ according to the way in which we wish to treat the one-electron problem.

It is clear now that the Hamiltonian E possesses a high degree of degeneracy, which may be regarded as of two kinds. First, there is that which arises simply from the fact that the Hamiltonian for the individual electrons does not depend on spin or on spatial orientation—the energy $E'(a^i)$ is really independent of m_l and m_s. This kind of degeneracy we have already studied in connection with one-electron spectra. Second, there is the degeneracy that goes with the fact that the N individual sets of quantum numbers of the *complete set* of $4N$ necessary to specify the state ϕ can be associated in any way with the electrons 1, 2, 3, ... to give a ϕ which belongs to the same total energy. This degeneracy arises because the Hamiltonian is a symmetric function of the coordinates of the several electrons, i.e. because of the dynamical equivalence of the electrons, and is therefore known as *equivalence degeneracy*.

A convenient notation for the different ϕ's associated with a particular set of quantum numbers is now desirable. We shall let A stand for the complete ordered set $a^1 a^2 \ldots a^N$ of one-electron quantum numbers, and shall understand by $\phi(A)$ the particular function

$$\phi(A) = u_1(a^1)\,u_2(a^2)\,u_3(a^3)\ldots u_N(a^N), \tag{4}$$

in which the first electron is associated with a^1, the second with a^2, the third with a^3, and so on. Then let the operator P stand for a particular permutation of the electron indices relative to the quantum numbers in $\phi(A)$. A particular P can be specified by the scheme

$$P: \begin{pmatrix} 1\,2\,3\,4\,5 \ldots \\ 2\,5\,3\,1\,4 \ldots \end{pmatrix}, \tag{5}$$

which means that after the permutation has been carried out the coordinates of the second electron appear where those of the first were, the fifth replaces the second, the third remains, and so on. Thus with P as in (5), the result of operation by P on $\phi(A)$ is

$$P\phi(A) = u_2(a^1)\, u_5(a^2)\, u_3(a^3)\, u_1(a^4)\, u_4(a^5) \ldots. \tag{5'}$$

With N electrons, there are $N!$ such permutations.* Any state associated with the quantum numbers $a^1 a^2 a^3 \ldots$ is thus of the form

$$\chi(A) = \sum_P C_P\, P\, \phi(A), \tag{6}$$

where the C_P's are an arbitrary set of coefficients.

The next problem is that of determining the particular values of C_P which correspond to the states of the atom as observed in nature. Before considering this question we may note that unless a number of the C_P's vanish, the occurrence of the summation in (6) makes it impossible to speak of a definite electron, say the first, as having a particular set of quantum numbers. There are N electrons and N sets of quantum numbers, but there is not a definite correlation of electron identities with quantum numbers.

3. The dynamical equivalence of the electrons.

We recognize immediately that all physical observables are symmetric functions of the coordinates of all electrons in the system (in fact, of all electrons in the world). If this were not so, the particular electron which occupied an unsymmetrical position in the form of some observable would be observably distinguishable from the rest, contrary to experiment.

There is, however, another phase to the complete dynamical equivalence of the electrons. If a particular electron occupied an essentially unsymmetrical position in a ψ function representing the state of a system, that electron would behave differently from the rest, even under the action of a symmetrical disturbance. We therefore postulate that a physical ψ function be such that all electrons have exactly the same properties—the same probabilities of being at various places, of having various momenta and spins, etc. The idea of the last sentence is formulated by the requirement that the mean value of any algebraic or differential function f of the electronic coordinates be the same as that of the function Pf, where P represents a permutation of the electrons.

* If two or more sets of quantum numbers are identical, these do not all lead to distinct eigenfunctions; the permutations are considered distinct nevertheless.

The form of an observable has physical significance as such; hence a physical observable must be a really symmetric function of the electrons. A ψ function has physical significance only through the probabilities which may be calculated from it. Hence we do not postulate that ψ be actually symmetric, but only that ψ have sufficient symmetry that

$$\int \bar{\psi} f \psi = \int \bar{\psi} (Pf) \psi \tag{1}$$

for all permutations P and operators f. Applying the inverse permutation P^{-1} to all factors in the integral on the right, which does not change its value, we obtain

$$\int \bar{\psi} f \psi = \int \overline{P^{-1}\psi} f P^{-1}\psi. \tag{2}$$

Now it seems obvious that if $\int \bar{\psi} f \psi = \int \bar{\phi} f \phi$ for all operators f, then $\phi = e^{i\delta}\psi$, where δ is a real constant; for if ϕ and ψ were otherwise related we could find some operator to bring out the distinction.* Hence we require that

$$P\psi = e^{i\delta}\psi, \tag{3}$$

for any permutation P.

Let us write our eigenfunction as $\psi_{ijk\ldots}$, the function which is obtained from this by interchanging the coordinates of electrons i and j as $\psi_{jik\ldots}$, etc. (3) then requires that $\psi_{jik\ldots} = e^{i\delta}\psi_{ijk\ldots}$. By interchanging i and j again we see that $\psi_{ijk\ldots} = e^{i\delta}\psi_{jik\ldots} = e^{2i\delta}\psi_{ijk\ldots}$, and hence that $\psi_{jik\ldots} = \pm\psi_{ijk\ldots}$. In the same way, the interchange of any two electrons in ψ merely multiplies the

* An immediate proof of this may be given for the case in which the coordinates take on only discrete values so that the integral becomes a summation, as for the case of the spin coordinates. A formal generalization of this proof to the case of continuous coordinates may be made by use of the δ function. Denote the coordinates by $x_1, x_2, x_3, \ldots$. Then if $f = \delta(x_1 - a_1)\,\delta(x_2 - a_2)\ldots$, the equation

$$\int \bar{\psi} f \psi = \int \bar{\phi} f \phi$$

becomes

$$\bar{\psi}_{a_1, a_2, \ldots}\,\psi_{a_1, a_2, \ldots} = \bar{\phi}_{a_1, a_2, \ldots}\,\phi_{a_1, a_2, \ldots}$$

or

$$\bar{\psi}\psi = \bar{\phi}\phi \qquad \text{at all points.} \tag{a}$$

If we take $f = \delta(x_1 - a_1)\,\delta(x_2 - a_2)\ldots\partial/\partial x_i$, we find that

$$\bar{\psi}_{a_1, a_2, \ldots} \frac{\partial}{\partial a_i}\psi_{a_1, a_2, \ldots} = \bar{\phi}_{a_1, a_2, \ldots} \frac{\partial}{\partial a_i}\phi_{a_1, a_2, \ldots},$$

or

$$\bar{\psi}\frac{\partial\psi}{\partial x_i} = \bar{\phi}\frac{\partial\phi}{\partial x_i} \qquad \text{at all points.} \tag{b}$$

Dividing (b) by (a) gives

$$\frac{\partial}{\partial x_i}\log\psi = \frac{\partial}{\partial x_i}\log\phi \qquad \text{at all points,}$$

or

$$\log\psi = \log(\phi g),$$

where g is a function of $x_1, x_2, \ldots$ which does not involve x_i (for any i) and which is therefore a constant. Then

$$\psi = \phi g + 2\pi n i,$$

or since ψ is a continuous function which vanishes at infinity, to satisfy (a)

$$\psi = \phi e^{-i\delta}.$$

function by ± 1. Since the functional form of ψ is completely determined by the position of the indices, let the interchange of the first two indices multiply ψ by $c_{12}(=\pm 1)$, the interchange of the second and the third indices by c_{23}, of the first and third by c_{13}, etc. Then

$$\psi_{jik\dots}=c_{12}\psi_{ijk\dots}.$$

But $$\psi_{jik\dots}=c_{23}\psi_{jki\dots}=c_{23}c_{13}\psi_{ikj\dots}=c_{23}c_{13}c_{23}\psi_{ijk\dots}.$$

Hence, since $c_{23}^2=1$, $$c_{12}=c_{13}.$$

In the same way all the c's may be shown to be equal, so that *a ψ function for a physical state must be either completely symmetric or completely antisymmetric in all the electrons*; i.e. the interchange of *any* two electrons must leave the function unchanged in the one case, or multiply it by -1 in the other.*

In the particular case of eigenfunctions $2^6 6$ of the Hamiltonian E, the choice

$$C_P^s=1 \qquad \text{for all } P \tag{4}$$

is seen to give us a symmetric state, while

$$C_P^a=(-1)^p \quad \text{for all } P \tag{5}$$

gives us an antisymmetric state. Here p has the parity of the permutation P.†

The antisymmetric eigenfunction (5) can be written as a determinant

$$\Phi(A)=\mathscr{A}\,\phi(A)=\frac{1}{\sqrt{N!}}\sum_P(-1)^pP\,\phi(A)=\frac{1}{\sqrt{N!}}\begin{vmatrix} u_1(a^1) & u_2(a^1) & \dots & u_N(a^1) \\ u_1(a^2) & u_2(a^2) & \dots & u_N(a^2) \\ \dots & \dots & \dots & \dots \\ u_1(a^N) & u_2(a^N) & \dots & u_N(a^N)\end{vmatrix}; \tag{6}$$

in this form the antisymmetry follows from a known property of determinants.

From any eigenfunction ψ, not necessarily of the form $2^6 4$, we may build a symmetric eigenfunction and an antisymmetric eigenfunction by the use of the operators

$$\left.\begin{aligned}\mathscr{S}&=\frac{1}{\sqrt{N!}}\sum_P P\\ \mathscr{A}&=\frac{1}{\sqrt{N!}}\sum_P(-1)^pP,\end{aligned}\right\} \tag{7}$$

* See PAULI, Handbuch der Physik, **24/1**, 2d ed., p. 191, for a group-theoretical discussion of this restriction.

† Any permutation can be regarded as made up of a number of simple interchanges, i.e. of permutations in which $(N-2)$ indices are left unaltered and the other two simply interchanged. The number of such interchanges for any given permutation is uniquely even or odd, and the permutation is accordingly said to be even or odd. p is an even number for an even permutation, an odd number for an odd permutation.

which we shall call the symmetrizer and the antisymmetrizer, respectively. These operators may be studied by purely operational methods. From the definitions we have the relations

$$\left.\begin{aligned} \mathscr{S}^2 - \sqrt{N!}\,\mathscr{S} &= 0, \\ \mathscr{A}^2 - \sqrt{N!}\,\mathscr{A} &= 0, \\ \mathscr{A}\mathscr{S} = \mathscr{S}\mathscr{A} &= 0. \end{aligned}\right\} \tag{8}$$

From the first two it follows that the eigenvalues of $\mathscr{S}$ and $\mathscr{A}$ are $\sqrt{N!}$ and 0. From the third it follows that if ψ is an eigen-ψ for $\sqrt{N!}$ of $\mathscr{A}$ it belongs to the zero eigenvalue of $\mathscr{S}$, and if an eigen-ψ for $\sqrt{N!}$ of $\mathscr{S}$ it belongs to the zero eigenvalue of $\mathscr{A}$.

Now a symmetric ψ is one for which $P\psi = \psi$ for all P. If $\mathscr{S}\psi = \sqrt{N!}\,\psi$, ψ is seen to be symmetric, since $\mathscr{S}P = P\mathscr{S} = \mathscr{S}$. But $\mathscr{S}(\mathscr{S}\psi) = \sqrt{N!}\,\mathscr{S}\psi$; hence for any ψ, $\mathscr{S}\psi$ is symmetric. That $\mathscr{S}\psi$ is essentially the only symmetric linear combination of the $N!$ $P\psi$'s is shown by the following argument. Suppose that $\Sigma\, C_P P\psi$ is symmetric, i.e. that $P\Sigma\, C_P P\psi = \Sigma\, C_P P\psi$ for all P. Then $\mathscr{S}\Sigma\, C_P P\psi = \sqrt{N!}\,\Sigma\, C_P P\psi$ by (7). But $\mathscr{S}\,\Sigma\, C_P P\psi = (\Sigma\, C_P)\mathscr{S}\psi$. Hence

$$\sum_P C_P P\psi = \frac{\Sigma\, C_P}{\sqrt{N!}}\,\mathscr{S}\psi$$

if $\Sigma\, C_P\, P\psi$ is symmetric.

Similarly, from the fact that $\mathscr{A}P = P\mathscr{A} = (-1)^p\mathscr{A}$ we see that $\mathscr{A}\psi$ is the unique antisymmetric linear combination of the $N!$ $P\psi$'s.

Hence $\mathscr{S}$ produces a symmetric, $\mathscr{A}$ an antisymmetric eigenfunction if it is possible to do so. If it is not possible to do so the functions $\mathscr{S}\psi$ or $\mathscr{A}\psi$ vanish.

The factor $1/\sqrt{N!}$ is placed in $\mathscr{S}$ and $\mathscr{A}$ so that if all $N!$ of the $P\psi$'s are mutually orthogonal, then $\mathscr{S}\psi$ and $\mathscr{A}\psi$ will be normalized if ψ is. Thus the state (6) is normalized provided that the N sets a^i of quantum numbers are all distinct.

Now we may show that if an atom is at any time in an antisymmetric state, it will always remain in an antisymmetric state. For according to $5^2 3$, if an atom has at a certain moment the antisymmetric eigenfunction ψ^a, the rate of change of its eigenfunction with time is given by $H\psi^a$, which itself is antisymmetric. Similarly, if the atom is at a certain moment in a symmetric state it will always remain in a symmetric state. A symmetric observable operating on a symmetric state gives a symmetric state; operating on an antisymmetric state it gives an antisymmetric state.

4. The Pauli exclusion principle.

In the empirical study of atomic spectra which preceded the development of quantum mechanics, Pauli* discovered a simple generalization applying to all atomic spectra. He noticed that if four quantum numbers be assigned to each electron, then states of atoms in which two electrons have all four quantum numbers the same do not occur. This was a rather perplexing discovery at the time it was made because it preceded the hypothesis of electron spin, so that an interpretation of the fourth quantum number was lacking. Although the interpretation of this quantum number in terms of electron spin was soon given, the curious fact remains that the two individual sets of four quantum numbers of each of two electrons cannot be alike. This is known as *Pauli's exclusion principle.*

We see at once that the quantum mechanics provides a natural place for the introduction of this principle. We found in the last section two systems of functions appropriate to equivalent particles such that which ever one actually occurs in nature must be the one that always occurs, since transitions between states of different type are impossible. It is now evident that the antisymmetric system of states satisfies the exclusion principle, since if any two individual sets in $3^6 6$ are identical the determinant vanishes. Nothing of the sort occurs for symmetric states. Hence Pauli's empirical principle is introduced into the theory by the requirement that *the ψ function describing the state of a system must be antisymmetric in all electrons.* This requirement is substantiated in other ways—for example by its equivalence for free particles to the experimentally verified Fermi electron-gas theory of metals. For this reason particles having antisymmetric eigenfunctions are said to obey *Fermi statistics.* Particles having symmetric eigenfunctions are said to obey Bose statistics—photons are particles of this nature.

We shall, as a matter of convenience, reserve capital Greek letters ($\Psi, \Phi, X, \Upsilon, \ldots$) for states satisfying the exclusion principle, i.e. for completely antisymmetric states.

The simplest case in which the Pauli principle makes a restriction is in the lowest energy levels of helium. One would expect that the lowest levels would be given by states in which each electron is in a $1s$ state. This gives four possible sets of quantum numbers, according to the four choices of the values of m_l and m_s,

$$A\colon (0^+0^+);\quad B\colon (0^+0^-);\quad C\colon (0^-0^+);\quad D\colon (0^-0^-);$$

where the sign of m_s is written as a superscript to the value of m_l. When one forms antisymmetric functions $3^6 6$ from these sets one sees that A and D are ruled out by the exclusion principle, while B and C do not give independent

* Pauli, Zeits. für Phys. **31**, 765 (1925).

functions. Thus there is left but one set, say B, and since for it the sum of the values of m_l and of m_s are both zero, it is a term of the type $^1S_0$ (see § 1^7).

It is to be observed that, although we have found a natural place for the Pauli principle in the theory, we have not a theoretical reason for the particular choice of the antisymmetric system. This is one of the unsolved problems of quantum mechanics. Presumably a more fundamental theory of the interaction of two equivalent particles will provide a better understanding of the matter, but such a theory has not been given as yet.

In the zero-order scheme* which we have been considering, certain otherwise acceptable states, namely those in which two electrons have the same individual set of quantum numbers, are prohibited by the exclusion principle on the ground that one cannot find a non-vanishing *antisymmetric* function characterized by those quantum numbers. In the same way in other schemes, in which one has no criterion so simple as the identity of the individual sets, one will find certain otherwise acceptable states prohibited. There is however one important theorem concerning this exclusion which we can prove here, namely:

If $\boldsymbol{J}$ *is a symmetric angular momentum, and if in any scheme the state* $\gamma j m$ *is allowed for one* m*, it is allowed for all* m. (1)

This means that if we can find a non-vanishing antisymmetric function characterized by the quantum numbers $\gamma j m^\circ$, say $\Psi(\gamma j m^\circ)$, then we can find non-vanishing antisymmetric functions $\Psi(\gamma j m)$ for all m. The proof is very simple; consider the equation $3^3 3$:

$$(J_x \pm iJ_y)\Psi(\gamma j m) = \hbar\sqrt{(j\mp m)(j\pm m+1)}\,\Psi(\gamma j m\pm 1).$$

Here J_x+iJ_y is assumed to be a symmetric operator, and $\Psi(\gamma j m)$ an antisymmetric function. Then $\Psi(\gamma j m+1)$ will also be an antisymmetric function which will not vanish unless J_x+iJ_y has a zero eigenvalue for the state $\gamma j m$. But J_x+iJ_y is easily shown not to have a zero eigenvalue unless $m=j$, for if J_x+iJ_y has a zero eigenvalue, so also has $(J_x-iJ_y)(J_x+iJ_y)$. But

$$(J_x-iJ_y)(J_x+iJ_y) = \boldsymbol{J}^2 - J_z^2 - J_z\hbar$$

has the eigenvalue $[j(j+1)-m(m+1)]\hbar^2$, which is zero only if $m=j$ (or $-j-1$). A similar consideration holds for the operator J_x-iJ_y, which steps the m value down.

* We shall call any scheme in which the state of a system is specified by giving a complete set of one-electron quantum numbers as in $3^6 6$ a *zero-order scheme*, since such states are eigenstates only of the central-field problem, but furnish a useful starting point for perturbation calculations.

5. Conventions concerning quantum numbers. Closed shells.

In the central-field approximation the state of an N-electron atom is specified by a *complete set* of quantum numbers, which consists of a list of N *individual sets* of quantum numbers. Each individual set contains four quantum numbers of the one-electron problem, usually the quantum numbers $n\,l\,m_s\,m_l$ or $n\,l\,j\,m$. Now the Pauli principle requires that all individual sets be distinct, and it is this requirement which gives rise to the occurrence of closed shells of electrons. In a complete set there can be only two individual sets with the same n, l, and m_l, since m_s is restricted to the two values $\pm\frac{1}{2}$. For a given nl there can be but $2l+1$ different values of m_l, so that altogether only $2(2l+1)$ individual sets may have the same nl.* When a complete set contains the maximum number of individual sets of a given nl, we say it contains a *closed shell* or *complete shell* of that type. It is to be noticed that for a closed shell a negative value of m_l and m_s occurs for every positive value so that the sums of the m_l and of the m_s values are zero.

We have seen that the antisymmetry of the eigenfunction makes it impossible to speak of a definite electron as having a particular individual set of quantum numbers. However, before the theory was fully developed, spectroscopists formed the habit of speaking of a state of the atom as involving, say, two $1s$ electrons and one $2p$ electron; and there is no harm in continuing this convenient mode of expression if it is understood that in such a statement the term 'nl electron' refers to the occurrence, in the complete set, of an individual set having quantum numbers nl. In the same way we may say that an atom has two electrons in the $1s$ shell and one electron in the $2p$ shell. Here, as was done early in spectroscopic theory, we let the nl values label the different *shells*, complete or incomplete.

Since the energy in the one-electron problem depends only on nl, the energy in an unperturbed state of the central-field approximation depends only on the distribution of the electrons among the shells, i.e. on the list of nl values of the individual sets. This list of nl values is said to specify the *electron configuration*. Then the first-order perturbation theory will need to consider only that part of the perturbation matrix which joins states belonging to the same configuration, and *we may to the first order treat the energy-level problem configuration by configuration*. In specifying a configuration of an atom a notation of the following type is used:

$$\text{Ti II} \quad 1s^2\,2s^2\,2p^6\,3s^2\,3p^6\,3d^2\,4p, \tag{1}$$

which indicates an atom (really the ion Ti^+) with $Z=21$, $N=20$, closed $1s$, $2s$, $2p$, $3s$, and $3p$ shells, two electrons in the $3d$ shell, and one in the $4p$.

* If we calculate this using the jm scheme: j may be $l+\frac{1}{2}$ or $l-\frac{1}{2}$, the first of which has $2l+2$ values of m, the second $2l$—total $4l+2$ sets as above.

Now the order of listing of the individual sets in a complete set does not affect the state in an essential way, but two different orders may lead to eigenfunctions ($3^6 6$) with opposite signs, i.e. with different choices of the arbitrary phase. For this reason we must for definiteness list the quantum numbers in a specified way. In the first place, the sets will be listed shell by shell, all the sets of a given shell being placed together. The order of listing the shells will in each case be specified by the order of nl values in the configuration designation of the type (1). Within a shell, we shall for convenience adopt a standard order of listing the quantum numbers. In the $nlm_s m_l$ scheme, the sets will be listed in decreasing order of m_l values, the set with $m_s = +\frac{1}{2}$ being placed before that with $m_s = -\frac{1}{2}$, in case both have the same m_l. In the $nljm$ scheme, the sets will be listed in decreasing order of m values, first all sets with $j = l + \frac{1}{2}$, then those with $j = l - \frac{1}{2}$. Having agreed on this standard order, the sets for a closed shell need not be listed explicitly, and only the $m_s m_l$ or jm values need be listed for the incomplete shells. In the $m_s m_l$ scheme we shall use a notation of the type

$$\text{Ne I} \quad 1s^2\, 2s^2\, 2p^5\, 3d\, (1^+\, 1^-\, 0^-\, -1^+\, -1^-\, 2^-), \tag{2}$$

where the numbers specify the m_l values and the superscripts the m_s values first of the five $2p$ electrons, then of the $3d$.

6. Matrix components for $\Sigma_i f(i)$.*

We need to know the matrix components, in terms of eigenfunctions based on the central-field approximation, of certain symmetric functions of the coordinates of the N electrons. The symmetric functions which actually occur in atomic theory are fortunately of two very simple types. The first is that in which there exists a function of the coordinates of a single electron which is written down once with each electron's coordinates taken in turn as the argument and the results added. Let F be such a quantity. Then we may write

$$F = \sum_{i=1}^{N} f(i), \tag{1}$$

where $f(i)$ operates only on the coordinates of the ith electron. It is easy to reduce the matrix components of this type of quantity to those for the motion of a single electron in the central field.

The matrix component $(A|F|B)$ connecting states of the antisymmetric type whose eigenfunctions are given by $3^6 6$ is given by

$$(A|F|B) = \int \bar{\Phi}(A)\, F\, \Phi(B) = \frac{1}{N!} \sum_i \sum_P \sum_{P'} (-1)^{p+p'} \int P\, \bar{\phi}(A) f(i)\, P'\, \phi(B), \tag{2}$$

where the integral means integration and summation over the $3N$ positional and the N spin coordinates respectively, as in § 5^3.

* CONDON, Phys. Rev. **36**, 1121 (1930). These matrix elements follow also from the considerations of JORDAN and WIGNER, Zeits. für Phys. **47**, 631 (1928) on second quantization.

When we are considering only antisymmetric states, since all the individual sets of quantum numbers are distinct, it is immaterial whether we regard P as a permutation of electron indices relative to quantum numbers, or of quantum numbers relative to electron indices. The latter point of view will usually be the most convenient. We may think of P as an operator acting on the quantum number of each electron and turning it into a different quantum number. We can thus denote the quantum number of the ith electron in $P\phi(A)$ by Pa^i.

The summand in (2),

$$\int P\,\bar{\phi}(A) f(i)\, P'\,\phi(B) = \int_1 \bar{u}_1(Pa^1)\, u_1(P'b^1) \cdot \int_2 \bar{u}_2(Pa^2)\, u_2(P'b^2)$$

$$\ldots \int_i \bar{u}_i(Pa^i) f(i)\, u_i(P'b^i) \ldots \int_N \bar{u}_N(Pa^N)\, u_N(P'b^N), \quad (3)$$

contains integrals simply of products of eigenfunctions of single electrons in every place except the ith. Therefore it vanishes unless $(N-1)$ of the individual sets in A agree with those in B, since the non-vanishing of the summand requires that

$$P'b^1 = Pa^1, \quad P'b^2 = Pa^2, \quad \ldots, \quad P'b^N = Pa^N, \quad (4)$$

except that $P'b^i$ need not equal Pa^i. It follows therefore that *all matrix components* $(A|F|B)$ *of a quantity of the type* F *vanish if* A *and* B *differ in regard to more than one individual set of quantum numbers.*

Continuing the calculation, we may suppose that A is the same as B except for one individual set, which is the set a^k of A when arranged in conventional order. Denote by B' the set obtained when the individual sets of B are arranged in the order

$$B': \quad a^1 a^2 \ldots a^{k-1} b^k a^{k+1} \ldots a^N.$$

This will not in general be the conventional order for the set B but we can pass to the set B by an $\left\{\begin{matrix}\text{even}\\ \text{odd}\end{matrix}\right\}$ number of interchanges. We then have

$$\Phi(B) = \{\pm\}\Phi(B'), \quad \text{and hence} \quad (A|F|B) = \{\pm\}(A|F|B'). \quad (5)$$

We may calculate $(A|F|B')$ quite simply. In order that the summand corresponding to (3) should not vanish, P' must be the same permutation as P, and must be such as to put the unmatched individual quantum numbers with the ith electron's coordinates. Since P' is the same as P the sign of each term in (2) will be positive. For a given i there are $(N-1)!$ P's satisfying this condition, so we obtain

$$(N-1)! \int \bar{u}_i(a^k) f(i)\, u_i(b^k)$$

for the sum over P and P' in (2). The same argument applies for each value of i from 1 to N and gives the same result, since a definite integral does not depend on the name of the integrated argument. The summation over i thus multiplies the above result by N, giving a total of $N!$, which cancels the $1/N!$ of (2). Hence we obtain finally

$$(A|F|B) = \pm \int \bar{u}_i(a^k) f(i)\, u_i(b^k). \tag{6}$$

In the case of the diagonal element $(A|F|A)$, the argument is similar. Here P' must equal P or the summand corresponding to (3) vanishes. Hence

$$(A|F|A) = \frac{1}{N!} \sum_i \sum_P \int \bar{u}_i(Pa^i) f(i)\, u_i(Pa^i).$$

Now of the permutations P there will be $(N-1)!$ for which $Pa^i = a^j$ $(j=1, \ldots, N)$. Hence on summing over P we obtain

$$(A|F|A) = \frac{1}{N} \sum_i \sum_j \int_i \bar{u}_i(a^j) f(i)\, u_i(a^j) = \sum_{j=1}^{N} \int \bar{u}_1(a^j) f(1)\, u_1(a^j), \tag{7}$$

since the value of the integral is independent of i.

To summarize—*the matrix component $(A|F|B)$ of the quantity*

$$F = \sum_{i=1}^{N} f(i),$$

(*a*) *vanishes if B differs from A by more than one individual set;*

(*b*) *has the value*

$$(A|F|B) = \pm (a^k|f|b^k) \tag{8}$$

if all the sets in A agree with all those in B except that $a^k \neq b^k$. The sign is positive or negative according to the parity of the permutation which changes the conventional order of B into one in which sets in B which match those in A all stand at the same places in the lists;

(*c*) *and if $B = A$, the diagonal element is*

$$(A|F|A) = \sum_{i=1}^{N} (a^i|f|a^i). \tag{9}$$

This completes the reduction of matrix elements of F in the N-electron problem to those of f in the one-electron problem. From the form of the results it is evident that if f is a diagonal matrix in the one-electron problem, then F is diagonal in the N-electron problem. Thus the sums of the z components of spin and of orbital angular momentum are each represented by diagonal matrices in the N-electron problem if the representation is based on the $nlm_s m_l$ scheme.

7. Matrix components for $\Sigma_{i,j}\, g(i,j)$.

The second type of symmetric function which occurs in atomic theory is that in which one has a symmetric function of the coordinates of two electrons, $g(i,j) \equiv g(j,i)$ which is summed over all pairs of electrons. We suppose that $i \neq j$, for terms of the type $\Sigma_i g(i, i)$, if present, constitute a quantity of the type F whose matrix components have already been calculated. It will now be shown how the reduction from the N- to the 2-electron problem can be made in this case.

Corresponding to $6^6 1$, a quantity G is defined by

$$G = \sum_{i>j=1}^{N} g(i,j). \tag{1}$$

The most important quantity of this type is the Coulomb-repulsion term which occurs in the Hamiltonian $1^6 1$, in which $g(i,j) = e^2/r_{ij}$. The matrix component of G joining the states A and B becomes

$$(A|G|B) = \frac{1}{N!} \sum_{i>j} \sum_{PP'} (-1)^{p+p'} \int_i \int_j \bar{u}_i(Pa^i)\, \bar{u}_j(Pa^j)\, g(i,j)\, u_i(P'b^i)\, u_j(P'b^j)$$
$$. \delta(Pa^1, P'b^1)\, \delta(Pa^2, P'b^2) \ldots \delta(Pa^N, P'b^N), \tag{2}$$

where the continued product of delta functions contains a term for each index except i and j. In order that there be some non-vanishing terms in the expression one sees that B can differ from A by at most *two* of the individual sets. We shall have three cases, those in which B differs from A by two sets, by one, and by none.

Considering first *the case in which A and B differ by two individual sets*, let us deal with the eigenstate B' in which $(N-2)$ of the sets are matched with those in A with the latter in conventional order:

$$\begin{array}{llllllll} A: & a^1 & a^2 & \ldots & a^k & \ldots & a^t & \ldots\ a^N \\ B': & a^1 & a^2 & \ldots & b^k & \ldots & b^t & \ldots\ a^N. \end{array} \tag{3}$$

Then B' is generally not in its conventional order, so we shall have to multiply by plus or minus one according to the parity of the permutation from the conventional order for B to this order to obtain the correct sign for the matrix component.

Consider now the expression (2) for a definite i and j. Since we deal with a double integral we see that for each P there are two P''s which contribute non-vanishing terms to the summand. We must have P such that $Pa^i = a^k$ or a^t and $Pa^j = a^t$ or a^k. Then P' must agree with P with regard to the $(N-2)$ electron indices other than i and j. There are $(N-2)!$ of the P's for which $Pa^i = a^k$ and $Pa^j = a^t$. For each of these we get two terms: first we may have $P'b^i = b^k$, $P'b^j = b^t$ and P' otherwise the same as P; then $P' \equiv P$, hence this term is positive—second, we can have $P'b^i = b^t$ and $P'b^j = b^k$ but P' otherwise the same as P; here P' and P differ by one interchange, so this is a negative term. Similarly there are $(N-2)!$ of the P's for which $Pa^i = a^t$ and $Pa^j = a^k$. For each of these we can have $P' \equiv P$, giving a positive term, and P' differing from P by one interchange, giving a negative term. Altogether then the summation over P and P' gives

$$\frac{1}{N(N-1)}\sum_{i>j}\Big[\iint \bar{u}_i(a^k)\,\bar{u}_j(a^t)\,g(i,j)\,u_i(b^k)\,u_j(b^t) - \iint \bar{u}_i(a^k)\,\bar{u}_j(a^t)\,g(i,j)\,u_i(b^t)\,u_j(b^k) + \iint \bar{u}_i(a^t)\,\bar{u}_j(a^k)\,g(i,j)\,u_i(b^t)\,u_j(b^k) - \iint \bar{u}_i(a^t)\,\bar{u}_j(a^k)\,g(i,j)\,u_i(b^k)\,u_j(b^t)\Big].$$

Since these terms are definite integrals, the first and third are equal and the second and fourth are equal. Moreover, the sum over i, j merely multiplies by the number of terms in the sum, which is $N(N-1)/2$, so altogether we have

$$(A|G|B) = \pm\Big[\iint \bar{u}_i(a^k)\,\bar{u}_j(a^t)\,g(i,j)\,u_i(b^k)\,u_j(b^t) - \iint \bar{u}_i(a^k)\,\bar{u}_j(a^t)\,g(i,j)\,u_i(b^t)\,u_j(b^k)\Big], \tag{4}$$

the $\pm$ sign being determined by the parity of the permutation necessary to change the conventional order of B into that of (3).

Next *if B differs from A by but one set,* we can take a permutation on the conventional order in B which will match up corresponding individual sets with the conventional order in A,

$$\begin{aligned} A&:\quad a^1\ a^2\ \ldots\ a^k\ \ldots\ a^N, \\ B'&:\quad a^1\ a^2\ \ldots\ b^k\ \ldots\ a^N. \end{aligned} \tag{5}$$

By arguments similar to those already used, it is then easy to show that

$$(A|G|B) = \pm\sum_{a^t}\Big[\iint \bar{u}_i(a^k)\,\bar{u}_j(a^t)\,g(i,j)\,u_i(b^k)\,u_j(a^t) - \iint \bar{u}_i(a^k)\,\bar{u}_j(a^t)\,g(i,j)\,u_i(a^t)\,u_j(b^k)\Big], \tag{6}$$

in which a^t runs over the $(N-1)$ individual sets that are common to A and B.

Finally, *the diagonal element* may be shown by similar arguments to be given by

$$(A|G|A) = \sum_{k>t=1}^{N}\Big[\iint \bar{u}_i(a^k)\,\bar{u}_j(a^t)\,g(i,j)\,u_i(a^k)\,u_j(a^t) - \iint \bar{u}_i(a^k)\,\bar{u}_j(a^t)\,g(i,j)\,u_i(a^t)\,u_j(a^k)\Big]. \tag{7}$$

In (7) we shall call the integrals with positive sign *direct integrals* and those with negative sign *exchange integrals*.

If $g(i,j)$ is independent of spin, then in the nlm_sm_l scheme the sum over the spin coordinates σ_i, σ_j implied in the $\iint$ sign in (7) may be carried out at once. This gives a factor unity on the direct integral and a factor $\delta(m_s^k, m_s^t)$ on the exchange integral, so we have exchange integrals only for electrons of like spins. Inspection of (4) and (6) shows that if $g(i,j)$ is independent of spin there are no matrix components connecting states A and B if these differ in regard to their Σm_s. This is consistent with the fact that in this case G commutes with S_z, the z component of resultant spin.

8. Matrix components of electrostatic interaction.

The most important quantity of the type considered in the preceding section is the Coulomb-repulsion term

$$Q = \sum_{i>j=1}^{N} q(i,j) = \sum_{i>j=1}^{N} e^2/r_{ij}$$

which occurs in the Hamiltonian of the many-electron atom. We wish now to carry out a computation of the integrals involved in this problem in the $nlm_s m_l$ scheme. The most general integral occurring is one in which four different individual sets, a, b, c, d, are involved. Hence we shall have to calculate

$$(a\,b|q|c\,d) = \int\int \bar{u}_1(a)\,\bar{u}_2(b)\frac{e^2}{r_{12}}u_1(c)\,u_2(d). \tag{1}$$

The single-electron eigenfunctions are given by $4^5 2$, which in the present notation becomes

$$u_1(a) = \frac{1}{r_1}R_1(n^a l^a)\,\Theta_1(l^a m_l^a)\,\Phi_1(m_l^a)\,\delta(\sigma_1, m_s^a).$$

Thus (1) can be factored into the product of a sextuple integral over the six positional coordinates and a double sum over the two spin coordinates—

$$\sum_{\sigma_1\sigma_2}\delta(\sigma_1, m_s^a)\,\delta(\sigma_1, m_s^c)\,\delta(\sigma_2, m_s^b)\,\delta(\sigma_2, m_s^d) = \delta(m_s^a, m_s^c)\,\delta(m_s^b, m_s^d) \tag{2}$$

—so the spin components of a and c, and of b and d, must be the same or the integral vanishes.

For further progress we make use of the relation

$$\frac{1}{r_{12}} = \frac{1}{\sqrt{r_1^2 + r_2^2 - 2r_1 r_2 \cos\omega}},$$

where ω is the angle between the radii vectores of electrons 1 and 2 from the origin. This can be developed in a series of Legendre polynomials

$$\frac{1}{r_{12}} = \sum_{k=0}^{\infty}\frac{r_<^k}{r_>^{k+1}}P_k(\cos\omega) \tag{3}$$

in which $r_<$ is the lesser and $r_>$ the greater of r_1 and r_2. Therefore the part of (1) independent of spins can be written

$$e^2\sum_{k=0}^{\infty}\left\{\int_0^\infty\int_0^\infty \frac{r_<^k}{r_>^{k+1}} R_1(n^a l^a)\,R_2(n^b l^b)\,R_1(n^c l^c)\,R_2(n^d l^d)\,dr_1 dr_2\right.$$
$$\cdot\int_0^\pi\int_0^{2\pi}\int_0^\pi\int_0^{2\pi} P_k(\cos\omega)\,\Theta_1(l^a m_l^a)\,\Theta_2(l^b m_l^b)\,\Theta_1(l^c m_l^c)\,\Theta_2(l^d m_l^d)$$
$$\left.\cdot\,\bar{\Phi}_1(m_l^a)\,\bar{\Phi}_2(m_l^b)\,\Phi_1(m_l^c)\,\Phi_2(m_l^d)\sin\theta_1\sin\theta_2\,d\theta_1 d\varphi_1 d\theta_2 d\varphi_2\right\}. \tag{4}$$

The second factor, depending on the spherical harmonics, can be further expanded by making use of the addition theorem ($4^3 22$). This expansion permits us to write the angle factor occurring in the kth term of (4) as

$$\frac{4\pi}{2k+1}\sum_{m=-k}^{k}\int_0^\pi \Theta_1(k\,m)\,\Theta_1(l^a\,m_l^a)\,\Theta_1(l^c\,m_l^c)\sin\theta_1\,d\theta_1\int_0^{2\pi}\Phi_1(m)\,\bar{\Phi}_1(m_l^a)\,\Phi_1(m_l^c)\,d\varphi_1$$

$$\cdot\int_0^\pi \Theta_2(k\,m)\,\Theta_2(l^b\,m_l^b)\,\Theta_2(l^d\,m_l^d)\sin\theta_2\,d\theta_2\int_0^{2\pi}\Phi_2(m)\,\bar{\Phi}_2(m_l^b)\,\Phi_2(m_l^d)\,d\varphi_2.$$

The φ integrals can be evaluated at once. They give

$$\frac{1}{2\pi}\,\delta(m, m_l^a - m_l^c)\,\delta(m, m_l^d - m_l^b).$$

Hence in the summation over m everything vanishes unless

$$m_l^a + m_l^b = m_l^c + m_l^d;$$

that is, there are no matrix components connecting states which differ in the value of the z component of the total orbital angular momentum. This is consistent with the commutation of Q with L_z. For states which do not differ in this value, only one term of the sum over m remains, namely that for

$$m = m_l^a - m_l^c = m_l^d - m_l^b. \tag{5}$$

When this condition is satisfied, the angle factor reduces to

$$c^k(l^a\,m_l^a, l^c\,m_l^c)\,c^k(l^d\,m_l^d, l^b\,m_l^b),$$

where c is defined by the equation

$$c^k(l\,m_l, l'\,m_l') = \sqrt{\frac{2}{2k+1}}\int_0^\pi \Theta(k\,m_l - m_l')\,\Theta(l\,m_l)\,\Theta(l'\,m_l')\sin\theta\,d\theta. \tag{6}$$

This expression is not symmetric in $l\,m_l$ and $l'\,m_l'$, but from $4^3 18$,

$$c^k(l\,m_l, l'\,m_l') = (-1)^{m_l - m_l'}\,c^k(l'\,m_l', l\,m_l). \tag{7}$$

If we now introduce the abbreviation

$$R^k(n^a l^a\,n^b l^b, n^c l^c\,n^d l^d)$$
$$= e^2\int_0^\infty\int_0^\infty \frac{r_<^k}{r_>^{k+1}}\,R_1(n^a l^a)\,R_2(n^b l^b)\,R_1(n^c l^c)\,R_2(n^d l^d)\,dr_1 dr_2, \tag{8}$$

we obtain as the final value of our matrix component

$$(a\,b|q|c\,d) = \delta(m_s^a, m_s^c)\,\delta(m_s^b, m_s^d)\,\delta(m_l^a + m_l^b, m_l^c + m_l^d)$$
$$\cdot\sum_{k=|m|}^{\infty} c^k(l^a\,m_l^a, l^c\,m_l^c)\,c^k(l^d\,m_l^d, l^b\,m_l^b)\,R^k(n^a l^a\,n^b l^b, n^c l^c\,n^d l^d). \tag{9}$$

The range of k may equally well be written from 0 to ∞, since the c^k vanish identically for $k < |m|$.

The values of the R^k here depend on our initial choice of central field and hence must be calculated independently for each different choice. But the

c's are essentially definite integrals of three associated Legendre polynomials and may be calculated once for all. Gaunt* has given a general formula for the integral of the product of the three functions $\Theta(l\,m)$, $\Theta(l'\,m')$, $\Theta(l''\,m+m')$, for $m \geqslant 0$, $m' \geqslant 0$. Our integral is seen to be reducible to this form by means of $4^3 18$. In the first place c^k vanishes unless k, l, and l' satisfy the so-called *triangular condition*, which requires that k, l, and l' equal the sides of a triangle of even perimeter. Expressed analytically, in order that c^k have a value different from zero, k must satisfy the conditions

$$\begin{aligned} &k+l+l' = 2g \qquad (g \text{ integral}) \\ &|l-l'| \leqslant k \leqslant l+l'. \end{aligned} \tag{10}$$

These conditions limit the range of k to a few values in any practical case and serve to limit the seemingly infinite series occurring in (9) to the sum of a very few terms.

Gaunt's formula gives, for $m \geqslant 0$, $m' \geqslant 0$, $2g = l+l'+l''$,

$$\int_0^\pi \Theta(l\,m)\,\Theta(l'\,m')\,\Theta(l''\,m+m') \sin\theta\, d\theta = \frac{(-1)^{g-l-m'}(2g-2l')!\,g!}{(g-l)!\,(g-l')!\,(g-l'')!\,(2g+1)!}$$
$$\cdot \sqrt{\frac{(2l+1)(2l'+1)(2l''+1)\,(l''-m-m')!\,(l+m)!\,(l'+m')!\,(l'-m')!}{2\,(l''+m+m')!\,(l-m)!}}$$
$$\cdot \sum_t (-1)^t \frac{(l''+m+m'+t)!\,(l+l'-m-m'-t)!}{(l''-m-m'-t)!\,(l-l'+m+m'+t)!\,(l'-m'-t)!\,t!}, \tag{11}$$

where, in the summation, t takes on all integral values consistent with the factorial notation, the factorial of a negative number being meaningless. This formula cannot be summed except in certain special cases. Because of its cumbersome nature, a table of values is needed for actual computations. Table 1^6 gives the values of the c's for s, p, d, and f electrons.

The particular matrix elements which occur in the calculation of the diagonal element of Q recur so frequently that it is convenient to introduce the special notation:

$$\left.\begin{aligned} (a\,b|q|a\,b) &= J(a,b) \\ (a\,b|q|b\,a) &= K(a,b). \end{aligned}\right\} \tag{12}$$

According to (9), we may write

$$J(a,b) = \sum_{k=0}^{\infty} a^k(l^a\,m_l^a, l^b\,m_l^b)\,F^k(n^a l^a, n^b l^b)$$

and

$$K(a,b) = \delta(m_s^a, m_s^b) \sum_{k=0}^{\infty} b^k(l^a\,m_l^a, l^b\,m_l^b)\,G^k(n^a l^a, n^b l^b). \tag{13}$$

Here a^k and b^k are defined in terms of our previous c^k by

$$\left.\begin{aligned} a^k(l^a\,m_l^a, l^b\,m_l^b) &= c^k(l^a\,m_l^a, l^a\,m_l^a)\,c^k(l^b\,m_l^b, l^b\,m_l^b) \\ b^k(l^a\,m_l^a, l^b\,m_l^b) &= [c^k(l^a\,m_l^a, l^b\,m_l^b)]^2, \end{aligned}\right\} \tag{14}$$

* Gaunt, Trans. Roy. Soc. **A228**, 151 (1929)—in particular, equation (9), p. 194.

while F^k and G^k are special cases of the R^k of (8), namely

$$F^k(n^a l^a, n^b l^b) = R^k(n^a l^a\, n^b l^b, n^a l^a\, n^b l^b)$$

$$= e^2 \int_0^\infty \int_0^\infty \frac{r_<^k}{r_>^{k+1}} R_1^2(n^a l^a)\, R_2^2(n^b l^b)\, dr_1 dr_2, \quad \text{(15a)}$$

$$G^k(n^a l^a, n^b l^b) = R^k(n^a l^a\, n^b l^b, n^b l^b\, n^a l^a)$$

$$= e^2 \int_0^\infty \int_0^\infty \frac{r_<^k}{r_>^{k+1}} R_1(n^a l^a)\, R_1(n^b l^b)\, R_2(n^a l^a)\, R_2(n^b l^b)\, dr_1 dr_2$$

$$= 2e^2 \int_0^\infty dr_2 \int_0^{r_2} \frac{r_1^k}{r_2^{k+1}} R_1(n^a l^a)\, R_1(n^b l^b)\, R_2(n^a l^a)\, R_2(n^b l^b)\, dr_1. \quad \text{(15b)}$$

Tables 1[6] and 2[6] give the values of b^k and a^k for s, p, d, and f electrons. To avoid the occurrence of fractional coefficients for the F's and G's in the evaluation of J and K matrix components, we define

$$F_k = F^k/D_k \quad \text{and} \quad G_k = G^k/D_k,$$

where D_k is the denominator occurring in Tables 1[6] and 2[6]. The formulas of the succeeding chapters will usually be expressed in terms of F_k and G_k.

It is seen that the R's which are obtained in taking a matrix element between two states of the same configuration will always reduce to these F's and G's. For a pair of equivalent electrons, F's and G's of the same k are equal:

$$F^k(n^a l^a, n^a l^a) = G^k(n^a l^a, n^a l^a).$$

We also note, from its definition, that F^k is essentially positive and a decreasing function of k. Although from its definition we cannot make such a statement, we shall find G^k to share these properties when we calculate its values in special cases.

We shall use later the fact that

$$I(n^a\, l^a\, m_s^a\, m_l^a, n^b\, l^b\, m_s^b\, m_l^b) = I(n^a\, l^a\, -m_s^a\, -m_l^a, n^b\, l^b\, -m_s^b\, -m_l^b), \quad \text{(16)}$$

where I equals J or K.

9. Specialization for closed shells.

If we are dealing with configurations involving closed shells of electrons, certain simplifications in the matrix components follow which are of importance for atomic theory. This case will now be investigated.

First, let us consider a diagonal element for a quantity of the type F. In 6[6]9 the summation extends over all the individual sets of quantum numbers. Consider separately the sum over the $2(2l+1)$ individual sets associated with the nl shell. This has the value

$$\sum_{m_s=-\frac{1}{2}}^{+\frac{1}{2}} \sum_{m_l=-l}^{+l} \sum_\sigma \int_0^\infty \int_0^\pi \int_0^{2\pi} R(nl)\, \Theta(l\, m_l)\, \bar{\Phi}(m_l)\, \delta(\sigma, m_s) f\, R(nl)$$
$$\cdot\, \Theta(l\, m_l)\, \Phi(m_l)\, \delta(\sigma, m_s) \sin\theta\, dr\, d\theta\, d\varphi.$$

If f is a function of the coordinates $r\,\theta\,\varphi\,\sigma$ only and not an operator, this becomes

$$\int_0^\infty \int_0^\pi \int_0^{2\pi} [f(r\,\theta\,\varphi+\tfrac{1}{2})+f(r\,\theta\,\varphi-\tfrac{1}{2})]\,R^2(nl)\left(\sum_{m_l}\frac{1}{2\pi}\,\Theta^2(l\,m_l)\right)\sin\theta\,dr\,d\theta\,d\varphi.$$

Now the sum in parentheses has the value $(2l+1)/4\pi$; this follows from the addition theorem $4^3 22$ by writing $\theta_1=\theta_2$, $\varphi_1=\varphi_2$, $\omega=0$, and $P_l(1)=1$. Hence our final result is

$$2\,(2l+1)\int_0^\infty\left(\frac{1}{4\pi}\int_0^\pi\int_0^{2\pi}\tfrac{1}{2}\,[f(r\,\theta\,\varphi+\tfrac{1}{2})+f(r\,\theta\,\varphi-\tfrac{1}{2})]\sin\theta\,d\theta\,d\varphi\right)R^2(nl)\,dr, \quad (1)$$

that is, the matrix component of the function f averaged for all directions in space, calculated as if the eigenfunctions of the states in the nl shell were

TABLE 1^6. $c^k(lm_l,\,l'm_l')$ *and* $b^k(lm_l,\,l'm_l')$.

We write $c_k=\pm\sqrt{x/D_k}$, where D_k depends only on l and l'. In the table are listed only the sign preceding the radical and the value of x, D_k being given at the head of each column. Since $b^k=(c^k)^2$, $b^k=+x/D_k$. Note that $c^k(l'm_l',\,lm_l)=(-1)^{m_l-m_l'}c^k(lm_l,\,l'm_l')$.

$c^k(lm_l,\,l'm_l')$ *for* $l+l'$ *odd*

$l\,l'$	m_l	m_l'	$k=$ 1	3	5
$s\,p$	0	±1	$-\sqrt{1/3}$		
	0	0	$+\ 1$		
$s\,f$	0	±3		$-\sqrt{1/7}$	
	0	±2		$+\ 1$	
	0	±1		$-\ 1$	
	0	0		$+\ 1$	
$p\,d$	±1	±2	$-\sqrt{6/15}$	$+\sqrt{3/245}$	
	±1	±1	$+\ 3$	$-\ 9$	
	±1	0	$-\ 1$	$+\ 18$	
	0	±2	0	$+\ 15$	
	0	±1	$-\ 3$	$-\ 24$	
	0	0	$+\ 4$	$+\ 27$	
	±1	∓2	0	$+\ 45$	
	±1	∓1	0	$-\ 30$	
$d\,f$	±2	±3	$-\sqrt{15/35}$	$+\sqrt{10/315}$	$-\sqrt{1/1524{\cdot}6}$
	±2	±2	$+\ 5$	$-\ 20$	$+\ 5$
	±2	±1	$-\ 1$	$+\ 24$	$-\ 15$
	±2	0	0	$-\ 20$	$+\ 35$
	±1	±3	0	$+\ 25$	$-\ 7$
	±1	±2	$-\ 10$	$-\ 15$	$+\ 24$
	±1	±1	$+\ 8$	$+\ 2$	$-\ 50$
	±1	0	$-\ 3$	$+\ 2$	$+\ 80$
	0	±3	0	$+\ 25$	$-\ 28$
	0	±2	0	0	$+\ 63$
	0	±1	$-\ 6$	$-\ 9$	$-\ 90$
	0	0	$+\ 9$	$+\ 16$	$+\ 100$
	±2	∓3	0	0	$-\ 210$
	±2	∓2	0	0	$+\ 126$
	±2	∓1	0	$+\ 10$	$-\ 70$
	±1	∓3	0	0	$-\ 84$
	±1	∓2	0	$+\ 25$	$+\ 112$
	±1	∓1	0	$-\ 15$	$-\ 105$

TABLE 1[6]—*continued.*

$c^k(lm_l, l'm'_l)$ *for* $l+l'$ *even*

ll'	m_l	m'_l	$k=$ 0	2	4	6
ss	0	0	+1			
sd	0	±2		$+\sqrt{1/5}$		
	0	±1		− 1		
	0	0		+ 1		
pp	±1	±1	+1	$-\sqrt{1/25}$		
	±1	0	0	+ 3		
	0	0	+1	+ 4		
	±1	∓1	0	− 6		
pf	±1	±3		$+\sqrt{45/175}$	$-\sqrt{1/189}$	
	±1	±2		− 30	+ 3	
	±1	±1		+ 18	− 6	
	±1	0		− 9	+ 10	
	0	±3		0	− 7	
	0	±2		+ 15	+ 12	
	0	±1		− 24	− 15	
	0	0		+ 27	+ 16	
	±1	∓3		0	− 28	
	±1	∓2		0	+ 21	
	±1	∓1		+ 3	− 15	
dd	±2	±2	+1	$-\sqrt{4/49}$	$+\sqrt{1/441}$	
	±2	±1	0	+ 6	− 5	
	±2	0	0	− 4	+ 15	
	±1	±1	+1	+ 1	− 16	
	±1	0	0	+ 1	+ 30	
	0	0	+1	+ 4	+ 36	
	±2	∓2	0	0	+ 70	
	±2	∓1	0	0	− 35	
	±1	∓1	0	− 6	− 40	
ff	±3	±3	+1	$-\sqrt{25/225}$	$+\sqrt{9/1089}$	$-\sqrt{1/7361 \cdot 64}$
	±3	±2	0	+ 25	− 30	+ 7
	±3	±1	0	− 10	+ 54	− 28
	±3	0	0	0	− 63	+ 84
	±2	±2	+1	0	− 49	+ 36
	±2	±1	0	+ 15	+ 32	− 105
	±2	0	0	− 20	− 3	+ 224
	±1	±1	+1	+ 9	+ 1	− 225
	±1	0	0	+ 2	+ 15	+ 350
	0	0	+1	+ 16	+ 36	+ 400
	±3	∓3	0	0	0	− 924
	±3	∓2	0	0	0	+ 462
	±3	∓1	0	0	+ 42	− 210
	±2	∓2	0	0	+ 70	+ 504
	±2	∓1	0	0	− 14	− 378
	±1	∓1	0	− 24	− 40	− 420

spherically symmetric, and multiplied by the total number of electrons in the shell.

Let us consider next the diagonal elements of the quantity

$$Q = \sum_{i>j} q(i,j) = \sum_{i>j} e^2/r_{ij}$$

TABLE 2^6. $a^k(l\,m_l,\,l'\,m_l')$.

The value of this coefficient is independent of the signs of m_l and m_l'. As in the preceding table, we print the common denominator of several related values but once at the beginning of each group. For $l=0$, $a^k(0\,0,\,l'\,m_l')=\delta(k,0)$ for all $l'\,m_l'$; for $k=0$, $a^0(l\,m_l,\,l'\,m_l')=1$ for all values of the arguments; in the table we give values only for $l,\,l',\,k>0$. Note that

$$a^k(l'\,m_l',\,l\,m_l)=a^k(l\,m_l,\,l'\,m_l').$$

$l\,l'$	$\lvert m_l\rvert$	$\lvert m_l'\rvert$	$k=2$	4
$p\,p$	1	1	1/25	
	1	0	−2	
	0	0	4	
$p\,d$	1	2	2/35	
	1	1	−1	
	1	0	−2	
	0	2	−4	
	0	1	2	
	0	0	4	
$p\,f$	1	3	5/75	
	1	2	0	
	1	1	−3	
	1	0	−4	
	0	3	−10	
	0	2	0	
	0	1	6	
	0	0	8	
$d\,d$	2	2	4/49	1/441
	2	1	−2	−4
	2	0	−4	6
	1	1	1	16
	1	0	2	−24
	0	0	4	36

$l\,l'$	$\lvert m_l\rvert$	$\lvert m_l'\rvert$	$k=2$	4	6
$d\,f$	2	3	10/105	3/693	
	2	2	0	−7	
	2	1	−6	1	
	2	0	−8	6	
	1	3	−5	−12	
	1	2	0	28	
	1	1	3	−4	
	1	0	4	−24	
	0	3	−10	18	
	0	2	0	−42	
	0	1	6	6	
	0	0	8	36	
$f\,f$	3	3	25/225	9/1089	1/7361·64
	3	2	0	−21	−6
	3	1	−15	3	15
	3	0	−20	18	−20
	2	2	0	49	36
	2	1	0	−7	−90
	2	0	0	−42	120
	1	1	9	1	225
	1	0	12	6	−300
	0	0	16	36	400

which represents the mutual Coulomb repulsion of the electrons. This element is given by the expression (cf. $7^6$7 and $8^6$12)

$$\left.\begin{aligned}(A|Q|A)=&\sum_{k>t=1}^{N}\Big[\iint\bar{v}_1(a^k)\,\bar{v}_2(a^t)\,q(1,2)\,v_1(a^k)\,v_2(a^t)\,d\tau_1 d\tau_2\\&-\delta(m_s^k,m_s^t)\iint\bar{v}_1(a^k)\,\bar{v}_2(a^t)\,q(1,2)\,v_1(a^t)\,v_2(a^k)\,d\tau_1 d\tau_2\Big]\\=&\sum_{k>t=1}^{N}[J(k,t)-K(k,t)],\end{aligned}\right\}\quad(2)$$

where $d\tau=\sin\theta\,dr\,d\theta\,d\varphi$ and $v(a)$ is the spin-free eigenfunction

$$R(n^a l^a)\,\Theta(l^a m_l^a)\,\Phi(m_l^a).$$

Consider now that part of the sum in (2) which results when a^k has the fixed value $nlm_s m_l$ and a^t runs over all the individual sets of the $n'l'$ closed shell. We do not assume nl to be different from $n'l'$. If $nl=n'l'$ there occurs one term in this sum, that for which $a^t\equiv a^k$, which does not occur in (2); but the summand of (2) is seen to vanish for $a^k \equiv a^t$ so we have introduced only a vanishing term. This sum is, in detail,

$$\sum_{m_s'} \sum_{m_l'=-l'}^{+l'} \Big[\iint q(1,2)\, R_1^2(nl)\, R_2^2(n'l')\, \Theta_1^2(l\,m_l)\, \Theta_2^2(l'\,m_l')\, (2\pi)^{-2}\, d\tau_1 d\tau_2$$
$$- \delta(m_s, m_s') \iint q(1,2)\, R_1(nl)\, R_1(n'l')\, R_2(nl)\, R_2(n'l')\, \Theta_1(l\,m_l)$$
$$. \bar{\Phi}_1(m_l)\, \Theta_1(l'\,m_l')\, \Phi_1(m_l')\, \Theta_2(l\,m_l)\, \Phi_2(m_l)\, \Theta_2(l'\,m_l')\, \bar{\Phi}_2(m_l')\, d\tau_1 d\tau_2 \Big]. \quad (3)$$

The summation over m_s' gives a factor two in the first term and a factor one in the second. The summation over m_l' may be carried out by the use of the addition theorem (4[3]22). This gives for our sum

$$2 \iint q(1,2)\, R_1^2(nl)\, R_2^2(n'l')\, \Theta_1^2(l\,m_l) \left(\frac{2l'+1}{4\pi}\right) \left(\frac{1}{2\pi}\right) d\tau_1 d\tau_2$$
$$- \iint q(1,2)\, R_1(nl)\, R_1(n'l')\, R_2(nl)\, R_2(n'l')\, \Theta_1(l\,m_l)\, \bar{\Phi}_1(m_l)$$
$$. \Theta_2(l\,m_l)\, \Phi_2(m_l) \left(\frac{2l'+1}{4\pi}\right) P_{l'}(\cos\omega)\, d\tau_1 d\tau_2. \quad (4)$$

Now writing
$$q(1,2) = \frac{e^2}{r_{12}} = e^2 \sum_\kappa \frac{r_<^\kappa}{r_>^{\kappa+1}} P_\kappa(\cos\omega),$$

each of these integrals can be written as a sum over κ of integrals involving $e^2 (r_<^\kappa / r_>^{\kappa+1})\, P_\kappa(\cos\omega)$ in place of $q(1,2)$.

In the direct integral one may choose a new coordinate system for the variables θ_2, φ_2 with the direction θ_1, φ_1 as pole, in which case $\omega = \theta_2$. Then all terms vanish except for $\kappa = 0$, in which case the integration over θ_2, φ_2 gives a factor 4π. The integration over θ_1, φ_1 may then be performed directly, giving for the value of the direct integral (cf. 8[6]15)

$$2\,(2l'+1)\, F^0(nl, n'l'). \quad (5)$$

The exchange integral, on using the same expansion, becomes

$$- \sum_\kappa e^2 \iint \frac{r_<^\kappa}{r_>^{\kappa+1}} R_1(nl)\, R_1(n'l')\, R_2(nl)\, R_2(n'l')\, \Theta_1(l\,m_l)\, \bar{\Phi}_1(m_l)$$
$$. \Theta_2(l\,m_l)\, \Phi_2(m_l) \left(\frac{2l'+1}{4\pi}\right) P_{l'}(\cos\omega)\, P_\kappa(\cos\omega)\, d\tau_1 d\tau_2. \quad (6)$$

To evaluate this we expand the product $P_{l'}(\cos\omega)\, P_\kappa(\cos\omega)$ in a series of Legendre polynomials $P_\lambda(\cos\omega)$. This we may do since the $P_\lambda(\cos\omega)$ form a complete set of functions in the interval 0 to π. The coefficient of $P_\lambda(\cos\omega)$ in this expansion is given by

$$\frac{2\lambda+1}{2} \int_0^\pi P_\lambda(\cos\omega)\, P_{l'}(\cos\omega)\, P_\kappa(\cos\omega)\, \sin\omega\, d\omega = \frac{2\lambda+1}{2}\, c_{\lambda l' \kappa},$$

where we introduce for the integral of the product of three Legendre polynomials the notation

$$\int_0^\pi P_u(\cos\theta)\, P_v(\cos\theta)\, P_w(\cos\theta) \sin\theta\, d\theta = C_{uvw}. \tag{7}$$

C_{uvw} is expressible in terms of the c's of $8^6 6$ by the relation

$$C_{uvw} = \frac{2\, c^u(v\,0, w\,0)}{\sqrt{(2v+1)(2w+1)}}, \tag{8}$$

or the corresponding forms obtained by permutation of $u\,v\,w$, since C is symmetric in the three indices. This integral has the value*

$$C_{uvw} = \frac{2(g!)^2\,(u+v-w)!\,(v+w-u)!\,(w+u-v)!}{(2g+1)!\,(g-u)!^2\,(g-v)!^2\,(g-w)!^2}, \tag{9}$$

where $2g = u+v+w$. The integral vanishes unless the triangular conditions are satisfied (cf. $8^6 10$).

When we substitute the relation

$$P_{l'}(\cos\omega)\, P_\kappa(\cos\omega) = \sum_\lambda \frac{2\lambda+1}{2}\, C_{\lambda l' \kappa}\, P_\lambda(\cos\omega),$$

the exchange integral (6) becomes a double sum over κ and λ. In each summand the dependence on θ_2, φ_2 is through the factor

$$\Theta_2(l\,m_l)\,\Phi_2(m_l)\, P_\lambda(\cos\omega).$$

Expanding $P_\lambda(\cos\omega)$ by the addition theorem, we find that the integral of this over the whole range of θ_2, φ_2 vanishes unless $\lambda = l$, in which case it has the value

$$\frac{4\pi}{2l+1}\,\Theta_1(l\,m_l)\,\Phi_1(m_l).$$

The integration over θ_1, φ_1 can now be performed directly to give simply a constant factor. Thus we are left with (cf. $8^6 15$)

$$-\frac{2l'+1}{2} \sum_\kappa C_{ll'\kappa}\, G^\kappa(nl, n'l') \tag{10}$$

for the value of the exchange integral.

Hence the contribution (3) of the nlm_sm_l electron and the $n'l'$ closed shell to the diagonal element of Q is given by

$$(2l'+1)\,[2\,F^0(nl, n'l') - \tfrac{1}{2} \sum_\kappa C_{ll'\kappa}\, G^\kappa(nl, n'l')]. \tag{11}$$

The most striking and important property of this result is that it is entirely independent of the values of m_l and m_s.

We may now obtain the contribution to (2) of all pairs of electrons in the same nl closed shell by multiplying (11) for $n'l' = nl$ by the number $2\,(2l+1)$ of electrons in the shell, and dividing by two since each pair of electrons is to be counted only once. This gives for the contribution to the diagonal

* Hobson, *Spherical Harmonics*, p. 87;
Gaunt, Trans. Roy. Soc. **A228**, 195 (1929).

element $(A|Q|A)$ of all pairs of individual sets belonging to the same closed shell the value

$$(2l+1)^2\,[2\,F^0(nl,nl)-\tfrac{1}{2}\sum_\kappa C_{ll\kappa}\,F^\kappa(nl,nl)]. \qquad (12)$$

Similarly if there is more than one closed shell in the complete set A there will be terms in the sum in which k refers to the closed shell nl and t refers to the closed shell $n'l'$. The contribution of such terms is

$$2\,(2l+1)\,(2l'+1)\,[2\,F^0(nl,n'l')-\tfrac{1}{2}\sum_\kappa C_{ll'\kappa}\,G^\kappa(nl,n'l')]. \qquad (13)$$

10. One electron outside closed shells.

We are now in a position to investigate the nature of the approximation which permits us to treat the spectra of the alkali-like atoms as arising from the motion of one electron in an effective central field as was done in § 8^5. We consider for this purpose the configurations in which all the electrons but one are in closed shells. The one electron not in a closed shell will be referred to as the *valence electron*.

The first approximation to the energy of any configuration, according to § 8^2, is given by the roots of a secular equation in which appear the matrix components of the perturbation energy taken with regard to all the states of the configuration. We label the states by a set of quantum numbers for the closed shells and an individual set $n\,l\,m_s\,m_l$ for the valence electron. The degeneracy is only in m_l and m_s. Since for each closed shell Σm_l and Σm_s vanish, the Σm_l and Σm_s for the complete set is just the m_l and m_s of the valence electron. But we have seen that the electrostatic interaction energy is diagonal with regard to Σm_l and Σm_s, so there are no non-diagonal matrix components due to this term in the Hamiltonian.

As to the spin-orbit interaction, it is a quantity of the type F, so the results of § 6^6 apply. The conventional order of listing quantum numbers will make the valence electron in set B come at the same place in the list of individual sets as it does in set A, so the sign in $6^6 8$ is positive. For the configurations under consideration it is possible for B to differ from A only in regard to the m_l and m_s of the valence electron, so the non-diagonal element that occurs is just what would occur for the spin-orbit interaction if there were but one electron in the whole atom. The diagonal elements of the spin-orbit interaction are given by $6^6 9$ where, as in § 4^5, f is of the form $\xi(r_i)\boldsymbol{L}_i\cdot\boldsymbol{S}_i$. The diagonal matrix components of this quantity are of the form $\zeta_{nl}m_s m_l$ for a single electron in the state $n\;l\;m_s\;m_l$. The sum of such quantities for all the electrons in a closed shell is obviously zero. Therefore the only term left in applying $6^6 9$ is that contributed by the valence electron, which is exactly what it would have been if the others had not been there, so the whole spin-orbit interaction is as in the one-electron problem.

Turning now to the diagonal elements of the electrostatic interaction we use the results of the preceding section. For each closed shell there will be a contribution of the form $9^6 12$ and for each pair of closed shells there will be the contributions of the form $9^6 13$. These two parts refer to the interactions of the electrons contained in the closed shells and so are the same for all states of the atom under consideration. Therefore they do not affect the relative positions of the different energy levels.

The remaining part of the diagonal element of Q is the part which depends on the quantum numbers of the valence electron. For this part k in $7^6 7$ refers to the valence electron while the index t runs over all the other individual sets. This gives a sum of contributions of the form $9^6 11$ for the interaction of the valence electron with each closed shell. The sum is independent of the $m_s m_l$ of the valence electron.

From the definition of $F^0(nl, n'l')$ the direct integral representing the interaction of the valence electron with the $n'l'$ shell is

$$2(2l'+1)\int_0^\infty \int_0^\infty \frac{e^2}{r_>} R_1^2(nl)\, R_2^2(n'l')\, dr_1 dr_2.$$

The integration with respect to r_2 can be carried out to give

$$V_{n'l'}(r_1) = 2(2l'+1)\left[\int_0^{r_1} \frac{e^2}{r_1} R_2^2(n'l')\, dr_2 + \int_{r_1}^\infty \frac{e^2}{r_2} R_2^2(n'l')\, dr_2\right], \tag{1}$$

so the direct integral can be written in the form

$$\int V_{n'l'}(r_1)\, R_1^2(nl)\, dr_1. \tag{2}$$

In other words the direct integral is just the interaction of the valence electron with the classical potential field due to the spherically symmetrical charge density $2(2l'+1)\, R^2(n'l')/r^2$.

The exchange integral has no such simple interpretation, which is consistent with the fact that it has no classical analogue.

The situation is thus as follows. To treat an atom with one electron outside closed shells we may assume all electrons to move in an effective central field $U(r)$. The one-electron central-field problem has to be solved for this field to find the unperturbed energy levels and radial eigenfunctions. Then the N-electron problem will be like this one-electron problem so far as the spin-orbit interaction and doublet splitting is concerned. It will be unlike it in that states will occur only for which the valence-electron quantum numbers are not those used up by the closed shells. The first-order perturbation (except the spin-orbit terms) will then be, when the valence electron is in an nl state, a sum of terms which it is convenient to regard in three groups.

First, there are the terms which refer only to the closed shells and which are therefore common to all of the states of the atom and so do not affect relative energy levels. These include not only a term like $9^6 12$ for each closed shell and one like 9[6]13 for each pair of closed shells, but also for each closed shell a term like 9[6]1 which, according to 1[6]4, has the special form

$$2(2l'+1)\int_0^\infty \left[-U(r)-\frac{Ze^2}{r}\right] R^2(n'l')\,dr \qquad (3)$$

for the $n'l'$ shell.

Second, there are the terms which are integrals of the form of a function of r multiplied by the square of the radial eigenfunction for the valence electron. Altogether they are

$$\int_0^\infty \left[-U(r)-\frac{Ze^2}{r}+\Sigma V_{n'l'}(r)\right] R^2(nl)\,dr, \qquad (4)$$

where the quantities $V_{n'l'}(r)$ are defined by (1) and the summation is over the $n'l'$ of all the closed shells.

Third, there are the terms of the exchange integrals between the valence electron and the electrons in the closed shells. In this category there is one sum like the second part of 9[6]11 for each of the closed shells.

The doublet structure has been shown to be the same as in the one-electron problem, so that it is simply the combined effects of terms like (4) and the exchange integrals which determine the relative positions of the different terms. These results give us a little more insight into the way the approximation depends on the choice of the function $U(r)$. The value of (4) is a functional of $U(r)$ of a rather complicated sort; it depends on $U(r)$ not only explicitly but also in that the $R(nl)$ depends on $U(r)$ and so also do the $V_{n'l'}(r)$ through involving the $R(n'l')$. A good choice of $U(r)$ would be one that makes (4) have a small value for as many of the values of nl as possible.

Now (4) is of the form which would arise if one were trying to approximate to the solutions for the potential energy

$$-\frac{Ze^2}{r}+\sum_{n'l'} V_{n'l'}(r) \qquad (5)$$

having started with the known exact solutions for the field $U(r)$. This suggests that one might take (5) as a new $U(r)$ for a second trial and then determine the eigenfunctions which belong to it and in turn calculate new $V_{n'l'}(r)$ from these as a means of approximating to the best potential-energy function. Detailed consideration of methods of obtaining $U(r)$ and results based on special central fields will be given in Chapter XIV.

11. Odd and even states.

In the course of the developments which follow we shall see that of all the ordinary quantum numbers, only those referring to the total angular momentum (J and M) are rigorously constants of the motion for a free atom. There is, however, another classification of states of the atom which is rigorously a constant of the motion. Each state may be characterized rigorously as 'even' or 'odd,' a fact of great importance in spectroscopy in that all the dipole radiation of the atom is concerned with transitions between states of opposite parity.

Let $\mathscr{P}$ be an observable* defined by the result of its operation on a Schrödinger ψ function of the coordinates $x_i y_i z_i \sigma_i$ $(i=1, \ldots, N)$ of the electrons as follows:

$$\mathscr{P}\psi_{x_i y_i z_i \sigma_i} = \psi_{-x_i - y_i - z_i \sigma_i}. \tag{1}$$

From this definition it is easily seen that with an observable which is a function of the position vectors $\boldsymbol{r}_i$, the momentum vectors $\boldsymbol{p}_i$, and the angular momenta $\boldsymbol{L}_i$ and $\boldsymbol{S}_i$, $\mathscr{P}$ has the following 'commutation' property

$$\mathscr{P}f(\boldsymbol{r}_i, \boldsymbol{p}_i, \boldsymbol{L}_i, \boldsymbol{S}_i) = f(-\boldsymbol{r}_i, -\boldsymbol{p}_i, \boldsymbol{L}_i, \boldsymbol{S}_i)\mathscr{P}. \tag{2}$$

Now the Hamiltonian for the free atom is a function only of $\boldsymbol{r}_i^2$, $\boldsymbol{r}_i \cdot \boldsymbol{r}_j$, $\boldsymbol{p}_i^2$, $\boldsymbol{L}_i$, and $\boldsymbol{S}_i$ and therefore commutes with $\mathscr{P}$:

$$[\mathscr{P}, H] = 0. \tag{3}$$

Hence stationary states of the atom may be taken also as eigenstates of $\mathscr{P}$. From its definition it is clear that $\mathscr{P}^2$ is the identical operator and so the eigenvalues of $\mathscr{P}$ are ± 1 (cf. Problem 1, § 1^2). States belonging to the eigenvalue $+1$ are called even states, to the eigenvalue -1, odd.

We shall now see that the eigenfunctions we have built up for the central-field approximation are also eigenstates of $\mathscr{P}$. It is evident geometrically that in polar coordinates

$$\mathscr{P}\psi_{r_i \theta_i \varphi_i \sigma_i} = \psi_{r_i\, \pi - \theta_i\, \pi + \varphi_i\, \sigma_i}. \tag{4}$$

Hence if the function is of the form of a radial function multiplied by a spherical harmonic only the latter is affected by $\mathscr{P}$. On $\Phi(m)$ the operation produces

$$\mathscr{P}\Phi(m) = e^{im\pi}\Phi(m) = (-1)^m \Phi(m),$$

while on $\Theta(lm)$ the result is

$$\mathscr{P}\Theta(lm) = (-1)^{l-m}\Theta(lm);$$

so that the whole effect on any spherical harmonic of order l is multiplication by $(-1)^l$. Then if $\mathscr{P}$ is applied to an eigenfunction like $3^6 6$, the eigenfunction is multiplied by $(-1)^{\Sigma l}$. Hence the state is even or odd according to whether Σl is even or odd. This is the origin of the even-odd terminology for the eigenstates of $\mathscr{P}$ belonging to ± 1 respectively.

Since $\mathscr{P}$ anticommutes with the electric moment $\boldsymbol{P}$ defined in $4^4 9$, it follows from $2^2 23$ that the electric moment can have non-vanishing matrix components only between states of opposite parity. Similarly $\mathfrak{N}$ and $\boldsymbol{M}$ commute with $\mathscr{P}$ and so will have non-vanishing matrix components only between states of the same parity.

* Real because we shall shortly find that the eigenvalues are real and the eigenstates orthogonal, or from the definition because clearly

$$\int \bar{\psi}_x(1)\mathscr{P}\psi_x(2)\,dx = \overline{\int \bar{\psi}_x(2)\mathscr{P}\psi_x(1)\,dx}$$

(notation of § 3^2).

CHAPTER VII

THE RUSSELL-SAUNDERS CASE: ENERGY LEVELS

We are now prepared to build on the results of preceding chapters by studying atomic configurations having more than one electron outside closed shells. For such configurations there are generally a large number of energy levels giving great complexity to the spectrum. In the first approximation of the perturbation theory the purely radial terms, $\sum_i \left(-\frac{Ze^2}{r_i} - U(r_i) \right)$, contribute quantities which have the same values for all levels belonging to a configuration. Therefore they do not affect the relative position of the levels of any one configuration. Because of this fact their consideration is best incorporated with the general study of the central-field problem which we take up in Chapter XIV. The Coulomb interaction of the electrons, $\sum_{i>j} e^2/r_{ij}$, is different for different states of the same configuration and thus serves partly to remove the degeneracy of the states belonging to the same configuration that exists in the simple central-field model of the preceding chapter. This is also true of the spin-orbit interaction terms, $\sum_i \xi(r_i)\, \boldsymbol{L}_i \cdot \boldsymbol{S}_i$.

Experience shows that in many atoms the spin-orbit interaction is small compared with the Coulomb interaction. For this reason it is useful to develop an intermediate approximation in which the Coulomb interaction is taken into account but the spin-orbit interaction neglected, or treated as small compared with the Coulomb interaction. This approximation is called the Russell-Saunders case in recognition of the pioneer work of Russell and Saunders* in which the main features of the theory of complex spectra were first recognized. In this chapter we study the scheme of levels belonging to a particular electron configuration. The next chapter is concerned with methods of finding the eigenfunctions for this approximation in terms of the zero-order functions for the central-field problem. Chapter IX completes the general study of the Russell-Saunders case by developing the theory of line strengths in this approximation.

The theory covered in these three chapters, while quite general, will be in a form which is suitable for actual calculation only for configurations involving a very few electrons outside of closed shells. Chapter XIII will show how the theory may be adapted to the calculation of configurations involving almost closed shells, i.e. shells with just a few electrons missing.

* RUSSELL and SAUNDERS, Astrophys. J., **61**, 38 (1925).

1. The LS-coupling scheme.*

The Hamiltonian without spin-orbit interaction terms commutes with all components of the resultant orbital angular momentum $\boldsymbol{L}=\boldsymbol{L}_1+\boldsymbol{L}_2+\ldots+\boldsymbol{L}_N$ and of the resultant spin angular momentum $\boldsymbol{S}=\boldsymbol{S}_1+\boldsymbol{S}_2+\ldots+\boldsymbol{S}_N$; this is most directly seen from the considerations in the first part of § 8[3]. Hence it commutes with all components of the total angular momentum $\boldsymbol{J}=\boldsymbol{S}+\boldsymbol{L}$. Therefore this Hamiltonian will have no matrix components connecting states labelled by two different precise values of $\boldsymbol{S}^2$, $\boldsymbol{L}^2$, $\boldsymbol{J}^2$, S_z, L_z, or J_z. We introduce quantum numbers S, L, J, M_S, M_L, M according to the scheme

$$\begin{aligned} \boldsymbol{S}^{2\prime} &= S(S+1)\hbar^2 & S_z' &= M_S\hbar \\ \boldsymbol{L}^{2\prime} &= L(L+1)\hbar^2 & L_z' &= M_L\hbar \\ \boldsymbol{J}^{2\prime} &= J(J+1)\hbar^2 & J_z' &= M\hbar \end{aligned} \tag{1}$$

for convenience in labelling states which correspond to precise values of these observables. Only four of these six observables are independent; one usually takes either the set SLM_SM_L or the set $SLJM$ to label the energy states. From 3[3]8 we obtain the important result that in the SLM_SM_L scheme the energy is independent of M_S and M_L. Since a definite state $SLJM$ is a linear combination of states SLM_SM_L with various M_SM_L but the same SL, this has as a consequence that in the $SLJM$ scheme the energy is independent of J and M.

The representation of states developed in the preceding chapter is based on a set of states specified by a particular set of $4N$ quantum numbers. We have now to transform to a new set of states in which the energy is diagonal to a certain degree of approximation. In whatever manner this is done each of the new states will likewise need to be specified by $4N$ quantum numbers, for this is the number of degrees of freedom of the system. In the first approximation of the perturbation scheme we neglect matrix components of the Hamiltonian which connect states belonging to different configurations. Therefore to this approximation the energy states can be labelled precisely by the configuration. If there are N' electrons not in closed shells and $N-N'$ in closed shells, the configuration label amounts to a specification of $4(N-N')+2N'$ quantum numbers, for all the quantum numbers occurring in a closed shell are prescribed and the nl values of the N' individual sets out of closed shells are also prescribed. Therefore in addition to the configuration label we need $2N'$ more quantum numbers with which to label states.

* In accordance with the recommendations of Russell, Shenstone, and Turner [Phys. Rev. **33**, 900 (1929)], we shall use capital letters S, L, J, M_S, M_L, M for quantum numbers which refer to the resultant momenta of several or all of the electrons in an atom or ion; the corresponding small letters will be used only for quantum numbers which refer to a single electron.

In case $N' = 2$ the four quantum numbers needed are provided by SL and either JM or $M_S M_L$. For $N' > 2$ these are not in general sufficient and we shall need to find additional quantum numbers in order to give a unique specification of the states. This can be accomplished in many cases (§ 2^8) by introducing the concept of parent terms.

Any set of states in which $\boldsymbol{L}^2$ and $\boldsymbol{S}^2$ are diagonal will be referred to as an *LS-coupling scheme*. On the older vector model of the atom the orbital angular momenta of the individual electrons were regarded as coupled together to give a resultant orbital angular momentum, and the individual spins were coupled to give a resultant spin. Then the $\boldsymbol{S}$ and $\boldsymbol{L}$ which resulted might be coupled under certain circumstances to give a resultant angular momentum $\boldsymbol{J}$ for the whole atom.

A set of $(2S+1)(2L+1)$ states belonging to a definite configuration and to a definite L and S will be called a *term*. If we are working with an SLM_SM_L scheme, then $|M_L| \leqslant L$ and $|M_S| \leqslant S$, giving $(2S+1)(2L+1)$ states in the term. Working in an $SLJM$ scheme, we have $|S-L| \leqslant J \leqslant S+L$ and $|M| \leqslant J$ for each J; this counts up to the same total number of states in the term. The quantity $2S+1$ is called the *multiplicity*: it is the number of J values and hence the number of *levels* that occur in the term if $L \geqslant S$. Even if $L < S$ we ascribe to a term this multiplicity and say that its multiplicity is not fully developed. The notation for a term, which has become standard, is to write $S, P, D, F, \ldots$ for $L = 0, 1, 2, 3, \ldots$ and to write the numerical value of the multiplicity as a superscript at the left of the symbol for the L value. Thus ^5P ('quintet P') indicates a term in which $L = 1$ and $S = 2$. This is in accord with the usage of § 4^5 for the doublets in one-electron spectra. If more than one term of the same kind occurs in a given configuration, we shall for the present distinguish them by an arbitrary extra letter as $a\,^5P$, $b\,^5P$, ...; in § 2^8 we shall examine the possibilities of a more significant way of characterizing them.

We have now to consider the problem of determining which terms occur in a given configuration.* By the methods of the preceding chapter we can write down the N individual sets comprising each complete set which occurs in the configuration. Each complete set will consist of the individual sets for electrons both in closed shells and out. As it will shortly appear that the number and nature of the closed shells is without effect on the following discussion, we shall in what follows use the word configuration to refer to the class of all configurations which have the same nl values for those electrons not in closed shells.

Having made a list of the complete sets which belong to a configuration

* Pauli, Zeits. für Phys. **31**, 765 (1925);
Goudsmit, *ibid.* **32**, 794 (1925).

we may classify them at once by the values of $\Sigma m_l = M_L$ and $\Sigma m_s = M_S$. To make the argument concrete let us consider $np\, n'p\,(=p\,p)$, the configuration in which two p electrons with differing n values occur outside closed shells. It will be convenient to use the notation of $5^6 2$ for the m_l and m_s values of the two electrons, the superscript $\pm$ sign indicating the m_s value for the individual set whose m_l value is given by the integer to which it is attached. The resulting table of complete sets for pp classified by M_L and M_S values is

pp		$M_S = 1$	$M_S = 0$	$M_S = -1$
M_L	2	$(1^+ 1^+)$	$(1^+ 1^-)(1^- 1^+)$	$(1^- 1^-)$
	1	$(1^+ 0^+)(0^+ 1^+)$	$(1^+ 0^-)(0^+ 1^-)(1^- 0^+)(0^- 1^+)$	$(1^- 0^-)(0^- 1^-)$
	0	$(1^+ -1^+)(0^+ 0^+)$ $(-1^+ 1^+)$	$(1^+ -1^-)(0^+ 0^-)(-1^+ 1^-)$ $(1^- -1^+)(0^- 0^+)(-1^- 1^+)$	$(1^- -1^-)(0^- 0^-)$ $(-1^- 1^-)$
	-1	$(-1^+ 0^+)(0^+ -1^+)$	$(-1^+ 0^-)(0^+ -1^-)(-1^- 0^+)(0^- -1^+)$	$(-1^- 0^-)(0^- -1^-)$
	-2	$(-1^+ -1^+)$	$(-1^+ -1^-)(-1^- -1^+)$	$(-1^- -1^-)$

. (2)

Because of the occurrence of a state for $M_L = 2$ and $M_S = 1$ there must be a term having the values $L = 2$ and $S = 1$, that is ^3D, and this will have a state in each cell of the table. Of the two states in the cell $M_L = 2$ and $M_S = 0$, only one is accounted for by the ^3D term, so there must be a ^1D term to account for the other. This accounts for all states with $M_L = 2$. Similarly from the presence of two states in the cell $M_L = 1$ and $M_S = 1$ we infer the occurrence of a ^3P term. Continuing this process we see that the terms are ^1S, ^3S, ^1P, ^3P, ^1D, and ^3D.

These terms are the same as one finds on the vector-coupling picture according to which S as the resultant of two spins each of magnitude one-half can have a value zero or unity, and L as the resultant of two orbital angular momenta each of magnitude unity can have the value zero, one, or two. However, the vector-coupling picture does not tell us which of the terms are eliminated by the exclusion principle in case the configuration involves equivalent electrons.

Let us see what effect the exclusion principle has in the case just considered if the n's of the two p electrons become equal, that is, for the configuration p^2. We have to rule out all complete sets in which identical individual sets occur, and must count but once sets which are now the same except for different order of listing of the individual sets. From $4^6 1$ we see that the exclusion principle excludes or permits whole terms so that its operation does not spoil the possibility of setting up an LS-coupling scheme. The $M_S M_L$ table now is

p^2		M_S 1	M_S 0	M_S -1
M_L	2		$(1^+ 1^-)$	
	1	$(1^+ 0^+)$	$(1^+ 0^-)(1^- 0^+)$	$(1^- 0^-)$
	0	$(1^+ -1^+)$	$(1^+ -1^-)(1^- -1^+)(0^+ 0^-)$	$(1^- -1^-)$
	-1	$(0^+ -1^+)$	$(0^+ -1^-)(0^- -1^+)$	$(0^- -1^-)$
	-2		$(-1^+ -1^-)$	

, (3)

so the terms are ${}^1S$, ${}^3P$ and ${}^1D$. A total of 36 states for pp is reduced by the exclusion principle to 15 for p^2.

Similar tables for other configurations may be made to find the terms occurring, although for several electrons outside closed shells the tables become quite extensive. Because of the symmetry which obtains it is necessary only to construct that part of the table which corresponds to $M_L \geqslant 0$ and $M_S \geqslant 0$.

2. Term energies.

Now let us calculate the first-order perturbation energy for the terms of a given configuration in the approximation in which we neglect spin-orbit interaction. There are no matrix components connecting states of different M_L and M_S; therefore the secular equation for the whole configuration factors into a chain of secular equations, one of which is associated with each value of $M_S M_L$. We make use of the diagonal-sum rule ($2^2 21$) which states that the sum of the roots of a secular equation is equal to the sum of the diagonal matrix components occurring in it. Since all the levels of a term have the same energy, this rule allows us to write a set of linear equations, one for each $M_S M_L$ cell, equating the sum of the energies of the terms having states in that cell to the sum of the diagonal matrix elements of the perturbation energy taken with regard to the complete sets in that cell. If, for brevity, we write ${}^3D$ for the first-order energy of the ${}^3D$ term, and write $(1^+ 1^+)$ for the diagonal matrix element of the Hamiltonian associated with this complete set, we obtain, for pp,

$$
\begin{aligned}
{}^3D &= (1^+ 1^+) \\
{}^1D + {}^3D &= (1^+ 1^-) + (1^- 1^+) \\
{}^3P + {}^3D &= (1^+ 0^+) + (0^+ 1^+) \\
{}^1P + {}^3P + {}^1D + {}^3D &= (1^+ 0^-) + (0^+ 1^-) + (1^- 0^+) + (0^- 1^+) \\
{}^3S + {}^3P + {}^3D &= (0^+ 0^+) + (1^+ -1^+) + (-1^+ 1^+) \\
{}^1S + {}^3S + {}^1P + {}^3P + {}^1D + {}^3D &= (0^+ 0^-) + (-1^+ 1^-) + (1^+ -1^-) + (0^- 0^+) + (-1^- 1^+) + (1^- -1^+),
\end{aligned}
$$

and additional equations from the remaining cells which are not needed except as checks in actual calculation.*

From these equations we readily solve for the term energies in terms of the diagonal matrix components:

$$\begin{aligned} {}^3D &= (1^+1^+) \\ {}^1D &= (1^+1^-)+(1^-1^+)-(1^+1^+) \\ {}^3P &= (1^+0^+)+(0^+1^+)-(1^+1^+), \text{ etc.} \end{aligned} \tag{1}$$

Corresponding calculations for the configuration p^2 give the formulas, analogous to (1),

$$\begin{aligned} &{}^1D = (1^+1^-) \qquad {}^3P = (1^+0^+) \\ &{}^1S = (1^+-1^-)+(1^--1^+)+(0^+0^-)-(1^+0^+)-(1^+1^-). \end{aligned} \tag{2}$$

If a particular kind of term occurs more than once in the configuration, the only way in which the energy of these terms occurs in the set of linear equations is as the sum of the energies of all like terms; hence this method gives only the sum. In order to find the individual energies one needs then actually to solve a secular equation involving the non-diagonal matrix elements. A case of such a calculation is taken up in § 7[8].

After the calculations for a particular configuration have given formulas like (1) and (2), these may be still further simplified by making use of some special properties of the diagonal matrix components from the preceding chapter. From the way in which these formulas are obtained it always turns out that the number of matrix components occurring with a plus sign is one greater than the number with a minus sign. Therefore any quantity which occurs as a constant added to all the diagonal matrix elements of a configuration will come up simply as that same constant in the expression for the energy of each of the terms. It is the electrostatic repulsion of the electrons which gives rise to the separation of the terms. From 7[6]7 we see that a particular diagonal element of $Q=\Sigma e^2/r_{ij}$ is given by

$$(A|Q|A) = \sum_{a>b=1}^{N} [(a\,b|e^2/r_{12}|a\,b)-(a\,b|e^2/r_{12}|b\,a)], \tag{3}$$

where the summation extends over all pairs of individual sets in the complete set. The sum in (3) breaks up into (*a*) pairs between individual sets in closed shells, (*b*) pairs between an individual set not in a closed shell with those in closed shells, both discussed in § 9[6], and (*c*) pairs between individual sets neither of which belongs to a closed shell. The parts (*a*) and (*b*) will be constant for the whole configuration. Part (*c*) is what gives structure

* We should like to call attention here to a slightly different formulation by VAN VLECK [Phys. Rev. 45, 405 (1934)], in terms of a vector model proposed by Dirac, of the problem of obtaining the first-order electrostatic energies. This formulation furnishes a procedure for obtaining the energies of the terms of any given multiplicity without first determining the energies of the terms of higher multiplicity.

to the configuration apart from its absolute location on the energy scale. It is simply the sum $\sum_{a>b=1}^{N'} [J(a,b)-K(a,b)]$ over the N' electrons outside of closed shells, in the notation of $8^6 12$. From formulas $8^6 13$ and the tables of a's and b's, it is a piece of straightforward computation to express the energy of each term with the F and G integrals. So many coefficients vanish that the summations in $8^6 13$ seldom involve more than two or three terms in cases of spectroscopic interest.

Let us illustrate the method by considering p^2 in detail. On writing E_0 for the constant part which arises from pairs (a) and (b) we have

$$(1^+ 1^-) = E_0 + F_0(np, np) + F_2(np, np)$$
$$(1^+ 0^+) = E_0 + F_0(np, np) - 5F_2(np, np)$$

and similar expressions for the other diagonal elements occurring in (2). Here we have used the fact that $G_k(np, np) = F_k(np, np) = F^k(np, np)/D_k$, where D_k is the denominator occurring in Tables 1^6 and 2^6. Hence finally the electrostatic contribution to the energies becomes

$$p^2 \begin{cases} {}^1S = E_0 + F_0 + 10F_2 \\ {}^1D = E_0 + F_0 + F_2 \\ {}^3P = E_0 + F_0 - 5F_2. \end{cases} \quad (4)$$

Although we cannot know the numerical values of the integrals $F_0(np, np)$ and $F_2(np, np)$ without basing the calculation on a definite assumption for $U(r)$ we see from the form of this result that, whatever the particular $U(r)$ chosen, the first-order theory predicts an interval ratio

$$({}^1S - {}^1D) : ({}^1D - {}^3P) = 3:2, \quad (5)$$

with the ${}^1S$ as the highest term.

This form of calculation thus provides relations between the term energies which can be compared with experiment independently of the more difficult problem of finding a good central field. Such predictions and their comparison with experimental data are given in § 5^7.

3. The Landé interval rule.*

The next problem in the theory of the energy levels in the Russell-Saunders case is that of allowing for the spin-orbit interaction, regarding this as small compared with the electrostatic interaction. The electrostatic interaction breaks the configuration up into separate terms, each characterized by a given L and S, but leaves these terms degenerate with regard to J. The spin-orbit interaction has the form

$$H^I = \sum_i H_i^I = \sum_i \xi(r_i)\, \boldsymbol{L}_i \cdot \boldsymbol{S}_i. \quad (1)$$

* Landé, Zeits. für Phys. **15**, 189; **19**, 112 (1923).

From 8^34 we see that this commutes with $\boldsymbol{J}$; this permits the levels to be labelled by J values even when the spin-orbit interaction is not neglected. But H^I does not commute with $\boldsymbol{S}$ and $\boldsymbol{L}$ so that these no longer furnish accurate quantum numbers. In case the spin-orbit interaction is small compared to the electrostatic, however, we need only consider the effect of the diagonal matrix components in a scheme of states labelled by $SLJM$. In this way, a given term will be split by the spin-orbit interaction into a close group of levels characterized still by S and L to a good approximation and distinguished by different values of J. For the individual levels we use the standard notation, ${}^{2S+1}L_J$. If we wish to specify a state we write the value of M as a right superscript, ${}^{2S+1}L_J^M$. In the SLM_SM_L scheme we shall designate a state by ${}^{2S+1}L$, M_S, M_L.

The complete matrix of H^I for all the states in the configuration includes components that are non-diagonal with respect to the terms. These are responsible for the breakdown of Russell-Saunders coupling, i.e. for the loss of validity of the characterization by S and L. The details of such effects are considered in Chapter XI; here we confine attention to the diagonal elements. From the fact that $\boldsymbol{r}_i^2$ commutes with $\boldsymbol{L}_i$, we see that $\xi(r_i)\boldsymbol{L}_i$ is a vector of type $\boldsymbol{T}$ with respect to $\boldsymbol{L}_i$, to $\boldsymbol{L}$, and to $\boldsymbol{J}$. Hence the diagonal element of the ith term of (1) is given by the first equation of 12^32 if we make the correlation

$$\boldsymbol{P}\to\xi(r_i)\boldsymbol{L}_i \qquad j_1\to S \qquad j\to J$$
$$\boldsymbol{Q}\to\boldsymbol{S}_i \qquad j_2\to L \qquad m\to M$$

as

$$(\gamma SLJM|\xi(r_i)\,\boldsymbol{L}_i\cdot\boldsymbol{S}_i|\gamma SLJM)$$
$$=\sum_{\gamma'}(\gamma L\vdots\xi(r_i)L_i\vdots\gamma' L)(\gamma' S\vdots S_i\vdots\gamma S)\left\{\frac{J(J+1)-L(L+1)-S(S+1)}{2}\right\}. \quad (2)$$

In this expression everything is independent of J and M except the factor in braces. Hence the whole dependence on J of each term of H^I is given by this factor. Since $\boldsymbol{L}\cdot\boldsymbol{S}=\frac{1}{2}(\boldsymbol{J}^2-\boldsymbol{L}^2-\boldsymbol{S}^2)$ this factor is recognized as the matrix element $(SLJM|\boldsymbol{L}\cdot\boldsymbol{S}|SLJM)$, so that we may write

$$\Gamma(\gamma SLJM)=(\gamma SLJM|H^I|\gamma SLJM)$$
$$=\zeta(\gamma SL)\left\{\frac{J(J+1)-L(L+1)-S(S+1)}{2}\right\} \quad (3)$$
$$=\zeta(\gamma SL)(SLJM|\boldsymbol{L}\cdot\boldsymbol{S}|SLJM),$$

where $\zeta(\gamma SL)$ is a factor depending only on S, L, and γ (which specifies the configuration, etc.) and which is entirely independent of J. The spin-orbit perturbation of the level γSLJ is thus given in the first approximation by (3).

From this result we see that the energy interval between levels differing by unity in their J value is

$$\Gamma(\gamma\, S\, L\, J)-\Gamma(\gamma\, S\, L\, J-1)=J\,\zeta(\gamma\, S\, L); \quad (4)$$

that is, the interval between two adjacent levels of a term is proportional to the higher J value of the pair. This is known as the *Landé interval rule.* It has a quite general validity and is of great use in empirical analysis of a spectrum. As an example we may take the ^5D term of the configuration $3d^6\,4s^2$ which is the normal term of Fe I:

Level	Energy (cm^{-1})	Interval	ζ observed
$^5D_4$	0·000		
		−415·934	−103·9
$^5D_3$	415·934		
		−288·067	− 96·1
$^5D_2$	704·001		
		−184·125	− 92·1
$^5D_1$	888·126		
		− 89·942	− 89·9
$^5D_0$	978·068		

Here 'ζ observed' is the observed interval divided by J, which would be constant if the rule were exactly obeyed.

The spin-orbit perturbation given by (3) has the property that the mean perturbation of all the states of the term vanishes, that is, the mean perturbation of the levels when weighted by the factor $(2J+1)$ is zero. This weighted mean of the energies of the levels of a term is called its *centre of gravity,* which is thus equal to the energy of the term as it would be in the absence of spin-orbit interaction. To prove this statement we merely multiply (3) by $(2J+1)$ and sum J from $|L-S|$ to $L+S$, using the summation formulas

$$\sum_{J=0}^{X} J = \tfrac{1}{2}X(X+1), \quad \sum_{J=0}^{X} J^2 = \tfrac{1}{6}X(X+1)(2X+1), \quad \sum_{J=0}^{X} J^3 = \tfrac{1}{4}X^2(X+1)^2. \quad (5)$$

Here, as in the case of one-electron spectra studied in Chapter v, there is a possibility of a small finite displacement of a term of a configuration containing s electrons through the occurrence of an infinite value of ζ multiplied by a factor that is formally zero. In such a case it is necessary to have recourse to the relativistic theory as was done in connection with the ^2S levels in hydrogen.

4. Absolute term intervals.*

We have obtained the Landé interval rule by a simple use of the first equation of $12^3 2$. In order to obtain the absolute term intervals in terms of the one-electron parameters

$$\zeta_{nl} = \hbar^2\,(nl|\xi(r)|nl), \quad (1)$$

we do not need to evaluate $3^7 2$ in greater detail (which as will be seen in Chapter xi is in general a rather complicated procedure) but can obtain all

* GOUDSMIT, Phys. Rev. **31**, 946 (1928).

the information we desire from the diagonal-sum rule except when two or more terms of the same kind occur in a configuration. The intervals in this latter case are considered in § 7[8].

In order to use the diagonal-sum rule we shall need to know the diagonal elements of the matrix of H^I in the SLM_SM_L scheme and in the zero-order nlm_sm_l scheme. In the SLM_SM_L scheme

$$\begin{aligned}(\gamma\, SLM_SM_L|H^I|\gamma\, SLM_SM_L) &= \sum_{JM}(\gamma\, SLM_SM_L|\gamma\, SLJM)(\gamma\, SLJM|H^I|\gamma SLJM)(\gamma\, SLJM|\gamma SLM_SM_L)\\ &= \zeta(\gamma\, SL)\,(SLM_SM_L|\mathbf{L}\cdot\mathbf{S}|SLM_SM_L)\\ &= M_LM_S\,\zeta(\gamma\, SL),\end{aligned} \tag{2}$$

the last form being obtained from 7[3]3. In the zero-order scheme, since H^I is a quantity of type F, the diagonal element for the state $A = a^1 a^2 \ldots a^N$ is given by

$$(A|H^I|A) = \sum_{i=1}^{N}(a^i|H_i^I|a^i) = \sum_{i=1}^{N}\zeta_{n^il^i}m_l^im_s^i. \tag{3}$$

The manner of using the diagonal sum rule is this: The zero-order states for a configuration are classified according to the values of M_L and M_S, and the terms which have components in each cell noted as in § 1[7]. By the diagonal-sum rule, for a given value of M_L and M_S, the sum of the elements (3) for the zero-order states is equal to the sum of the elements (2) for the terms which have such components. This gives for each cell an equation expressing certain of the $\zeta(SL)$ in terms of the ζ_{nl}, and there will be a sufficient number of equations to determine all the $\zeta(SL)$ unless two or more terms of a kind occur, in which case only the sum of the $\zeta(SL)$ for these terms may be obtained.

As a simple example consider the configuration $np\,n'p$ which has been discussed in § 1[7]. From 1[7]2 we see that $(1^+\,1^+)$ is the state* ^3D, 1, 2; hence we obtain the equation

$$2\,\zeta(^3D) = \tfrac{1}{2}\,(\zeta_{np} + \zeta_{n'p})$$

or

$$\zeta(^3D) = \tfrac{1}{4}\,(\zeta_{np} + \zeta_{n'p}).$$

The states ^3D, 1, 1 and ^3P, 1, 1 are linear combinations of the states $(1^+\,0^+)$ and $(0^+\,1^+)$; hence

$$\zeta(^3D) + \zeta(^3P) = \tfrac{1}{2}(\zeta_{np} + \zeta_{n'p})$$

or

$$\zeta(^3P) = \tfrac{1}{4}(\zeta_{np} + \zeta_{n'p}).$$

If we attempt to write a similar equation for a cell with either M_L or $M_S = 0$, we obtain of course only the result $0 = 0$, but the new terms occurring in those cells are either S terms or singlets, for which ζ has no real significance. We thus have obtained the result that

$$\zeta(np\,n'p\,^3D) = \zeta(np\,n'p\,^3P),$$

* See preceding section for notation.

which is directly amenable to comparison with experiment, and says that the total width of the ^3D should be $\frac{5}{3}$ that of the ^3P.

It will be noted that in this case the lowest J value of each term has the lowest energy, since the ζ_{nl} are all essentially positive (§ 4[5]). This positive sign for $\zeta(SL)$ is characteristic of all configurations which contain few electrons—the resulting order with the lowest J value lowest is said to be *normal*. A term for which $\zeta(SL)$ is negative so that the lowest J value is highest is said to be *inverted*, and the intervals are specified as negative.

Since the situation mentioned at the end of the preceding section in which a single level has a spin-orbit displacement cannot occur unless the configuration contains an s electron, none of the singlet or S levels of $np\,n'p$ can possibly have a spin-orbit displacement in the Russell-Saunders case.

5. Formulas and experimental comparison.*

We now turn to a systematic comparison with experimental data of the formulas for the electrostatic energies and Landé term intervals in the Russell-Saunders approximation. For the electrostatic energies in the simplest cases the theory predicts certain interval ratios as in 2[7]5. In other cases where there are more terms and more F's and G's, we treat the latter as adjustable parameters. This enables us to see if the formulas agree with the data in any sense: if they do, then it still has to be remembered that the F's and G's are not really independent. The actual values of F's and G's have to be obtained from some particular choice of the $U(r)$ underlying the central-field approximation. This part of the problem is considered in Chapter XIV.

Departures from these first-order formulas we may regard as due to two causes. First, the spin-orbit interaction may be so large that the LS coupling is broken down. Second, even though the spin-orbit interaction be small, the neglected matrix components of the electrostatic interaction which connect different configurations may be important. The first point is considered in Chapter XI, the second in Chapter XV. From § 11[6] we know already that there are no matrix components of the Hamiltonian between configurations of opposite parity. Hence all perturbations between Russell-Saunders configurations occur between terms of the same S and L values and of the same parity.

<u>*ls*</u>

$$^1L = F_0 + G_l$$
$$^3L = F_0 - G_l \qquad\qquad \zeta(^3L) = \tfrac{1}{2}\zeta_l$$

Here $F_0 = F_0(n's, nl)$, $G_l = G_l(n's, nl)$; these arguments need not be specified

* CONDON and SHORTLEY, Phys. Rev. **37**, 1025 (1931).

unless there is ambiguity. These configurations are considered in detail in § 3^{11}.

p^2

$$^1S = F_0 + 10F_2$$
$$^1D = F_0 + F_2$$
$$^3P = F_0 - 5F_2 \qquad \zeta(^3P) = \tfrac{1}{2}\zeta_p$$

As F_2 is essentially positive, the terms in order of increasing energy are ^3P, ^1D, ^1S, with intervals in the ratio given by 2:3. This is to be compared with the experimental values*

Atom	Configuration	$(^1S - {}^1D)/(^1D - {}^3P)$
Theory	np^2	1·50
C I	$2p^2$	1·13
N II	$2p^2$	1·14
O III	$2p^2$	1·14
Si I	$3p^2$	1·48
Ge I	$4p^2$	1·50
Sn I	$5p^2$	1·39
La II	$6p^2$	18·43
Pb I	$6p^2$	0·62

The departures in case of C I, N II, O III are probably due to action of the configuration $2p\,3p$ above. The same remark holds for Si I, so the good agreement here is probably accidental. In Ge I, Sn I and Pb I there is appreciable breakdown of LS coupling (see § 3^{11}). In La II the configuration is high and lies in the midst of many possible perturbing configurations.

p^3

$$^2P = 3F_0 \qquad \zeta(^2P) = 0$$
$$^2D = 3F_0 - 6F_2 \qquad \zeta(^2D) = 0$$
$$^4S = 3F_0 - 15F_2$$

Here again we can eliminate the unknown F_2 and compare observed with theoretical ratios:

Atom	Configuration	$(^2P - {}^2D)/(^2D - {}^4S)$
Theory	np^3	0·667
N I	$2p^3$	0·500
O II	$2p^3$	0·509
S II	$3p^3$	0·651
As I	$4p^3$	0·715
Sb I	$5p^3$	0·908
Bi I	$6p^3$	1·121

The deviations in N I and O II are probably connected with disturbance by $2p^2\,3p$, the lowest term of which in N I is only four times the over-all separation of $2p^3$ above this configuration. Similar remarks hold for S II. In the last three examples departure from LS coupling is important (§ 3^{11}).

* We wish here to acknowledge our indebtedness to the excellent compilation of energy levels by Bacher and Goudsmit (*Atomic Energy States*, McGraw-Hill, 1932) for most of the experimental data used in this book in energy-value comparisons.

The first-order formulas give $\zeta(^2D)=0$ and $\zeta(^2P)=0$, so the actual term intervals must be referred to perturbations or breakdown of Russell-Saunders coupling.

p^4

The formulas are the same as for p^2 with $6F_0$ written in place of F_0. The data are

Atom	Configuration	$(^1S-{}^1D)/(^1D-{}^3P)$
Theory	np^4	1·50
O I	$2p^4$	1·14
Te I	$4p^4$	1·50

In O I there is probably perturbation by $2p^3\,3p$ and in Te I the coupling departure is great, so the agreement is fortuitous (see § 3^{13}, where Te I is shown to agree well with the theory for intermediate coupling).

p^2s

$$\begin{aligned} {}^2S &= F_0+10F_2-\ G_1 & & \\ {}^2P &= F_0-\ 5F_2+\ G_1 & \qquad \zeta(^2P) &= \tfrac{2}{3}\zeta_p \\ {}^4P &= F_0-\ 5F_2-2G_1 & \qquad \zeta(^4P) &= \tfrac{1}{3}\zeta_p \\ {}^2D &= F_0+\ \ F_2-\ G_1 & \qquad \zeta(^2D) &= 0 \end{aligned}$$

where of course the G_1 is $G_1(np, n's)$. Evidently $(^2S-{}^2D)/(^2D-P)=\frac{3}{2}$, where $P=\frac{1}{3}(2\,{}^4P+{}^2P)$. There are only two cases, Sb III $5s\,5p^2$ and C II $2s\,2p^2$; here the order of terms is right, but the ratios are badly off: 3·58 and 0·481 respectively. In the case of Sb III the $5s^2\,6s\,{}^2S$ and the $5s^2\,5d\,{}^2D$ are nearby and located so as to increase the ${}^2S-{}^2D$ interval. In C II, $2s^2\,3s\,{}^2S$ is located so as to decrease the ${}^2S-{}^2D$ interval.

In Sb III $5s\,5p^2$ we find 2140 and 2400 (cm^{-1}) for the two values of $\zeta(^4P)$ inferred from the two intervals; this is roughly half of $\zeta(^2P)=4500$, as it should be, while $\zeta(^2D)$ is 507 which, although not zero, is quite small relative to the others. Similarly in C II the $\zeta(^4P)$ is 14·3 and 11·4 from the intervals, roughly half of $\zeta(^2P)=27·4$, while $\zeta(^2D)=-0·8$ which is very much smaller.

p^3s

$$\begin{aligned} {}^1P,\ {}^3P &= F_0 \qquad\quad +0,\ -2G_1 & \qquad \zeta(^3P) &= 0 \\ {}^1D,\ {}^3D &= F_0-\ 6F_2+0,\ -2G_1 & \qquad \zeta(^3D) &= 0 \\ {}^3S,\ {}^5S &= F_0-15F_2+G_1,-3G_1 & & \end{aligned}$$

where $\quad F_0=3\,F_0(np,np)+3\,F_0(n's,np);\quad F_2=F_2(np,np);\quad G_1=G_1(n's,np).$

This configuration for O I is discussed in § 2^8.

pp

$$\begin{aligned} {}^1S,\ {}^3S &= F_0+10F_2\pm(G_0+10G_2) & & \\ {}^1P,\ {}^3P &= F_0-\ 5F_2\mp(G_0-\ 5G_2) & \qquad \zeta(^3P) &= \tfrac{1}{4}(\zeta_{np}+\zeta_{n'p}) \\ {}^1D,\ {}^3D &= F_0+\ \ F_2\pm(G_0+\ \ G_2) & \qquad \zeta(^3D) &= \tfrac{1}{4}(\zeta_{np}+\zeta_{n'p}) \end{aligned}$$

We expect that G_2 is always small enough that G_0 determines the sign of the expressions in the G's. Thus the theory predicts an alternation: singlet above triplet for S, then below for P and then above for D. This phenomenon occurs quite generally in two-electron configurations.* We may eliminate the G's by dealing with the arithmetic means of corresponding singlets and triplets, and find $(S-D)/(D-P)=\frac{3}{2}$ for the interval ratio of the means. The means do not even lie in this order in the known cases which are in C I, N II and O III, probably because of perturbation by p^2.

Here the $\zeta(^3D)$ and $\zeta(^3P)$ should be equal; this is fairly well borne out by the data:

	$\zeta(^3D)$	$\zeta(^3P)$
C I $2p\,3p$	11·2	10·2
	10·6	12·5
N II $2p\,3p$	32·1	29·2
	30·4	35·2
O III $2p\,3p$	73·4	65·3
	68·2	82·0

The table gives the value of ζ inferred from each of the intervals in the triplet, so comparison vertically tests the Landé interval rule and comparison horizontally tests the theoretical equality of $\zeta(^3D)$ and $\zeta(^3P)$. The $2p\,4p$ configuration in these same elements shows the same degree of agreement with the theory.

pd

$$^1P,\,^3P=F_0+7F_2\pm(G_1+63G_3) \qquad \zeta(^3P)=-\tfrac{1}{4}\zeta_p+\tfrac{3}{4}\zeta_d$$
$$^1D,\,^3D=F_0-7F_2\mp(3G_1-21G_3) \qquad \zeta(^3D)=\tfrac{1}{12}\zeta_p+\tfrac{5}{12}\zeta_d$$
$$^1F,\,^3F=F_0+2F_2\pm(6G_1+3G_3) \qquad \zeta(^3F)=\tfrac{1}{6}\zeta_p+\tfrac{1}{3}\zeta_d$$

These show the singlet-triplet alternation as in pp. Using means eliminates the G's, giving a theoretical ratio $(P-F)/(F-D)=\frac{5}{9}$. The means do not come in this order in C I, N II and O III, nor in Yt II, La II, Ge I $4p\,5d$, where the configurations are high and surrounded by perturbing configurations. In Ge I $4p\,4d$ and Zr III $5p\,4d$ the means are in the theoretical order but the ratios are 0·28 and 3·58 respectively instead of 0·555.

The configuration $4p\,3d$ can be followed† in the long isoelectronic sequence Ca I to Ni IX. The arrangement of the levels is shown in Fig. 1^7, from which it is evident that the configuration is strongly perturbed in the low stages of ionization, as is shown especially by the occurrence of the ^1P below the ^3P. This lowers the P mean enough that the ratio $(P-F)/(F-D)$ is negative for Ca I and Sc II. In the higher stages of ionization the behaviour is more in accord with the formulas, but here the intervals inside terms are great enough to indicate appreciable breakdown of Russell-Saunders coupling.

* Russell and Meggers, Sci. Papers Bur. Standards, **22**, 364 (1927).
† Cady, Phys. Rev. **43**, 322 (1933).

The ^3P is partially inverted so a detailed study of the Landé intervals would be without meaning. The theoretical ζ's given in the table imply the relation

$$\zeta(^3P) = 5\,\zeta(^3D) - 4\,\zeta(^3F),$$

which makes it clear why the ^3P intervals are so small since $\zeta(^3F)$ is enough larger than $\zeta(^3D)$ to make $\zeta(^3P)$ almost zero.

Similarly in C I $2p\,3d$ the relation between the ζ's indicates a reason why the ^3P is inverted, although the conclusion is uncertain because there are large deviations from the interval rule.

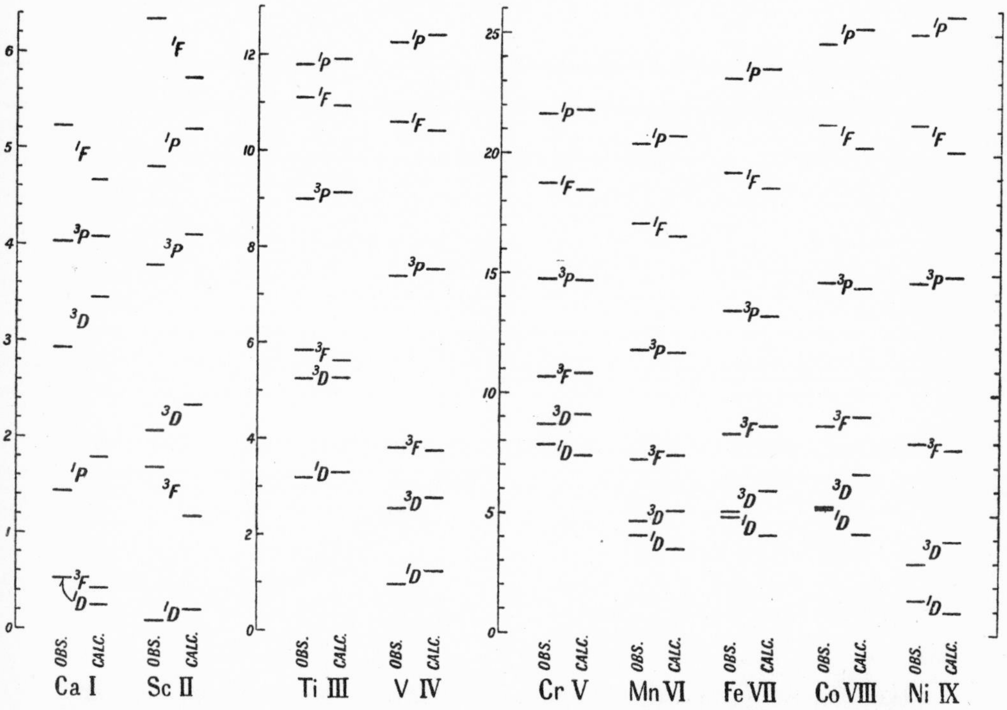

Fig. 1[7]. The $4p\,3d$ configurations in the isoelectronic sequence from Ca I to Ni IX. The calculated terms are obtained by choosing values of F_0, F_2, G_1, and G_3 which give approximately the best fit of the six observed levels. (Scales in thousands of wave numbers.)

<u>*pf*</u>

$$^1D,\,^3D = F_0 + 12F_2 \pm (3G_2 + 36G_4) \qquad \zeta(^3D) = -\tfrac{1}{6}\zeta_p + \tfrac{2}{3}\zeta_f$$
$$^1F,\,^3F = F_0 - 15F_2 \mp (15G_2 - 9G_4) \qquad \zeta(^3F) = \tfrac{1}{24}\zeta_p + \tfrac{11}{24}\zeta_f$$
$$^1G,\,^3G = F_0 + 5F_2 \pm (45G_2 + G_4) \qquad \zeta(^3G) = \tfrac{1}{8}\zeta_p + \tfrac{3}{8}\zeta_f$$

Only one example is known: La II $4f\,6p$. The order of the singlet-triplet means agrees with theory but the ratio $(D-G)/(G-F)$ is 1·05 instead of the theoretical 0·35 ($=7/20$). The interval rule is not satisfied, ^3D being partly inverted.

pds

Here we have the first instance of a configuration in which a term of a particular kind occurs more than once so that the diagonal sum rule does not suffice to give complete formulas. The terms are ${}^{2,2,4}PDF$. The formulas give the energies of the means of the pairs of doublets, indicated by ${}^{2}_{\mathrm{m}}L$.

$$
\begin{aligned}
{}^{2}_{\mathrm{m}}P &= F_0 + 7F_2 & \zeta({}^{2}_{\mathrm{m}}P) &= -\tfrac{1}{6}\zeta_p + \tfrac{1}{2}\zeta_d \\
{}^{4}P &= F_0 + 7F_2 - (G_1^{pd} + 63G_3^{pd}) - G_1^{sp} - G_2^{sd} & \zeta({}^{4}P) &= -\tfrac{1}{6}\zeta_p + \tfrac{1}{2}\zeta_d \\
{}^{2}_{\mathrm{m}}D &= F_0 - 7F_2 & \zeta({}^{2}_{\mathrm{m}}D) &= \tfrac{1}{18}\zeta_p + \tfrac{5}{18}\zeta_d \\
{}^{4}D &= F_0 - 7F_2 + (3G_1^{pd} - 21G_3^{pd}) - G_1^{sp} - G_2^{sd} & \zeta({}^{4}D) &= \tfrac{1}{18}\zeta_p + \tfrac{5}{18}\zeta_d \\
{}^{2}_{\mathrm{m}}F &= F_0 + 2F_2 & \zeta({}^{2}_{\mathrm{m}}F) &= \tfrac{1}{9}\zeta_p + \tfrac{2}{9}\zeta_d \\
{}^{4}F &= F_0 + 2F_2 - (6G_1^{pd} + 3G_3^{pd}) - G_1^{sp} - G_2^{sd} & \zeta({}^{4}F) &= \tfrac{1}{9}\zeta_p + \tfrac{2}{9}\zeta_d
\end{aligned}
$$

in which $\zeta({}^{2}_{\mathrm{m}}L)$ is the mean of the ζ's for the two ${}^{2}L$'s, and

$$
\begin{aligned}
F_0 &= F_0(ns, n'p) + F_0(ns, n''d) + F_0(n'p, n''d) \\
F_2 &= F_2(n'p, n''d) \\
G_1^{sp} &= G_1(ns, n'p) & G_2^{sd} &= G_2(ns, n''d) \\
G_1^{pd} &= G_1(n'p, n''d) & G_3^{pd} &= G_3(n'p, n''d)
\end{aligned}
$$

The predicted order of the doublet means is expressed in

$$({}^{2}_{\mathrm{m}}P - {}^{2}_{\mathrm{m}}F)/({}^{2}_{\mathrm{m}}F - {}^{2}_{\mathrm{m}}D) = \tfrac{5}{9}.$$

The order is DPF in Sc I $3d\,4s\,4p$ which is probably strongly perturbed by $3d^2\,4p$. In Yt I and Zr II $4d\,5s\,5p$ the order is correct but with large departures from the ratio, probably mainly because of perturbation by $4d^2\,5p$.

d^2

$$
\begin{aligned}
{}^{1}S &= F_0 + 14F_2 + 126F_4 & & \\
{}^{3}P &= F_0 + 7F_2 - 84F_4 & \zeta({}^{3}P) &= \tfrac{1}{2}\zeta_d \\
{}^{1}D &= F_0 - 3F_2 + 36F_4 & & \\
{}^{3}F &= F_0 - 8F_2 - 9F_4 & \zeta({}^{3}F) &= \tfrac{1}{2}\zeta_d \\
{}^{1}G &= F_0 + 4F_2 + F_4 & &
\end{aligned}
$$

Regarding the F_2 and F_4 as adjustable parameters we have here four term intervals to be expressed in terms of two parameters. A convenient graphical way of doing this is to plot the observed term values as ordinates against the coefficient of F_2 in the formulas as abscissas. If the F_4 coefficients were all zero, the points should lie on a straight line whose slope is F_2 and whose intercept is F_0. As they are not we must draw a line on the graph so that the ordinate differences between the points and the line are as closely proportional to the coefficients of F_4 as possible. Of course the parameters can be chosen by some analytic process like least squares, but the graphical method is accurate enough and gives a better idea of the nature of the fit.

The values of the parameters chosen so as to give the best agreement with the data for $3d^2$ in the isoelectronic sequence Sc II to Ni IX, as investigated by Cady,* are shown in Fig. 2^7. They vary roughly linearly with the degree of ionization as they should. The values are something like 80 per cent. larger than what is given by using hydrogenic radial functions for the one-electron $3d$ states in $8^6$15a, the hydrogenic values being $F_2(3d^2) = 203Z$ and $F_4(3d^2) = 14{\cdot}7Z$ in cm^{-1}. The ^1S term is much too low in every case, so it was left out of account in determining the parameters. The accuracy with which the other terms of these configurations are represented by the theoretical formulas in terms of these parameter values is shown in Fig. 3^7. There is no obvious reason for the discrepancies in Yt II. The Zr III $4d^2$ configuration cannot be represented at all with possible values of F_2 and F_4, probably because of strong perturbation by $4d\,5s$. Similarly La II $5d^2$ seems to be strongly perturbed by $5d\,6s$.

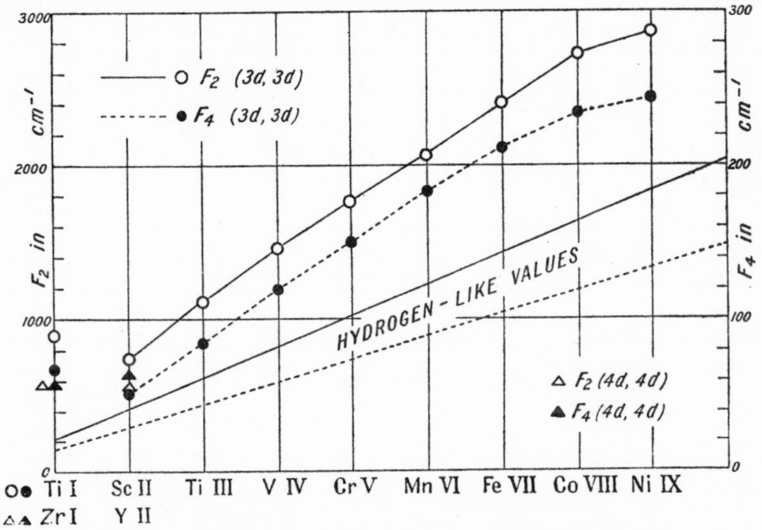

Fig. 2^7. Parameter values in the configuration d^2.

The interval rule and the relation $\zeta(^3P) = \zeta(^3F)$ are quite accurately fulfilled in the long isoelectronic sequence studied by Cady.

$\underline{d^2 s}$

$$
\begin{array}{ll}
^2S = F_0 + 14F_2 + 126F_4 - G_2 & \\
^2P = F_0 + 7F_2 - 84F_4 + G_2 & \zeta(^2P) = \tfrac{2}{3}\zeta_d \\
^4P = F_0 + 7F_2 - 84F_4 - 2G_2 & \zeta(^4P) = \tfrac{1}{3}\zeta_d \\
^2D = F_0 - 3F_2 + 36F_4 - G_2 & \zeta(^2D) = 0 \\
^2F = F_0 - 8F_2 - 9F_4 + G_2 & \zeta(^2F) = \tfrac{2}{3}\zeta_d \\
^4F = F_0 - 8F_2 - 9F_4 - 2G_2 & \zeta(^4F) = \tfrac{1}{3}\zeta_d \\
^2G = F_0 + 4F_2 + F_4 - G_2 & \zeta(^2G) = 0
\end{array}
$$

Here $F_0 = F_0(nd, nd) + 2F_0(nd, n's)$; $F_{2,4} = F_{2,4}(nd, nd)$; $G_2 = G_2(nd, n's)$.

* Cady, *loc. cit.*

We observe that the separations of ^2S, $\frac{1}{3}(^2P+2\,^4P)$, ^2D, $\frac{1}{3}(^2F+2\,^4F)$, and ^2G are the same as those of ^1S, ^3P, ^1D, ^3F, ^1G in d^2.

Neglecting the ^2S which is much too low we may fit Ti II $3d^2\,4s$ and Yt I $4d^2\,5s$ with the values

	Ti II	Yt I
F_2	1015	434
F_4	59	32

as shown in Fig. 4^7. These configurations overlap and interact with $3d^3$ and $4d^3$ respectively. In Zr II $4d^2\,5s$ the ^2S is not so badly off, although the

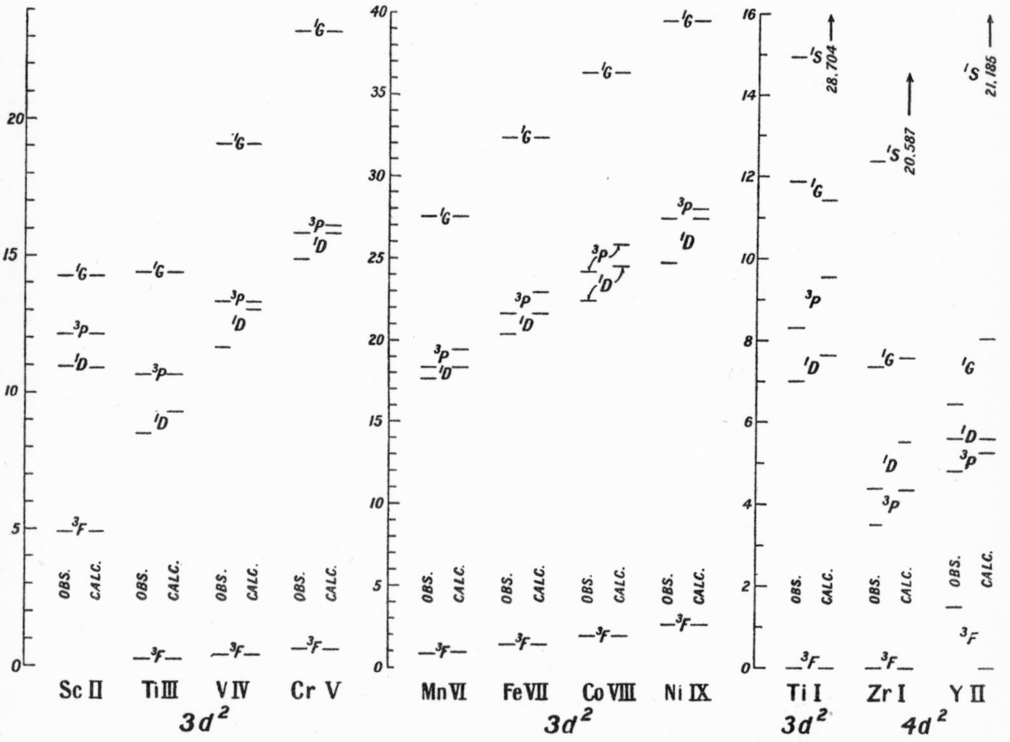

Fig. 3^7. The configuration d^2. (Scales in thousands of wave numbers.)

agreement in general is very poor. The best parameters are $F_2=905$ and $F_4=55$.

According to the formulas $^2P-{^4P}$ should equal $^2F-{^4F}$. This is only roughly true in the first two cases and not at all true in the third:

	Ti II	Yt I	Zr II
$(^2P-{^4P})$	6620	3964	−1877
$(^2F-{^4F})$	4558	4357	5404

Further reference to these cases is made in Chapter XV, where effects of configuration interaction are discussed.

dd

$$^1S,\ ^3S = F_0 + 14F_2 + 126F_4 \pm (G_0 + 14G_2 + 126G_4)$$
$$^1P,\ ^3P = F_0 + 7F_2 - 84F_4 \mp (G_0 + 7G_2 - 84G_4)$$
$$^1D,\ ^3D = F_0 - 3F_2 + 36F_4 \pm (G_0 - 3G_2 + 36G_4)$$
$$^1F,\ ^3F = F_0 - 8F_2 - 9F_4 \mp (G_0 - 8G_2 - 9G_4)$$
$$^1G,\ ^3G = F_0 + 4F_2 + F_4 \pm (G_0 + 4G_2 + G_4)$$
$$\zeta(^3P) = \zeta(^3D) = \zeta(^3F) = \zeta(^3G) = \tfrac{1}{4}\zeta_{nd} + \tfrac{1}{4}\zeta_{n'd}$$

The best example of this configuration is Sc II $3d\,4d$. As before the ^1S is much too low and was left out of account. The parameters used are $F_2 = 107$,

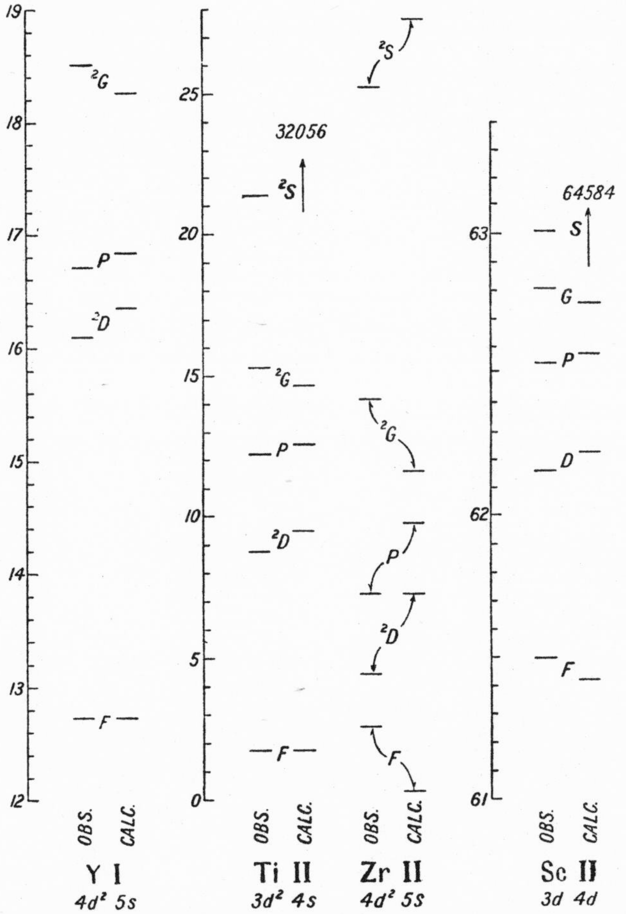

Fig. 4[7]. Configurations d^2s and dd. (Scales in thousands of wave numbers.)

$F_4 = 6$; Fig. 4[7] shows the relation of observed and calculated singlet-triplet means. The five singlet-triplet differences are represented by $G_0 = 2230$,

$G_2 = 36{\cdot}7$ and $G_4 = 3{\cdot}5$ in the manner shown in the table:

	$^1S-{}^3S$	$^1P-{}^3P$	$^1D-{}^3D$	$^1F-{}^3F$	$^1G-{}^3G$
Observed	3872	−4275	4384	−3935	4864
Calculated	6368	−4385	4492	−3810	4760

Other known instances are Yt II $4d\,5d$, Zr III $4d\,5d$ and La II $5d\,6d$ where the agreement is not as good; the singlet-triplet differences are entirely in disagreement in La II.

d^3

This configuration gives two ^2D terms, so we can obtain only their mean by the diagonal sum rule:

$$\begin{array}{llll}
{}^2P = 3F_0 - 6F_2 - 12F_4 & & \zeta({}^2P) = & \tfrac{2}{3}\zeta_d \\
{}^4P = 3F_0 & - 147F_4 & \zeta({}^4P) = & \tfrac{1}{3}\zeta_d \\
{}^2_{\mathrm{m}}D = 3F_0 + 5F_2 + 3F_4 & & \zeta({}^2_{\mathrm{m}}D) = & \tfrac{1}{6}\zeta_d \\
{}^2F = 3F_0 + 9F_2 - 87F_4 & & \zeta({}^2F) = & -\tfrac{1}{6}\zeta_d \\
{}^4F = 3F_0 - 15F_2 - 72F_4 & & \zeta({}^4F) = & \tfrac{1}{3}\zeta_d \\
{}^2G = 3F_0 - 11F_2 + 13F_4 & & \zeta({}^2G) = & \tfrac{3}{10}\zeta_d \\
{}^2H = 3F_0 - 6F_2 - 12F_4 & & \zeta({}^2H) = & \tfrac{1}{5}\zeta_d
\end{array}$$

It is interesting to note that ^2P and ^2H have the same first-order energy. This configuration is discussed in detail in § 7^8.

df

$$\begin{array}{ll}
{}^1P,\,{}^3P = F_0 + 24F_2 + 66F_4 \pm (G_1 + 24G_3 + 330G_5) & \zeta({}^3P) = -\tfrac{1}{2}\zeta_d + \zeta_f \\
{}^1D,\,{}^3D = F_0 + 6F_2 - 99F_4 \mp (3G_1 + 42G_3 - 165G_5) & \zeta({}^3D) = \tfrac{1}{2}\zeta_f \\
{}^1F,\,{}^3F = F_0 - 11F_2 + 66F_4 \pm (6G_1 + 19G_3 + 55G_5) & \zeta({}^3F) = \tfrac{1}{8}\zeta_d + \tfrac{3}{8}\zeta_f \\
{}^1G,\,{}^3G = F_0 - 15F_2 - 22F_4 \mp (10G_1 - 35G_3 - 11G_5) & \zeta({}^3G) = \tfrac{7}{40}\zeta_d + \tfrac{13}{40}\zeta_f \\
{}^1H,\,{}^3H = F_0 + 10F_2 + 3F_4 \pm (15G_1 + 10G_3 + G_5) & \zeta({}^3H) = \tfrac{1}{5}\zeta_d + \tfrac{3}{10}\zeta_f
\end{array}$$

The only known case is La II $5d\,4f$, a case in which knowledge of the theoretical formulas was of use in the analysis.* The singlet-triplet means are well represented on taking $F_0 = 21400$, $F_2 = 115$, $F_4 = 16$:

	P	D	F	G	H
Observed	25214	20534	21467	19038	23680
Calculated	25216	20506	21191	19323	22598

The G parameters were fitted to the observed separations by least squares, the values being $G_1 = 357{\cdot}6$, $G_3 = 29{\cdot}7$, $G_5 = 3{\cdot}78$ with the following representation of the data for the singlet-triplet differences:

	P	D	F	G	H
Observed	4421	−3279	6112	−4879	9690
Calculated	4638	−3397	5837	−4987	11330

This gives a beautiful example of the alternation in sign of the singlet-triplet difference with increasing L. The triplet intervals do not follow the Landé rule sufficiently well for a comparison of the ζ's.

* Condon and Shortley, *loc. cit.* p. 1042.

f^2 and ff

The formulas for ff are

$$\begin{array}{ll}
{}^1S,\ {}^3S = F_0+60F_2+198F_4+1716F_6 \pm (G_0+60G_2+198G_4+1716G_6) \\
{}^1P,\ {}^3P = F_0+45F_2+\ 33F_4-1287F_6 \mp (G_0+45G_2+\ 33G_4-1287G_6) \\
{}^1D,\ {}^3D = F_0+19F_2-\ 99F_4+\ 715F_6 \pm (G_0+19G_2-\ 99G_4+\ 715G_6) \\
{}^1F,\ {}^3F = F_0-10F_2-\ 33F_4-\ 286F_6 \mp (G_0-10G_2-\ 33G_4-\ 286G_6) \\
{}^1G,\ {}^3G = F_0-30F_2+\ 97F_4+\ \ 78F_6 \pm (G_0-30G_2+\ 97G_4+\ \ 78G_6) \\
{}^1H,\ {}^3H = F_0-25F_2-\ 51F_4-\ \ 13F_6 \mp (G_0-25G_2-\ 51G_4-\ \ 13G_6) \\
{}^1I,\ {}^3I = F_0+25F_2+\ \ 9F_4+\ \ \ \ F_6 \pm (G_0+25G_2+\ \ 9G_4+\ \ \ \ G_6)
\end{array}$$

$$\zeta(^3P)=\zeta(^3D)=\zeta(^3F)=\zeta(^3G)=\zeta(^3I)=\tfrac{1}{4}\zeta_{nf}+\tfrac{1}{4}\zeta_{n'f}$$

The formulas for f^2 are obtained by omitting the expressions in the G's and noting that the allowed terms are: ^1S, ^3P, ^1D, ^3F, ^1G, ^3H, ^1I. The intervals are given by $\zeta(^3P)=\zeta(^3F)=\zeta(^3H)=\frac{1}{2}\zeta_f$.

No case of ff is known at present. But one f^2 is known: La II $4f^2$. Using least squares the F parameters were determined as $F_2=93{\cdot}33$, $F_4=21{\cdot}58$, $F_6=0{\cdot}262$. The comparison with observation is shown in Fig. 5[7]. The intervals in the triplets of La II $4f^2$ are badly perturbed.

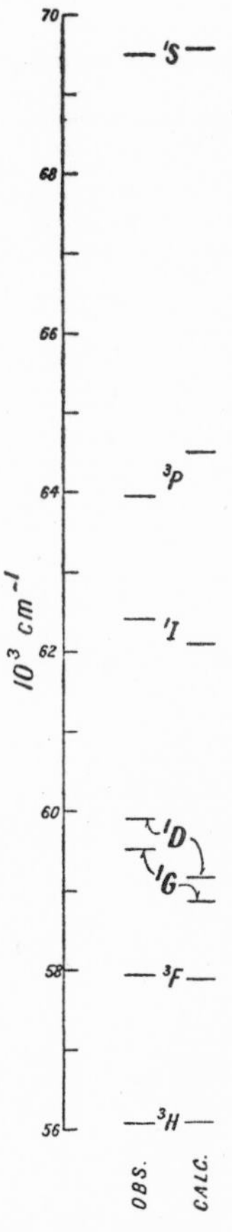

Fig. 5[7]. The $4f^2$ configuration of La II.

The upshot of all these comparisons is clearly that, although the first-order formulas agree with the data in many cases and explain some qualitative facts like alternation in sign of singlet-triplet differences, the second-order perturbations are usually quite important.

6. Terms in the nl^x configurations.

It is of interest to consider the configurations in which all of the electrons outside of closed shells belong to the same shell. These are commonly the type associated with the lowest levels of an atom (compare § 1[14]). The number of different individual sets possible in the nl shell is $N_l=2(2l+1)$, so without the Pauli principle the number of different complete sets possible with x electrons in the nl shell would be N_l^x. The exclusion principle however effects a strong reduction in this number, since it requires that all individual sets in a complete set be different and that two complete sets containing the same individual sets in a different order not count as different.

Therefore the number of states associated with the configuration nl^x is

$$\frac{N_l(N_l-1)\dots(N_l-x+1)}{x!}=\binom{N_l}{x},$$

the binomial coefficient which gives the coefficient of z^x in $(1+z)^{N_l}$. This number is symmetric with regard to $x=N_l/2$ since $\binom{N_l}{x}=\binom{N_l}{N_l-x}$ and has a maximum value for $x=N_l/2$. Therefore the complexity of these configurations increases with increase in x to the middle of the shell and then decreases symmetrically for larger x up to $x=N_l$.

The Russell-Saunders terms corresponding to the configurations s^x and p^x have been considered already in the course of the preceding section. For the configurations d^x and f^x the calculation of the terms by the method of § 2 becomes rather lengthy. Formulas for the term energies as expressions in the F and G integrals have not been worked out,* but the kind of terms to be expected are given in Table 1^7 which, for completeness, includes the s^x and p^x configurations already discussed.† The values of $\zeta(SL)$ as given by Goudsmit (*loc. cit.* § 4^7) for p and d electrons are also included.

TABLE 1^7. *Russell-Saunders terms and Landé factors for nl^x configurations.*

The small numerals under the term designation show the number of terms of the kind occurring in the configuration. The fractions in parentheses give the values of α in the equations $\zeta(SL)=\alpha\zeta_l$ for $x<2l+1$ and $\zeta(SL)=-\alpha\zeta_l$ for $x>2l+1$. Where two terms of a kind occur, the mean value of α is given.

s	^{2}S			
s^2	^{1}S			
$p,\ p^5$	$^{2}P(1)$			
$p^2,\ p^4$	^{1}SD	$^{3}P(\frac{1}{2})$		
p^3	$^{2}PD(0)$	^{4}S		
$d,\ d^9$	$^{2}D(1)$			
$d^2,\ d^8$	^{1}SDG	$^{3}PF(\frac{1}{2}\ \frac{1}{2})$		
$d^3,\ d^7$	$^{2}PDFGH(\frac{2}{3}\ \frac{1}{6}\ -\frac{1}{6}\ \frac{3}{10}\ \frac{1}{5})$ 2 (under D)	$^{4}PF(\frac{1}{3}\ \frac{1}{3})$		
$d^4,\ d^6$	$^{1}SDFGI$ 2 2 2 (under S D G)	$^{3}PDFGH(\frac{1}{4}\ -\frac{1}{12}\ \frac{1}{24}\ \frac{3}{20}\ \frac{1}{10})$ 2 2 (under P F)	$^{5}D(\frac{1}{4})$	
d^5	$^{2}SPDFGHI(0)$ 3 2 2 (under D F G)	$^{4}PDFG(0)$	^{6}S	
$f,\ f^{13}$	^{2}F			
$f^2,\ f^{12}$	$^{1}SDGI$	^{3}PFH		
$f^3,\ f^{11}$	$^{2}PDFGHIKL$ 2 2 2 2	$^{4}SDFGI$		
$f^4\ f^{10}$	$^{1}SDFGHIKLN$ 2 4 4 2 3 2	$^{3}PDFGHIKLM$ 3 2 4 3 4 2 2	$^{5}SDFGI$	
$f^5,\ f^9$	$^{2}PDFGHIKLMNO$ 4 5 7 6 7 5 5 3 2	$^{4}SPDFGHIKLM$ 2 3 4 4 3 3 2	^{6}PFH	
$f^6,\ f^8$	$^{1}SPDFGHIKLMNQ$ 4 6 4 8 4 7 3 4 2 2	$^{3}PDFGHIKLMNO$ 6 5 9 7 9 6 6 3 3	$^{5}SPDFGHIKL$ 3 2 3 2 2	^{7}F
f^7	$^{2}SPDFGHIKLMNOQ$ 2 5 7 10 10 9 9 7 5 4 2	$^{4}SPDFGHIKLMN$ 2 2 6 5 7 5 5 3 3	$^{6}PDFGHI$	^{8}S

* The term energies for d^4 are given by OSTROFSKY, Phys. Rev. **46**, 604 (1934).

† RUSSELL, Phys. Rev. **29**, 782 (1927);
GIBBS, WILBER, and WHITE, *ibid.* **29**, 790 (1927).

The configurations $3d^x$ occur in the elements from Sc I to Zn I, the so-called iron group; $4d^x$ and $5d^x$ occur in the elements Y I to Cd I (the palladium group) and Lu I to Hg I (the platinum group) respectively. The configurations $4f^x$ occur in the rare earths, whose spectra are not yet analysed in any detail. Even in the case of the iron group where quite elaborate analyses have been made the results are as yet far from complete. For example, in Fe I, the normal configuration $3d^6$ gives rise to sixteen terms according to the table, but reference to Bacher and Goudsmit's tables shows that only the ^5D, ^3F, and ^3G terms are known. Similarly of the sixteen terms expected from $3d^5$ only one, ^6S, is known in Mn I.

It is easy to find the largest L value of the terms of highest multiplicity. For $x \leqslant 2l+1$, it will be possible to have an individual set in which all m_s are positive, satisfying the Pauli principle by taking different m_l values. Hence the terms of highest multiplicity have $s = x/2$. The largest m_l is clearly given by taking $m_{l1} = l$, $m_{l2} = l-1$, ... in which case

$$L = \Sigma m_l = \sum_{y=1}^{x} (l - y + 1) = xl - \tfrac{1}{2}x(x-1).$$

Thus for a d shell this gives ^1S, ^2D, ^3F, ^4F, ^5D, and ^6S in agreement with the table as x ranges from 0 to 5. For $x \geqslant 2l+1$ we can have m_s positive for the first $2l+1$ individual sets and for these $\Sigma m_l = 0$, for the others we must take m_s negative and take the remaining m_l's as large as possible. This means that from the middle of the shell on the highest multiplicity diminishes by unity for each added electron and the associated L values run through the same sequence as in the first half of the shell.

It is an empirical fact, noted by Hund,* that the term of largest S and among these the term of largest L is lowest in energy.

It is easy to find an expression for the $\zeta(\gamma SL)$ of the term of largest S and L in these configurations by the method of § 4[7], since this term occurs as the sole occupant of the cell of largest M_S and M_L in the table of complete sets. The result is that for this term

$$\zeta(\gamma SL) = \pm \frac{1}{2S} \zeta_{nl},$$

the plus sign applying for $x < (2l+1)$ and the minus sign for $x > (2l+1)$. This result affords an interpretation of the trend of the observed $\zeta(\gamma SL)$ in the elements of the iron group in which the $3d^x$ configurations occur. These are plotted against x in Fig. 6[7]. We see that they remain fairly constant in the first half of the shell, then reverse sign and increase rapidly in magnitude in the second half of the shell.

Using the formula just derived we can infer the value of ζ_{3d} in each

* Hund, *Linienspektren*, p. 124.

element from the observed $\zeta(\gamma SL)$. The values so obtained are also plotted in the same figure. They lie on a smooth curve showing a steady increase with atomic number. This is just what we should expect because with increased atomic number the radial function $R(3d)$ will be drawn in toward the origin, giving more weight to the regions where $\left(\frac{1}{r}\frac{dU}{dr}\right)$ is large and consequently increasing the value of ζ_{3d}. A similar situation holds in the palladium group where the $4d^x$ configurations occur, but the result is rougher because the data are not as complete and the departures from Russell-Saunders coupling are greater.

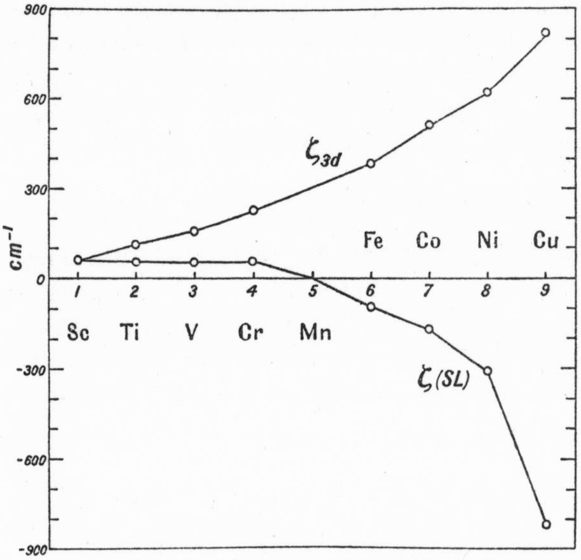

Fig. 6^7. Term-interval parameters in the $3d^x$ configurations of the iron-group elements.

PROBLEM

Show that the term of highest L value for nl^x is a singlet with $L = xl - \frac{1}{4}x(x-2)$ if x is even, or a doublet with $L = xl - \frac{1}{4}(x-1)^2$ if x is odd ($x \leqslant 2l+1$).

7. The triplet terms of helium.

Our discussion of the relativistic theory of the one-electron problem in § 5^5 showed how intimately connected is the spin-orbit interaction energy with other relativistic effects. We have allowed for this interaction approximately thus far by introducing the terms

$$\sum_i \xi(r_i)\,\boldsymbol{L}_i\cdot\boldsymbol{S}_i \quad \text{with} \quad \xi(r_i) = \frac{1}{2\mu^2c^2}\left(\frac{1}{r_i}\frac{\partial U(r_i)}{\partial r_i}\right)$$

in the Hamiltonian, and in § 5^7 we have seen that this simple approximation is useful in correlating a large amount of empirical data on intervals between levels of a term. In Chapter XI we shall see that is also quite suitable for discussion of a wider class of cases in which the spin-orbit interaction energy becomes comparable with the electrostatic interaction of electrons not in closed shells.

But there are cases, particularly for light atoms, in which a more refined discussion of the spin terms is necessary. The classical instance is that of the triplet terms of helium. Data are known for $2\,^3P$, $3\,^3P$, $3\,^3D$ and $4\,^3D$. These terms are narrow and inverted, and the departures from the Landé interval rule are so great that they were at first thought to be doublet terms. Thus for $2p\,^3P$ the interval ratio is 1:14 ($-0{\cdot}07\,\text{cm}^{-1}:-0{\cdot}99\,\text{cm}^{-1}$) instead of the Landé value of 2:1. These facts have been the subject of a number of papers* and are definitely related to the inadequacy of the usual approximation for spin-orbit interaction. The question has been approached in two ways. One is to adopt the model of the electron as a small magnet and to set up the additional terms for its magnetic interaction with the nucleus and other moving electrons by classical mechanics. The other is to take the relativistic classical formula for the interaction of two moving charges and try to adapt it to quantum mechanics by methods suggested by Dirac's theory for one electron. The first approach is that originally adopted by Heisenberg, his work being earlier than the relativistic theory of Dirac.

The spin contributions to the Hamiltonian, according to Heisenberg, may be written for the two-electron problem as

$$\begin{aligned}&\xi(r_1)\,\boldsymbol{L}_1\cdot\boldsymbol{S}_1+\xi(r_2)\,\boldsymbol{L}_2\cdot\boldsymbol{S}_2\\&\quad-\frac{e^2}{\mu c^2}\left\{\frac{(\boldsymbol{r}_1-\boldsymbol{r}_2)\times(\boldsymbol{v}_1-\boldsymbol{v}_2)}{r_{12}^3}-\frac{1}{2}\frac{(\boldsymbol{r}_1-\boldsymbol{r}_2)\times\boldsymbol{v}_1}{r_{12}^3}\right\}\cdot\boldsymbol{S}_1\\&\quad-\frac{e^2}{\mu c^2}\left\{\frac{(\boldsymbol{r}_2-\boldsymbol{r}_1)\times(\boldsymbol{v}_2-\boldsymbol{v}_1)}{r_{12}^3}-\frac{1}{2}\frac{(\boldsymbol{r}_2-\boldsymbol{r}_1)\times\boldsymbol{v}_2}{r_{12}^3}\right\}\cdot\boldsymbol{S}_2\\&\quad+\frac{e^2}{\mu^2c^2}\left\{\frac{\boldsymbol{S}_1\cdot\boldsymbol{S}_2\,r_{12}^2-3\boldsymbol{S}_1\cdot(\boldsymbol{r}_2-\boldsymbol{r}_1)\boldsymbol{S}_2\cdot(\boldsymbol{r}_2-\boldsymbol{r}_1)}{r_{12}^5}\right\}.\end{aligned}\qquad(1)$$

In this expression $\xi(r)$ is to be calculated simply for the Coulomb field of the nucleus, so the first line contains terms which we have already considered in the theory for hydrogen. The next two lines represent the interaction of each electron magnet with the field produced by the other electron, $\boldsymbol{v}_1$ and $\boldsymbol{v}_2$ being the velocities of the two electrons. Finally the last line is the direct interaction of the magnetic dipoles of the two electrons. The terms of the last three lines are given by purely classical considerations by associating

* Heisenberg, Zeits. für Phys. **39**, 499 (1926);
Gaunt, Proc. Roy. Soc. **A122**, 153 (1929); Trans. Roy. Soc. **228**, 151 (1929);
Oppenheimer, Phys. Rev. **35**, 461 (1930);
Breit, Phys. Rev. **36**, 383 (1930).

with each electron a magnetic moment $-e\boldsymbol{S}/\mu c$. The problem of spin-orbit interaction in a particular spectrum is thus reduced to the calculation of the perturbation energy associated with these terms.

If we rearrange the terms in the second two lines we see that they can be expressed in terms of the products $\boldsymbol{L}_1\cdot\boldsymbol{S}_1$, $\boldsymbol{L}_2\cdot\boldsymbol{S}_2$, $\boldsymbol{L}_2\cdot\boldsymbol{S}_1$ and $\boldsymbol{L}_1\cdot\boldsymbol{S}_2$, together with terms in which mixed expressions like $(\boldsymbol{r}_1\times\boldsymbol{v}_2)\cdot\boldsymbol{S}_1$ occur. The whole term in $\boldsymbol{L}_1\cdot\boldsymbol{S}_1$ is found to be

$$\frac{e^2}{\mu^2c^2}\left[\frac{Z}{r_1^3}-\frac{1}{2r_{12}^3}\right]\boldsymbol{L}_1\cdot\boldsymbol{S}_1,$$

on using the definition of $\xi(r_1)$ in terms of the Coulomb field. This shows in a rough way how part of the mutual interaction of electrons comes in to reduce the full effect of the nuclear field in the terms of the form $\boldsymbol{L}_i\cdot\boldsymbol{S}_i$ and affords a slight justification for the use of $U(r)$ in place of the Coulomb field for calculating spin-orbit interaction in complex spectra. But the whole situation is at present quite unsatisfactory and calls for more accurate treatment.

The approach through Dirac's equations does not make use of the magnetic moment in the model. It starts by replacing the ordinary Coulomb interaction by the interaction law of classical electron theory

$$\frac{e^2}{r}\left(1-\frac{\boldsymbol{v}_1\cdot\boldsymbol{v}_2}{c^2}\right)$$

which includes the magnetic interaction of the moving electrons as well as their electrostatic interaction. The velocities $\boldsymbol{v}_1/c$ and $\boldsymbol{v}_2/c$ are replaced by the matrix vectors $\boldsymbol{\beta}_1$ and $\boldsymbol{\beta}_2$ of the Dirac theory and this term incorporated into a Hamiltonian which also includes Dirac's relativistic Hamiltonian written down for each of the two electrons. This procedure is entirely provisional and is beset with numerous difficulties, so we shall not give further details. When the Hamiltonian so formed is reduced by approximate elimination of the small Dirac functions that go with negative values of the rest-mass term it is in close correspondence with the one obtained from the classical model. The differences are akin to those discussed in § 5^5 in relation to the S-levels of the one-electron problem.

The most careful computation of the triplet structure is that made by Breit, who obtains a quite satisfactory agreement with experiment in the cases of $2\,^3P$ of He I and Li II.

CHAPTER VIII

THE RUSSELL-SAUNDERS CASE: EIGENFUNCTIONS

In this chapter we discuss not only the calculation of eigenfunctions in LS coupling in terms of the zero-order functions, but the question of the meaning of quantum numbers referring to less than the whole atom for antisymmetric states. In § 1 we find that operators may be rigorously defined to characterize the resultant momenta of all the electrons in a given shell in the configuration. This enables us in § 2 to introduce a characterization of LS coupling terms by parents and grandparents and in § 3 to calculate the Landé intervals of a term from those of its ancestors. §§ 4, 5, and 6 are devoted to calculations of eigenfunctions proper; while § 7 discusses the calculation from these eigenfunctions of the separate energies and Landé splittings in cases where two terms of the same character occur in a configuration, with detailed results for the two ^2D's of d^3.

1. Vector coupling in antisymmetric states.*

In this section we consider the meaning of vector coupling in antisymmetric states, and the extent to which the matrix methods of Chapter III are available for use in connection with such states. The situation is roughly as follows. In the antisymmetric state characterized by the quantum numbers $n^1 l^1 m_s^1 m_l^1$, $n^2 l^2 m_s^2 m_l^2$, ...; of what operator is m_l^1 the eigenvalue? Clearly not the operator L_z-of-the-first-electron (unless all the m_l's are equal). But if $n^1 l^1$ differs from all the other nl's in the configuration, it is an eigenstate of the operator L_z-of-the-$n^1 l^1$-electron. If $n^2 l^2$ also differs from all the other nl's, we may add the two L's to obtain a resultant L and M_L and the two S's to obtain a resultant S and M_S for the $n^1 l^1$ and $n^2 l^2$ electrons by the formulas of § 14^3. Then in terms of these states the matrices of L's and S's of the $n^1 l^1$ and $n^2 l^2$ electrons would be given by $9^3 11$ and $10^3 2$. But if $n^1 l^1 \equiv n^2 l^2$, we can no longer define an operator $\boldsymbol{L}$-of-the-$n^1 l^1$-electron, since no operator will distinguish between the two electrons in an antisymmetric coupled state. It is, however, still sensible to define a resultant $\boldsymbol{L}$ for the two $n^1 l^1$ electrons, but this operator will not be the sum of two commuting angular momenta, and will not have the allowed values determined by the addition of a vector l^1 to a vector l^1. Hence if a group of equivalent electrons occurs in a configuration, we must be content to work with the whole group as a unit in our vector coupling, not trying to define the angular momentum of less than the whole group. These ideas are given precision by

* SHORTLEY, Phys. Rev. **40**, 197 (1932). For a discussion from the point of view of second quantization see JOHNSON, Phys. Rev. **43**, 627 (1933).

the following consideration of the coupling of two inequivalent groups of electrons.

We consider a group of N_{I} electrons $(1 \ldots N_{\text{I}})$ in a given configuration I. Let us denote the allowed terms for this configuration by the quantum numbers $\gamma^{\text{I}} S^{\text{I}} L^{\text{I}}$, and the corresponding antisymmetric states by

$$X_{\text{I}}(\gamma^{\text{I}} S^{\text{I}} L^{\text{I}} M_S^{\text{I}} M_L^{\text{I}}) \equiv X_{\text{I}}(\lambda^{\text{I}}). \tag{1}$$

Such a state will be a linear combination* of products of the N_{I} one-electron functions of the type

$$X_{\text{I}}(\lambda^{\text{I}}) = \sum_{m_s^i m_l^i} \sum_P C_{m_s^i m_l^i, P}\, \phi_{P1}(n^1 l^1 m_s^1 m_l^1)\, \phi_{P2}(n^2 l^2 m_s^2 m_l^2) \ldots. \tag{2}$$

The phases in (1) for the states of a given term are supposed to be chosen so that the matrices $(\lambda_{\text{I}}^{\text{I}}|\boldsymbol{L}_{\text{I}}|\lambda'^{\text{I}}_{\text{I}})$† and $(\lambda_{\text{I}}^{\text{I}}|\boldsymbol{S}_{\text{I}}|\lambda'^{\text{I}}_{\text{I}})$ of $\boldsymbol{L}_{\text{I}} = \boldsymbol{L}_1 + \boldsymbol{L}_2 + \ldots + \boldsymbol{L}_{N_{\text{I}}}$ and $\boldsymbol{S}_{\text{I}} = \boldsymbol{S}_1 + \boldsymbol{S}_2 + \ldots + \boldsymbol{S}_{N_{\text{I}}}$ have the values given by $3^3 7$.

Consider also a second group, of N_{II} electrons $(N_{\text{I}}+1, N_{\text{I}}+2, \ldots, N_{\text{I}}+N_{\text{II}})$ in a configuration II *whose nl values are all different from any in* I.‡ We denote the antisymmetric states of this group by

$$X_{\text{II}}(\gamma^{\text{II}} S^{\text{II}} L^{\text{II}} M_S^{\text{II}} M_L^{\text{II}}) \equiv X_{\text{II}}(\lambda^{\text{II}}). \tag{3}$$

The products

$$X_{\text{I}}(\lambda^{\text{I}})\, X_{\text{II}}(\lambda^{\text{II}}) \tag{4}$$

of functions of type (1) and functions of type (3) are functions of all $N = N_{\text{I}} + N_{\text{II}}$ electrons for which the matrix elements of $\boldsymbol{L}_{\text{I}}$ have the values

$$(\lambda_{\text{I}}^{\text{I}} \lambda_{\text{II}}^{\text{II}}|\boldsymbol{L}_{\text{I}}|\lambda'^{\text{I}}_{\text{I}} \lambda'^{\text{II}}_{\text{II}}) = (\lambda_{\text{I}}^{\text{I}}|\boldsymbol{L}_{\text{I}}|\lambda'^{\text{I}}_{\text{I}})\, \delta(\lambda^{\text{II}}, \lambda'^{\text{II}}). \tag{5}$$

Let us now define a new operator $\boldsymbol{L}^{\text{I}}$ to represent, not the sum of the angular momenta of electrons $1 \ldots N_{\text{I}}$, but the sum of the angular momenta of the electrons in configuration I. Explicitly, we define $\boldsymbol{L}^{\text{I}}$ as that operator which when acting on a state ϕ of configuration I+II, expressed as a product of one-electron functions as in $2^6 5'$, has the same effect as the operator $\boldsymbol{L}_a + \boldsymbol{L}_b + \ldots$, where $a, b, \ldots$ are the indices in ϕ of the electrons in configuration I. Here II is any configuration inequivalent to I. When acting on a state of a configuration which cannot be represented as I+II, $\boldsymbol{L}^{\text{I}}$ vanishes. We make corresponding definitions of $\boldsymbol{L}^{\text{II}}$, $\boldsymbol{S}^{\text{I}}$, and $\boldsymbol{S}^{\text{II}}$.

For states formed by the exchange of electrons between groups in (4), the matrix elements of $\boldsymbol{L}^{\text{I}}$ have the values

$$X_{Q\text{I}}(\lambda^{\text{I}})\, X_{Q\text{II}}(\lambda^{\text{II}})\, \boldsymbol{L}^{\text{I}}\, X_{Q'\text{I}}(\lambda'^{\text{I}})\, X_{Q'\text{II}}(\lambda'^{\text{II}}) = \delta(Q, Q')\, (\lambda^{\text{I}} \lambda^{\text{II}}|\boldsymbol{L}^{\text{I}}|\lambda'^{\text{I}} \lambda'^{\text{II}}), \tag{6}$$

where

$$(\lambda^{\text{I}} \lambda^{\text{II}}|\boldsymbol{L}^{\text{I}}|\lambda'^{\text{I}} \lambda'^{\text{II}}) = (\lambda_{\text{I}}^{\text{I}} \lambda_{\text{II}}^{\text{II}}|\boldsymbol{L}_{\text{I}}|\lambda'^{\text{I}}_{\text{I}} \lambda'^{\text{II}}_{\text{II}}). \tag{7}$$

Here QI is the set of electrons which the permutation Q of electrons $1 \ldots N$ puts in place of electrons $1 \ldots N_{\text{I}}$. Q is restricted to be such a permutation that the electrons in QI and QII are ordered. We shall for convenience call

* We shall see in §§ 4^8, 5^8, 6^8 how to determine these linear combinations.

† The symbol $\lambda'^{\text{I}}_{\text{I}}$ indicates that group $_{\text{I}}$ of electrons $(1 \ldots N_{\text{I}})$ have quantum numbers λ'^{I}.

‡ Two configurations related in this way will be said to be 'inequivalent.'

such a permutation an *exchange*. Formula (6) follows from the fact that the resultant of operation by $\boldsymbol{L}^{\mathrm{I}}$ on its operand in (6) is the same as the result of operation by $\boldsymbol{L}_{Q'\mathrm{I}}$ on this operand. Matrix components of $\boldsymbol{L}^{\mathrm{I}}$ between states of two different configurations are of course all zero.

The state $$X_{Q\mathrm{I}}(\lambda^{\mathrm{I}})\, X_{Q\mathrm{II}}(\lambda^{\mathrm{II}}) \equiv Q\, X_{\mathrm{I}}(\lambda^{\mathrm{I}})\, X_{\mathrm{II}}(\lambda^{\mathrm{II}}) \tag{8}$$
is thus an eigenstate of $(\boldsymbol{S}^{\mathrm{I}})^2$, $(\boldsymbol{L}^{\mathrm{I}})^2$, S_z^{I}, L_z^{I}, with the quantum numbers S^{I}, L^{I}, M_S^{I}, M_L^{I}. A similar statement holds for the corresponding operators referring to configuration II. We may easily construct an antisymmetric linear combination of these eigenstates:

$$\Upsilon(\lambda^{\mathrm{I}}\lambda^{\mathrm{II}}) = \frac{1}{\sqrt{N!/N_{\mathrm{I}}!N_{\mathrm{II}}!}} \sum_Q (-1)^q Q\, X_{\mathrm{I}}(\lambda^{\mathrm{I}})\, X_{\mathrm{II}}(\lambda^{\mathrm{II}}) = \mathscr{A}'\, X_{\mathrm{I}}(\lambda^{\mathrm{I}})\, X_{\mathrm{II}}(\lambda^{\mathrm{II}}), \tag{9}$$

where the summation runs over all the $N!/N_{\mathrm{I}}!N_{\mathrm{II}}!$ exchanges Q, and q has the parity of Q. Since the states in the summand of (9) are all orthogonal, Υ is normalized, and is hence an allowed antisymmetric state of configuration I+II. The matrix components of $\boldsymbol{L}^{\mathrm{I}}$ between states of the type (9) are seen from (6) to be given by

$$\overline{\Upsilon}(\lambda^{\mathrm{I}}\lambda^{\mathrm{II}})\, \boldsymbol{L}^{\mathrm{I}}\, \Upsilon(\lambda'^{\mathrm{I}}\lambda'^{\mathrm{II}}) = (\lambda^{\mathrm{I}}\lambda^{\mathrm{II}}|\boldsymbol{L}^{\mathrm{I}}|\lambda'^{\mathrm{I}}\lambda'^{\mathrm{II}}). \tag{10}$$

Hence, the states Υ of (9) have all the properties of the states $\phi(j_1 j_2 m_1 m_2)$ of 6[3]1 with respect to the angular momenta $\boldsymbol{L}^{\mathrm{I}}$, $\boldsymbol{L}^{\mathrm{II}}$, and again with respect to $\boldsymbol{S}^{\mathrm{I}}$, $\boldsymbol{S}^{\mathrm{II}}$. Therefore we can use the formulas of § 14[3] to add $\boldsymbol{L}^{\mathrm{I}}$ and $\boldsymbol{L}^{\mathrm{II}}$ or $\boldsymbol{S}^{\mathrm{I}}$ and $\boldsymbol{S}^{\mathrm{II}}$, or both, to obtain states

$$\Psi(\gamma^{\mathrm{I}} S^{\mathrm{I}} L^{\mathrm{I}}, \gamma^{\mathrm{II}} S^{\mathrm{II}} L^{\mathrm{II}}, S\, L\, M_S M_L), \tag{11}$$

where $$\boldsymbol{S} = \boldsymbol{S}_{\mathrm{I}} + \boldsymbol{S}_{\mathrm{II}} = \boldsymbol{S}^{\mathrm{I}} + \boldsymbol{S}^{\mathrm{II}}, \quad \boldsymbol{L} = \boldsymbol{L}_{\mathrm{I}} + \boldsymbol{L}_{\mathrm{II}} = \boldsymbol{L}^{\mathrm{I}} + \boldsymbol{L}^{\mathrm{II}}.$$

The expressions of § 10[3] for the matrix components of $\boldsymbol{S}^{\mathrm{I}}$, $\boldsymbol{L}^{\mathrm{I}}$, $\boldsymbol{S}^{\mathrm{II}}$, $\boldsymbol{L}^{\mathrm{II}}$ will accordingly hold for this system of states.

For any quantity $F = F_1 + F_2 + \ldots + F_N$, let us similarly define F^{I} as that operator which behaves like $F_a + F_b + \ldots$ when acting on a state of configuration I+II, and which vanishes otherwise. Then the action of F^{I} on a state of type (8) gives

$$F^{\mathrm{I}} X_{Q\mathrm{I}}(\lambda^{\mathrm{I}})\, X_{Q\mathrm{II}}(\lambda^{\mathrm{II}}) = \sum_{\lambda^{\mathrm{III}}} X_{Q\mathrm{I}}(\lambda^{\mathrm{III}})\, X_{Q\mathrm{II}}(\lambda^{\mathrm{II}})\, (\lambda^{\mathrm{III}}\lambda^{\mathrm{II}}|F^{\mathrm{I}}|\lambda^{\mathrm{I}}\lambda^{\mathrm{II}}) + \sum_{\alpha} X(\alpha)\, (\alpha|F^{\mathrm{I}}|\lambda^{\mathrm{I}}\lambda^{\mathrm{II}}), \tag{12}$$

where $$(\lambda^{\mathrm{III}}\lambda^{\mathrm{II}}|F^{\mathrm{I}}|\lambda^{\mathrm{I}}\lambda^{\mathrm{II}}) = (\lambda_{\mathrm{I}}^{\mathrm{III}}\lambda_{\mathrm{II}}^{\mathrm{II}}|F_{\mathrm{I}}|\lambda_{\mathrm{I}}^{\mathrm{I}}\lambda_{\mathrm{II}}^{\mathrm{II}}), \tag{13}$$
and α is a state of a configuration which cannot be written in the form III+II.* Similarly the action of F^{II} on this state gives

$$F^{\mathrm{II}} X_{Q\mathrm{I}}(\lambda^{\mathrm{I}})\, X_{Q\mathrm{II}}(\lambda^{\mathrm{II}}) = \sum_{\lambda^{\mathrm{IV}}} X_{Q\mathrm{I}}(\lambda^{\mathrm{I}})\, X_{Q\mathrm{II}}(\lambda^{\mathrm{IV}})\, (\lambda^{\mathrm{I}}\lambda^{\mathrm{IV}}|F^{\mathrm{II}}|\lambda^{\mathrm{I}}\lambda^{\mathrm{II}}) + \sum_{\beta} X(\beta)\, (\beta|F^{\mathrm{II}}|\lambda^{\mathrm{I}}\lambda^{\mathrm{II}}), \tag{14}$$

* For example, if configuration I is d^2 and II p^2, a typical III is df, while a typical α is $d\,p^3$. F^{I} has of course no components between configurations differing in more than one electron.

where β is a state of a configuration which cannot be written in the form I+IV.

From these formulas we see that the general matrix component of $F = F^{\mathrm{I}} + F^{\mathrm{II}}$ connecting states of configurations I+II and III+IV is

$$\bar{X}_{Q\mathrm{I}}(\lambda^{\mathrm{III}})\,\bar{X}_{Q\mathrm{II}}(\lambda^{\mathrm{IV}})\,F\,X_{Q'\mathrm{I}}(\lambda^{\mathrm{I}})\,X_{Q'\mathrm{II}}(\lambda^{\mathrm{II}})$$
$$= \delta(Q,Q')\,\delta(\lambda^{\mathrm{IV}},\lambda^{\mathrm{II}})\,(\lambda^{\mathrm{III}}\lambda^{\mathrm{II}}|F^{\mathrm{I}}|\lambda^{\mathrm{I}}\lambda^{\mathrm{II}}) + \delta(Q,Q')\,\delta(\lambda^{\mathrm{III}},\lambda^{\mathrm{I}})\,(\lambda^{\mathrm{I}}\lambda^{\mathrm{IV}}|F^{\mathrm{II}}|\lambda^{\mathrm{I}}\lambda^{\mathrm{II}}). \quad (15)$$

Hence the matrix components of F between two antisymmetric states of the type (9) are

$$(\lambda^{\mathrm{III}}\lambda^{\mathrm{IV}}|F|\lambda^{\mathrm{I}}\lambda^{\mathrm{II}}) = \delta(\lambda^{\mathrm{IV}},\lambda^{\mathrm{II}})\,(\lambda^{\mathrm{III}}\lambda^{\mathrm{II}}|F^{\mathrm{I}}|\lambda^{\mathrm{I}}\lambda^{\mathrm{II}}) + \delta(\lambda^{\mathrm{III}},\lambda^{\mathrm{I}})\,(\lambda^{\mathrm{I}}\lambda^{\mathrm{IV}}|F^{\mathrm{II}}|\lambda^{\mathrm{I}}\lambda^{\mathrm{II}}). \quad (16)$$

From this, (13), and (10) we see that in evaluating these matrix components and those joining corresponding coupled states of type (11), we may treat F as $F^{\mathrm{I}} + F^{\mathrm{II}}$, where F^{I} commutes with $\boldsymbol{S}^{\mathrm{II}}$, $\boldsymbol{L}^{\mathrm{II}}$, and F^{II} with $\boldsymbol{S}^{\mathrm{I}}$, $\boldsymbol{L}^{\mathrm{I}}$. If F is a vector of type $\boldsymbol{T}$ (§ 8[3]) with respect to either $\boldsymbol{L}$ or $\boldsymbol{S}$, and commutes with the other, the matrix components of F^{I} and F^{II} between states of type (11) will be given by the formulas of § 11[3].

To summarize, we have found that the matrix components of a quantity of type F between an antisymmetric state of configuration I+II and an antisymmetric state of III+IV may be reduced to those of F_{I} between antisymmetric states of I and III and of F_{II} between antisymmetric states of II and IV in just the way we should if the uncoupled states of I+II and III+IV had been taken of type (4), without exchange of electrons in the process of antisymmetrizing as in (9). Here configurations I and II must be inequivalent, and III and IV must be inequivalent.

2. Genealogical characterization of LS-coupling terms.

As pointed out in § 1[7], a term is not in general completely characterized by its configuration and SL value since a given configuration may contain several terms of the same sort. We shall now investigate the allowed terms from a vector-coupling standpoint with a view to obtaining where possible a more general characterization and an introduction of the important concept of parentage.

If we divide the electrons of a configuration into two inequivalent groups as in the preceding paragraph, we see that the allowed terms of the configuration may be obtained as in 1[8]11 by coupling in all possible ways the $\boldsymbol{S}$ and $\boldsymbol{L}$ vectors of the groups. This means that *we can obtain the allowed terms for any configuration by using Table* 1[7] *of allowed terms for groups of equivalent electrons*, and that these terms will be characterized by the terms of the constituent groups. Where a group contains three or more equivalent d or f electrons, we cannot distinguish between the terms occurring more than once in this group except by giving the actual eigenfunctions, since it is impossible to divide such a group.

Thus for the hypothetical configuration $p^2 d^2$, the allowed terms are

$$\begin{array}{lll} p^2\,(^1S)\,d^2\,(^1S)\ ^1S & p^2\,(^3P)\,d^2\,(^1S)\ ^3P & p^2\,(^1D)\,d^2\,(^1S)\ ^1D \\ \qquad\quad (^3P)\ ^3P & \qquad\quad (^3P)\ ^{1,3,5}SPD & \qquad\quad (^3P)\ ^3PDF \\ \qquad\quad (^1D)\ ^1D & \qquad\quad (^1D)\ ^3PDF & \qquad\quad (^1D)\ ^1SPDFG \\ \qquad\quad (^3F)\ ^3F & \qquad\quad (^3F)\ ^{1,3,5}DFG & \qquad\quad (^3F)\ ^3PDFGH \\ \qquad\quad (^1G)\ ^1G & \qquad\quad (^1G)\ ^3FGH & \qquad\quad (^1G)\ ^1DFGHI. \end{array} \tag{1}$$

Although there are six each of ^1D's, ^3P's, and ^3F's occurring here, each is individually characterized by giving the terms of p^2 and d^2 from which it arises.

When one electron is added to an 'ion,' the terms of the ion are called the *parents* of the terms of the atom. Thus if we consider the addition of nonequivalent s, p, and d electrons respectively to the sd configuration of an ion, we obtain the terms

$$\begin{array}{lll} \underline{s\,d\,s} & \underline{s\,d\,p} & \underline{s\,d\,d} \\ s\,d\,(^1D)\,s\ ^2D & s\,d\,(^1D)\,p\ ^2PDF & s\,d\,(^1D)\,d\ ^2SPDFG \\ s\,d\,(^3D)\,s\ ^2D,\ ^4D & s\,d\,(^3D)\,p\ ^2PDF,\ ^4PDF & s\,d\,(^3D)\,d\ ^2SPDFG,\ ^4SPDFG. \end{array} \tag{2}$$

The groups of terms of the same multiplicity which arise from the addition of s, p, and d electrons are called *monads*, *triads*, and *pentads* respectively* (in general *polyads*). Thus, in the example above, the configuration sdp contains a doublet triad having as parent $sd\,(^1D)$ and a doublet and a quartet triad having $sd\,(^3D)$ as parent.

The occurrence of these close groups of three or five terms was noticed empirically soon after the beginning of the analysis of complex spectra. Hence the eigenfunctions which arise in the convenient theoretical designation by parentage must to a certain approximation be the actual eigenfunctions which diagonalize the electrostatic interaction, even when more than one term of a kind occurs. The reason for this is illustrated by the plot, in Fig. 1^8, of the energy-level spectrum of O I. This spectrum appears essentially as the sum of three one-electron spectra converging to the three terms of the low $2p^3$ configuration of O II as limits. The terms built on each of the parent levels have essentially the energies of a single valence electron moving in a central field due to the ion in a particular state. For this spectrum the central-field approximation which regards the electrostatic interaction as a small perturbation on each configuration is quite invalid; the electrostatic interaction of the core is larger than the separation between configurations in the central-field problem. Large interactions between configurations are

* They are designated in this way even when the number of terms is not as large as indicated, e.g. $(^1P)\,d$ gives a pentad containing only three terms ^2PDF.

expected, not only among the low configurations plotted, but between these and configurations high above the ionization potential which have other terms of O II as parents; for the agreement with the formulas of § 5^7 of $2p^3\,3s$ is no better or worse than the agreement of the O II $2p^3$ parent, and the perturbations of this parent are caused by higher states of O II. But whatever

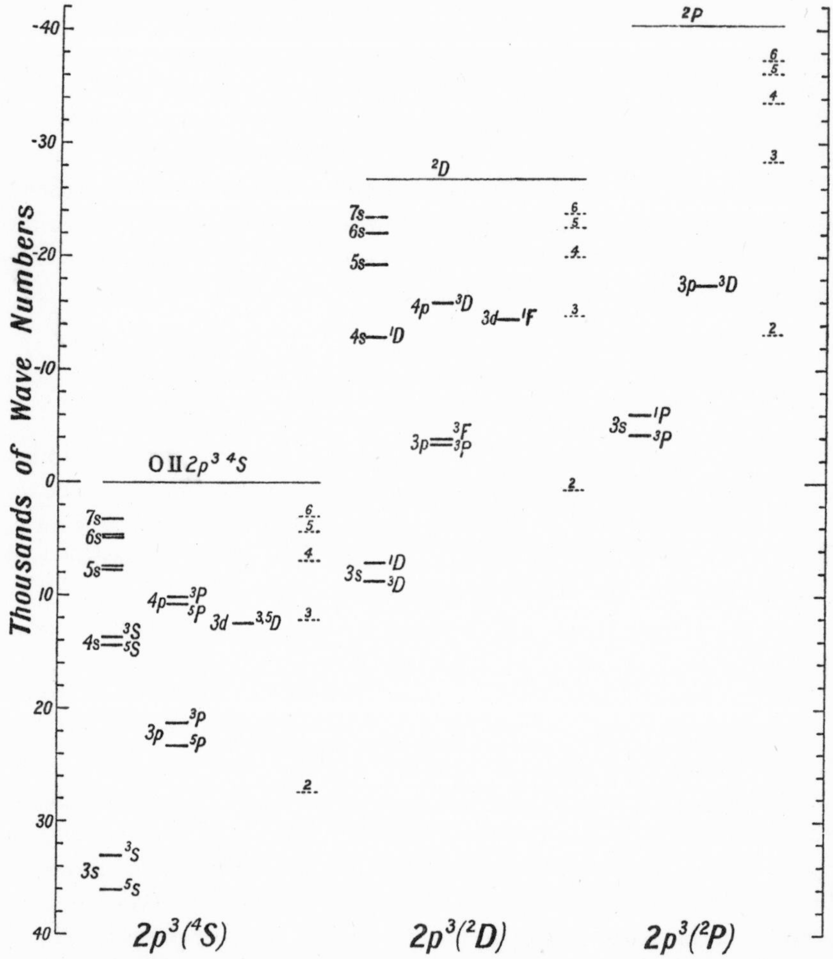

Fig. 1^8. The energy levels of oxygen I. This shows all observed levels except $2p^4$ (76,000-110,000 cm^{-1}) and $2s\,2p^5\,{}^3P$ ($-13{,}000$ cm^{-1}). Hydrogen levels are indicated by broken lines.

the perturbation of O II p^3, it is certain that the actual eigenstates of the p^3 levels are to a good approximation the parents of the plotted levels of O I. The electrostatic interaction of the valence electron with the ion causes only the small separation of the singlet and triplet, or triplet and quintet polyads.

In many of the complex spectra, such as those of Ti, V, and Zr, where the overlapping of configurations is not so exaggerated as in O I, the levels of a given configuration will still be found to occur in groups (polyads) having approximately the relative locations of the terms of the parent configuration. The parentage is not exact, however, for as soon as the intervals within the polyads—due to the electrostatic interaction of the valence electron with the ion—becomes comparable with the separation of the polyads we may expect an interaction between the similar terms arising from different parents.

3. Landé intervals for terms of coupled groups.*

With the results of § 1[8] we may easily obtain a formula for the intervals in the terms of a configuration I+II in terms of the intervals in the in-equivalent configurations I and II separately. From 3[7]2 we find that

$$\zeta(\gamma SL) = \sum_{i=1}^{N} (\gamma L \vdots \xi(r_i) L_i \vdots \gamma L)\,(\gamma S \vdots S_i \vdots \gamma S). \tag{1}$$

We have omitted the sum over γ' because each quantum number in γ' will in general either refer to the spins, in which case $\xi(r_i)\boldsymbol{L}_i$ will be diagonal with respect to it, or will not refer to the spins, in which case $\boldsymbol{S}_i$ will be diagonal. If now the term γSL is derived by the addition of the term $S^{\mathrm{I}}L^{\mathrm{I}}$ of configuration I and the term $S^{\mathrm{II}}L^{\mathrm{II}}$ of configuration II, we may split the spin-orbit interaction into two parts, one of which refers to I, the other to II. This breaks the sum (1) into two parts

$$\begin{aligned}\zeta(S^{\mathrm{I}}L^{\mathrm{I}}S^{\mathrm{II}}L^{\mathrm{II}}SL) &= \sum_{\mathrm{I}} (S^{\mathrm{I}}L^{\mathrm{I}}S^{\mathrm{II}}L^{\mathrm{II}}L \vdots \xi(r_i)L_i \vdots S^{\mathrm{I}}L^{\mathrm{I}}S^{\mathrm{II}}L^{\mathrm{II}}L)\,(S^{\mathrm{I}}L^{\mathrm{I}}S^{\mathrm{II}}L^{\mathrm{II}}S \vdots S_i \vdots S^{\mathrm{I}}L^{\mathrm{I}}S^{\mathrm{II}}L^{\mathrm{II}}S)\\ &+ \sum_{\mathrm{II}}(\ldots).\end{aligned} \tag{2}$$

In accordance with the discussion at the end of § 1[8], we may evaluate each of these parts independently without troubling about the fact that the eigenfunction is antisymmetric. In the first sum, $\xi(r_i)\boldsymbol{L}_i$ and $\boldsymbol{S}_i$ may then be assumed to commute with L^{II} and S^{II}, in which case the factors are each given by 11[3]8 if we correlate

$$\begin{array}{ll} \boldsymbol{Q} \to \xi(r_i)\boldsymbol{L}_i \text{ or } \boldsymbol{S}_i & \qquad j_1 \to L^{\mathrm{I}} \text{ or } S^{\mathrm{I}} \\ j \to L \text{ or } S & \qquad j_2 \to L^{\mathrm{II}} \text{ or } S^{\mathrm{II}}. \end{array} \tag{3}$$

This reduces the first sum to a known multiple of

$$\sum_{\mathrm{I}} (L^{\mathrm{I}} \vdots \xi(r_i) L_i \vdots L^{\mathrm{I}})\,(S^{\mathrm{I}} \vdots S_i \vdots S^{\mathrm{I}}) = \zeta(S^{\mathrm{I}}L^{\mathrm{I}}).$$

* GOUDSMIT and HUMPHREYS, Phys. Rev. 31, 960 (1928).

The second sum is similarly reduced, to obtain finally

$$\zeta(S^{\text{I}}L^{\text{I}}\,S^{\text{II}}\,L^{\text{II}}\,S\,L) = \frac{L(L+1)-L^{\text{II}}(L^{\text{II}}+1)+L^{\text{I}}(L^{\text{I}}+1)}{2L(L+1)} \times \frac{S(S+1)-S^{\text{II}}(S^{\text{II}}+1)+S^{\text{I}}(S^{\text{I}}+1)}{2S(S+1)}\,\zeta(S^{\text{I}}\,L^{\text{I}})$$
$$+\frac{L(L+1)-L^{\text{I}}(L^{\text{I}}+1)+L^{\text{II}}(L^{\text{II}}+1)}{2L(L+1)} \times \frac{S(S+1)-S^{\text{I}}(S^{\text{I}}+1)+S^{\text{II}}(S^{\text{II}}+1)}{2S(S+1)}\,\zeta(S^{\text{II}}\,L^{\text{II}}). \quad (4)$$

By this formula, the values of ζ for any configuration may be expressed in terms of those for configurations of the type nl^x, which were given in Table 1[7]. In particular, if a single nl electron is added to an ion, $\zeta(S^{\text{II}}\,L^{\text{II}}) = \zeta_{nl}$. Formula (4) gives the values of ζ immediately for two-electron configurations $nl\,n'l'$ if we take

$$L^{\text{I}}=l,\quad L^{\text{II}}=l',\quad S^{\text{I}}=S^{\text{II}}=\tfrac{1}{2},\quad \zeta(S^{\text{I}}\,L^{\text{I}})=\zeta_{nl},\quad \text{and}\quad \zeta(S^{\text{II}}\,L^{\text{II}})=\zeta_{n'l'};$$

as we shall show in § 6[8], this gives the correct answer even if the electrons are equivalent.

4. Calculation of eigenfunctions by direct diagonalization.*

We wish now to show how to determine the eigenfunctions of states in LS coupling in terms of states in the zero-order nlm_sm_l scheme.

States characterized by $SLJM$ are given directly in terms of states characterized by SLM_SM_L by the formulas of § 14[3]. States characterized by SLM_SM_L for configuration I+II are given in terms of those of configurations I and II by the general procedure of § 1[8]; this is discussed in detail in § 6[8]. These considerations reduce the problem essentially to the determination of eigenfunctions in the SLM_SM_L scheme for configurations consisting wholly of equivalent electrons. In order to use the formulas of § 14[3] in the way we have sketched, it is however necessary that the relative phases of the states of a given term be such that the matrices of $\boldsymbol{L}$ and $\boldsymbol{S}$ have the values given by 3[3]7.

The direct method of obtaining eigenfunctions characterized by SLM_SM_L is to diagonalize the matrices of $\boldsymbol{L}^2$ and $\boldsymbol{S}^2$ in the zero-order nlm_sm_l scheme. Since such a diagonalization does not give states with phases chosen in any particular way, it is necessary, if we wish states characterized by $SLJM$, to diagonalize also the matrix of $\boldsymbol{J}^2$ (or, for greater convenience, of $\boldsymbol{L}\cdot\boldsymbol{S}$). Because the process is laborious and the phases arbitrary, this method has little practical application. We shall, nevertheless, consider the determination of the matrices of $\boldsymbol{L}^2$, $\boldsymbol{S}^2$, and $\boldsymbol{L}\cdot\boldsymbol{S}$ in the nlm_sm_l scheme because

* Johnson, Phys. Rev. 39, 197 (1932).

our discussion in Chapter XII of transformations for configurations containing almost closed shells will depend on the properties of these matrices. Because the simultaneous diagonalization of these three matrices is an excellent illustration of the process of matrix diagonalization described in § 7[2], we shall give detailed results for the configuration p^3.

The matrix components of $\boldsymbol{L}^2$, $\boldsymbol{S}^2$, and $\boldsymbol{L}\cdot\boldsymbol{S}$ in the nlm_sm_l scheme may be easily expressed in terms of the one-electron components (cf. 3[3]4 and 7[3]3)

$$\begin{aligned}(a|L_x|b) &= (n^a l^a m_s^a m_l^a|L_x|n^b l^b m_s^b m_l^b)\\ &= \delta(n^a l^a m_s^a, n^b l^b m_s^b)\,\delta(m_l^a \pm 1, m_l^b)\,\tfrac{1}{2}\hbar\sqrt{(l^b \pm m_l^b)(l^b \mp m_l^b + 1)}, \quad \text{(1a)*}\end{aligned}$$

$$(a|S_x|b) = \delta(n^a l^a m_l^a, n^b l^b m_l^b)\,\delta(m_s^a \pm 1, m_s^b)\,\tfrac{1}{2}\hbar, \qquad \text{(1b)*}$$

$$\begin{aligned}(a|\boldsymbol{L}\cdot\boldsymbol{S}|b) &= \hbar^2\,\delta(n^a l^a m^a, n^b l^b m^b)\{\delta(a,b)\, m_s^a m_l^a\\ &\quad + \tfrac{1}{2}\,\delta(m_s^a, m_s^b \pm 1)\sqrt{(l^a - m^a + \tfrac{1}{2})(l^a + m^a + \tfrac{1}{2})}\} \qquad \text{(1c)*}\end{aligned}$$

by the formulas of §§ 6[6], 7[6] and the relation 3[3]6. In this way the following results are obtained. (The sign is in all cases to be chosen positive or negative according to the even- or odd-ness of the permutation which changes the set A' from its standard order to the order in which sets in A' which match those in A all stand at the same places in the lists, with a' corresponding to a, b' to b).

Matrix of $\boldsymbol{L}^2$. We get non-diagonal elements of $\boldsymbol{L}^2$ only if A' differs from A in two individual sets, say that A has the sets a, b while A' has a', b'. This component has the value

$$(A|\boldsymbol{L}^2|A') = \pm 4\{(a|L_x|a')(b|L_x|b') - (a|L_x|b')(b|L_x|a')\}\,\delta(M, M'). \qquad \text{(3a)}$$

The diagonal element of $\boldsymbol{L}^2$ is

$$(A|\boldsymbol{L}^2|A) = \hbar^2 M_L^2 + \hbar^2 \sum_a \{l^a(l^a+1) - (m_l^a)^2\} - 4\sum_{a<b}(a|L_x|b)^2, \qquad \text{(3b)}$$

where a and b run over all the individual sets of A.

Matrix of $\boldsymbol{S}^2$. We get non-diagonal elements of $\boldsymbol{S}^2$ only if A' differs from A in two individual sets, say a and b. This component has the value

$$(A|\boldsymbol{S}^2|A') = \pm 4\{(a|S_x|a')(b|S_x|b') - (a|S_x|b')(b|S_x|a')\}\,\delta(M, M'). \qquad \text{(4a)}$$

The diagonal element of $\boldsymbol{S}^2$ is

$$(A|\boldsymbol{S}^2|A) = \hbar^2(M_S^2 + \tfrac{1}{2}N - N'), \qquad \text{(4b)}$$

where N is the number of electrons and N' is the number of pairs of individual sets with the same $n\,l\,m_l$, each pair counted only once.

* It is to be noted that for any of these quantities

$$(n^a\, l^a\, m_s^a\, m_l^a| \quad |n^b\, l^b\, m_s^b\, m_l^b) = (n^a\, l^a\; -m_s^a\; -m_l^a| \quad |n^b\, l^b\; -m_s^b\; -m_l^b). \qquad (2)$$

This property will be of use in Chapter XII.

Matrix of $\boldsymbol{L}\cdot\boldsymbol{S}$. If A' differs from A in regard to two individual sets, a and b, we obtain

$$(A|\boldsymbol{L}\cdot\boldsymbol{S}|A') = \pm 2\{(a|L_x|a')(b|S_x|b') + (b|L_x|b')(a|S_x|a') - (a|L_x|b')(b|S_x|a') - (b|L_x|a')(a|S_x|b')\}\,\delta(M,M'). \quad (5a)$$

If A' differs from A in regard to one individual set, a,

$$(A|\boldsymbol{L}\cdot\boldsymbol{S}|A') = \pm\{(a|\boldsymbol{L}\cdot\boldsymbol{S}|a') - 2\sum_b (a|L_x|b)(b|S_x|a') - 2\sum_c (a'|L_x|c)(c|S_x|a)\}\,\delta(M,M'), \quad (5b)$$

where b and c run over all the individual sets common to A and A'. It is worth noting in connection with this formula that one obtains no value unless a, a' are of the form $(m_l)^-$, $(m_l-1)^+$. The second term in the braces vanishes except for one value of b, namely $(m_l-1)^-$; the third term vanishes except for one value of c, namely $(m_l)^+$. When either of the last two terms does not vanish (i.e. when the appropriate b or c exists in the set A), it has just the value $-(a|\boldsymbol{L}\cdot\boldsymbol{S}|a')$. The diagonal element of $\boldsymbol{L}\cdot\boldsymbol{S}$ is given by

$$(A|\boldsymbol{L}\cdot\boldsymbol{S}|A) = M_S M_L \hbar^2. \quad (5c)$$

With these formulas it is rather easy to calculate the matrices of $\boldsymbol{L}^2$, $\boldsymbol{S}^2$ and $\boldsymbol{L}\cdot\boldsymbol{S}$ in the zero-order scheme. Only those sets of the configuration which are outside of closed shells need be considered in calculating these matrices. The non-diagonal elements are obviously not affected by the presence of closed shells. The part of a diagonal element which results from a closed shell is easily seen to be zero [using 3^75 in the case of $(A|\boldsymbol{L}^2|A)$].

Illustration. The configuration p^3. This method of calculation will be illustrated in detail for the case of the configuration p^3. This configuration leads to the five levels ${}^4S_{\frac{3}{2}}$, ${}^2P_{\frac{1}{2}}$, ${}^2P_{\frac{3}{2}}$, ${}^2D_{\frac{3}{2}}$, ${}^2D_{\frac{5}{2}}$. We shall consider only states characterized by $M=\frac{1}{2}$—this will give us one state of each level. The zero-order states for $M=\frac{1}{2}$ are given by

		M_S	M_L		
A	$(1^-, 0^+, 0^-)$	$-\frac{1}{2}$	1	α	
B	$(1^+, 1^-, -1^-)$	$-\frac{1}{2}$	1	α	
C	$(1^+, 0^+, -1^-)$	$\frac{1}{2}$	0	β	(6a)
D	$(1^+, 0^-, -1^+)$	$\frac{1}{2}$	0	β	
E	$(1^-, 0^+, -1^+)$	$\frac{1}{2}$	0	β	

If we make a plot such as 1^72 we see that the terms ${}^2P$ and ${}^2D$ have states in both box α and box β, while ${}^4S$ has a state only in β. Hence we may infer that the eigenfunctions for ${}^4S_{\frac{3}{2}}$ will not involve states A and B. In this scheme, the matrices of $\boldsymbol{S}^2$, $\boldsymbol{L}^2$, and $\boldsymbol{L}\cdot\boldsymbol{S}$ have the form (cf. § 7^2 for notation; the lines in the matrices indicate the separation between α and β)

$$\|(A|\boldsymbol{S}^2|B)\| = \left\|\begin{array}{c|cc|ccc} & A & B & C & D & E \\ \hline A & 3 & 0 & & 0 & \\ B & 0 & 3 & & & \\ \hline C & & & 7 & 4 & 4 \\ D & & 0 & 4 & 7 & 4 \\ E & & & 4 & 4 & 7 \end{array}\right\| \cdot \tfrac{1}{4}\hbar^2, \quad (6b)$$

$$\|(A|\boldsymbol{L}^2|B)\| = \begin{array}{c|cc|ccc|} & A & B & C & D & E \\ \hline A & 4 & -2 & & 0 & \\ B & -2 & 4 & & & \\ \hline C & & & 2 & -2 & 0 \\ D & & 0 & -2 & 4 & -2 \\ E & & & 0 & -2 & 2 \\ \hline \end{array} \cdot \hbar^2, \tag{6c}$$

$$\|(A|\boldsymbol{L}\cdot\boldsymbol{S}|B)\| = \begin{array}{c|cc|ccc|} & A & B & C & D & E \\ \hline A & -1 & 0 & \sqrt{2} & -\sqrt{2} & 0 \\ B & 0 & -1 & 0 & \sqrt{2} & -\sqrt{2} \\ \hline C & \sqrt{2} & 0 & & & \\ D & -\sqrt{2} & \sqrt{2} & & 0 & \\ E & 0 & -\sqrt{2} & & & \\ \hline \end{array} \cdot \tfrac{1}{2}\hbar^2. \tag{6d}$$

So far as the diagonalization of $\boldsymbol{L}^2$ and $\boldsymbol{S}^2$ are concerned α and β are quite independent. Consider the matrix of $\boldsymbol{S}^2$. For α the eigenvalues of $\boldsymbol{S}^2$ which may occur are those corresponding to the two doublets, namely $\frac{3}{4}$ and $\frac{3}{4}$. For β we have in addition to these the quartet eigenvalue $\frac{15}{4}$. A transformation which diagonalizes this matrix is found by the method of § 7^2:

$$\|(A|{}^2M)\| = \begin{array}{c|cc|ccc|} & {}^2M & {}^2N & {}^2Q & {}^2R & {}^4S \\ \hline A & \sqrt{6} & 0 & & 0 & \\ B & 0 & \sqrt{6} & & & \\ \hline C & & & 1 & \sqrt{3} & \sqrt{2} \\ D & & 0 & 1 & -\sqrt{3} & \sqrt{2} \\ E & & & -2 & 0 & \sqrt{2} \\ \hline \end{array} \cdot \frac{1}{\sqrt{6}}. \tag{6e}$$

(Because of the degeneracy this transformation is not uniquely determined for either pair of doublets.)

When the matrix (6c) of $\boldsymbol{L}^2$ is transformed to this scheme in which $\boldsymbol{S}^2$, S_z, L_z are diagonal by the relation (cf. 7^27)

$$\|({}^2M|\boldsymbol{L}^2|{}^2N)\| = \|({}^2M|A)\|\cdot\|(A|\boldsymbol{L}^2|B)\|\cdot\|(B|{}^2N)\|, \tag{6f}$$

it takes the form

$$\|({}^2M|\mathbf{L}^2|{}^2N)\| = \begin{array}{c|cc|ccc|} & {}^2M & {}^2N & {}^2Q & {}^2R & {}^4S \\ \hline {}^2M & 4 & -2 & & & \\ {}^2N & -2 & 4 & & 0 & \\ \hline {}^2Q & & & 3 & -\sqrt{3} & 0 \\ {}^2R & & 0 & -\sqrt{3} & 5 & 0 \\ {}^4S & & & 0 & 0 & 0 \\ \hline \end{array} \cdot \hbar^2, \tag{6g}$$

which is diagonal with respect to S, as required. The eigenvalues of $\mathbf{L}^2$ for both the stepmatrices α, $S=\frac{1}{2}$ and β, $S=\frac{1}{2}$ are now known to be those corresponding to P and D, namely 2 and 6. The transformation which diagonalizes (6g) is given by

$$\|({}^2M|{}^2P_\alpha)\| = \begin{array}{c|cc|ccc|} & {}^2P_\alpha & {}^2D_\alpha & {}^2P_\beta & {}^2D_\beta & {}^4S_\beta \\ \hline {}^2M & \sqrt{2} & \sqrt{2} & & & \\ {}^2N & \sqrt{2} & -\sqrt{2} & & 0 & \\ \hline {}^2Q & & & \sqrt{3} & -1 & 0 \\ {}^2R & & 0 & 1 & \sqrt{3} & 0 \\ {}^2S & & & 0 & 0 & 2 \\ \hline \end{array} \cdot \tfrac{1}{2}. \tag{6h}$$

The new states are now in the SLM_SM_L scheme, the values of M_S and M_L being indicated by α or β.

The transformation from the zero-order scheme to the SLM_SM_L scheme may be obtained by the multiplication (cf. 7^26) of (6e) by (6h):

$$\|(A|{}^2P_\alpha)\| = \|(A|{}^2M)\| \cdot \|({}^2M|{}^2P_\alpha)\|; \tag{6i}$$

$$\|(A|{}^2P_\alpha)\| = \begin{array}{c|cc|ccc|} & {}^2P_\alpha & {}^2D_\alpha & {}^2P_\beta & {}^2D_\beta & {}^4S_\beta \\ \hline A & \sqrt{3} & \sqrt{3} & & & \\ B & \sqrt{3} & -\sqrt{3} & & 0 & \\ \hline C & & & \sqrt{3} & 1 & \sqrt{2} \\ D & & 0 & 0 & -2 & \sqrt{2} \\ E & & & -\sqrt{3} & 1 & \sqrt{2} \\ \hline \end{array} \cdot \frac{1}{\sqrt{6}}. \qquad \alpha\begin{cases} M_S = -\frac{1}{2} \\ M_L = 1 \end{cases} \quad \beta\begin{cases} M_S = \frac{1}{2} \\ M_L = 0 \end{cases} \tag{6j}*$$

* We have chosen the phases so that ${}^2P_{\alpha \text{ and } \beta}$, ${}^2D_{\alpha \text{ and } \beta}$ have the proper relative phases as determined by the method of the next section, and so that ${}^2P^{\frac{1}{2}}_{\frac{1}{2} \text{ and } \frac{3}{2}}$, ${}^2D^{\frac{1}{2}}_{\frac{3}{2} \text{ and } \frac{5}{2}}$ are the proper linear combinations of these states as given by § 14^3. In this way we can make full use of these matrices without having to rewrite them with more useful phases.

When the matrix (6d) of $\boldsymbol{L}\cdot\boldsymbol{S}$ is transformed to this scheme, it becomes

$$\|({}^2P_\alpha|\boldsymbol{L}\cdot\boldsymbol{S}|{}^2D_\alpha)\| = \begin{array}{c|cc|ccc|} & {}^2P_\alpha & {}^2D_\alpha & {}^2P_\beta & {}^2D_\beta & {}^4S_\beta \\ \hline {}^2P_\alpha & -1 & 0 & \sqrt{2} & 0 & 0 \\ {}^2D_\alpha & 0 & -1 & 0 & \sqrt{6} & 0 \\ \hline {}^2P_\beta & \sqrt{2} & 0 & & & \\ {}^2D_\beta & 0 & \sqrt{6} & & 0 & \\ {}^4S_\beta & 0 & 0 & & & \\ \hline \end{array} \cdot \tfrac{1}{2}\hbar^2, \qquad (6\text{k})$$

which is diagonal with respect to L and S, as required. The eigenvalues of this matrix are calculated from the formula $\boldsymbol{L}\cdot\boldsymbol{S} = \frac{1}{2}(\boldsymbol{J}^2 - \boldsymbol{L}^2 - \boldsymbol{S}^2)$, which gives -1, $\frac{1}{2}$, $-\frac{3}{2}$, 1, and 0 for ${}^2P_{\frac{1}{2}}$, ${}^2P_{\frac{3}{2}}$, ${}^2D_{\frac{3}{2}}$, ${}^2D_{\frac{5}{2}}$, and ${}^4S_{\frac{3}{2}}$ respectively. The transformation to these states is given by

$$\|({}^2P_\alpha|{}^2P_{\frac{1}{2}})\| = \begin{array}{c|ccccc|} & {}^2P_{\frac{1}{2}} & {}^2P_{\frac{3}{2}} & {}^4S_{\frac{3}{2}} & {}^2D_{\frac{3}{2}} & {}^2D_{\frac{5}{2}} \\ \hline {}^2P_\alpha & -\sqrt{10} & \sqrt{5} & 0 & 0 & 0 \\ {}^2D_\alpha & 0 & 0 & 0 & -3 & \sqrt{6} \\ \hline {}^2P_\beta & \sqrt{5} & \sqrt{10} & 0 & 0 & 0 \\ {}^2D_\beta & 0 & 0 & 0 & \sqrt{6} & 3 \\ {}^4S_\beta & 0 & 0 & \sqrt{15} & 0 & 0 \\ \hline \end{array} \cdot \frac{1}{\sqrt{15}}. \qquad (6\text{l})$$

The transformation from the zero-order scheme to the SLJM scheme is then given by the product of (6j) and (6l):

$$\|(A|{}^2P_{\frac{1}{2}})\| = \begin{array}{c|ccccc|} & {}^2P_{\frac{1}{2}} & {}^2P_{\frac{3}{2}} & {}^4S_{\frac{3}{2}} & {}^2D_{\frac{3}{2}} & {}^2D_{\frac{5}{2}} \\ \hline A & -\sqrt{10} & \sqrt{5} & 0 & -3 & \sqrt{6} \\ B & -\sqrt{10} & \sqrt{5} & 0 & 3 & -\sqrt{6} \\ \hline C & \sqrt{5} & \sqrt{10} & \sqrt{10} & \sqrt{2} & \sqrt{3} \\ D & 0 & 0 & \sqrt{10} & -\sqrt{8} & -\sqrt{12} \\ E & -\sqrt{5} & -\sqrt{10} & \sqrt{10} & \sqrt{2} & \sqrt{3} \\ \hline \end{array} \cdot \frac{1}{\sqrt{30}}. \quad (M = \tfrac{1}{2}) \qquad (6\text{m})^*$$

* See footnote to (6j), p. 224.

It is interesting now to verify directly the electrostatic energies which were obtained in § 5[7] by the diagonal-sum rule. From the results of §§ 7[6] and 8[6] we find for the matrix of electrostatic interaction in the zero-order scheme

$$\|(A|Q|B)\| = \begin{array}{c|cc|ccc|} & A & B & C & D & E \\ \hline A & 1 & -1 & & & \\ B & -1 & 1 & & 0 & \\ \hline C & & & 2 & 1 & 2 \\ D & & 0 & 1 & 3 & 1 \\ E & & & 2 & 1 & 2 \\ \hline \end{array} \cdot(-3F_2)+3F_0. \tag{7}$$

When this matrix is transformed by (6m) to the $SLJM$ scheme, it becomes

$$\|({}^2P_{\frac{1}{2}}|Q|{}^2P_{\frac{3}{2}})\| = \begin{array}{c|ccccc|} & {}^2P_{\frac{1}{2}} & {}^2P_{\frac{3}{2}} & {}^4S_{\frac{3}{2}} & {}^2D_{\frac{3}{2}} & {}^2D_{\frac{5}{2}} \\ \hline {}^2P_{\frac{1}{2}} & 0 & & & & \\ {}^2P_{\frac{3}{2}} & & 0 & & & \\ {}^4S_{\frac{3}{2}} & & & 5 & & \\ {}^2D_{\frac{3}{2}} & & & & 2 & \\ {}^2D_{\frac{5}{2}} & & & & & 2 \\ \hline \end{array} \cdot(-3F_2)+3F_0, \tag{8}$$

which agrees with the results of § 5[7]. The same eigenvalues would have been obtained by transforming (7) to the SLM_SM_L scheme by (6j).

5. Calculation of eigenfunctions using angular-momentum operators.*

We shall now give a convenient method of computing SL eigenfunctions proposed by Gray and Wills. Having found the eigenfunction $\Psi(SLM_SM_L)$ for $M_S=S$, $M_L=L$, this method makes use of the symmetric operator $\mathscr{L}=L_x-iL_y$ to give the eigenfunction $\Psi(S\,L\,M_S\,M_L-1)$ by 3[3]3, and the operator $\mathscr{S}=S_x-iS_y$ to give $\Psi(S\,L\,M_S-1\,M_L)$, and so forth. In this way the whole of an SL term is found with the proper relative phases for all states, so that the results of § 14[3] may be used directly to obtain the $SLJM$ states. The first state is found purely from orthogonality considerations; hence the states of *different* terms have no particular phase relation. We shall need to discuss only the transformation to the SLM_SM_L states.

In order to carry out the operations mentioned we shall need to know the result of operating with $\mathscr{L}$ and $\mathscr{S}$ on a state expressed in terms of the antisymmetric zero-order functions $\Phi(m_s^1 m_l^1, m_s^2 m_l^2, \ldots)$. Since $\mathscr{L}=\mathscr{L}_1+\mathscr{L}_2+\ldots$, we have from the general relation

$$(J_x-iJ_y)\,\psi(j\,m)=\hbar\sqrt{(j+m)(j-m+1)}\,\psi(j\,m-1) \tag{1}$$

* Gray and Wills, Phys. Rev. **38**, 248 (1931).

the result

$$\hbar^{-1}\mathscr{L}\,\phi_1(m_s^1 m_l^1)\,\phi_2(m_s^2 m_l^2)\ldots = \sqrt{(l^1+m_l^1)(l^1-m_l^1+1)}\,\phi_1(m_s^1 m_l^1-1)\,\phi_2(m_s^2 m_l^2)\ldots$$
$$+\sqrt{(l^2+m_l^2)(l^2-m_l^2+1)}\,\phi_1(m_s^1 m_l^1)\,\phi_2(m_s^2 m_l^2-1)\ldots+\ldots. \tag{2}$$

Since $\mathscr{L}$ is a symmetric operator, it commutes with the operator $\mathscr{A}$ of $3^6 6$. When we apply $\mathscr{A}$ to (2) we find

$$\hbar^{-1}\mathscr{L}\,\Phi(m_s^1 m_l^1, m_s^2 m_l^2, \ldots) = \sqrt{(l^1+m_l^1)(l^1-m_l^1+1)}\,\Phi(m_s^1 m_l^1-1, m_s^2 m_l^2, \ldots)$$
$$+\sqrt{(l^2+m_l^2)(l^2-m_l^2+1)}\,\Phi(m_s^1 m_l^1, m_s^2 m_l^2-1, \ldots)+\ldots. \tag{3}$$

With this formula the result of operation by $\mathscr{L}$ on any state is easily calculated.

The result of operation by $\mathscr{S}$ is very similar, the m_s values being successively decreased by 1 in place of the m_l values. The coefficients which occur are limited to 1 and 0, 1 for $m_s=\frac{1}{2}$, 0 for $m_s=-\frac{1}{2}$, so that operation by $\mathscr{S}$ gives just the sum of the states resulting when successive m_l^+ are changed to m_l^-. The operator $\mathscr{J}$, if one has occasion to apply it, is just the sum of $\mathscr{L}$ and $\mathscr{S}$.

Illustration—the configuration d^3. As an illustration of this method of calculation, let us obtain some of the eigenfunctions for the configuration d^3; in particular let us obtain one eigenfunction for each of the two ^2D's to use in § 7^8 for the separation of the energies of these terms. A classification of the zero-order states of d^3 according to M_S, M_L down to $\frac{1}{2}$, 2 together with the terms which have components in each partition is given below (cf. $1^7 2$):

M_S	M_L				M_S	M_L				
$\frac{1}{2}$	5	A	$(2^+ 2^- \; 1^+)$	$[^2H]$						(d^3)
$\frac{1}{2}$	4	B	$(2^+ 2^- \; 0^+)$	^2H						
		C	$(2^+ 1^+ \; 1^-)$	^2G						
$\frac{1}{2}$	3	E	$(2^+ 2^- \; -1^+)$	^2H	$\frac{3}{2}$	3	D	$(2^+ 1^+ \; 0^+)$	$[^4F]$	
		F	$(2^+ 1^+ \; 0^-)$	^2G						
		G	$(2^+ 1^- \; 0^+)$	^4F						
		H	$(2^- 1^+ \; 0^+)$	^2F						(4)
$\frac{1}{2}$	2	J	$(2^+ 2^- \; -2^+)$	^2H	$\frac{3}{2}$	2	I	$(2^+ 1^+ \; -1^+)$	$[^4F]$	
		K	$(2^+ 1^+ \; -1^-)$	^2G						
		L	$(2^+ 1^- \; -1^+)$	^4F						
		M	$(2^- 1^+ \; -1^+)$	^2F						
		N	$(2^+ 0^+ \; 0^-)$	^2D						
		O	$(1^+ 1^- \; 0^+)$	^2D						

Our problem is, then, to obtain two (orthogonal) eigenfunctions for ^2D, $\frac{1}{2}$, 2. Before beginning the calculation proper, let us calculate the action of $\mathscr{L}$ and $\mathscr{S}$ on certain of the above states [omitting the factor $\hbar^{-1}$, cf. (3)]:

$$\begin{aligned}
&\mathscr{L}A = -2C+\sqrt{6}B && \mathscr{L}G = 2O-\sqrt{6}N+\sqrt{6}L\\
&\mathscr{L}B = -2H+2G+\sqrt{6}E && \mathscr{L}H = -2O+\sqrt{6}M\\
&\mathscr{L}C = -\sqrt{6}G+\sqrt{6}F && \\
&\mathscr{L}E = -2M+2L+2J && \mathscr{S}D = H+G+F\\
&\mathscr{L}F = \sqrt{6}N+\sqrt{6}K && \mathscr{S}I = M+L+K.
\end{aligned} \tag{5}$$

Now the state $^2H, \frac{1}{2}, 5$ is just A. From (1) we find that

$$\mathscr{L}(^2H, \tfrac{1}{2}, 5) = \sqrt{10}\,(^2H, \tfrac{1}{2}, 4).$$

But $$\mathscr{L}(^2H, \tfrac{1}{2}, 5) = \mathscr{L}A = -2C + \sqrt{6}B$$

from (5). Hence $$^2H, \tfrac{1}{2}, 4 = 10^{-\frac{1}{2}}[\sqrt{6}B - 2C].$$

The other ^2H's are obtained in succession:

$$\begin{aligned}
&^2H, \tfrac{1}{2}, 5 = A \\
&^2H, \tfrac{1}{2}, 4 = 10^{-\frac{1}{2}}[\sqrt{6}B - 2C] \\
&^2H, \tfrac{1}{2}, 3 = 30^{-\frac{1}{2}}[\sqrt{6}E - 2F + 4G - 2H] \\
&^2H, \tfrac{1}{2}, 2 = 30^{-\frac{1}{2}}[J - K + 3L - 2M - 3N + \sqrt{6}O].
\end{aligned} \tag{6a}$$

The state $^2G, \frac{1}{2}, 4$ is a linear combination of B and C orthogonal to $^2H, \frac{1}{2}, 4$. The choice of its phase is purely arbitrary.

$$\begin{aligned}
&^2G, \tfrac{1}{2}, 4 = 10^{-\frac{1}{2}}[2B + \sqrt{6}C] \\
&^2G, \tfrac{1}{2}, 3 = 20^{-\frac{1}{2}}[\sqrt{6}E + 3F - G - 2H] \\
&^2G, \tfrac{1}{2}, 2 = \sqrt{\tfrac{3}{140}}[2J + 3K + L - 4M + 4N + \sqrt{\tfrac{2}{3}}O].
\end{aligned} \tag{6b}$$

We have now $^4F, \frac{3}{2}, 3 = D$. By operation with $\mathscr{S}$ on this we obtain $^4F, \frac{1}{2}, 3$:

$$\mathscr{S}(^4F, \tfrac{3}{2}, 3) = \sqrt{3}\,(^4F, \tfrac{1}{2}, 3) = \mathscr{S}D.$$

$$\begin{aligned}
&^4F, \tfrac{1}{2}, 3 = 3^{-\frac{1}{2}}[F + G + H] \\
&^4F, \tfrac{1}{2}, 2 = 3^{-\frac{1}{2}}[K + L + M].
\end{aligned} \tag{6c}$$

$^2F, \frac{1}{2}, 3$ is determined by its orthogonality to the above three states for $M_S = \frac{1}{2}$, $M_L = 3$:

$$\begin{aligned}
&^2F, \tfrac{1}{2}, 3 = 12^{-\frac{1}{2}}[-\sqrt{6}E + F + G - 2H] \\
&^2F, \tfrac{1}{2}, 2 = 12^{-\frac{1}{2}}[-2J + K - L + \sqrt{6}O].
\end{aligned} \tag{6d}$$

We have now four states of the partition $M_S = \frac{1}{2}$, $M_L = 2$. The other states of this partition are both ^2D's; any state orthogonal to the four above is a ^2D and there are two linearly independent such states. Let us choose one at random and a second orthogonal to it—

$$\begin{aligned}
&\text{a}\ ^2D, \tfrac{1}{2}, 2 = \tfrac{1}{2}\ [-J - K + L + N] \\
&\text{b}\ ^2D, \tfrac{1}{2}, 2 = 84^{-\frac{1}{2}}[-5J + 3K + L - 4M - 3N - 2\sqrt{6}O].
\end{aligned} \tag{6e}$$

This procedure may be continued to obtain any desired state. Any other configuration is handled in the same way. A check is continuously furnished by the fact that the states must come out properly orthonormal.

6. Calculation of eigenfunctions from vector-coupling formulas.

Let us first consider how we may obtain the eigenfunctions resulting from the coupling of the terms of two inequivalent groups of electrons as in § 1[8].

Let $X_{\text{I}}(S^{\text{I}}L^{\text{I}}M_S^{\text{I}}M_L^{\text{I}})$ represent one of the antisymmetric eigenfunctions of a given term of group I. This is supposed to be known in terms of the zero-order functions $\Phi_{\text{I}}(A) = \Phi_{\text{I}}(a^1, a^2, \ldots, a^{N_1})$ for the configuration in question.

Similarly let $X_{II}(S^{II}\,L^{II}\,M_S^{II}\,M_L^{II})$ be one of the antisymmetric eigenfunctions of a term of the second group. The product of these two states will be a known linear combination of the form

$$X_I(S^I\,L^I\,M_S^I\,M_L^I)\,X_{II}(S^{II}\,L^{II}\,M_S^{II}\,M_L^{II}) = \sum_{A,B} C_{A,B}\,\Phi_I(a^1, a^2, \ldots, a^{N_I})\,\Phi_{II}(b^1, b^2, \ldots, b^{N_{II}}). \quad (1)$$

This is antisymmetrized as in 1[8]9 by application of the operator $\mathscr{A}'$. It is easily seen that

$$\mathscr{A}'\,\Phi_I(a^1, a^2, \ldots, a^{N_I})\,\Phi_{II}(b^1, b^2, \ldots, b^{N_{II}}) = \Phi(a^1, a^2, \ldots, a^{N_I}, b^1, b^2, \ldots, b^{N_{II}}),$$

an antisymmetric zero-order function for configuration I+II; hence the antisymmetric function

$$\Upsilon(S^I\,L^I\,M_S^I\,M_L^I,\, S^{II}\,L^{II}\,M_S^{II}\,M_L^{II}) = \sum_{A,B} C_{A,B}\,\Phi(a^1, a^2, \ldots, b^{N_{II}}). \quad (2)$$

In other words, these antisymmetric functions are obtained by simply replacing the products of Φ's for I and II by Φ's for I+II.

Now let us see how to apply the formulas of § 14[3] to the coupling of S^I and S^{II} and of L^I and L^{II}. Let us write the function Υ *purely symbolically* as a product of a function referring to the S's by a function referring to the L's:

$$\Upsilon(S^I\,L^I\,M_S^I\,M_L^I,\, S^{II}\,L^{II}\,M_S^{II}\,M_L^{II}) = \tau(S^I\,S^{II}\,M_S^I\,M_S^{II})\,\omega(L^I\,L^{II}\,M_L^I\,M_L^{II}). \quad (3)$$

Let $T(S^I\,S^{II}\,S\,M_S)$ designate the linear combination of the τ's which is given by § 14[3] for the coupling of S^I and S^{II}; let $\Omega(L^I\,L^{II}\,L\,M_L)$ similarly refer to the coupling of L^I and L^{II}.* Then symbolically, the LS-coupling states are given by

$$\Psi(S^I\,L^I,\, S^{II}\,L^{II},\, S\,L\,M_S\,M_L) = T(S^I\,S^{II}\,S\,M_S)\,\Omega(L^I\,L^{II}\,L\,M_L), \quad (4)$$

where the right side of this equation is to be interpreted according to (3).

Two non-equivalent electrons.†

The eigenfunctions for a configuration consisting of just two non-equivalent electrons are given by these considerations if group I is one of the electrons, group II the other. $X_I\,X_{II}$ of (1) is then just

$$\phi_1(n^1\,l^1\,m_s^1\,m_l^1)\,\phi_2(n^2\,l^2\,m_s^2\,m_l^2),$$

and Υ of (2) just

$$\Phi(n^1\,l^1\,m_s^1\,m_l^1,\, n^2\,l^2\,m_s^2\,m_l^2),$$

which we write as in (3) as $\tau(m_s^1\,m_s^2)\,\omega(m_l^1\,m_l^2)$. Making the correlation $s^1 = j_1$, $s^2 = j_2$, we find for the T's:

$$\begin{aligned} T(S=0,\ M_S=0) &= 2^{-\frac{1}{2}}[\tau(+,-) - \tau(-,+)] \\ T(\ 1,\quad 1) &= \tau(+,+) \\ T(\ 1,\quad 0) &= 2^{-\frac{1}{2}}[\tau(+,-) + \tau(-,+)] \\ T(\ 1,\ -1) &= \tau(-,-). \end{aligned} \quad (5)$$

* Note that we are *not* here following our convention of capital letters for antisymmetric functions, small letters for all others.

† Bartlett, Phys. Rev. 38, 1623 (1931).

If we consider in particular the configuration $np\,n'p$, with the correlation $l^1=j_1$, $l^2=j_2$, we find for the Ω's:

$$\begin{array}{lll} \Omega(L=0, & M_L=0)=3^{-\frac{1}{2}}[\omega(1,\,-1)-\omega(0,0)+\omega(-1,1)] & \\ \Omega(\quad 1, & \quad 1)=2^{-\frac{1}{2}}[\omega(1,0)-\omega(0,1)] & \\ \Omega(\quad 1, & \quad 0)=2^{-\frac{1}{2}}[\omega(1,\,-1)-\omega(-1,1)] & (6)\\ \Omega(\quad 1, & -1)=2^{-\frac{1}{2}}[\omega(0,\,-1)-\omega(-1,0)] & \\ \Omega(\quad 2, & \text{etc.} & \end{array}$$

The SL eigenfunctions (4) are then the symbolic products of T and Ω:

$$\begin{array}{l} \Psi({}^1S,\ \ 0,0)=6^{-\frac{1}{2}}[\Phi(1^+\,-1^-)-\Phi(1^-\,-1^+)-\Phi(0^+0^-)+\Phi(0^-0^+)+\Phi(-1^+1^-)-\Phi(-1^-1^+)] \\ \Psi({}^3S,\ \ 1,0)=3^{-\frac{1}{2}}[\Phi(1^+\,-1^+)-\Phi(0^+0^+)+\Phi(-1^+1^+)] \\ \Psi({}^3S,\ \ 0,0)=6^{-\frac{1}{2}}[\Phi(1^+\,-1^-)+\Phi(1^-\,-1^+)-\Phi(0^+0^-)-\Phi(0^-0^+)+\Phi(-1^+1^-)+\Phi(-1^-1^+)] \\ \Psi({}^3S,\,-1,0)=3^{-\frac{1}{2}}[\Phi(1^-\,-1^-)-\Phi(0^-0^-)+\Phi(-1^-1^-)] \\ \Psi({}^1P,\qquad \text{etc.} \end{array} \qquad (pp)\ \ (7)$$

For the important configurations ls, we have

$$\Omega(L,M_L)=\omega(M_L,0). \qquad \text{(for all } M_L) \quad (8)$$

Hence from (5)

$$\begin{array}{l} \Psi({}^1L,\ \ 0,M_L)=2^{-\frac{1}{2}}[\Phi(M_L^+,0^-)-\Phi(M_L^-,0^+)] \\ \Psi({}^3L,\ \ 1,M_L)=\Phi(M_L^+,0^+) \\ \Psi({}^3L,\ \ 0,M_L)=2^{-\frac{1}{2}}[\Phi(M_L^+,0^-)+\Phi(M_L^-,0^+)] \\ \Psi({}^3L,\,-1,M_L)=\Phi(M_L^-,0^-). \end{array} \qquad (ls)\ \ (9)$$

Addition of an electron to an ion.

Let us consider the addition of an s electron to the configuration pp to obtain pps. In particular, let us find the eigenfunctions for the two ${}^2S$'s which result, namely $pp\,({}^1S)\,s\,{}^2S$ and $pp\,({}^3S)\,s\,{}^2S$. The states $pp\,{}^1S$ and ${}^3S$ are given by (7). The first of the ${}^2S$'s is given merely by adding 0^+ or 0^- to the single state of $pp\,{}^1S$ and antisymmetrizing:

$$\begin{array}{l} \Psi(pp\,({}^1S)\,s\,{}^2S,\,\pm\tfrac{1}{2},0)=6^{-\frac{1}{2}}[\Phi(1^+\,-1^-0^\pm)-\Phi(1^-\,-1^+0^\pm) \\ \qquad -\Phi(0^+0^-0^\pm)+\Phi(0^-0^+0^\pm)+\Phi(-1^+1^-0^\pm)-\Phi(-1^-1^+0^\pm)]. \end{array} \quad (10a)$$

For the other ${}^2S$ we must add $S^{\mathrm{I}}=1$ to $S^{\mathrm{II}}=\frac{1}{2}$ to obtain $S=\frac{1}{2}$:

$$\begin{array}{l} \mathrm{T}(S=\tfrac{1}{2},\ M_S=\ \ \tfrac{1}{2})=3^{-\frac{1}{2}}[\sqrt{2}\,\tau(1,-)-\tau(0,+)] \\ \mathrm{T}(S=\tfrac{1}{2},\ M_S=-\tfrac{1}{2})=3^{-\frac{1}{2}}[\tau(0,-)-\sqrt{2}\,\tau(-1,+)]. \end{array} \quad (11)$$

Hence

$$\begin{array}{l} \Psi(pp\,({}^3S)\,s\,{}^2S,\tfrac{1}{2},0)=18^{-\frac{1}{2}}[2\,\Phi(1^+\,-1^+0^-)-2\,\Phi(0^+0^+0^-)+2\,\Phi(-1^+1^+0^-)-\Phi(1^+\,-1^-0^+) \\ \qquad -\Phi(1^-\,-1^+0^+)+\Phi(0^+0^-0^+)+\Phi(0^-0^+0^+)-\Phi(-1^+1^-0^+)-\Phi(-1^-1^+0^+)] \\ \Psi(pp\,({}^3S)\,s\,{}^2S,-\tfrac{1}{2},\ \text{etc.} \end{array} \quad (10b)$$

This state is orthogonal to the other ${}^2S$, $\frac{1}{2}$, 0, as required.

We may characterize the two ${}^2S$'s of this same configuration in another way by considering the addition of a non-equivalent p electron to ps as parent. Let us obtain the eigenfunctions for $ps\,({}^3P)\,p\,{}^2S$ and $ps\,({}^1P)\,p\,{}^2S$.

For the first we must add $S^{\text{I}}=1$ to $S^{\text{II}}=\frac{1}{2}$ as in (11), and $L^{\text{I}}=1$ to $L^{\text{II}}=1$ to obtain $L=0$ as follows:

$$\Omega(L=0, M_L=0)=3^{-\frac{1}{2}}[\omega(1,-1)-\omega(0,0)+\omega(-1,1)]. \tag{12}$$

The symbolic product of this with (11) gives, when we use the functions (9),

$$\begin{aligned}\Psi(p\,s\,(^3P)\,p\,^2S, \tfrac{1}{2}, 0)=18^{-\frac{1}{2}}[&2\,\Phi(1^+\,0^+\,-1^-)-2\,\Phi(0^+\,0^+\,0^-)+2\,\Phi(-1^+\,0^+\,1^-)-\Phi(1^+\,0^-\,-1^+)\\ &-\Phi(1^-\,0^+\,-1^+)+\Phi(0^+\,0^-\,0^+)+\Phi(0^-\,0^+\,0^+)-\Phi(-1^+\,0^-\,1^+)-\Phi(-1^-\,0^+\,1^+)]\end{aligned}$$

$$\Psi(p\,s\,(^3P)\,p\,^2S, -\tfrac{1}{2}, \quad \text{etc.} \tag{13a}$$

The states of the second ^2S are given by (12) and (9):

$$\begin{aligned}\Psi(p\,s\,(^1P)\,p\,^2S, \pm\tfrac{1}{2}, 0)=6^{-\frac{1}{2}}[&\Phi(1^+\,0^-\,-1\pm)-\Phi(1^-\,0^+\,-1\pm)-\Phi(0^+\,0^-\,0\pm)\\ &+\Phi(0^-\,0^+\,0\pm)+\Phi(-1^+\,0^-\,1\pm)-\Phi(-1^-\,0^+\,1\pm)].\end{aligned} \tag{13b}$$

The quantum numbers in these zero-order functions refer to the electrons np, $n''s$, $n'p$ respectively, while those of (10) refer to np, $n'p$, $n''s$. This difference of arrangement must be taken into account in a comparison of the two sets of states. We see that the states (13) are quite different from the states (10); but since there are only two possible states ^2S, $\frac{1}{2}$, 0, they must be obtainable from (10) by a unitary transformation. This transformation is in fact the following—

$$\begin{aligned}\Psi(p\,s\,(^3P)\,p\,^2S, \tfrac{1}{2}, 0)&=-\tfrac{1}{2}\sqrt{3}\,\Psi(p\,p\,(^1S)\,s\,^2S, \tfrac{1}{2}, 0)+\tfrac{1}{2}\Psi(p\,p\,(^3S)\,s\,^2S, \tfrac{1}{2}, 0)\\ \Psi(p\,s\,(^1P)\,p\,^2S, \tfrac{1}{2}, 0)&=-\tfrac{1}{2}\Psi(p\,p\,(^1S)\,s\,^2S, \tfrac{1}{2}, 0)-\tfrac{1}{2}\sqrt{3}\,\Psi(p\,p\,(^3S)\,s\,^2S, \tfrac{1}{2}, 0).\end{aligned} \tag{14}$$

Two equivalent electrons.

As noted in § 4[6], we may look at the exclusion principle from the following point of view: We build up non-antisymmetric states characterized by quantum numbers $n_1l_1m_{s1}m_{l1}$, $n_2'l_2'm_{s2}'m_{l2}'$, $S_{12}L_{12}$, $n_3''l_3''m_{s3}''m_{l3}''$, …, SLM_SM_L. (The subscripts here refer as usual to the definite electron which has the quantum numbers indicated by the superscripts.) Then, by application of the operator $\mathscr{A}$, these states are antisymmetrized to obtain states still characterized by SLM_SM_L, but not necessarily by any other definite quantum numbers. If certain of the electrons are equivalent, two things happen. First, the antisymmetrized states are no longer normalized and some of them vanish—these are forbidden; second, some of the antisymmetrized states are linearly dependent—this reduces the number of distinct eigenfunctions. In this way we may obtain all the eigenfunctions for any configuration by coupling vectors and then antisymmetrizing. In general, however, for configurations containing more than two equivalent electrons, this is less easy than the method of § 5[8] which takes full cognizance of the exclusion principle from the start.

For the case of two equivalent electrons, however, it is easy to see the effect of the exclusion principle and to obtain the eigenfunctions by this method. Let us, according to the formulas of § 14[3], couple the s's and l's of two equivalent electrons to obtain the state $\psi(n_1l_1\,n_2l_2\,SLM_SM_L)$. This state is literally the product of an orbital function $\psi_r(n_1l_1\,n_2l_2\,LM_L)$ and a spin function $\psi_\sigma(s_1\,s_2\,SM_S)$. By 14^37 interchange of electrons 1 and 2 merely multiplies ψ_r by $(-1)^{2l-L}=(-1)^L$ and ψ_σ by $(-1)^{2s-S}=(-1)^{S+1}$. Hence ψ is already either antisymmetric or symmetric according to whether $L+S$ is even or odd. The symmetric states are forbidden; the antisymmetric states are all distinct since they all refer to different SLM_SM_L and hence are all allowed. *The allowed terms for* nl^2 *are hence* ^1S, ^3P, ^1D, ^3F, ..., $^1(2l)$, as we have found in special cases in Chapter VII. Since the states ψ are normalized, we may write

$$\Psi(nl\,nl\,SLM_SM_L)=\psi(n_1l_1\,n_2l_2\,SLM_SM_L) \tag{15}$$

if SL is an allowed term. Since the operator $\mathscr{A}$ has the eigenvalue $\sqrt{2!}$ when applied to an antisymmetric two-electron function (§ 3^6), $\Psi=2^{-\frac{1}{2}}\mathscr{A}\psi$. It is $\mathscr{A}\psi$ which we obtain if we set $n=n'$ in the eigenfunctions for the configuration $nl\,n'l$. Hence we can obtain the eigenfunctions for nl^2 from those for $nl\,n'l$ by setting $n=n'$ and dividing by $\sqrt{2}$ to normalize. Thus we obtain from (7)

$$\Psi(np^2\,{}^1S,0,0)=3^{-\frac{1}{2}}[\Phi(1^+,-1^-)-\Phi(0^+,0^-)-\Phi(1^-,-1^+)], \tag{16}$$

while if we set $n=n'$ in ^3S, these states all vanish.

Because of the relation (15), we have the important result that *we can obtain the matrix component of a quantity* $F=F_1+F_2$ *between two allowed states of* nl^2 *by giving quantum numbers* $n_1s_1l_1$ *to electron* 1 *and* $n_2s_2l_2$ *to electron* 2. This permits us to use formulas of Chapter III for such configurations in the same way as we found in § 1^8 that we could use them for non-equivalent electrons. For the matrix component between an allowed state of nl^2 and a state of $nl\,n'l'$ we have

$$\begin{aligned}
&\bar{\Psi}(nl\,nl\,SLM_SM_L)\,F\,\Psi(nl\,n'l'\,S'L'M_S'M_L')\\
&\quad=\bar{\psi}(n_1l_1\,n_2l_2\,SLM_SM_L)\,F\,2^{-\frac{1}{2}}[\psi(n_1l_1\,n_2'l_2'\,S'L'M_S'M_L')\\
&\qquad\qquad-\psi(n_2l_2\,n_1'l_1'\,S'L'M_S'M_L')]\\
&\quad=2^{-\frac{1}{2}}\bar{\psi}(n_1l_1\,n_2l_2\,SLM_SM_L)\,F_2\,\psi(n_1l_1\,n_2'l_2'\,S'L'M_S'M_L')\\
&\qquad-2^{-\frac{1}{2}}\bar{\psi}(n_1l_1\,n_2l_2\,SLM_SM_L)\,F_1\,\psi(n_2l_2\,n_1'l_1'\,S'L'M_S'M_L');
\end{aligned}$$

on interchanging electrons 1 and 2 in the second term, this becomes

$$=\sqrt{2}\,\bar{\psi}(n_1l_1\,n_2l_2\,SLM_SM_L)\,F_2\,\psi(n_1l_1\,n_2'l_2'\,S'L'M_S'M_L'). \tag{17}$$

A similar result holds for states characterized by J and M. Hence if we use the formulas of Chapter III to find matrix components between configurations nl^2 and $nl\,n'l'$ by assigning definite quantum numbers to the electrons, *we must set those components from forbidden states of* nl^2 *equal to zero, and multiply the rest by* $\sqrt{2}$.

No considerations of this sort have been given for more than two equivalent electrons, so that it is probably not possible to apply matrix methods to calculations within groups of more than two equivalent electrons.

7. Separation of the ^2D's of d^3.*

When two terms of a kind occur in a given configuration, we have seen that the diagonal-sum rule will not determine their separate energies or Landé splittings. These terms may be separated, however, by finding the complete matrix of electrostatic interaction Q by the formulas of § 8^6 if we know a set of LS-coupling eigenfunctions. Since Q is diagonal with respect to SLM_SM_L and independent of M_SM_L, it will only be necessary to use that part of the matrix of Q which refers to a given SLM_SM_L.

Thus if we wish to separate the two ^2D's of d^3, we need only use the second-order matrix connecting the two $^2D, \frac{1}{2}, 2$'s of 5^86e. This matrix is found to be

	a $^2D, \frac{1}{2}, 2$	b $^2D, \frac{1}{2}, 2$
a $^2D, \frac{1}{2}, 2$	$3F_0+7F_2+63F_4$	$3\sqrt{21}\,(F_2-5F_4)$
b $^2D, \frac{1}{2}, 2$	$3\sqrt{21}\,(F_2-5F_4)$	$3F_0+3F_2-57F_4$

, (1)

which has for its eigenvalues

$$\epsilon=3F_0+5F_2+3F_4\pm\sqrt{193F_2^2-1650F_2F_4+8325F_4^2}. \tag{2}$$

This formula then gives the separate energies of the two ^2D terms. The eigenfunctions for these terms, for $M_S=\frac{1}{2}$, $M_L=2$, may now be obtained. If we write the eigenfunction for either of them as

$$\alpha\Psi(\text{a}\,^2D, \tfrac{1}{2}, 2)+\beta\Psi(\text{b}\,^2D, \tfrac{1}{2}, 2),$$

we find

$$\alpha=\left[\left(\frac{3F_0+7F_2+63F_4-\epsilon}{3\sqrt{21}(F_2-5F_4)}\right)^2+1\right]^{-\frac{1}{2}}$$
$$\beta=-\frac{3F_0+7F_2+63F_4-\epsilon}{3\sqrt{21}(F_2-5F_4)}\alpha. \tag{3}$$

Since (1) and (2) are independent of M_S and M_L, the same linear combination of a ^2D and b ^2D obtains for any state of the term.

Let us now consider the absolute Landé intervals for these terms. Since $\Psi(^2D, \frac{1}{2}, 2)=\Psi(^2D^{\frac{5}{2}}_{\frac{5}{2}})$, we have from 3^73

$$(^2D, \tfrac{1}{2}, 2|H^I|^2D, \tfrac{1}{2}, 2)=\zeta(^2D). \tag{4}$$

Equation (3) gives the eigenfunction for ^2D, $\frac{1}{2}$, 2 in terms of zero-order states each characterized by $M_S=\frac{1}{2}$, $M_L=2$. Since these states all belong to the same configuration and have the same M_S and M_L, no two of them can differ in regard to just one individual set. Hence H^I, a quantity of type F, can

* UFFORD and SHORTLEY, Phys. Rev. **42**, 167 (1932).

have no matrix components connecting two of these states, so that we can calculate (4) entirely in terms of the diagonal components $4^7$3. In this way we find that

$$\zeta(^2D) = (\tfrac{1}{2}\alpha^2 - \tfrac{1}{3}\sqrt{21}\alpha\beta - \tfrac{1}{6}\beta^2)\,\zeta_d$$
$$= \tfrac{1}{6}\left[1 \mp \frac{59F_2 - 435F_4}{\sqrt{193F_2^2 - 1650F_2F_4 + 8325F_4^2}}\right]\zeta_d. \quad (5)$$

We shall now see how these calculations compare with the experimental data. The first instance of d^3 occurs in Ti II $3d^3$, which although completely mixed up with d^2s gives a fairly good fit for the terms exclusive of the ^2D's with $F_2 = 845$, $F_4 = 54$ $(3F_0 = 17{,}750)$; see Fig. 2^8. If we put these values in (2), we find for the energies of the two ^2D's 12,820 and 31,450 cm^{-1}. The only observed ^2D is at 12,710 cm^{-1}, which agrees excellently with our lower ^2D. The position of the second ^2D is predicted 10,000 cm^{-1} higher than any other level of the configuration, which may account for its not being found. The intervals in this configuration also fit the formulas fairly well. This is seen from the following list of values of ζ_d, calculated from the observed intervals (beginning with the widest) in the terms:

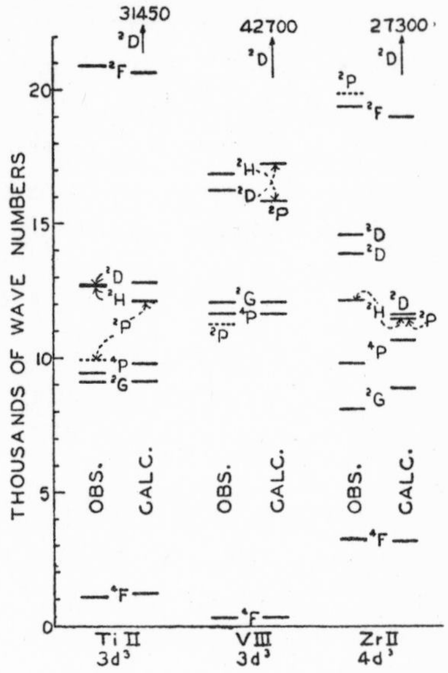

Fig. 2^8. The d^3 configuration.

^4F	85·6, 88·6, 81·0	^2H	88·9
^2G	89·2	^2D	81·0
^4P	146·7, 64·1	^2F	102·7
^2P	125·0		

If the formulas were accurately followed, these values would be all equal. It is pleasing to find the ^2F inverted, as the theory predicts. If we take 89 cm^{-1} as the correct value of ζ_d, we calculate from (5) the lower ^2D interval as 142 cm^{-1}, while the observed value is 129·4. The interval for the higher ^2D is calculated as $-67{\cdot}7$ cm^{-1}.

In the $3d^3$ of V III, all the terms except ^2F and one ^2D are known. The configuration d^3 has the peculiarity that the calculated electrostatic energies for ^2H and ^2P are equal. In this instance these energies are not at all equal, so that we cannot depend much on ^2H and ^2P. However, if we choose the F's to make the other three terms ^4F, ^4P, ^2G fit exactly, the ^2H energy is fairly good while that of the ^2P is not. The values of the constants which are obtained in this way are $F_2=1171$, $F_4=83$ $(3F_0=23{,}891)$. These constants give for the values of the two ^2D's, 17,300 and 42,700 cm^{-1}. The first of these is agreeably close to the ^2D found by White at 16,317 cm^{-1}; the second is predicted 25,800 cm^{-1} higher than any other level of the configuration, which has a spread of only 16,900 cm^{-1} as analysed! Hence it is not surprising that this second ^2D was not found. The intervals as usual agree only roughly; ^4F fits the interval rule fairly well, and if we use the ζ given by this level (which is about an average for the configuration), the calculated ^2D intervals are 248 and -110 cm^{-1}; the first of these is to be compared with the observed interval of 147 cm^{-1}.

For the $4d^3$ of Zr II, in which two ^2D's are reported, we obtain an approximate fit of all terms except ^2P with $F_2=683$, $F_4=36$ $(3F_0=16{,}000)$. With these values the calculated ^2D's lie at 11,750 and 27,300 cm^{-1}, with separations of 593 and -309 cm^{-1}, respectively. These separations are calculated using the ^2H, which gives an average value for ζ, as standard. The observed ^2D's lie at 13,869 and 14,559 cm^{-1} with separations of 734 and 435 cm^{-1}. Hence we must infer that, if these are correctly classified, one of them is very strongly perturbed. In this configuration again, d^3 is completely mixed up with d^2s.*

These are all the instances of d^3 which are sufficiently analysed for comparison with the theory. In general we have seen that the lower ^2D corresponds well with the one usually observed; the second ^2D is predicted extremely high and inverted.

* The interaction between these configurations and the ^2D assignments have been considered in detail by Ufford, Phys. Rev. **44**, 732 (1933), who finds that he can account for the discrepancies of Fig. 2[8] satisfactorily. The better agreement of ^2H over ^2P in the above approximation is explained by the fact that ^2P is strongly perturbed by ^2P's in neighbouring configurations, whereas there is no nearby configuration which contains a ^2H term to perturb the ^2H of d^3.

CHAPTER IX

THE RUSSELL-SAUNDERS CASE: LINE STRENGTHS

In this chapter we shall develop the theory of radiation as outlined in Chapter IV to obtain formulas for the strengths of the various spectral lines for atoms obeying Russell-Saunders coupling. In § 1 we obtain some general results concerning the possible configuration changes in radiative processes which are true for any coupling. In § 2 formulas for the relative strengths of lines in a multiplet are obtained as a special application of the results of § 11³. The problem of the relative strengths of the different multiplets arising from a transition between a particular pair of configurations is treated in the next two sections. In § 5 we develop for quadrupole multiplets the formulas giving relative strengths of the lines analogous to those for dipole multiplets in § 2.

1. Configuration selection rules.

From the results of previous sections we may immediately obtain two important selection rules. Since the electric-dipole moment $\boldsymbol{P}$ is a quantity which anticommutes with the parity operator $\mathscr{P}$ of § 11⁶, it follows that $\boldsymbol{P}$ has no matrix components between states of the same parity. Hence *all spectral lines due to electric-dipole radiation arise from transitions between states of opposite parity.* This rule was discovered by Laporte* and is usually known as the *Laporte rule.* Its importance lies in the fact that it remains valid even in complicated cases where it is no longer possible to assign configurations uniquely to the energy levels, and enables the spectroscopist to characterize such levels as uniquely even or odd. The symbol ° is attached to the quantum numbers of odd terms when it is desired to express the parity explicitly.

Since $\boldsymbol{P}$ is a quantity of type F considered in § 6⁶, it will have non-vanishing matrix components only between states which differ in regard to at most one individual set of quantum numbers. The non-vanishing matrix component connecting states A and B which differ in regard to one individual set is just the corresponding one-electron matrix element connecting the non-identical individual sets of A and B. From the theory of one-electron spectra we know that this matrix component of $\boldsymbol{P}$ vanishes unless $\Delta l = \pm 1$, where Δl is the difference of the l's in the two non-identical individual sets. Hence if the energy levels are accurately characterized by configuration assignments, *the configuration change occurring in dipole*

* LAPORTE, Zeits. für Phys. **23**, 135 (1924); see also RUSSELL, Science, **51**, 512 (1924).

radiation is by just the nl of one electron, the change in l being restricted to ± 1. Such transitions are called one-electron jumps. Transitions are observed in which there is an apparent change in two of the *nl* values (two-electron jumps), but this is to be connected with a breakdown in the association of configurations with levels.

With regard to quadrupole and magnetic-dipole radiation we observe from $4^4 9$ that $\mathfrak{R}$ and $\boldsymbol{M}$ are also quantities of type F. The restriction to one-electron jumps also applies to them. But these quantities commute with the parity operator $\mathscr{P}$, so that they have non-vanishing matrix components only between states of the same parity, which is just the opposite of the Laporte rule for dipole radiation.

2. Line strengths in Russell–Saunders multiplets.

The ensemble of all lines connected with transitions from one term characterized by L and S to another characterized by L' and S' we shall call a *multiplet*. We wish now to consider the theory of the strengths of the lines of a multiplet. We confine our attention at present to electric-dipole radiation.

Since $\boldsymbol{P}$ commutes with $\boldsymbol{S}$, it follows at once that the matrix of $\boldsymbol{P}$ can contain no components connecting states of different S. Because of this, spectroscopists find that they can conveniently divide the terms of most atoms into *systems* of different multiplicities, e.g., a singlet system, triplet system, and quintet system. The lines connecting terms in different systems are in general missing or weak. For example, only one line connecting the singlet terms with the triplet terms of helium has ever been observed, and that with considerable difficulty. This is then the first important selection rule of the Russell-Saunders case: *intersystem combinations are forbidden.* In passing we note that this is also true of the electric-quadrupole and the magnetic-dipole radiations, since the corresponding moments also commute with the spin.

The line strengths are given by formulas $7^4 5$ if we write γSLJ for αj. We have the selection rule $\Delta J = \pm 1, 0$. Since $\boldsymbol{P}$ commutes with $\boldsymbol{S}$, the dependence on J of the factors $(\gamma SLJ \vdots P \vdots \gamma' S' L' J')$ occurring on the right of $7^4 5$ is given by formulas $11^3 8$ if we correlate j_1 with S and j_2 with L. It follows from these formulas that the selection rule on L is the same as that on J, namely $\Delta L = \pm 1, 0$.

The factors $(\gamma SL \vdots P \vdots \gamma' SL')$ occurring on the right of $11^3 8$ will be independent of S since the states $\Psi(\gamma SLM_S M_L)$ may be expressed as a sum of products of a function of the electronic spin coordinates only and a function of the positional coordinates only, and the operator $\boldsymbol{P}$ acts on the positional coordinates only (cf. the discussion of § 11^3). Hence we may write in general

$$\mathbf{S}(\gamma SLJ, \gamma' SL'J') = f(SLJ, SL'J') \, |(\gamma L \vdots P \vdots \gamma' L')|^2, \qquad (1)$$

where the whole dependence on S, J, and J' is contained in the function $f(SLJ, SL'J')$, which may be evaluated from 7^45 and 11^38.

First let us consider the multiplets for which $L' = L-1$. For these we have:

$$\mathsf{S}(\gamma\, S\, L\, J, \gamma'\, S\, L{-}1\, J{+}1) = (-1)^2 \frac{(J+S-L+1)(L+S-J)(J+S-L+2)(L+S-J-1)}{4(J+1)} |(\gamma L\vdots P\vdots \gamma' L-1)|^2$$

$$\mathsf{S}(\gamma\, S\, L\, J, \gamma'\, S\, L{-}1\, J) = (2J+1) \frac{(J+L-S)(J+S-L+1)(S+L+1+J)(S+L-J)}{4J(J+1)} |(\gamma L\vdots P\vdots \gamma' L-1)|^2$$

$$\mathsf{S}(\gamma\, S\, L\, J, \gamma'\, S\, L{-}1\, J{-}1) = \frac{(J+L-S-1)(J+L-S)(S+L+J+1)(S+L+J)}{4J} |(\gamma L\vdots P\vdots \gamma' L-1)|^2. \qquad (2a)$$

The sum of these expressions is given by $13^35'$:

$$\sum_{J'} \mathsf{S}(\gamma\, S\, L\, J, \gamma'\, S\, L{-}1\, J') = (2J+1)\, L\, (2L-1)\, |(\gamma\, L\vdots P\vdots \gamma'\, L-1)|^2 \qquad (3)$$

—that is, the total strength of the lines of the multiplet originating in the level γSLJ is proportional to the statistical weight $(2J+1)$ of the initial level. Since the formulas for strengths of lines are symmetric between initial and final states, we may regard γSLJ as the final state in the preceding formulas; in which case they give at once the strengths for a multiplet $L' = L+1$ and the sum rule says that the sum of the strengths of the lines ending in the same final level γSLJ is proportional to the statistical weight of that level.

For the multiplets in which $L' = L$ we have:

$$\mathsf{S}(\gamma\, S\, L\, J, \gamma'\, S\, L\, J{+}1) = (-1)^2 \frac{(J-S+L+1)(J+S-L+1)(S+L+J+2)(S+L-J)}{4(J+1)} |(\gamma L\vdots P\vdots \gamma' L)|^2$$

$$\mathsf{S}(\gamma\, S\, L\, J, \gamma'\, S\, L\, J) = (2J+1) \frac{[J(J+1) - S(S+1) + L(L+1)]^2}{4J(J+1)} |(\gamma L\vdots P\vdots \gamma' L)|^2 \qquad (2b)$$

$$\mathsf{S}(\gamma\, S\, L\, J, \gamma'\, S\, L\, J{-}1) = (-1)^2 \frac{(J-S+L)(J+S-L)(S+L+J+1)(S+L+1-J)}{4J} |(\gamma L\vdots P\vdots \gamma' L)|^2.$$

From $13^35'$ we learn, as before, that the sum,

$$\sum_{J'} \mathsf{S}(\gamma S\, L\, J, \gamma'\, S\, L\, J') = (2J+1)\, L\, (L+1)\, |(\gamma\, L\vdots P\vdots \gamma'\, L)|^2, \qquad (4)$$

of these three quantities is proportional to $2J+1$. The corresponding sum for the case $L' = L+1$ has the value

$$\sum_{J'} \mathsf{S}(\gamma\, S\, L\, J, \gamma'\, S\, L{+}1\, J') = (2J+1)(L+1)(2L+3)\, |(\gamma\, L\vdots P\vdots \gamma'\, L{+}1)|^2. \quad (5)$$

Thus for any Russell-Saunders multiplet we have derived the general result *that the sum of the strengths of the lines having a given initial level is proportional to the statistical weight* $(2J+1)$ *of that initial level, and that the sum of the strengths of the lines having a given final level is proportional to the statistical weight of that final level.* This *sum rule* was discovered empirically by Ornstein, Burger, and Dorgelo.*

* BURGER and DORGELO, Zeits. für Phys. **23**, 258 (1924); ORNSTEIN and BURGER, *ibid.* **24**, 41 (1924).

Finally we may sum these formulas over the various values of J associated with the initial term. Since the statistical weight of the whole term is $(2S+1)(2L+1)$, we find that the total strength of all the lines in the multiplet is given by

$$\begin{aligned}\mathsf{S}(\gamma\, S\, L, \gamma'\, S\, L+1) &= (2S+1)(2L+1)(L+1)(2L+3)\,|(\gamma\, L \vdots P \vdots \gamma'\, L+1)|^2 \\ \mathsf{S}(\gamma\, S\, L, \gamma'\, S\, L) &= (2S+1)(2L+1)\, L\, (L+1)\,|(\gamma\, L \vdots P \vdots \gamma'\, L)|^2 \qquad (6) \\ \mathsf{S}(\gamma\, S\, L, \gamma'\, S\, L-1) &= (2S+1)(2L+1)\, L\, (2L-1)\,|(\gamma\, L \vdots P \vdots \gamma'\, L-1)|^2,\end{aligned}$$

for the three kinds of change in L. These formulas are symmetric in the initial and final terms. They bear a striking resemblance to formulas $7^4 5$ which express the total strength of all components of a line.

Now that we have seen how the relative intensities of the lines in a multiplet follow from the general formulas of Chapter III, let us consider the application to spectra and something of the historical development in this field.

The simplest result concerns the relative strengths of the lines ${}^2S_{\frac{1}{2}}$-${}^2P_{\frac{1}{2}}$ and ${}^2S_{\frac{1}{2}}$-${}^2P_{\frac{3}{2}}$. Since there is but one level in the S state, the sum rule tells us that these should have strengths in the ratio 1:2. This is actually the case for the yellow D lines of Na I and is quite generally true for the S-P doublets of the alkalis. We consider in § 5^{15} the explanation of some departures from this result.

In the ${}^3P$-${}^3S$ triplets the sum rule predicts that the strengths of the lines are as 5:3:1 and this was found to be experimentally the case in Zn and Cd. Generally the ratios of the strengths of the three lines ${}^{2S+1}P_{S+1}$-, ${}^{2S+1}P_{S}$-, ${}^{2S+1}P_{S-1}$-${}^{2S+1}S_S$ are as $(2S+3):(2S+1):(2S-1)$. This was studied* for quintets, sextets and octets in Cr I and Mn I, with the results shown in the following table (the calculated relative intensities are obtained by multiplying the above strengths by the σ^4 factors of § 7^4):

Multiplet	Wave-lengths in Å	Relative intensity
Cr: $a\ {}^5S \rightarrow z\ {}^5P^\circ$	5208, 5206, 5204	7:5:3 Calculated 7:5·04:3·15 Observed
Mn: $z\ {}^6P^\circ \rightarrow b\ {}^6S$	6021, 6016, 6013	8:6:4 Calculated 8:6·15:4·32 Observed
Mn: $z\ {}^8P^\circ \rightarrow a\ {}^8S$	4823, 4783, 4754	10:8·28:6·35 Calculated 10:8:6·15 Observed

Doublets and combinations with S terms are the only multiplets in which the strengths are fully determined by the sum rules alone. Experimentally it was found, however, that the sum rules are valid quite generally. The complete intensity formulas (2) were derived by the aid of the correspondence principle before quantum mechanics simultaneously by Kronig, Sommerfeld and Hönl, and Russell, and quantum mechanically first by

* DORGELO, Zeits. für Phys. **22**, 170 (1924).

Dirac.* Before the complete formulas were obtained, Sommerfeld and Heisenberg† found by the correspondence principle an important qualitative characteristic of them, namely, that the strongest lines are those in which the change in J is the same as that in L. These are called the principal lines of the multiplet. Among the principal lines the strongest is that with the largest J of the initial level, the strengths decreasing with decreasing J. The other lines are called *satellites*. For multiplets in which L changes the satellites may be classified into those of first order, in which J does not change; and of second order, in which the J change is opposite to the L change. The satellites of second order are, as a class, weaker than those of first order. Among the satellites the strength is a maximum for intermediate values of J. These qualitative characteristics of the multiplet intensity formulas are naturally of great importance in helping to recognize the character of a multiplet when a spectrum is being analysed.

For practical work it is convenient to have tables showing the relative strengths of multiplet lines in the most important cases. These have been calculated by Russell (*loc. cit.*) and by White and Eliason.‡ The tables of White and Eliason are given here in Table 1^9. These are essentially tables of relative values of $f(SLJ, SL'J')$ [cf. (1)] for all J and J' consistent with a given L, L', and S. The largest value is in each case set equal to 100; the others are sharply rounded off. Since in the theory of hyperfine structure and in jj coupling we need tables of this same function with half-integral values of the arguments L and L', such tables are also included.

We have taken care to speak of the strengths of spectral lines rather than intensities in order to place on the experimentalist the burden of inferring the values of the strengths from his observed intensities.§ Among the difficulties that arise is the lack of a statistical equilibrium in the atoms of the source ‖ and a partial reabsorption of the light emitted in one part of the source by the atoms in another part of the source. The latter difficulty is diminished by use of low-pressure sources, but is hard to get rid of entirely. Many of the experimental data on intensities are not suitable for comparison with the theory because data are lacking with which to calculate strengths from observed intensities.

* Kronig, Zeits. für Phys. **31**, 885 (1925);
Sommerfeld and Hönl, Sitz. der Preuss. Akad. 1925, p. 141;
Russell, Proc. Nat. Acad. Sci. **11**, 314 (1925);
Dirac, Proc. Roy. Soc. **A111**, 302 (1926).

† Sommerfeld and Heisenberg, Zeits. für Phys. **11**, 131 (1922).

‡ White and Eliason, Phys. Rev. **44**, 753 (1933).

§ Important references are:
Dorgelo, Zeits. für Phys. **13**, 206 (1923);
Harrison, J. Opt. Soc. Am. **19**, 267 (1929).

‖ On this point see an interesting study of conditions in the neon glow discharge by Ladenburg, Rev. Mod. Phys. **5**, 243 (1933); see also Schütz, Ann. der Phys. **18**, 746 (1933).

TABLE 1[9].

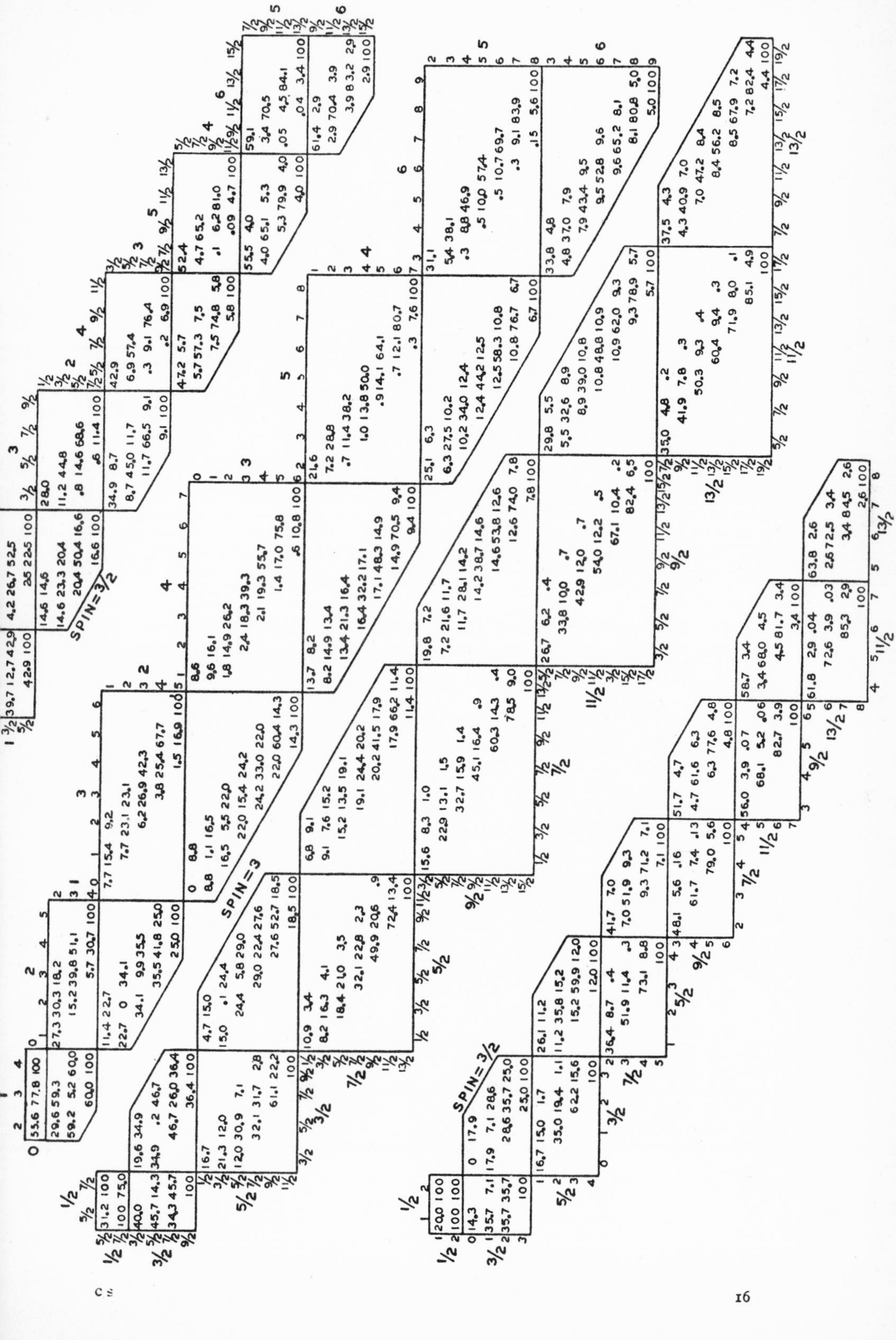

TABLE 1^9—*continued.*

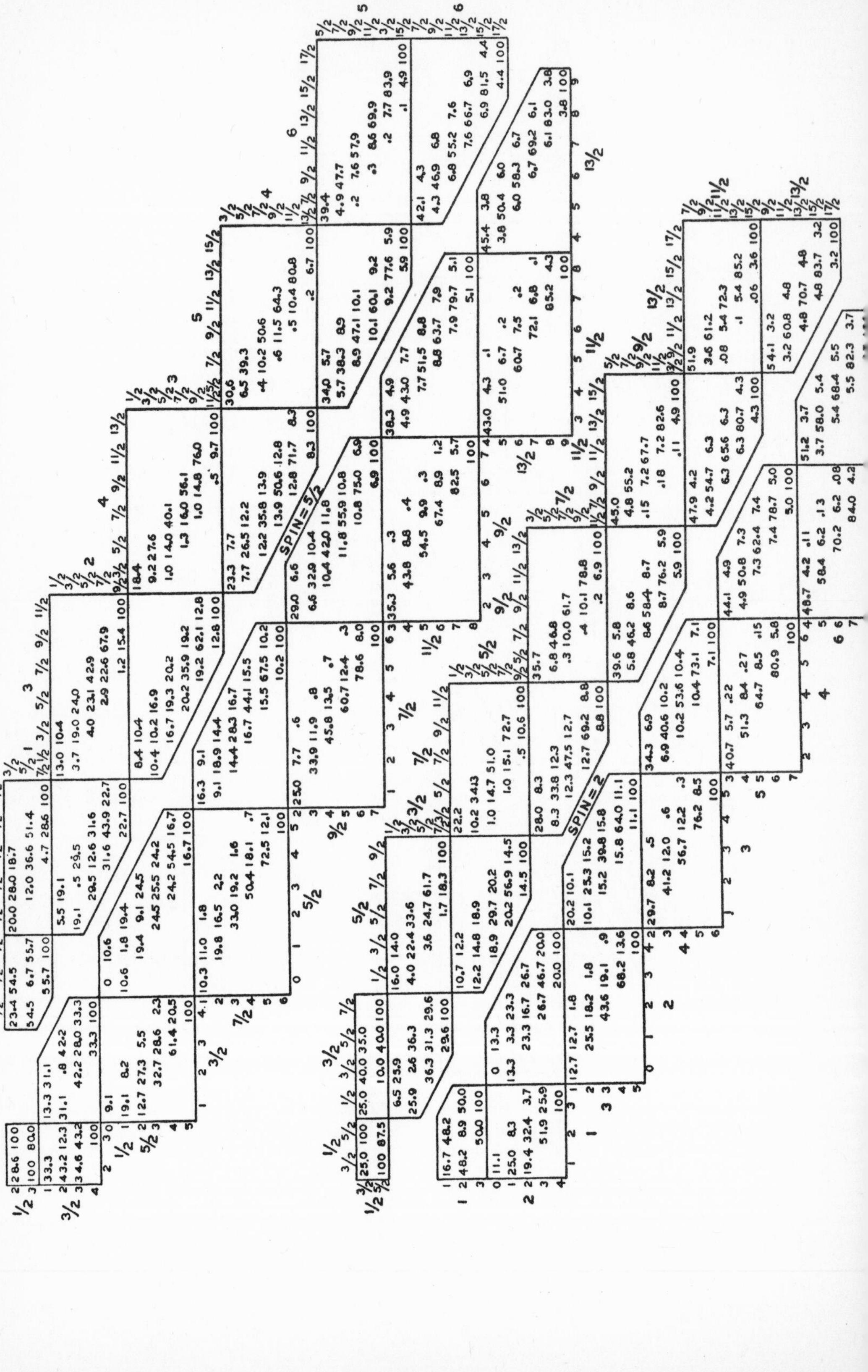

TABLE 1[9]—*continued.*

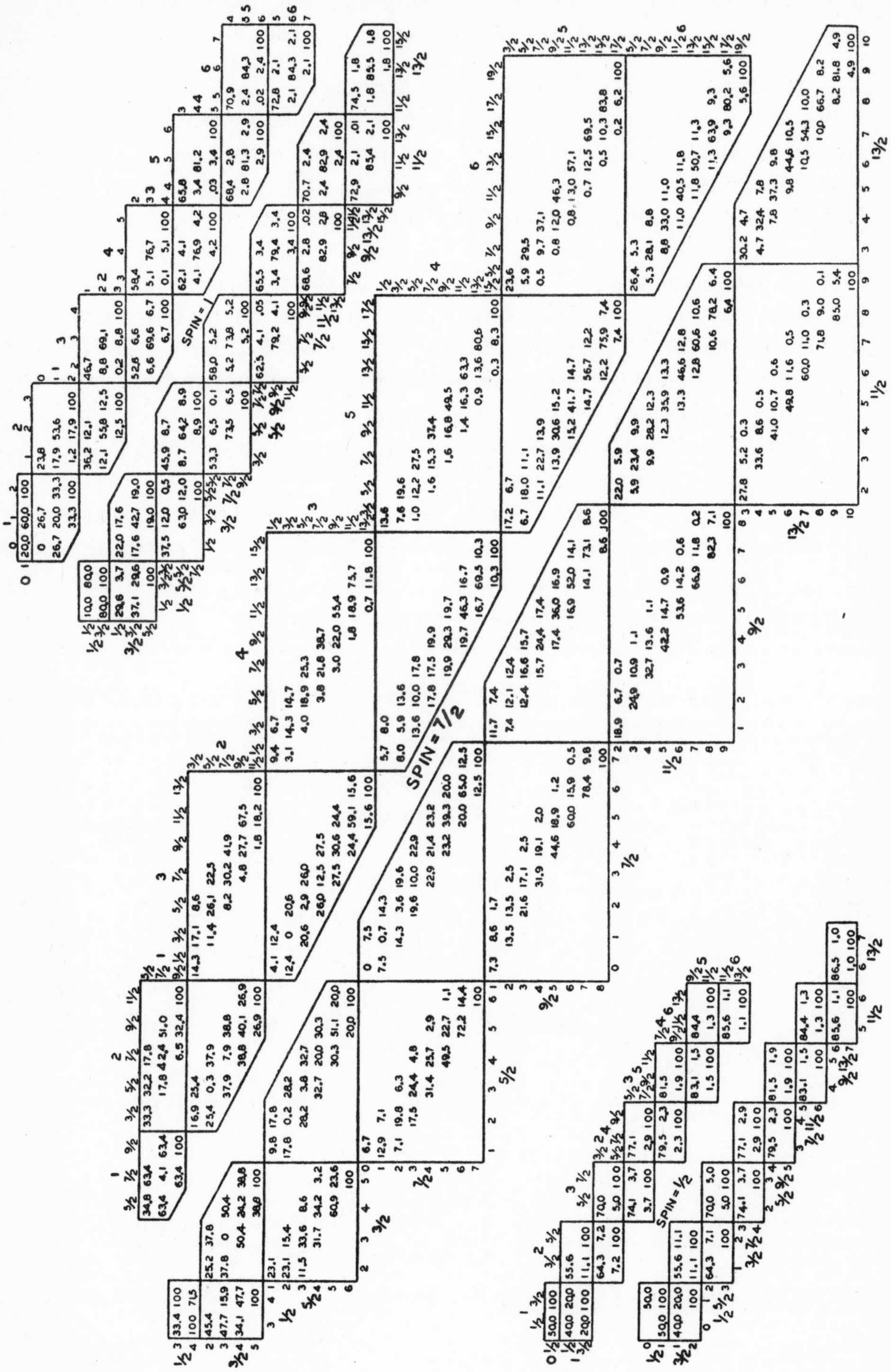

For further comparisons between theory and experiment we may refer to the work of Frerichs, of van Milaan, and of Harrison.*

3. Multiplet strengths in a transition array.†

We now consider the problem of finding the relative strengths of the multiplets which result from transitions occurring between two configurations. Formulas 2^96 express the total strength of a multiplet in terms of $|(\gamma L\vdots P\vdots\gamma' L')|^2$, so our problem is reduced to the evaluation of these quantities.

If the two configurations are expressible in the form I+II and I+IV respectively, we may use the results of § 1^8, in particular 1^816, to reduce our problem to that of the relative strengths of the multiplets in the transitions from configuration II to configuration IV. If the two configurations are not expressible in this form, we cannot proceed by matrix methods (except for two-electron configurations, cf. 6^817) but may use the method of § 4^9.‡

The quantity we must evaluate is

$$|(\gamma^{\rm I}\, S^{\rm I}\, L^{\rm I}\,\gamma^{\rm IV}\, S^{\rm IV}\, L^{\rm IV}\, L\vdots P\vdots\gamma'^{\rm I}\, S'^{\rm I}\, L'^{\rm I}\gamma^{\rm II}\, S^{\rm II}\, L^{\rm II}\, L')|^2. \qquad (1)$$

According to 1^816, this vanishes unless $\gamma^{\rm I}\, S^{\rm I}\, L^{\rm I}\equiv\gamma'^{\rm I}\, S'^{\rm I}\, L'^{\rm I}$; i.e. transitions are forbidden between terms built on different terms of configuration I. If the terms of configuration I are the same, (1) reduces to

$$|(\gamma^{\rm I}\, S^{\rm I}\, L^{\rm I}\,\gamma^{\rm IV} S^{\rm IV} L^{\rm IV}\, L\vdots P^{\rm II}\vdots\gamma^{\rm I}\, S^{\rm I}\, L^{\rm I}\,\gamma^{\rm II}\, S^{\rm II}\, L^{\rm II}\, L')|^2. \qquad (2)$$

From the considerations of § 1^8, we see that we can evaluate this expression by means of 11^38, treating $\boldsymbol{P}^{\rm II}$ as a vector which commutes with $\boldsymbol{L}^{\rm I}$ and $\boldsymbol{S}^{\rm II}$. Hence this vanishes unless $S^{\rm IV}=S^{\rm II}$, in which case it is the same function of $L^{\rm I}\, L^{\rm IV}\, L,\ L^{\rm I}\, L^{\rm II}\, L'$ as $|(\gamma\, S\, L\, J\vdots P\vdots\gamma'\, S\, L'\, J')|^2$ is of $S\, L\, J,\ S\, L'\, J'$. On comparing 7^45 with 2^96, we then see that the total strengths of the multiplets are given by

$$\begin{aligned}&\mathsf{S}(\gamma^{\rm I}\, S^{\rm I}\, L^{\rm I}\,\gamma^{\rm IV}\, S^{\rm II}\, L^{\rm IV}\, S\, L, \gamma^{\rm I}\, S^{\rm I}\, L^{\rm I}\,\gamma^{\rm II}\, S^{\rm II}\, L^{\rm II}\, S\, L')\\ &\qquad =(2S+1)\, f(L^{\rm I}\, L^{\rm IV}\, L,\ L^{\rm I}\, L^{\rm II}\, L')\,|(\gamma^{\rm IV}\, L^{\rm IV}\vdots P^{\rm II}\vdots\gamma^{\rm II}\, L^{\rm II})|^2, \qquad (3)\end{aligned}$$

where the function f is defined by 2^91. If we sum this expression over the various values of L and L' which are consistent with the given values of $L^{\rm I}$, $L^{\rm II}$, and $L^{\rm IV}$, we obtain, in analogy with 2^96,

$$\begin{aligned}&\mathsf{S}(\gamma^{\rm I}\, S^{\rm I}\, L^{\rm I}\,\gamma^{\rm IV}\, S^{\rm IV}\, L^{\rm IV}\, S, \gamma^{\rm I}\, S^{\rm I}\, L^{\rm I}\,\gamma^{\rm II}\, S^{\rm II}\, L^{\rm II}\, S)\\ &\qquad =\frac{(2S+1)(2L^{\rm I}+1)}{2S^{\rm II}+1}\mathsf{S}(\gamma^{\rm IV}\, S^{\rm IV}\, L^{\rm IV}, \gamma^{\rm II}\, S^{\rm II}\, L^{\rm II}). \qquad (4)\end{aligned}$$

* Frerichs, Ann. der Phys. **81**, 807 (1926);
van Milaan, Zeits. für Phys. **34**, 921 (1925); **38**, 427 (1926);
Harrison, J. Opt. Soc. Am. **17**, 389 (1928).

† Johnson, Proc. Nat. Acad. Sci. **19**, 916 (1933); Shortley, *ibid.* **20**, 591 (1934). Following Harrison, we shall call the totality of lines resulting from the transitions between two configurations a *transition array*.

‡ For example, these considerations are applicable to the transitions between $d^2\, s\, p$ and $d^2\, s^2$, but not to the transitions between $d^2\, s^2$ and $d^3\, s$.

The last **S** here is the total strength of the multiplet $\gamma^{\mathrm{IV}}\, S^{\mathrm{IV}}\, L^{\mathrm{IV}} \to \gamma^{\mathrm{II}}\, S^{\mathrm{II}}\, L^{\mathrm{II}}$ in transitions between configuration IV and configuration II; it of course contains a factor $\delta(S^{\mathrm{IV}}, S^{\mathrm{II}})$. We shall now discuss these formulas by means of illustrative examples.

Addition of an electron to an ion.

The set of all lines arising in transitions between two polyads (§ 2[8]) having the same parent configuration is known as a *supermultiplet*. The relative strengths of the multiplets in a supermultiplet are given by (3) if we let $\gamma^{\mathrm{I}}\, S^{\mathrm{I}}\, L^{\mathrm{I}}$ represent the parent term of the ion, $\gamma^{\mathrm{IV}}\, L^{\mathrm{IV}} = nl$ and $\gamma^{\mathrm{II}}\, L^{\mathrm{II}} = n'l'$, the nl values of the jumping electron in the initial and final states, and $S^{\mathrm{II}} = \frac{1}{2}$. In the first place we see that *transitions between polyads based on different parent terms are forbidden.* For an allowed supermultiplet,

$$\mathsf{S}(\gamma^{\mathrm{I}}\, S^{\mathrm{I}}\, L^{\mathrm{I}}\, nl\, S\, L, \gamma^{\mathrm{I}}\, S^{\mathrm{I}}\, L^{\mathrm{I}}\, n'l'\, S\, L') = (2S+1)\, f(L^{\mathrm{I}}\, l\, L, L^{\mathrm{I}}\, l'\, L')\, |(nl\vdots P\vdots n'l')|^2. \tag{5}$$

The last factor here is just the factor which arises in connection with the intensities in one-electron spectra and which was evaluated in 6[5]3. It vanishes unless $l' = l \pm 1$, in which case it has the value

$$(n\,l\vdots P\vdots n'\,l{-}1) = \mathsf{s}(n\,l, n'\,l{-}1) = \frac{-e}{\sqrt{4l^2-1}} \int_0^\infty r\, R(n\,l)\, R(n'\,l{-}1)\, dr. \tag{6}$$

The functions f are given by 2[9]2 according to the definition 2[9]1. *The relative strengths of the multiplets in a supermultiplet are thus the same functions of $L^{\mathrm{I}}\, l\, L$, $L^{\mathrm{I}}\, l'\, L'$ as the relative strengths of the lines in a multiplet are of $S\, L\, J$, $S\, L'\, J'$.* These relative strengths are thus given by Table 1[9] with this correlation.

For example, the relative strengths of the multiplets in the

$$\text{Ti I}\ d^2\,4s\,(^4F)\,5s\,^5F \to d^2\,4s\,(^4F)\,4p\,^5DFG$$

supermultiplet are the same as the relative strengths of the lines in a $^7S \to {}^7P$ transition, namely 50:70:90. Several carefully determined measurements of relative strengths of multiplets in supermultiplets have been made by Harrison* with good theoretical agreement. For example, for the above supermultiplet of titanium he finds the relative strengths 53:70:89. The maximum discrepancy of 6 per cent. between these and the above values is probably little more than the experimental error.

This relation for the relative strengths of the multiplets of a supermultiplet was first obtained by Kronig† from the correspondence principle in 1925. It was apparently independently inferred from the experimental data by Russell and Meggers‡ in 1927 when they called attention to the

* Harrison, J. Opt. Soc. Am. **19**, 109 (1929);
Harrison and Engwicht, *ibid.* **18**, 287 (1929).
† Kronig, Zeits. für Phys. **33**, 261 (1925).
‡ Russell and Meggers, Sci. Papers Bur. of Standards, **22**, 329 (1927).

striking similarity between the visually estimated relative multiplet intensities in a $dd \rightarrow dp$ triplet supermultiplet of Sc II and the theoretical relative strengths of the lines in a $^5D \rightarrow {}^5P$ multiplet.

Now let us consider the relative total strengths of the various supermultiplets in a transition array between two configurations formed by adding an electron to the same parent configuration. The total strength of a supermultiplet is obtained from (4) if we make the same correlation as we did in (3) to obtain the relative strengths of the multiplets. This immediately shows that *the strengths of the different supermultiplets in a transition array are proportional to* $(2S+1)(2L^{\mathrm{I}}+1)$, where L^{I} is the L value of the parent ion and S the multiplicity of the supermultiplet. For example, in the transition array $p^2s \rightarrow p^2p$, the relative total strengths of the four supermultiplets are 2:6:12:10, as seen from the following table:

		s		p	$(2S+1)(2L^{\mathrm{I}}+1)$
p^2	(^1S)	^2S	$\rightarrow$	^2P	2.1
	(^3P)	^2P	$\rightarrow$	^2SPD	2.3
		^4P	$\rightarrow$	^4SPD	4.3
	(^1D)	^2D	$\rightarrow$	^2PDF	2.5

Experimental data on relative strengths of supermultiplets are almost nonexistent. A comparison of the intensities of a quartet and a sextet supermultiplet is furnished by Seward* for the case of the quartet and sextet supermultiplets of Mn I $3d^6\,(^5D)\,4p \rightarrow 3d^6\,(^5D)\,5s$. The relative strengths of ^4F, ^4D, $^4P \rightarrow {}^4D$ are the same as the relative strengths of the lines in $^5P \rightarrow {}^5S$, namely 7:5:3. The multiplets ^6F, ^6D, $^6P \rightarrow {}^6D$ should be $\frac{6}{4}$ as strong as these, i.e. $10\frac{1}{2}:7\frac{1}{2}:4\frac{1}{2}$. When these values are corrected by the actual σ^4 factors of Mn I, the theoretical relative intensities of these six multiplets for excitation at infinite temperature are given by Seward as

382:354:259:1380:777:754.

The measured intensities, when corrected for the actual distribution among initial levels by an assumption of a Boltzmann distribution at the most favourable temperature, are given by Seward as

381:400:278:1328:740:819.

The agreement is quite good, considering the difficulty in temperature correction and intensity measurements for lines of rather different frequencies originating in levels of different energy.

We have thus solved completely the problem of reducing to the radial integral (6) the strengths in a transition array in which the jumping electron is not equivalent to any in the ion. It should be pointed out that this solves in particular the problem of transitions between two-electron configurations

* SEWARD, Phys. Rev. 37, 344 (1931).

neither of which is composed of equivalent electrons, since such transition arrays contain only two supermultiplets—a singlet and a triplet. For example, the relative strengths of the multiplets in either supermultiplet for $ps \rightarrow pp$ are the same as the relative strengths of the lines in ${}^3S \rightarrow {}^3P$, but the triplets have thrice the strength of the singlets. The line strengths in this array are given in Table 2^9. In transitions between two-electron configurations, at most one can be composed of equivalent electrons; from $6^8 17$ we see that the above considerations are true for the unexcluded multiplets of this case also, except that all absolute strengths must be multiplied by 2. Thus, e.g., the unexcluded multiplets of $ps \rightarrow p^2$ have the same relative strengths as the corresponding multiplets of $ps \rightarrow pp$; see Table 2^9.

TABLE 2^9. *Relative strengths of the lines in $ps \rightarrow pp$ and $ps \rightarrow p^2$ in LS coupling.*

Invariant sums are shown in parentheses. The signs of $\mathbf{S}^{\frac{1}{2}}$ are indicated.

ps \ *pp*	${}^1S_0$	${}^3P_0$	${}^1P_1$	${}^3S_1$	${}^3P_1$	${}^3D_1$	${}^1D_2$	${}^3P_2$	${}^3D_2$	${}^3D_3$	
${}^3P_0$				$+3\frac{1}{3}$	-10	$+16\frac{2}{3}$ (30)					(30)
${}^1P_1$	-10	(10)	$+30$			(30)	$+50$		(50)		(90)
${}^3P_1$		-10 (10)		-10	$+7{\cdot}5$	$+12{\cdot}5$ (30)		$-12{\cdot}5$	$+37{\cdot}5$ (50)		(90)
${}^3P_2$				$-16\frac{2}{3}$	$-12{\cdot}5$	$-0\frac{5}{6}$ (30)		$+37{\cdot}5$	$+12{\cdot}5$ (50)	$+70$ (70)	(150)
	(10)	(10)	(30)	(30)	(30)	(30)	(50)	(50)	(50)	(70)	

ps \ p^2	${}^1S_0$	${}^3P_0$	${}^3P_1$	${}^1D_2$	${}^3P_2$		
${}^3P_0$			-20 (20)			(20)	
${}^1P_1$	-20	(20)		$+100$		120	(180)
${}^3P_1$		-20 (20)	$+15$ (15)		-25 (125)	60	
${}^3P_2$			-25 (25)		$+75$ (75)	(100)	
	(20)	(20)	(60)	(100)	(100)		

The J-file sum rule.

We now derive a sum rule which applies in the case we have been considering in which the jumping electron is equivalent to none in the ion. For each multiplet the sum of the strengths of lines originating in a given level is proportional to its statistical weight $2J+1$. We may by formulas analogous to $2^9 3$, 4, 5 sum (5) over L' to find the total strength of the lines of the array which originate in the term $\gamma^{\rm I}\, L^{\rm I}\, S^{\rm I}\, nl\, S\, L$. This sum has the value

$$\begin{aligned} &(2S+1)(2L+1)\,l\,(2l-1)\,|(n\,l\vdots P\vdots n'\,l-1)|^2 && \text{if } l'=l-1 \\ &(2S+1)(2L+1)(l+1)(2l+3)\,|(n\,l\vdots P\vdots n'\,l+1)|^2 && \text{if } l'=l+1, \end{aligned} \tag{7}$$

which depends on the initial term only through its statistical weight $(2S+1)(2L+1)$. Hence *the total strength of the lines of the transition array originating in (or terminating on) any level is proportional to the statistical weight* $2J+1$ *of that level*. This statement, which we shall see in Chapter XI holds for any type of coupling, we shall call the *J-file sum rule* (we call any row or column of a transition array such as that of Table 2^9 a *J file*). We shall show in § 4^{11} that even if the jumping electron is equivalent to other electrons in *one* of the configurations, this sum rule still holds for the J files in *one* direction in the transition array.

More general configurations.

As an illustration of the use of formulas (3) and (4) for transition arrays in which the jumping electron is equivalent to one of the others, we may consider the strengths in the case $d^2\,s\,p \rightarrow d^2\,p^2$. By (3) and (4) any group of electrons which remains intact during the transition, such as d^2 here, may be eliminated, and the problem reduced to transitions between the remaining electrons, as to $s\,p \rightarrow p^2$. The allowed transitions for $d^2\,s\,p \rightarrow d^2\,p^2$ are

$d^2\,(^1S)$	$s\,p\,(^1P)\,^1P$	$\rightarrow p^2\,(^1S)\,^1S$	$\mathsf{S}=1$	$(^1P\rightarrow{}^1S)$	
	$\searrow$	$(^1D)\,^1D$	5	$(^1P\rightarrow{}^1D)$	
	$(^3P)\,^3P$ $\rightarrow$	$(^3P)\,^3P$	9	$(^1P\rightarrow{}^1P)$	
$d^2\,(^3P)$	$s\,p\,(^1P)\,^3SPD \rightarrow$	$p^2\,(^1S)\,^3P$	9	$(^3P\rightarrow{}^3S)$	
	$\searrow$	$(^1D)\,^3PDF$	45	$(^3P\rightarrow{}^3D)$	(8)
	$(^3P)\,^1SPD \rightarrow$	$(^3P)\,^1SPD$	9	$(^3P\rightarrow{}^3P)$	
	$^3SPD \rightarrow$	^3SPD	27	$(^3P\rightarrow{}^3P)$	
	$^5SPD \rightarrow$	^5SPD	45	$(^3P\rightarrow{}^3P)$	
$d^2\,(^1D)$ etc.					

It is convenient to consider together all multiplets of the same multiplicity which are based on the same terms of d^2, $s\,p$, and p^2, and which are analogous to supermultiplets. Each arrow in the above represents one such group. If we let $L^{\rm I}\,S^{\rm I}$ represent the term of d^2, $L^{\rm IV}\,S^{\rm IV}$ the term of $s\,p$, and $L^{\rm II}\,S^{\rm II}$ the term of p^2, (3) shows that the relative strengths of the multiplets in a group are the same as the relative strengths of the lines in the multiplet $SL\rightarrow SL'$, where $S=L^{\rm I}$, $L=L^{\rm IV}$, and $L'=L^{\rm II}$. These multiplets are indicated at the right of (8). The relative total strengths of the different groups are to be obtained from (4) in terms of the known relative strengths of the multiplets in the transition $s\,p\rightarrow p^2$—

$$\mathsf{S}(^1P\rightarrow{}^1S):\mathsf{S}(^1P\rightarrow{}^1D):\mathsf{S}(^3P\rightarrow{}^3P)=1:5:9.$$

In this way, e.g., for the last entry of (8), $2S+1=5$, $2L^{\mathrm{I}}+1=3$, $2S^{\mathrm{II}}+1=3$, $\mathsf{S}(sp\,{}^3P \to p^2\,{}^3P)=9$. Substitution in (4) gives 45 for the total strength of the group. Thus the relative strengths of the multiplets in a group are the same as the relative strengths of the lines in the multiplet in parentheses at the right of (8), and the relative total strengths of the different groups are given by the numbers in italics.*

4. Multiplet strengths obtained from spectroscopic stability.†

We have obtained very simply the strengths of the multiplets for configurations in which the jumping electron is not equivalent to any other electron in either configuration or is equivalent to one other electron in only one of the configurations. We have also been able in many cases to reduce the problem to a simpler one. But we have not been able to handle such transitions as $p^2 s \to p s^2$ or $p^2 s \to p^3$, to name two of the simplest. It is possible to determine these strengths from the principle of spectroscopic stability in much the same way as we determined the electrostatic energies from the diagonal-sum rule.

As shown in § 1^9, the only transitions possible are between configurations which differ in regard to but one electron, which has quantum numbers $n\,l$ in one configuration, and $n'\,l-1$ in the other. We calculate first, in the zero-order $nlm_s m_l$ scheme, the absolute squares of the matrix components of $\boldsymbol{P}$ connecting these two configurations. There are no components connecting states which differ in regard to the quantum numbers of more than one electron. By the results of § 6^6 a non-vanishing matrix component is simply the corresponding matrix component of the one-electron problem (§ 6^5). This is given by the formula

$$|(n\,l\,m_s\,m_l|e\boldsymbol{r}|n'\,l-1\,m_s'\,m_l')|^2 = \mathsf{s}^2\,\delta(m_s,m_s')\,\delta(m_l,m_l')\,(l^2-m_l^2) + \mathsf{s}^2\,\delta(m_s,m_s')\,\delta(m_l',m_l\mp 1)\,\tfrac{1}{2}\,(l\pm m_l)(l\pm m_l-1), \quad (1)$$

where s is given by $3^9 6$. Thus all the squared matrix components of $\boldsymbol{P}$ may be expressed as multiples of s^2.

Now let us consider the squared matrix components of $\boldsymbol{P}$ connecting these configurations when expressed in the SLM_SM_L scheme. All of these are expressed in terms of the quantities $|(\gamma\,S\,L\vdots P\vdots\gamma'\,S\,L')|^2$, which we are seeking, by means of the formulas $9^3 11$. The states of the zero-order scheme are related to those of the SLM_SM_L scheme by a unitary transformation which has non-vanishing components only between states of the same M_S

* If we had substituted in (4) the absolute strengths of the multiplets in the array $s\,p \to p^2$, we should have obtained the total *absolute* strengths of these groups in terms of the radial integral (6).

† Condon and Ufford, Phys. Rev. **44**, 740 (1933).

and M_L. Therefore we may apply the principle of spectroscopic stability $2^2 25$ to the squared matrix components occurring in any box of which the rows and columns are labelled by the same values of M_S and M_L in the two schemes. This enables us to equate the sum of the squared matrix components in a particular box in one scheme to the corresponding sum of the squares in the other scheme. In this way a set of equations is obtained which is usually sufficient to determine the quantities $|(\gamma\, S\, L \vdots P \vdots \gamma'\, S\, L')|^2$, from which we obtain the multiplet strengths by $2^9 6$. Analogous to the limitation of the diagonal-sum rule for energy, the method gives complete specification only when there is but one multiplet of each kind occurring. If there are several multiplets of the same kind, the method determines only the sum of the strengths.

We shall illustrate the method by consideration of the configuration array $p^3 \rightarrow p^2 s$ in Russell-Saunders coupling. Only one component occurs with $M_S = M'_S = \frac{3}{2}$ and $M_L = 0$, $M'_L = 1$. In the zero-order scheme this is the transition $(1^+ 0^+ -1^+) \rightarrow (1^+ 0^+ 0^+)$, for which the squared matrix component of $\boldsymbol{P}$ is s^2 by the formulas. In the SLM_SM_L scheme this is the transition $({}^4S, \frac{3}{2}, 0) \rightarrow ({}^4P, \frac{3}{2}, 1)$; the squared matrix component being, by $9^3 11$,

$$|(p^3\, {}^4S \vdots P \vdots p^2 s\, {}^4P)|^2.$$

Hence we infer that $$|(p^3\, {}^4S \vdots P \vdots p^2 s\, {}^4P)|^2 = \mathrm{s}^2.$$

This result may be checked from the component having $M_S = M'_S = \frac{3}{2}$, $M_L = M'_L = 0$.

In the same way we may pass to the part of the matrix associated with $M_S = M'_S = \frac{1}{2}$ and find in the zero-order scheme that the coefficients of s^2 in the squares of the non-vanishing matrix elements of $\boldsymbol{P}$ are given by

p^3 \ $p^2 s$	$(1^+ 1^- 0^+)$	$(1^+ 0^- 0^+)$	$(1^- 0^+ 0^+)$	$(1^+ 0^+ 0^-)$	$(0^+ 0^- 0^+)$	$(1^+ -1^- 0^+)$	$(1^- -1^+ 0^+)$	$(1^+ -1^+ 0^-)$
$(1^+ 1^- \; 0^+)$	1	0	1	1				
$(1^+ 1^- -1^+)$	1	0	0	0	0	0	1	1
$(1^+ 0^+ \; 0^-)$	0	1	0	1	1	0	0	0
$(1^+ 0^+ -1^-)$		0	0	1	0	1	0	0
$(1^+ 0^- -1^+)$		1	0	0	0	0	0	1
$(1^- 0^+ -1^+)$		0	1	0	0	0	1	0

The table need not be extended to negative values of M_L unless this is desired for checking. The corresponding portion of the table in the SLM_SM_L scheme, in which the numbers are the factors which multiply the corresponding quantity of the type $|(p^3\, {}^2D \vdots P \vdots p^2 s\, {}^2P)|^2$, is

p^3 \ $p^2 s$	$^2D, 2, \frac{1}{2}$	^2D	^2P	$^4P, 1, \frac{1}{2}$	^2D	^2P	^2S	$^4P, 0, \frac{1}{2}$
$^2D, 2, \frac{1}{2}$	4	2	6	0				
^2D	2	1	3	0	3	3	0	0
$^2P, 1, \frac{1}{2}$	6	3	1	0	1	1	1	0
^2D		3	1	0	0	4	0	0
^2P		3	1	0	4	0	1	0
$^4S, 0, \frac{1}{2}$		0	0	1	0	0	0	1

Applying the principle to each of the boxes, we obtain a set of equations which lead to results

		$p^2 s$ (or $p^4 s$)			
		^4P	^2D	^2P	^2S
p^3 (or $p^3 s^2$)	^4S	1	0	0	0
	^2D	0	$\frac{1}{4}$	$\frac{1}{4}$	0
	^2P	0	$\frac{1}{12}$	$\frac{3}{4}$	$\frac{2}{3}$

for the coefficient of s^2 in the value of $|(p^3\, SL \vdots P \vdots p^2 s\, SL')|^2$. The total multiplet strengths, given in terms of this quantity by formulas 2[9]6, are

TABLE 3[9]. *Relative multiplet strengths.*

The terms of the first configuration are placed on the left, of the second, at the top, of the boxes. Only allowed transitions are listed, and where two or more multiplets of a kind occur, the sum of the strengths is given. In the same transition array, values in boxes of different multiplicities are to be directly compared.

$p\,s^2 \rightarrow p^2 s$ or $p^5 \rightarrow p^4 s$

	^2S	^2P	^2D
^2P	1	9	5

$p^5 s \rightarrow p^4 s^2$ or $p\,s \rightarrow p^2$

	^3P
^3P	9

	^1D	^1S
^1P	5	1

$p^3 \rightarrow p^2 s$ or $p^3 s^2 \rightarrow p^4 s$

	^4P
^4S	12

	^2D	^2P	^2S
^2D	15	15	—
^2P	5	9	4

$p^2 s^2 \rightarrow p^3 s$ or $p^4 \rightarrow p^3 s$

	^3S	^3P	^3D
^3P	12	9	15

	^1P	^1D
^1S	4	—
^1D	5	15

$p^5 \rightarrow p^4 d$

	$3\,^2D$	$2\,^2P$	^2S
^2P	19	9	2

$d^2 p \rightarrow d^3$

	^4F	^4P
^4G	180	—
^4F	1260	—
$2\,^4D$	1080	520
^4P	—	240
^4S	—	320

	^2H	^2G	^2F	$2\,^2D$	^2P
^2H	396	44	—	—	—
$2\,^2G$	1584	576	180	—	—
$3\,^2F$	—	1000	420	470	—
$3\,^2D$	—	—	660	625	215
$3\,^2P$	—	—	—	705	255
^2S	—	—	—	—	70

$p^2 d \rightarrow p^3$

	^4S
^4P	24

	^2D	^2P
$2\,^2F$	42	—
$3\,^2D$	15	25
$2\,^2P$	3	9
^2S	—	2

$p^3 d \rightarrow p^4$

	^3P
$3\,^3D$	120
$2\,^3P$	45
^3S	15

	^1D	^1S
$2\,^1F$	63	—
$2\,^1D$	30	—
$2\,^1P$	7	20
^1S	—	0

given in Table 3^9, together with the relative strengths of the multiplets in other simple transition arrays which cannot be solved by the formulas of § 3^9.*

The method of this section has the disadvantage, not shared by that of § 3^9, of being capable of giving us only the amplitude of the electric-moment matrix. This is of little use for cases of departures from Russell-Saunders coupling as discussed in Chapter XI, since to effect a transformation of this matrix from LS coupling to an intermediate coupling we need to know the correct phases of the matrix elements in a definite scheme of states. Such a calculation of the electric-moment matrix may be made if we know the eigenfunctions of the LS-coupling states in terms of the zero-order states. In order to separate two like multiplets in cases not amenable to characterization by parentage we must also go to the eigenfunctions, just as we did in § 7^8 to separate the energies of the two ^2D's of d^3. Several of the arrays of Table 3^9 are taken from a paper by Ufford,† who calculated them by first obtaining the eigenfunctions.

5. Quadrupole multiplets.

Let us now obtain the strengths of the lines in a quadrupole multiplet, as we did for the dipole-multiplet in § 2^9. In § 6^4 we were able to find the dependence on M of the quadrupole-moment matrix components by building them up from the dipole matrix components of $9^3 11$: here we may repeat the process by applying § 11^3 to find the dependence of the matrix components on L and S in the SLJ scheme.

After the matrix components are found the strengths of the quadrupole lines may be found by $7^4 7$. This gives us a set of formulas (Table 4^9) for quadrupole multiplets analogous to $2^9 2$ for dipole multiplets. In the process of squaring, the sign of the matrix component is lost, so in Table 4^9 we give the sign in a separate column. If one wants to know one of the matrix components D, E, F occurring in $7^4 7$, he can find it from the strengths by dividing by the appropriate factor in $7^4 7$ and attaching the sign here given to the positive square root of the quotient. G, H, I are related to the matrix components of $\boldsymbol{r}_i$, where $\mathfrak{R} = -e\sum_i \boldsymbol{r}_i \boldsymbol{r}_i$, as follows:

$$G = -e\sum_i \tfrac{1}{4} \sum_{\alpha''} [(\alpha\, S\, L \vdots \boldsymbol{r}_i \vdots \alpha''\, S\, L)(\alpha''\, S\, L \vdots \boldsymbol{r}_i \vdots \alpha'\, S\, L) - (\alpha\, S\, L \vdots \boldsymbol{r}_i \vdots \alpha''\, S\, L+1)(\alpha''\, S\, L+1 \vdots \boldsymbol{r}_i \vdots \alpha\, S\, L) - (\alpha\, S\, L \vdots \boldsymbol{r}_i \vdots \alpha''\, S\, L-1)(\alpha''\, S\, L-1 \vdots \boldsymbol{r}_i \vdots \alpha\, S\, L)],$$

$$H = -e\sum_i \tfrac{1}{4} \sum_{\alpha''} [(\alpha\, S\, L \vdots \boldsymbol{r}_i \vdots \alpha''\, S\, L)(\alpha''\, S\, L \vdots \boldsymbol{r}_i \vdots \alpha'\, S\, L-1) + (\alpha\, S\, L \vdots \boldsymbol{r}_i \vdots \alpha''\, S\, L-1)(\alpha''\, S\, L-1 \vdots \boldsymbol{r}_i \vdots \alpha'\, S\, L-1)],$$

$$I = -e\sum_i \tfrac{1}{4} \sum_{\alpha''} (\alpha\, S\, L \vdots \boldsymbol{r}_i \vdots \alpha''\, S\, L-1)(\alpha''\, S\, L-1 \vdots \boldsymbol{r}_i \vdots \alpha'\, S\, L-2).$$

* A very complete set of tables of relative multiplet strengths is being prepared by Mr Goldberg, of the Harvard College Observatory; this will probably be published in the Astrophysical Journal in 1935.

† Ufford, Phys. Rev. **40**, 974 (1932).

The strengths of Table 4[9] were calculated by Rubinowicz*, who showed that they obey the sum rule of Ornstein, Burger and Dorgelo, namely, that the sum of the strengths of the lines in a multiplet from a given initial (or final) level is proportional to the statistical weight, $2J+1$, of that level.

TABLE 4[9]. *Strengths of quadrupole-multiplet lines.*

Here $P(J)$, $Q(J)$, and $R(J)$ are the factors used in 11[3]8 with the correlation $j\to J$, $j_1\to S$, $j_2\to L$.

L'	J'	Sign	$\mathbf{S}(\alpha\,S\,L\,J,\alpha'\,S\,L'\,J')$
L	$J-2$	+	$\frac{P(J-1)\,P(J)\,Q(J-2)\,Q(J-1)}{J\,(J-1)(2J-1)}\,G^2$
	$J-1$	−	$\frac{2\,P(J)\,Q(J-1)\,[R(J)-(J+1)]^2}{J\,(J-1)(J+1)}\,G^2$
	J	+	$\frac{2\{3\,R(J)\,[R(J)-1]-4J\,(J+1)\,L\,(L+1)\}^2\,(2J+1)}{3J\,(J+1)(2J-1)(2J+3)}\,G^2$
	$J+1$	−	$\frac{2\,P(J+1)\,Q(J)\,[R(J)+J]^2}{J\,(J+1)(J+2)}\,G^2$
	$J+2$	+	$\frac{P(J+1)\,P(J+2)\,Q(J)\,Q(J+1)}{(J+1)(J+2)(2J+3)}\,G^2$
$L-1$	$J-2$	−	$\frac{P(J-2)\,P(J-1)\,P(J)\,Q(J-1)}{J\,(J-1)(2J-1)}\,H^2$
	$J-1$	+	$\frac{2\,P(J-1)\,P(J)\,[R(J)-(J+1)(L+1)]^2}{J\,(J-1)(J+1)}\,H^2$
	J	+	$\frac{6\,P(J)\,Q(J)\,[R(J)+J(L+1)]^2\,(2J+1)}{J\,(J+1)(2J-1)(2J+3)}\,H^2$
	$J+1$	−	$\frac{2\,Q(J)\,Q(J+1)\,[R(J)-J\,(L+1)]^2}{J\,(J+1)(J+2)}\,H^2$
	$J+2$	+	$\frac{P(J+1)\,Q(J)\,Q(J+1)\,Q(J+2)}{(J+1)(J+2)(2J+3)}\,H^2$
$L-2$	$J-2$	+	$\frac{P(J-3)\,P(J-2)\,P(J-1)\,P(J)}{J\,(J-1)(2J-1)}\,I^2$
	$J-1$	+	$\frac{2\,P(J-2)\,P(J-1)\,P(J)\,Q(J)}{J\,(J-1)(J+1)}\,I^2$
	J	+	$\frac{6\,P(J-1)\,P(J)\,Q(J)\,Q(J+1)\,(2J+1)}{J\,(J+1)(2J-1)(2J+3)}\,I^2$
	$J+1$	−	$\frac{2\,P(J)\,Q(J)\,Q(J+1)\,Q(J+2)}{J\,(J+1)(J+2)}\,I^2$
	$J+2$	+	$\frac{Q(J)\,Q(J+1)\,Q(J+2)\,Q(J+3)}{(J+1)(J+2)(2J+3)}\,I^2$

There is very little experimental material available for testing these formulas. The relative line strengths in some of the $^2D\to{}^2S$ quadrupole

* RUBINOWICZ, Zeits. für Phys. **65**, 662 (1930).

doublets in the alkalis have been measured by Prokofjew* and found to be 3:2 in accordance with theory. The difficulty is that multiplets violating the Laporte rule may arise because the atoms are disturbed by electric fields in the source, in which case the line strengths are given by other formulas (§ 3^{17}, especially $3^{17}5$). The place where quadrupole lines are of importance is in the spectra of nebulae or rarefied stellar atmospheres where the mutual disturbance of neighbouring atoms is negligible. In fact the most interesting developments in this field grew out of Bowen's interpretation of the nebulium lines as quadrupole lines. This we consider in § 5^{11} because these lines are combinations between terms of different multiplicity and so owe their existence to the departure from Russell-Saunders coupling.

There are two good examples of quadrupole multiplets in the Fe II spectrum found by Merrill† in the spectrum of the star η-Carinae. They are $3d^5\,4s^2\,{}^6S \rightarrow 3d^6\,4s\,{}^6D$ and $3d^6\,4s\,{}^4G \rightarrow 3d^7\,{}^4F$. The intensity data are merely eye estimates from plates taken in Chile by Moore and Sanford. For the former there are five lines having the same initial level and ending on the five levels of the ${}^6D$ term. By the sum rule the strengths are proportional to $(2J+1)$ for the final level. If we take Merrill's intensities divided by numbers proportional to $\sigma^6(2J+1)$, the quotient should be constant. The values are

$$6{\cdot}2 \quad ? \quad 6{\cdot}6 \quad 6{\cdot}8 \quad 6{\cdot}4,$$

the ? being due to disturbance by another line. This is surprisingly good agreement, probably accidental in view of the lack of photometric measurements.

	${}^4F_{\frac{9}{2}}$	${}^4F_{\frac{7}{2}}$	${}^4F_{\frac{5}{2}}$	${}^4F_{\frac{3}{2}}$
${}^4G_{\frac{11}{2}}$	46·4	6·8		
	279·6	43·0		
	0·17	0·16		
${}^4G_{\frac{9}{2}}$	3·2	20·9	?	
	60·5	155·5	52·8	
	0·053	0·135		
${}^4G_{\frac{7}{2}}$	0·0	5·3	11·1	5·6
	5·4	71·1	99·0	39·5
	—	0·07	0·11	0·14
${}^4G_{\frac{5}{2}}$	0·0	0·0	6·35	?
	0·1	6·9	55·5	98·7
	—	—	0·09	

For the other multiplet the agreement is not nearly as good. In the table we give for each line the observed relative strengths (intensity divided by σ^6), the theoretical relative strengths, and the ratio, observed to theoretical. The latter should be constant in any one row as these lines have a common

* Prokofjew, Zeits. für Phys. **57**, 387 (1929).
† Merrill, Astrophys. J. **67**, 405 (1928).

initial level. The sum of the observed strengths from $^4G_{1\frac{1}{2}}$ divided by 12 is 4·4, which would be equal to the sum from $^4G_{\frac{7}{2}}$ divided by 8, which is 2·75, if the temperature were infinite. The observed ratio, if we assume a Maxwell-Boltzmann distribution of the atoms, corresponds to a temperature of 1600° K.

We now consider the absolute values of the quadrupole strengths. The methods developed in the preceding two sections could be taken over with proper generalizations to treat the corresponding problems for quadrupole lines. Such developments have not been carried out in detail because the possibility of experimental verification seems too remote. We confine attention to the case of one-electron spectra, which is directly analogous to the work of §§ 6^5 and 9^5. The problem is to express the coefficients, G, H, and I, of the formulas for line strengths in terms of the radial integral

$$\mathsf{s}_2 = -e\int_0^\infty r^2 R(nl)\, R(n'l')\, dr, \tag{1}$$

which obviously will measure the second-order moment of the charge distribution. This may be done by calculating, by direct integration, one of the matrix components $(n\,l\,m_s\,l|\quad|n'\,l'\,m_s\,l')$ in the nlm_sm_l scheme. Then, since for $m_s = +\frac{1}{2}$ and $m_l = l$ we have

$$\phi(l, m_s = \tfrac{1}{2}, m_l = l) = \psi(l, j = l+\tfrac{1}{2}, m = l+\tfrac{1}{2}),$$

we know that this is also equal to $(n\,l\,l+\frac{1}{2}\,l+\frac{1}{2}|\quad|n'\,l'\,l'+\frac{1}{2}\,l'+\frac{1}{2})$ in the $nljm$ scheme. The same matrix component can be expressed in terms of D, E, or F from $6^4 6$, so by appropriate choices D, E, and F and the coefficient of $\frac{1}{3}\mathfrak{Y}$ in $6^4$6a can all be found in terms of s_2. Then by combining $7^4 7$ with the table of strengths of this section we can find G, H, and I in terms of s_2, which is all we need. The work is a little tedious but straightforward. It is clear that for a one-electron spectrum $H = 0$, otherwise the parity selection rule would be violated. The values of G and I found in this way are

$$\begin{aligned} G &= \frac{-\mathsf{s}_2/2}{(2l-1)(2l+3)} \\ I &= \frac{\mathsf{s}_2}{4\,(2l-1)\sqrt{(2l+1)(2l-3)}}. \end{aligned} \tag{2}$$

Combining these with the table of strengths we have the means of expressing in absolute measure all the strengths of one-electron doublet lines.

It is convenient, as in $6^5 5$, to express the numerical coefficient in the quadrupole transition probability in atomic units. In $7^4 8$, if we measure σ with the Rydberg constant as unit, and $\mathsf{S}(A, B)$ in atomic units, $e^2\mathrm{a}^4$, we may write

$$\mathsf{A}(A, B) = \frac{\alpha^5}{2^6 5}\frac{\sigma^5\,\mathsf{S}(A, B)}{2J_A+1}\tau^{-1} = \frac{\sigma^5\,\mathsf{S}(A, B)}{2J_A+1}\,.\,2649\ \mathrm{sec}^{-1}. \tag{3}$$

The numerical coefficient is about one-millionth of that for dipole radiation ($6^5 5$). To exemplify the formulas just presented, and also to gain a better appreciation of the order of magnitude of quadrupole transition probabilities, we shall calculate $\mathsf{A}(A, B)$ for the $3d \rightarrow 1s$ doublet in hydrogen. First we find that the radial integral

$$s_2 = \int_0^\infty r^2 R(3d) R(1s) dr = \sqrt{2^{-15}.3^9.5};$$

hence $I^2 = 3^7/2^{19}$, from which, using the table of strengths,

$$\mathsf{S}(3d_{\frac{5}{2}} \rightarrow 1s_{\frac{1}{2}}) = 3.3^8/2^{13}; \quad \mathsf{S}(3d_{\frac{3}{2}} \rightarrow 1s_{\frac{1}{2}}) = 2.3^8/2^{13}.$$

These are in the ratio of the statistical weights as the sum rule demands. For this line we have $\sigma = \frac{8}{9}$ so the spontaneous emission transition probability of either line is

$$\mathsf{A}(3d \rightarrow 1s) = 616 \sec^{-1}.$$

From Table 5^5, the closely analogous dipole transition probability is

$$\mathsf{A}(3p \rightarrow 1s) = 1{\cdot}64 \times 10^8 \sec^{-1},$$

which is about $0{\cdot}27 \times 10^6$ times as great.

In hydrogenic ions of nuclear charge Z we have $\sigma \propto Z^2$ and $r \propto Z^{-1}$, so the dipole transition probability varies as Z^4 while the quadrupole transition probability varies as Z^6. In all actual atoms the quadrupole transition probability remains smaller than the dipole, although for $Z = 92$ the ratio is only 1:32. As a consequence of this trend quadrupole transitions are sometimes of appreciable intensity in X-ray spectra. For the higher values of Z the calculation should be made using the relativistic radial functions of § 5^5, but our result shows the trend in order of magnitude.

In view of the great ratio of dipole to quadrupole intensity it is out of the question to observe quadrupole lines in the hydrogen spectrum. In the alkalis, where the terms are well separated, this is possible and the $s \rightarrow d$ series is easily obtained in absorption. Stevenson* has calculated the transition probabilities from the normal S level to the lowest D doublet for the alkali metals using Hartree wave functions. Experimental values were obtained by the anomalous dispersion method by Prokofjew.† His values have to be divided by four because of a difference in the theoretical formulas for quadrupole and dipole dispersion.‡ The final values of 10^6 times the ratio of the spontaneous transition probability for the first $D \rightarrow S$ quadrupole line to that for the first $P \rightarrow S$ dipole line are

	Na	K	Rb	Cs
Observed	1·1	1·5	2·7	0·6
Calculated	3·5	2·5	2·9	

* Stevenson, Proc. Roy. Soc. **A128**, 591 (1930).

† Prokofjew, Zeits. für Phys. **57**, 387 (1929).

‡ This was pointed out by Rubinowicz and Blaton, Ergebnisse der exakten Naturwissenschaften, **11**, 216 (1932).

CHAPTER X

jj COUPLING

The most direct method of solution of the first-order perturbation problem would consist of the diagonalization of the matrix of electrostatic plus spin-orbit energy for a given configuration in the zero-order scheme. That procedure one might call 'impossible' for all but the very simplest configurations because of the high order of the resulting secular equations. The general solution is 'possible' for a great many more cases if one uses the $SLJM$ scheme, but in order to utilize this scheme one must, except in special cases, obtain the transformation to it. Because of the complexity of the problem and the 'impossibility' of obtaining the general solution for complicated configurations, it is desirable to obtain as much information as possible of an elementary, although approximate, character. In the preceding three chapters we have considered the important case in which the spin-orbit interaction is weak compared to the electrostatic; in this chapter it will be interesting to consider the less important case in which the electrostatic interaction is weak compared to the spin-orbit. In this way we shall know the character of the general solution at both extremes.

1. The jj-coupling scheme and the spin-orbit interaction.

We wish to determine the first-order energies which result from the spin-orbit interaction, when the electrostatic interaction is absent. Since the spin-orbit interaction

$$H^I = \sum_i H^I_i = \sum_i \xi(r_i)\, \boldsymbol{L}_i \cdot \boldsymbol{S}_i \tag{1}$$

is a quantity of type F, it is diagonal in the zero-order scheme in which $nljm$ are taken as the electronic quantum numbers. This, then, is the natural scheme to use when considering this interaction. The diagonal element for the state $A = a^1 \dots a^N$ is given by

$$\begin{aligned}(A|H^I|A) &= \sum_i (n^i l^i j^i m^i | H^I_i | n^i l^i j^i m^i) \\ &= \sum_i \zeta_{n^i l^i} \left\{ \frac{j^i(j^i+1) - l^i(l^i+1) - s^i(s^i+1)}{2} \right\}\end{aligned} \tag{2}$$

(cf. § 4^5), which depends only on the values of $n^i l^i j^i$.

Although H^I is so highly degenerate as to be diagonal in any scheme specified by the set $n^i l^i j^i$, it is most convenient to consider the particular scheme in which the j's have been combined to form resultant J and M. The scheme characterized by the set of $n^i l^i j^i$ and JM is known as the jj-coupling scheme. It may be here noted that although this designation utilizes more quantum numbers than the LS-coupling scheme, it still does not in general furnish a complete set for more than two electrons.

The process of finding the allowed levels in jj coupling is similar to that in LS coupling, depending in the same way on $4^6 1$. For example, consider the configuration $np\,n'p$. Here j and j' may take on the values $\frac{1}{2}$ and $\frac{3}{2}$. If $j=\frac{1}{2}$, $j'=\frac{1}{2}$, (m,m') may have the values $(\frac{1}{2},\frac{1}{2})$, $(\frac{1}{2},-\frac{1}{2})$, $(-\frac{1}{2},\frac{1}{2})$, and $(-\frac{1}{2},-\frac{1}{2})$. For the first of these states $M=1$, and since this is the highest M occurring for $j=\frac{1}{2}$, $j'=\frac{1}{2}$, there must be a level $J=1$. There are two states with $M=0$; the level $J=1$ has a state $M=0$ and there must in addition be a level with $J=0$. The one state with $M=-1$ is taken care of by $J=1$. Hence we infer the existence of the levels $np_{\frac{1}{2}}n'p_{\frac{1}{2}}$, $J=1, 0$. In a similar way the levels for the other values of j, j' are determined in the following table. In the body of the table are given (m, m'), while at the foot of each column are given the resulting J values.

$np\,n'p$		j, j': $\frac{3}{2}, \frac{3}{2}$	$\frac{3}{2}, \frac{1}{2}$	$\frac{1}{2}, \frac{3}{2}$	$\frac{1}{2}, \frac{1}{2}$
M	3	$(\frac{3}{2}, \frac{3}{2})$			
	2	$(\frac{3}{2}, \frac{1}{2})(\frac{1}{2}, \frac{3}{2})$	$(\frac{3}{2}, \frac{1}{2})$	$(\frac{1}{2}, \frac{3}{2})$	
	1	$(\frac{3}{2}, -\frac{1}{2})(-\frac{1}{2}, \frac{3}{2})(\frac{1}{2}, \frac{1}{2})$	$(\frac{1}{2}, \frac{1}{2})(\frac{3}{2}, -\frac{1}{2})$	$(\frac{1}{2}, \frac{1}{2})(-\frac{1}{2}, \frac{3}{2})$	$(\frac{1}{2}, \frac{1}{2})$
	0	$(\frac{3}{2}, -\frac{3}{2})(-\frac{3}{2}, \frac{3}{2})(\frac{1}{2}, -\frac{1}{2})(-\frac{1}{2}, \frac{1}{2})$	$(\frac{1}{2}, -\frac{1}{2})(-\frac{1}{2}, \frac{1}{2})$	$(\frac{1}{2}, -\frac{1}{2})(-\frac{1}{2}, \frac{1}{2})$	$(\frac{1}{2}, -\frac{1}{2})(-\frac{1}{2}, \frac{1}{2})$
	−1	$(\frac{1}{2}, -\frac{3}{2})(-\frac{3}{2}, \frac{1}{2})(-\frac{1}{2}, -\frac{1}{2})$	$(-\frac{1}{2}, -\frac{1}{2})(-\frac{3}{2}, \frac{1}{2})$	$(-\frac{1}{2}, -\frac{1}{2})(\frac{1}{2}, -\frac{3}{2})$	$(-\frac{1}{2}, -\frac{1}{2})$
	−2	$(-\frac{3}{2}, -\frac{1}{2})(-\frac{1}{2}, -\frac{3}{2})$	$(-\frac{3}{2}, -\frac{1}{2})$	$(-\frac{1}{2}, -\frac{3}{2})$	
	−3	$(-\frac{3}{2}, -\frac{3}{2})$			
J values:		3, 2, 1, 0	2, 1	2, 1	1, 0

(3)

These J values check with those obtained in § 1^7 for this configuration, namely ${}^1S_0$, ${}^3S_1$, ${}^1P_1$, ${}^3P_{0,1,2}$, ${}^1D_2$, ${}^3D_{1,2,3}$. They could equally well have been obtained by the usual vector-coupling picture: $j=\frac{3}{2}$ and $j'=\frac{3}{2}$ can combine to give a resultant $J=3, 2, 1, 0$, etc. Just as in LS coupling, the resultant levels cannot be obtained in this way if there are equivalent electrons. For the configuration np^2, (3) becomes modified as follows:

np^2		j, j': $\frac{3}{2}, \frac{3}{2}$	$\frac{3}{2}, \frac{1}{2}$	$\frac{1}{2}, \frac{1}{2}$
M	2	$(\frac{3}{2}, \frac{1}{2})$	$(\frac{3}{2}, \frac{1}{2})$	
	1	$(\frac{3}{2}, -\frac{1}{2})$	$(\frac{3}{2}, -\frac{1}{2})(\frac{1}{2}, \frac{1}{2})$	
	0	$(\frac{3}{2}, -\frac{3}{2})(\frac{1}{2}, -\frac{1}{2})$	$(\frac{1}{2}, -\frac{1}{2})(-\frac{1}{2}, \frac{1}{2})$	$(\frac{1}{2}, -\frac{1}{2})$
	−1	$(-\frac{3}{2}, \frac{1}{2})$	$(-\frac{3}{2}, \frac{1}{2})(-\frac{1}{2}, -\frac{1}{2})$	
	−2	$(-\frac{3}{2}, -\frac{1}{2})$	$(-\frac{3}{2}, -\frac{1}{2})$	
J values:		2, 0	2, 1	0

(4)

These J values agree with those previously found, namely ${}^1S_0$, ${}^3P_{0,1,2}$, ${}^1D_2$. Because of the symmetry which obtains, the half of these tables for $M \geqslant 0$ will usually give one all the information desired.

The spin-orbit energies for these states as calculated from (2) are

$$\left.\begin{array}{lllll} npn'p: & j=\frac{3}{2}, & j'=\frac{3}{2}: & \frac{1}{2}\zeta_{np}+\frac{1}{2}\zeta_{n'p} & (16) \\ & \frac{3}{2} & \frac{1}{2} & \frac{1}{2}\zeta_{np}-\zeta_{n'p} & (8) \\ & \frac{1}{2} & \frac{3}{2} & -\zeta_{np}+\frac{1}{2}\zeta_{n'p} & (8) \\ & \frac{1}{2} & \frac{1}{2} & -\zeta_{np}-\zeta_{n'p} & (4) \end{array}\right\} \quad (5)$$

$$\left.\begin{array}{lllll} np^2: & j=\frac{3}{2}, & j'=\frac{3}{2}: & \zeta_{np} & (6) \\ & \frac{3}{2} & \frac{1}{2} & -\frac{1}{2}\zeta_{np} & (8) \\ & \frac{1}{2} & \frac{1}{2} & -2\zeta_{np} & (1) \end{array}\right\} \quad (6)$$

(The numbers in parentheses at the right denote the degeneracies of the energy levels.) Thus we see that p^2 is split by the spin-orbit interaction into three equally spaced energy levels, while pp is split into four energy levels whose spacing depends on the relative values of ζ_{np} and $\zeta_{n'p}$. In each case the centre of gravity of the configuration remains at zero. This is a general property of the spin-orbit interaction, which follows from the theorem given in § 3^7 about the centre of gravity of a term. In § 3^7 we calculated the diagonal elements of H^I in the $LSJM$ scheme and found that their sum for each term, and hence for the whole configuration, was zero; by the diagonal-sum rule the sum of the eigenvalues of H^I must have this same value.

2. The addition of a weak electrostatic interaction.*

We shall now determine the effect of the addition of an electrostatic interaction which is sufficiently weak to be considered as a perturbation upon the degenerate spin-orbit levels. We must then, in $1^{10}3$ and $1^{10}4$, calculate that part of the matrix of electrostatic energy which refers to each column, and determine its eigenvalues. But the electrostatic energy is diagonal with respect to J and independent of M, so that there will occur one eigenvalue for each J, with a degeneracy of $2J+1$. Hence unless a given J occurs more than once per column, these eigenvalues can be determined by the diagonal-sum rule from the diagonal matrix elements.

We therefore need the diagonal elements of the electrostatic energy

$$Q = \sum_{i>j} q(i,j) = \sum_{i>j} e^2/r_{ij} \quad (1)$$

* INGLIS, Phys. Rev. **38**, 862 (1931).

in the *nljm* scheme. From 7^67 this element becomes

$$
\begin{aligned}
(A|Q|A) &= \sum_{k>t=1}^{N}\Big[\iint \bar\psi_1(n^kl^kj^km^k)\,\bar\psi_2(n^tl^tj^tm^t)\,q(1,2)\,\psi_1(n^kl^kj^km^k)\,\psi_2(n^tl^tj^tm^t)\,d\tau_1 d\tau_2 \\
&\qquad - \iint \bar\psi_1(n^kl^kj^km^k)\,\bar\psi_2(n^tl^tj^tm^t)\,q(1,2)\,\psi_1(n^tl^tj^tm^t)\,\psi_2(n^kl^kj^km^k)\,d\tau_1 d\tau_2\Big] \\
&= \sum_{k>t=1}^{N} T(n^kl^kj^km^k, n^tl^tj^tm^t). \qquad (2)
\end{aligned}
$$

Each of the ψ's of (2) must now be expressed by 4^58 in terms of the $\phi(nlm_sm_l)$, in which case we obtain a sum of terms of the type 8^61 which have already been evaluated. We shall content ourselves with giving Table 1^{10}, as calculated by Inglis, of the quantities $T(nljm, n'l'j'm')$ in terms of the $F^k(nl,n'l')$ and $G^k(nl,n'l')$ of 8^615, for s, p, and d electrons.

For example, for the configuration $np\,n'p$ the electrostatic energies are found from $1^{10}3$ and Table 1^{10} to be

$$
\begin{array}{lrl}
np\,n'p: & j=\tfrac32, j'=\tfrac32;\ J=3: & F_0+\ F_2-G_0-\ G_2 \\
& 2: & F_0-3F_2+G_0-3G_2 \\
& 1: & F_0+\ F_2-G_0-\ G_2 \\
& 0: & F_0+5F_2+G_0+5G_2 \\[4pt]
& j=\tfrac32, j'=\tfrac12;\ J=2: & F_0\qquad\quad -\ G_2 \\
& 1: & F_0\qquad\quad -5G_2 \\[4pt]
& j=\tfrac12, j'=\tfrac32,\ J=2: & F_0\qquad\quad -\ G_2 \\
& 1: & F_0\qquad\quad -5G_2 \\[4pt]
& j=\tfrac12, j'=\tfrac12,\ J=1: & F_0\qquad -G_0 \\
& 0: & F_0\qquad +G_0.
\end{array}
\qquad (3)
$$

In the approximation we are considering F_0 need not be small since it occurs uniformly in all diagonal elements and all energies. It serves merely to raise the whole configuration and may be considered as a constant independent of coupling. In the case of np^2 we have

$$
\begin{array}{lrl}
np^2: & j=\tfrac32, j'=\tfrac32;\ J=2: & F_0-3F_2 \\
& 0: & F_0+5F_2 \\[4pt]
& j=\tfrac32, j'=\tfrac12;\ J=2: & F_0-\ F_2 \\
& 1: & F_0-5F_2 \\[4pt]
& j=\tfrac12, j'=\tfrac12;\ J=0: & F_0.
\end{array}
\qquad (4)
$$

In Fig. 1^{10} these levels are compared schematically with those of Chapter VII for this configuration in *LS* coupling. The J values occur in the same order on each side, and as the parameters are continuously varied we shall see in § 3^{11} that the two sides of Fig. 1^{10} grow into each other in this order. That two levels of the same J value cannot in general cross when one

TABLE 1[1]. $T(nljm, n'l'j'm')$.

The numbers given are to be used as coefficients for the integrals at the top of the column. To each of these T's is to be added $F^0(nl, n'l')$, which is omitted from the table for convenience.

$l\,l'$	$j\ m$	$j'\ m'$	$-G^0$	$\frac{-G^1}{9}$	$\frac{-G^2}{25}$
$s\,s$	$\frac{1}{2}\ \pm\frac{1}{2}$	$\frac{1}{2}\ \pm\frac{1}{2}$	1		
	$\pm\frac{1}{2}$	$\mp\frac{1}{2}$	0		
$s\,p$	$\frac{1}{2}\ \pm\frac{1}{2}$	$\frac{3}{2}\ \pm\frac{3}{2}$		3	
	$\pm\frac{1}{2}$	$\pm\frac{1}{2}$		2	
	$\pm\frac{1}{2}$	$\mp\frac{3}{2}$		0	
	$\pm\frac{1}{2}$	$\mp\frac{1}{2}$		1	
	$\frac{1}{2}\ \pm\frac{1}{2}$	$\frac{1}{2}\ \pm\frac{1}{2}$		1	
	$\pm\frac{1}{2}$	$\mp\frac{1}{2}$		2	
$s\,d$	$\frac{1}{2}\ \pm\frac{1}{2}$	$\frac{5}{2}\ \pm\frac{5}{2}$			5
	$\pm\frac{1}{2}$	$\pm\frac{3}{2}$			4
	$\pm\frac{1}{2}$	$\pm\frac{1}{2}$			3
	$\pm\frac{1}{2}$	$\mp\frac{5}{2}$			0
	$\pm\frac{1}{2}$	$\mp\frac{3}{2}$			1
	$\pm\frac{1}{2}$	$\mp\frac{1}{2}$			2
	$\frac{1}{2}\ \pm\frac{1}{2}$	$\frac{3}{2}\ \pm\frac{3}{2}$			1
	$\pm\frac{1}{2}$	$\pm\frac{1}{2}$			2
	$\pm\frac{1}{2}$	$\mp\frac{3}{2}$			4
	$\pm\frac{1}{2}$	$\mp\frac{1}{2}$			3

$l\,l'$	$j\ m$	$j'\ m'$	$\frac{F^2}{25}$	$-G^0$	$\frac{-G^2}{25}$
$p\,p$	$\frac{3}{2}\ \pm\frac{3}{2}$	$\frac{3}{2}\ \pm\frac{3}{2}$	1	1	1
	$\pm\frac{3}{2}$	$\pm\frac{1}{2}$	−1	0	2
	$\pm\frac{1}{2}$	$\pm\frac{1}{2}$	1	1	1
	$\pm\frac{3}{2}$	$\mp\frac{3}{2}$	1	0	0
	$\pm\frac{3}{2}$	$\mp\frac{1}{2}$	−1	0	2
	$\pm\frac{1}{2}$	$\mp\frac{1}{2}$	1	0	0
	$\frac{3}{2}\ \pm\frac{3}{2}$	$\frac{1}{2}\ \pm\frac{1}{2}$			1
	$\pm\frac{1}{2}$	$\pm\frac{1}{2}$			2
	$\pm\frac{3}{2}$	$\mp\frac{1}{2}$			4
	$\pm\frac{1}{2}$	$\mp\frac{1}{2}$			3
	$\frac{1}{2}\ \pm\frac{1}{2}$	$\frac{1}{2}\ \pm\frac{1}{2}$		1	
	$\pm\frac{1}{2}$	$\mp\frac{1}{2}$		0	

$l\,l'$	$j\ m$	$j'\ m'$	$\frac{F^2}{175}$	$\frac{-G^1}{225}$	$\frac{-G^3}{1225}$
$p\,d$	$\frac{3}{2}\ \pm\frac{3}{2}$	$\frac{5}{2}\ \pm\frac{5}{2}$	10	90	15
	$\pm\frac{3}{2}$	$\pm\frac{3}{2}$	− 2	36	36
	$\pm\frac{3}{2}$	$\pm\frac{1}{2}$	− 8	9	54
	$\pm\frac{1}{2}$	$\pm\frac{5}{2}$	−10	0	50
	$\pm\frac{1}{2}$	$\pm\frac{3}{2}$	2	54	49
	$\pm\frac{1}{2}$	$\pm\frac{1}{2}$	8	54	24
	$\pm\frac{3}{2}$	$\mp\frac{5}{2}$	10	0	0
	$\pm\frac{3}{2}$	$\mp\frac{3}{2}$	− 2	0	45
	$\pm\frac{3}{2}$	$\mp\frac{1}{2}$	− 8	0	60
	$\pm\frac{1}{2}$	$\mp\frac{5}{2}$	−10	0	75
	$\pm\frac{1}{2}$	$\mp\frac{3}{2}$	2	0	10
	$\pm\frac{1}{2}$	$\mp\frac{1}{2}$	8	27	2
	$\frac{3}{2}\ \pm\frac{3}{2}$	$\frac{3}{2}\ \pm\frac{3}{2}$	7	9	9
	$\pm\frac{3}{2}$	$\pm\frac{1}{2}$	− 7	6	36
	$\pm\frac{1}{2}$	$\pm\frac{3}{2}$	− 7	6	36
	$\pm\frac{1}{2}$	$\pm\frac{1}{2}$	7	1	81
	$\pm\frac{3}{2}$	$\mp\frac{3}{2}$	7	0	180
	$\pm\frac{3}{2}$	$\mp\frac{1}{2}$	− 7	0	90
	$\pm\frac{1}{2}$	$\mp\frac{3}{2}$	− 7	0	90
	$\pm\frac{1}{2}$	$\mp\frac{1}{2}$	7	8	108
	$\frac{1}{2}\ \pm\frac{1}{2}$	$\frac{5}{2}\ \pm\frac{5}{2}$			25
	$\pm\frac{1}{2}$	$\pm\frac{3}{2}$			50
	$\pm\frac{1}{2}$	$\pm\frac{1}{2}$			75
	$\pm\frac{1}{2}$	$\mp\frac{5}{2}$			150
	$\pm\frac{1}{2}$	$\mp\frac{3}{2}$			125
	$\pm\frac{1}{2}$	$\mp\frac{1}{2}$			100
	$\frac{1}{2}\ \pm\frac{1}{2}$	$\frac{3}{2}\ \pm\frac{3}{2}$		75	
	$\pm\frac{1}{2}$	$\pm\frac{1}{2}$		50	
	$\pm\frac{1}{2}$	$\mp\frac{3}{2}$		0	
	$\pm\frac{1}{2}$	$\mp\frac{1}{2}$		25	

$l\,l'$	$j\ m$	$j'\ m'$	$\frac{F^2}{1225}$	$\frac{F^4}{441}$	$-G^0$	$\frac{-G^2}{1225}$	$\frac{-G^4}{441}$
$d\,d$	$\frac{5}{2}\ \pm\frac{5}{2}$	$\frac{5}{2}\ \pm\frac{5}{2}$	100	1	1	100	1
	$\pm\frac{5}{2}$	$\pm\frac{3}{2}$	− 20	−3	0	120	4
	$\pm\frac{5}{2}$	$\pm\frac{1}{2}$	− 80	2	0	60	9
	$\pm\frac{3}{2}$	$\pm\frac{3}{2}$	4	9	1	4	9
	$\pm\frac{3}{2}$	$\pm\frac{1}{2}$	16	−6	0	48	10
	$\pm\frac{1}{2}$	$\pm\frac{1}{2}$	64	4	1	64	4
	$\pm\frac{5}{2}$	$\mp\frac{5}{2}$	100	1	0	0	0
	$\pm\frac{5}{2}$	$\mp\frac{3}{2}$	− 20	3	0	0	14
	$\pm\frac{5}{2}$	$\mp\frac{1}{2}$	− 80	2	0	0	14
	$\pm\frac{3}{2}$	$\mp\frac{3}{2}$	4	9	0	0	0
	$\pm\frac{3}{2}$	$\mp\frac{1}{2}$	16	−6	0	108	5
	$\pm\frac{1}{2}$	$\mp\frac{1}{2}$	64	4	0	0	0
	$\frac{5}{2}\ \pm\frac{5}{2}$	$\frac{3}{2}\ \pm\frac{3}{2}$	70			30	1
	$\pm\frac{5}{2}$	$\pm\frac{1}{2}$	− 70			40	6
	$\pm\frac{3}{2}$	$\pm\frac{3}{2}$	− 14			36	4
	$\pm\frac{3}{2}$	$\pm\frac{1}{2}$	14			2	15
	$\pm\frac{1}{2}$	$\pm\frac{3}{2}$	− 56			27	10
	$\pm\frac{1}{2}$	$\pm\frac{1}{2}$	56			6	24
	$\pm\frac{5}{2}$	$\mp\frac{3}{2}$	70			0	56
	$\pm\frac{5}{2}$	$\mp\frac{1}{2}$	− 70			0	21
	$\pm\frac{3}{2}$	$\mp\frac{3}{2}$	− 14			0	35
	$\pm\frac{3}{2}$	$\mp\frac{1}{2}$	14			32	30
	$\pm\frac{1}{2}$	$\mp\frac{3}{2}$	− 56			12	20
	$\pm\frac{1}{2}$	$\mp\frac{1}{2}$	56			25	30
	$\frac{3}{2}\ \pm\frac{3}{2}$	$\frac{3}{2}\ \pm\frac{3}{2}$	49		1	49	
	$\pm\frac{3}{2}$	$\pm\frac{1}{2}$	− 49		0	98	
	$\pm\frac{1}{2}$	$\pm\frac{1}{2}$	49		1	49	
	$\pm\frac{3}{2}$	$\mp\frac{3}{2}$	49		0	0	
	$\pm\frac{3}{2}$	$\mp\frac{1}{2}$	− 49		0	98	
	$\pm\frac{1}{2}$	$\mp\frac{1}{2}$	49		0	0	

parameter is varied has been shown by von Neumann and Wigner.* The sums of the electrostatic energies in the two limits (weighted by $2J+1$) are equal as required by the diagonal-sum rule.

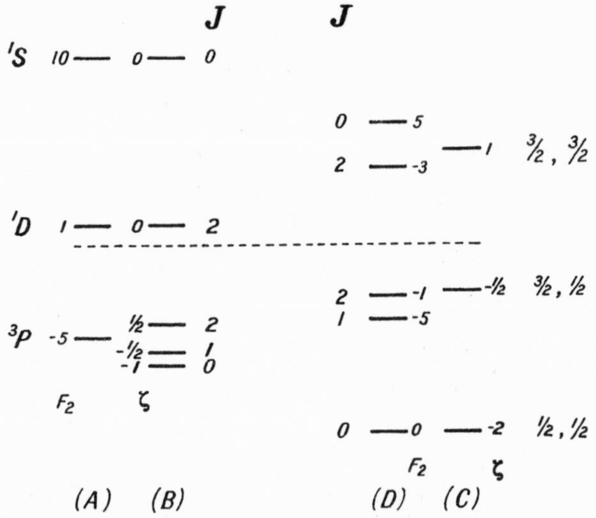

Fig. 1^{10}. Limiting cases for p^2: (*A*) No spin-orbit interaction. (*B*) Weak spin-orbit interaction. (*C*) No electrostatic interaction. (*D*) Weak electrostatic interaction.

3. Eigenfunctions.

The energy levels which result from the spin-orbit interaction are characterized for a given configuration by a set of j values, and are degenerate with respect to J and M. If a weak electrostatic interaction is applied, these energy levels are split into levels characterized by J values. Let us for convenience call the set of states characterized by a given set of j values a (jj-coupling) 'term,' in analogy with the terms in LS coupling. We shall use a notation in which we represent the three possible terms of the configuration $2p^3$ by $2p_{\frac{3}{2}}\, 2p_{\frac{1}{2}}^2$, $2p_{\frac{3}{2}}^2\, 2p_{\frac{1}{2}}$, and $2p_{\frac{3}{2}}^3$. The subscripts denote the j values of the electrons, the superscripts the number of electrons of the same nlj values.

The exclusion principle operates in jj coupling in a slightly different fashion than in LS coupling. All terms for which not more than $2j+1$ electrons have the same nlj value are allowed. However, not all the levels of a given term as given by coupling the j's vectorially are allowed unless all the nlj's are different. The eigenfunctions for jj coupling may be found by any of the methods used in Chapter VIII, but we can in general obtain more of them by coupling vectors than we could there. For example if, as in § 1^8, we couple the terms of two configurations to get terms of a new configuration,

* VON NEUMANN and WIGNER, Phys. Zeits. **30**, 467 (1929).

all the J's obtained by coupling the allowed J^{I}'s and J^{II}'s will be allowed and the eigenfunctions may be found as in § 6[8] provided that no nlj value occurs in both I and II. This is less stringent than the LS-coupling requirement that no nl value occurs in both configurations. Thus we may find the allowed J values for any term if we know the allowed J values for terms of the type nl_j^x. These are given in Table 2[10]. Thus we see that the three allowed levels of $np_{\frac{3}{2}}^2 np_{\frac{1}{2}}$ are

$$np_{\frac{3}{2}}^2(0)\, np_{\frac{1}{2}}(\tfrac{1}{2}),\ \tfrac{1}{2} \text{ and } np_{\frac{3}{2}}^2(2)\, np_{\frac{1}{2}}(\tfrac{1}{2}),\ \tfrac{3}{2} \text{ and } \tfrac{5}{2}.$$

Here in parentheses are given the J values of the groups of nlj-equivalent electrons, and at the end the resultant J values.

TABLE 2[10]. *Allowed J values for groups of nlj-equivalent electrons.*

$l_{\frac{1}{2}}^0, l_{\frac{1}{2}}^2$	0	$l_{\frac{3}{2}}^0, l_{\frac{3}{2}}^4$	0	$l_{\frac{5}{2}}^0, l_{\frac{5}{2}}^6$	0	$l_{\frac{7}{2}}^0, l_{\frac{7}{2}}^8$	0
$l_{\frac{1}{2}}^1$	$\frac{1}{2}$	$l_{\frac{3}{2}}^1, l_{\frac{3}{2}}^3$	$\frac{3}{2}$	$l_{\frac{5}{2}}^1, l_{\frac{5}{2}}^5$	$\frac{5}{2}$	$l_{\frac{7}{2}}^1, l_{\frac{7}{2}}^7$	$\frac{7}{2}$
		$l_{\frac{3}{2}}^2$	0, 2	$l_{\frac{5}{2}}^2, l_{\frac{5}{2}}^4$	0, 2, 4	$l_{\frac{7}{2}}^2, l_{\frac{7}{2}}^6$	0, 2, 4, 6
				$l_{\frac{5}{2}}^3$	$\frac{3}{2}, \frac{5}{2}, \frac{9}{2}$	$l_{\frac{7}{2}}^3, l_{\frac{7}{2}}^5$	$\frac{3}{2}, \frac{5}{2}, \frac{7}{2}, \frac{9}{2}, \frac{11}{2}, \frac{15}{2}$
						$l_{\frac{7}{2}}^4$	0, 2, 2, 4, 4, 5, 6, 8

In coupling two equivalent electrons, all J values are allowed if $j^1 \neq j^2$. If $j^1 = j^2$, the states resulting from the coupling are as in § 6[8] either antisymmetric or symmetric according to whether J is even or odd. Hence only the even J values are allowed if $j = j'$. The use of vector-coupling formulas for obtaining matrix elements between two states of the configuration l^2 is considered in detail in § 6[12]. Let us here consider, for use in the next section on line strengths, the matrix components

$$(nlj\, nlj'\, JM|F|nlj''\, n'l'j'''\, J'M') \equiv (jj'|F|j''j''') \tag{1}$$

of $F = F_1 + F_2$ connecting the states of nl^2 and $nl\, n'l'$. If $j = j'$, we see from a consideration similar to that which gave 6[8]17 that

$$(jj|F|j''j''') = \sqrt{2}\,\delta_{jj''}(j_1 j_2|F_2|j_1 j_2'''), \tag{2}$$

where the matrix component at the right is between states in which the individual electrons have definite quantum numbers, and is hence amenable to the use of the formulas of Chapter III. If $j \neq j'$, we see from the relations

$$\Psi(nlj\, nlj'\, JM) = 2^{-\frac{1}{2}}[\psi(n_1 l_1 j_1\, n_2 l_2 j_2'\, JM) - \psi(n_2 l_2 j_2\, n_1 l_1 j_1'\, JM)],$$

$$\Psi(nlj''\, n'l'j'''\, J'M') = 2^{-\frac{1}{2}}[\psi(n_1 l_1 j_1''\, n_2' l_2' j_2'''\, J'M') - \psi(n_2 l_2 j_2''\, n_1' l_1' j_1'''\, J'M')]$$

that

$$(jj'|F|j''j''') = \tfrac{1}{2}\delta(j,j'')\,[(j_2 j_1'|F_1|j_2 j_1''') + (j_1 j_2'|F_2|j_1 j_2''')] - \tfrac{1}{2}\delta(j',j'')\,[(j_1 j_2'|F_1|j_2' j_1''') + (j_2 j_1'|F_2|j_1' j_2''')].$$

In the last term here, the order of the quantum numbers is significant, and must be reversed by the use of the relation $14^{3}7$:

$$\psi(j_2 j_1' JM) = (-1)^{j+j'-J}\psi(j_1' j_2 JM). \tag{3}$$

In this way we find

$$(jj'|F|j''j''') = \delta(j,j'')(j_1 j_2'|F_2|j_1 j_2''') - \delta(j',j'')(-1)^{j+j'-J}(j_1' j_2|F_2|j_1' j_2'''). \quad (j \neq j') \tag{4}$$

4. Line strengths.*

Line strengths in *jj* coupling may be found by methods similar to those of Chapter IX for *LS* coupling. Their main application is in the rare-gas spectra (Chapter XIII), where only the strengths for two-electron configurations are needed. We shall first briefly consider transitions between the two two-electron configurations $n^1l^1 nl$ and $n^1l^1 n'l'$. If the jumping electron has different nlj from the other electron in both configurations, we may use the formulas $11^{3}8$ to evaluate the quantities

$$|(n^1l^1j^1 nlj\, J\vdots P\vdots n^1l^1j^1 n'l'j'\, J')|^2 \tag{1}$$

which are needed in $7^{4}5$ to give the strengths of the corresponding lines. In this way we obtain, as in $2^{9}1$,

$$\mathbf{S}(j^1 j\, J, j^1 j'\, J') = f(j^1 j\, J, j^1 j'\, J')\,|(nlj\vdots P\vdots n'l'j')|^2. \tag{2}$$

Hence the relative strengths of the lines in the 'multiplet' connecting the terms $j^1 j$ and $j^1 j'$ are the same as those in *LS* coupling for the multiplet $SL \to SL'$ with the correlation $S \to j^1$, $L \to j$, $L' \to j'$—except that j is always, L never, half-integral. They are thus given by Table 1^9 for half-integral L. The total strengths of the multiplets are given as in $3^{9}5$ by

$$\mathbf{S}(j^1 j, j^1 j') = (2j^1+1) f(slj, sl'j')\,|(nl\vdots P\vdots n'l')|^2, \tag{3}$$

which reduces all the intensities to the single matrix element $3^{9}6$. If one of the configurations is composed of equivalent electrons, say that $nl = n^1l^1$, we see from $3^{10}2$ that the allowed lines when $j = j^1$ have twice the strength of the corresponding lines for non-equivalent electrons, and from $3^{10}4$ that when $j \neq j^1$ either $\mathbf{S}(j^1 j\, J, j^1 j' J')$ or $\mathbf{S}(j j^1\, J, j^1 j' J')$, whichever occurs, has the same intensity as $\mathbf{S}(j^1 j\, J, j^1 j' J')$ for non-equivalent electrons. These calculations are illustrated by Table 3^{10} of the relative strengths of the lines in $ps \to pp$ and $ps \to p^2$ in *jj* coupling.

More generally, formula (2) is applicable, if we write J^1 for j^1, to transitions between the levels resulting from the addition of electrons of quantum numbers nlj and $n'l'j'$ to any ion in a state of resultant angular momentum J^1, if none of the electrons in the ion has quantum numbers nlj or $n'l'j'$.

* Bartlett, Phys. Rev. **35**, 229 (1930).

TABLE 3^{10}. *Relative strengths of the lines in* $p\,s \to p\,p$ *and* $p\,s \to p^2$ *in jj coupling.*

Invariant sums are shown in parentheses. The signs of $S^{\frac{1}{2}}$ are indicated.

$p\,p$

$p\,s$	$\frac{1}{2},\frac{1}{2}^0$	$\frac{3}{2},\frac{3}{2}^0$	$\frac{1}{2},\frac{1}{2}^1$	$\frac{1}{2},\frac{3}{2}^1$	$\frac{3}{2},\frac{1}{2}^1$	$\frac{3}{2},\frac{3}{2}^1$	$\frac{1}{2},\frac{3}{2}^2$	$\frac{3}{2},\frac{1}{2}^2$	$\frac{3}{2},\frac{3}{2}^2$	$\frac{3}{2},\frac{3}{2}^3$	
$\frac{1}{2},\frac{1}{2}^0$			-10	$+20$							(30)
						(30)					
$\frac{1}{2},\frac{1}{2}^1$	-10		$+20$	$+10$			$+50$				(90)
		(10)				(30)			(50)		
$\frac{3}{2},\frac{1}{2}^1$		-10			-5	$+25$		-25	$+25$		(90)
		(10)				(30)			(50)		
$\frac{3}{2},\frac{1}{2}^2$					-25	-5		$+25$	$+25$	$+70$	(150)
						(30)			(50)	(70)	
	(10)	(10)	(30)	(30)	(30)	(30)	(50)	(50)	(50)	(70)	

p^2

$p\,s$	$\frac{1}{2},\frac{1}{2}^0$	$\frac{3}{2},\frac{3}{2}^0$	$\frac{3}{2},\frac{1}{2}^1$	$\frac{3}{2},\frac{1}{2}^2$	$\frac{3}{2},\frac{3}{2}^2$	
$\frac{1}{2},\frac{1}{2}^0$			$+20$			(20)
			(20)			
$\frac{1}{2},\frac{1}{2}^1$	-20		$+10$	-50		80 ⎫
		(20)				⎬ (180)
$\frac{3}{2},\frac{1}{2}^1$		-20	-5	-25	$+50$	100 ⎭
		(20)	(15)		(125)	
$\frac{3}{2},\frac{1}{2}^2$			-25	$+25$	$+50$	(100)
			(25)		(75)	
	(20)	(20)	(60)	(100)	(100)	

CHAPTER XI

INTERMEDIATE COUPLING

We have calculated the first-order energy levels and line strengths in two limiting cases: the Russell-Saunders, or LS-coupling limit, in which the electrostatic energy predominates, and the jj-coupling limit, in which the spin-orbit energy predominates. Practically, the actual levels of all atoms lie between these limits, although many are very close to Russell-Saunders coupling, and a few, particularly heavy atoms and those containing almost closed shells, are close to jj-coupling.

In this chapter we shall obtain the first-order energies and line strengths in the general case in which the two interactions may be of any relative order of magnitude by calculating the complete energy matrix for states in the $SLJM$ scheme. In this scheme the electrostatic energy is completely diagonal (the diagonal elements being known from Chapter VII) and the spin-orbit interaction is diagonal with respect to J and M. The total energy is independent of M, hence the least value of $|M|$ which occurs for the configuration is usually the most convenient to consider, since every level will have such a state. In Chapter XII we shall see how to calculate the same energy matrix in the jj-coupling scheme which is particularly suitable for the considerations of the rare-gas spectra in Chapter XIII.

These calculations all neglect the interactions between configurations, which will be considered in Chapter XV.

1. Matrix of spin-orbit interaction for configurations consisting of coupled groups.*

Our first task is to obtain the complete matrix of the spin-orbit interaction

$$H^I = \sum_{i=1}^{N} \xi(r_i)\, \boldsymbol{L}_i \cdot \boldsymbol{S}_i \tag{1}$$

for the $SLJM$ scheme of a configuration. In § 3[8] we saw how we could obtain the diagonal elements of this matrix for configuration I + II in terms of the diagonal elements for the inequivalent configurations I and II separately. Such a calculation is however not restricted to the diagonal elements. If we make the correlation

$$\begin{array}{lll} \boldsymbol{P} \to \xi(r_i)\, \boldsymbol{L}_i & j_1 \to S & j \to J \\ \boldsymbol{Q} \to \boldsymbol{S}_i & j_2 \to L & m \to M, \end{array} \tag{2}$$

the dependence on J (and M) of the matrix elements

$$\sum_i (\gamma^{\rm I} S^{\rm I} L^{\rm I} \gamma^{\rm II} S^{\rm II} L^{\rm II}\, SLJM | \xi(r_i)\, \boldsymbol{L}_i \cdot \boldsymbol{S}_i | \gamma'^{\rm I} S'^{\rm I} L'^{\rm I} \gamma'^{\rm II} S'^{\rm II} L'^{\rm II}\, S'L'JM) \tag{3}$$

* Johnson, Phys. Rev. **38**, 1628 (1931).

is given by 12^32 in terms of the quantities

$$\sum_{\gamma''^{\mathrm{I}}\gamma''^{\mathrm{II}}}\sum_{i}(\gamma^{\mathrm{I}}S^{\mathrm{I}}L^{\mathrm{I}}\gamma^{\mathrm{II}}S^{\mathrm{II}}L^{\mathrm{II}}\,L\vdots\xi(r_i)\,L_i\vdots\gamma''^{\mathrm{I}}S^{\mathrm{I}}L'^{\mathrm{I}}\gamma''^{\mathrm{II}}S^{\mathrm{II}}L'^{\mathrm{II}}\,L')$$
$$(\gamma''^{\mathrm{I}}S^{\mathrm{I}}L'^{\mathrm{I}}\gamma''^{\mathrm{II}}S^{\mathrm{II}}L'^{\mathrm{II}}\,S\vdots S_i\vdots\gamma'^{\mathrm{I}}S'^{\mathrm{I}}L'^{\mathrm{I}}\gamma'^{\mathrm{II}}S'^{\mathrm{II}}L'^{\mathrm{II}}\,S'). \quad (4)$$

As in 3^81, this may be broken into two sums, over group I and II independently. Formulas 11^38 may then, with the correlation 3^83, be used to reduce the first sum to a known multiple of

$$\sum_{\gamma''^{\mathrm{I}}}\sum_{\mathrm{I}}(\gamma^{\mathrm{I}}L^{\mathrm{I}}\vdots\xi(r_i)\,L_i\vdots\gamma''^{\mathrm{I}}L'^{\mathrm{I}})(\gamma''^{\mathrm{I}}S^{\mathrm{I}}\vdots S_i\vdots\gamma'^{\mathrm{I}}S'^{\mathrm{I}}). \quad (5)$$

The second sum may be correspondingly reduced. Since the matrix elements of spin-orbit interaction for configuration I are known multiples of (5), this effectively expresses the matrix elements for configuration I+II in terms of those for I and II separately.*

Two-electron configurations.

These considerations completely solve the problem of finding the matrix of spin-orbit interaction for a two-electron configuration, since the matrix for a one-electron configuration is known from 4^56:

$$(nljm|\xi(r)\,\boldsymbol{L}\cdot\boldsymbol{S}|nljm)=\zeta_{nl}\tfrac{1}{2}\{j(j+1)-\tfrac{3}{4}-l(l+1)\}.$$

A comparison of this with 12^32 shows that

$$(nl\vdots\xi(r)\,L\vdots nl)(ns\vdots S\vdots ns)=\zeta_{nl}\hbar^2, \quad (6)$$

which gives the value of the quantity (5) for a single electron. Thus for a two-electron configuration the element

$$\sum_{i=1}^{2}(n^1l^1\,n^2l^2\,SLJM|\xi(r_i)\,\boldsymbol{L}_i\cdot\boldsymbol{S}_i|n^1l^1\,n^2l^2\,S'L'JM)$$
$$=g(SLJ,S'L'J)\sum_{i=1}^{2}\zeta_{n^il^i}(L\vdots L^i\vdots L')(S\vdots S^i\vdots S'), \quad (7)$$

where $g(SLJ,S'L'J)$ is the coefficient of (4) in (3) as given by 12^32 with the correlation (2), and

$$\left.\begin{matrix}(L\vdots L^1\vdots L)\\(L\vdots L^2\vdots L)\end{matrix}\right\}=\frac{L(L+1)\mp l^2(l^2+1)\pm l^1(l^1+1)}{2L(L+1)}\hbar$$

$$\left.\begin{matrix}(L\vdots L^1\vdots L-1)\\(L\vdots L^2\vdots L-1)\end{matrix}\right\}=\pm\hbar\sqrt{\frac{(L-l^1+l^2)(L+l^1-l^2)(l^1+l^2+L+1)(l^1+l^2-L+1)}{4L^2(2L-1)(2L+1)}}$$

$$(S\vdots S^1\vdots S)=(S\vdots S^2\vdots S)=(S\vdots S^1\vdots S-1)=-(S\vdots S^2\vdots S-1)=\tfrac{1}{2}\hbar \quad (8)$$

* For a given configuration, say I, to write that

$$\sum_{\mathrm{I}}(\gamma^{\mathrm{I}}\,S^{\mathrm{I}}\,L^{\mathrm{I}}\,J^{\mathrm{I}}\,M^{\mathrm{I}}|\xi(r_i)\,\boldsymbol{L}_i\cdot\boldsymbol{S}_i|\gamma'^{\mathrm{I}}\,S'^{\mathrm{I}}\,L'^{\mathrm{I}}\,J^{\mathrm{I}}\,M^{\mathrm{I}}) \quad (3^{\mathrm{I}})$$

equals the factor given by 12^32 times

$$\sum_{\gamma''^{\mathrm{I}}}\sum_{\mathrm{I}}(\gamma^{\mathrm{I}}\,S^{\mathrm{I}}\,L^{\mathrm{I}}\vdots\xi(r_i)\,L_i\vdots\gamma''^{\mathrm{I}}\,S^{\mathrm{I}}\,L'^{\mathrm{I}})(\gamma''^{\mathrm{I}}\,S^{\mathrm{I}}\,L'^{\mathrm{I}}\vdots S_i\vdots\gamma'^{\mathrm{I}}\,S'^{\mathrm{I}}\,L'^{\mathrm{I}}) \quad (4^{\mathrm{I}})$$

has in general a sense only symbolic, since no state characterized by $S^{\mathrm{I}}\,L'^{\mathrm{I}}$ may be allowed, although (3^{I}) has a value different from zero. The above procedure is justified by the fact, which follows from § 1^8, that the matrix component (3) is a linear combination of the components (3^{I}) and (3^{II}) which depends only on $S^{\mathrm{I}}\,L^{\mathrm{I}}\,S^{\mathrm{II}}\,L^{\mathrm{II}}$ and $S'^{\mathrm{I}}\,L'^{\mathrm{I}}\,S'^{\mathrm{II}}\,L'^{\mathrm{II}}$ and not on the structure of the inequivalent configurations I and II.

TABLE 1^{11}. *Matrices of spin-orbit interaction.*

sl

	$^3L_{l+1}$	^3L_l	^1L_l	$^3L_{l-1}$	
$^3L_{l+1}$	l				
^3L_l		-1	$\sqrt{l(l+1)}$		$\cdot\frac{1}{2}\zeta_l$
^1L_l		$\sqrt{l(l+1)}$	0		
$^3L_{l-1}$				$-(l+1)$	

p^3

	$^2P_{\frac{1}{2}}$	$^2P_{\frac{3}{2}}$	$^4S_{\frac{3}{2}}$	$^2D_{\frac{3}{2}}$	$^2D_{\frac{5}{2}}$	
$^2P_{\frac{1}{2}}$	0					
$^2P_{\frac{3}{2}}$		0	2	$\sqrt{5}$		
$^4S_{\frac{3}{2}}$		2	0	0		$\cdot\frac{1}{2}\zeta_p$
$^2D_{\frac{3}{2}}$		$\sqrt{5}$	0	0		
$^2D_{\frac{5}{2}}$					0	

p^2 and pp

———— $np\,n'p$ $\begin{cases}\alpha=\frac{1}{4}(\zeta_{np}+\zeta_{n'p})\\ \beta=\frac{1}{4}(\zeta_{np}-\zeta_{n'p})\end{cases}$

.... np^2 $\quad \alpha=\frac{1}{2}\zeta_{np}$

	$^3D_2$	$^1D_2$	$^3P_2$
$^3D_2$	$-\alpha$	$-\sqrt{6}\beta$	$\sqrt{3}\beta$
$^1D_2$	$-\sqrt{6}\beta$	0	$\sqrt{2}\alpha$
$^3P_2$	$\sqrt{3}\beta$	$\sqrt{2}\alpha$	α

	$^3P_0$	$^1S_0$
$^3P_0$	-2α	$-2\sqrt{2}\alpha$
$^1S_0$	$-2\sqrt{2}\alpha$	0

	$^3D_1$	$^3P_1$	$^1P_1$	$^3S_1$	
$^3D_1$	$-3\sqrt{3}\alpha$	$\sqrt{5}\beta$	$-\sqrt{10}\alpha$	0	
$^3P_1$	$\sqrt{5}\beta$	$-\sqrt{3}\alpha$	$-\sqrt{6}\beta$	4β	$\cdot\frac{1}{\sqrt{3}}$
$^1P_1$	$-\sqrt{10}\alpha$	$-\sqrt{6}\beta$	0	$2\sqrt{2}\alpha$	
$^3S_1$	0	4β	$2\sqrt{2}\alpha$	0	

	$^3D_3$
$^3D_3$	2α

p^2s

	$^2D_{\frac{5}{2}}$	$^4P_{\frac{5}{2}}$	
$^2D_{\frac{5}{2}}$	0	$\sqrt{2}$	$\cdot\frac{1}{2}\zeta_p$
$^4P_{\frac{5}{2}}$	$\sqrt{2}$	1	

	$^2D_{\frac{3}{2}}$	$^4P_{\frac{3}{2}}$	$^2P_{\frac{3}{2}}$	
$^2D_{\frac{3}{2}}$	0	$\sqrt{3}$	$-\sqrt{15}$	
$^4P_{\frac{3}{2}}$	$\sqrt{3}$	-2	$-\sqrt{5}$	$\cdot\frac{1}{6}\zeta_p$
$^2P_{\frac{3}{2}}$	$-\sqrt{15}$	$-\sqrt{5}$	2	

	$^4P_{\frac{1}{2}}$	$^2P_{\frac{1}{2}}$	$^2S_{\frac{1}{2}}$	
$^4P_{\frac{1}{2}}$	-5	$-\sqrt{2}$	$-4\sqrt{3}$	
$^2P_{\frac{1}{2}}$	$-\sqrt{2}$	-4	$-2\sqrt{6}$	$\cdot\frac{1}{6}\zeta_p$
$^2S_{\frac{1}{2}}$	$-4\sqrt{3}$	$-2\sqrt{6}$	0	

p^3s

	$^3D_3$
$^3D_3$	0

	$^3P_0$
$^3P_0$	0

	$^3D_2$	$^1D_2$	$^3P_2$	$^5S_2$	
$^3D_2$	0	0	$\sqrt{3}$	0	
$^1D_2$	0	0	$\sqrt{2}$	0	$\cdot\frac{1}{2}\zeta_p$
$^3P_2$	$\sqrt{3}$	$\sqrt{2}$	0	2	
$^5S_2$	0	0	2	0	

	$^3D_1$	$^3P_1$	$^1P_1$	$^3S_1$	
$^3D_1$	0	$\sqrt{5}$	$-\sqrt{10}$	0	
$^3P_1$	$\sqrt{5}$	0	0	-2	$\cdot\frac{1}{2\sqrt{3}}\zeta_p$
$^1P_1$	$-\sqrt{10}$	0	0	$2\sqrt{2}$	
$^3S_1$	0	-2	$2\sqrt{2}$	0	

TABLE 1^{11}—*continued.*

$\underline{d^2}$ and $\underline{d\,d}$

———— $nd\,n'd\begin{cases}\alpha=\frac{1}{4}(\zeta_{nd}+\zeta_{n'd})\\ \beta=\frac{1}{4}(\zeta_{nd}-\zeta_{n'd})\end{cases}$

············ nd^2 $\quad \alpha=\frac{1}{2}\zeta_{nd}$

	$^3G_4$	$^1G_4$	$^3F_4$
$^3G_4$	$-\alpha$	$-2\sqrt{5}\beta$	$\sqrt{5}\beta$
$^1G_4$	$-2\sqrt{5}\beta$	0	2α
$^3F_4$	$\sqrt{5}\beta$	2α	3α

	$^3G_3$	$^3F_3$	$^1F_3$	$^3D_3$
$^3G_3$	$-5\sqrt{7}\alpha$	$3\sqrt{3}\beta$	-6α	0
$^3F_3$	$3\sqrt{3}\beta$	$-\sqrt{7}\alpha$	$-2\sqrt{21}\beta$	8β
$^1F_3$	-6α	$-2\sqrt{21}\beta$	0	$4\sqrt{3}\alpha$
$^3D_3$	0	8β	$4\sqrt{3}\alpha$	$2\sqrt{7}\alpha$

$\cdot\frac{1}{\sqrt{7}}$

	$^3D_2$	$^3F_2$	$^1D_2$	$^3P_2$
$^3D_2$	$-\sqrt{5}\alpha$	$4\sqrt{2}\beta$	$-\sqrt{30}\beta$	$3\sqrt{7}\beta$
$^3F_2$	$4\sqrt{2}\beta$	$-4\sqrt{5}\alpha$	$-4\sqrt{3}\alpha$	0
$^1D_2$	$-\sqrt{30}\beta$	$-4\sqrt{3}\alpha$	0	$\sqrt{42}\alpha$
$^3P_2$	$3\sqrt{7}\beta$	0	$\sqrt{42}\alpha$	$\sqrt{5}\alpha$

$\cdot\frac{1}{\sqrt{5}}$

	$^3D_1$	$^3P_1$	$^1P_1$	$^3S_1$
$^3D_1$	-3α	$\sqrt{7}\beta$	$-\sqrt{14}\alpha$	0
$^3P_1$	$\sqrt{7}\beta$	$-\alpha$	$-\sqrt{2}\beta$	4β
$^1P_1$	$-\sqrt{14}\alpha$	$-\sqrt{2}\beta$	0	$2\sqrt{2}\alpha$
$^3S_1$	0	4β	$2\sqrt{2}\alpha$	0

	$^3P_0$	$^1S_0$
$^3P_0$	-2α	$-2\sqrt{6}\alpha$
$^1S_0$	$-2\sqrt{6}\alpha$	0

	$^3G_5$
$^3G_5$	4α

$\underline{dp}$

	$^3F_4$
$^3F_4$	$\zeta_d+\frac{1}{2}\zeta_p$

	$^3F_3$	$^1F_3$	$^3D_3$
$^3F_3$	$-2\zeta_d-\zeta_p$	$-2\sqrt{3}(2\zeta_d-\zeta_p)$	$2\sqrt{2}(\zeta_d-\zeta_p)$
$^1F_3$	$-2\sqrt{3}(2\zeta_d-\zeta_p)$	0	$\sqrt{6}(\zeta_d+\zeta_p)$
$^3D_3$	$2\sqrt{2}(\zeta_d-\zeta_p)$	$\sqrt{6}(\zeta_d+\zeta_p)$	$5\zeta_d+\zeta_p$

$\cdot\frac{1}{6}$

	$^3F_2$	$^3D_2$	$^1D_2$	$^3P_2$
$^3F_2$	$-8\sqrt{5}(2\zeta_d+\zeta_p)$	$4\sqrt{7}(\zeta_d-\zeta_p)$	$-2\sqrt{42}(\zeta_d+\zeta_p)$	0
$^3D_2$	$4\sqrt{7}(\zeta_d-\zeta_p)$	$-\sqrt{5}(5\zeta_d+\zeta_p)$	$-\sqrt{30}(5\zeta_d-\zeta_p)$	$9\sqrt{3}(\zeta_d-\zeta_p)$
$^1D_2$	$-2\sqrt{42}(\zeta_d+\zeta_p)$	$-\sqrt{30}(5\zeta_d-\zeta_p)$	0	$9\sqrt{2}(\zeta_d+\zeta_p)$
$^3P_2$	0	$9\sqrt{3}(\zeta_d-\zeta_p)$	$9\sqrt{2}(\zeta_d+\zeta_p)$	$3\sqrt{5}(3\zeta_d-\zeta_p)$

$\cdot\frac{1}{12\sqrt{5}}$

	$^3D_1$	$^3P_1$	$^1P_1$
$^3D_1$	$-5\zeta_d-\zeta_p$	$\sqrt{3}(\zeta_d-\zeta_p)$	$-\sqrt{6}(\zeta_d+\zeta_p)$
$^3P_1$	$\sqrt{3}(\zeta_d-\zeta_p)$	$-3\zeta_d+\zeta_p$	$-\sqrt{2}(3\zeta_d+\zeta_p)$
$^1P_1$	$-\sqrt{6}(\zeta_d+\zeta_p)$	$-\sqrt{2}(3\zeta_d+\zeta_p)$	0

$\cdot\frac{1}{4}$

	$^3P_0$
$^3P_0$	$-\frac{3}{2}\zeta_d+\frac{1}{2}\zeta_p$

are the factors given by $10^3 2$ with $S^1=\frac{1}{2}$, $S^2=\frac{1}{2}$, $S=1$. It follows from $6^8 15$ that these formulas hold for the allowed elements even if the two electrons are equivalent.

In this way the matrices of spin-orbit interaction for the two-electron configurations in Table 1^{11} may be readily calculated. In these matrices the phases of the eigenstates in the configuration $n^1l^1 n^2l^2$ are such that $\boldsymbol{L}=\boldsymbol{L}^1+\boldsymbol{L}^2$, $\boldsymbol{S}=\boldsymbol{S}^1+\boldsymbol{S}^2$, $\boldsymbol{J}=\boldsymbol{S}+\boldsymbol{L}$, with the vectors added in this order by the formulas of § 14^3. *We shall always use this system of phases for two-electron configurations.* According to this convention, the phases of a given term of $n^2l^2 n^1l^1$ are $(-1)^{l^1+l^2-L-S+1}$ times those of the corresponding terms of $n^1l^1 n^2l^2$.

More complex configurations may be calculated in a similar fashion. Thus in Table 1^{11} is included the configuration p^3s. This matrix was obtained in the way we have sketched from that given for p^3, which is calculated in the next section. In this matrix the phases of the eigenstates are such that an eigenstate of p^3s is obtained from those of p^3 with phases given by $4^8 6$j by adding $\boldsymbol{S}^{p^3}+\boldsymbol{S}^s=\boldsymbol{S}$, $\boldsymbol{L}^{p^3}+\boldsymbol{L}^s=\boldsymbol{L}$, $\boldsymbol{S}+\boldsymbol{L}=\boldsymbol{J}$ in this order. We shall always use such a system of phases when adding an electron to an ion or in general in coupling two groups.

2. Matrix of spin-orbit interaction obtained from the eigenfunctions.

For configurations containing more than two equivalent electrons we cannot obtain completely the matrix of spin-orbit interaction by the method of the preceding section. But if we know the eigenstates for the $SLJM$ scheme (Chapter VIII) and the matrix of spin-orbit interaction in the zero-order scheme, this matrix may readily be transformed to the $SLJM$ scheme in the usual way. The matrix of the spin-orbit interaction in the zero-order scheme is obtained from § 6^6 (cf. $4^8 1$ for notation and sign convention). One obtains a non-diagonal element only between states differing in regard to one individual set, say that a occurs in A while a' occurs in A'. The value of this component is

$$(A|H^I|A') = \pm\hbar^{-2}\zeta_{n^a l^a}(a|\boldsymbol{L}\cdot\boldsymbol{S}|a'). \tag{1a}$$

The diagonal element has the value

$$(A|H^I|A) = \sum_a \zeta_{n^a l^a} m_l^a m_s^a, \tag{1b}$$

where a runs over all the sets (outside of closed shells) of A.

From these formulas we may easily calculate the matrix of H^I in the zero-order scheme, where it is diagonal with respect to M. When transformed to the $SLJM$ scheme it becomes diagonal also with respect to J and independent of M.

Illustration: The configuration p^3. The $SLJM$ states for p^3 were obtained in 4^86m for $M=\frac{1}{2}$. In the zero-order scheme 4^86a the matrix of H^I has the value

$$\|(A|H^I|B)\| = \begin{array}{c|cc|ccc|} & A & B & C & D & E \\ \hline A & -1 & 0 & 0 & 0 & \sqrt{2} \\ B & 0 & 1 & \sqrt{2} & 0 & 0 \\ \hline C & 0 & \sqrt{2} & 2 & 0 & 0 \\ D & 0 & 0 & 0 & 0 & 0 \\ E & \sqrt{2} & 0 & 0 & 0 & -2 \\ \hline \end{array} \cdot \tfrac{1}{2}\zeta_{np}.$$

When this is transformed to the $SLJM$ scheme by 4^86m, we obtain the matrix given in Table 1^{11}.

3. Illustrations of the transition from LS to jj coupling.

sl*

By adding the electrostatic energies of § 5^7 to the spin-orbit matrices of Table 1^{11}, and solving the resulting secular equations, we find for the energy levels of sl in any coupling the values:

$$\begin{aligned} {}^3L_{l+1} &= F_0 - G_l + \tfrac{1}{2}l\zeta \\ \left.\begin{array}{l}{}^1L_l'\\ {}^3L_l'\end{array}\right\} &= F_0 \qquad -\tfrac{1}{4}\zeta \pm \sqrt{(G_l+\tfrac{1}{4}\zeta)^2+\tfrac{1}{4}l(l+1)\,\zeta^2} \\ {}^3L_{l-1} &= F_0 - G_l - \tfrac{1}{2}(l+1)\,\zeta. \end{aligned}$$

Here $\zeta=\zeta_{nl}$. The complete transition from LS coupling ($G_1 \gg \zeta$) to jj coupling ($\zeta \gg G_1$) is plotted in Fig. 1^{11} for the configuration sp. F_0 is an additive constant which does not influence the intervals between the levels and hence does not need here to be considered further. Apart from F_0, we see that the ratio ϵ/G_1 of the energy value to G_1 is a function only of $\chi=\frac{3}{4}\zeta/G_1$. For $\zeta=0$, the energies in these units are $+1$ and -1, while for $G_1\approx 0$, they are $\frac{2}{3}\chi$ and $-\frac{4}{3}\chi$. As $\chi\to\infty$, ϵ/G_1 approaches infinity. Hence we cannot show the whole transition by making a plot of ϵ/G_1 against χ. In order to keep the ordinates from going to infinity we plot instead $\dfrac{\epsilon/G_1}{\sqrt{1+\chi^2}}$ as ordinate, and in order to confine the abscissas to a finite range, we plot this against $\chi/(1+\chi)$ in Fig. 1^{11}. In this way we show the true interval ratios for all χ from 0 to ∞. At the left end, ϵ/G_1 is effectively plotted against $\frac{3}{4}\zeta/G_1$, while

* Houston, Phys. Rev. **33**, 297 (1929);
Condon and Shortley, Phys. Rev. **35**, 1342 (1930).

at the right end $\epsilon/\frac{3}{4}\zeta$ is effectively plotted against $1-(G_1/\frac{3}{4}\zeta)$. The factor $\frac{3}{4}$ is chosen to make the total splitting the same at the two ends.

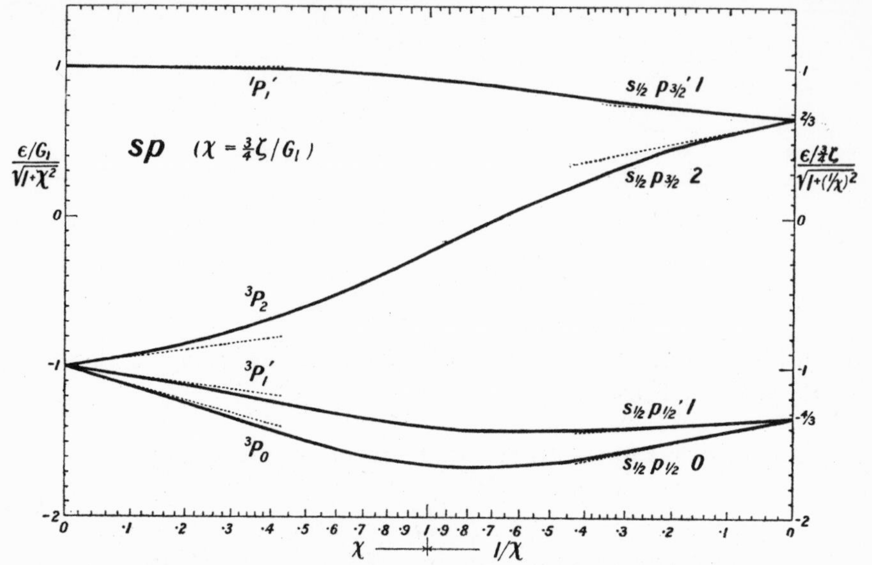

Fig. 1[11]. The configuration sp in intermediate coupling.

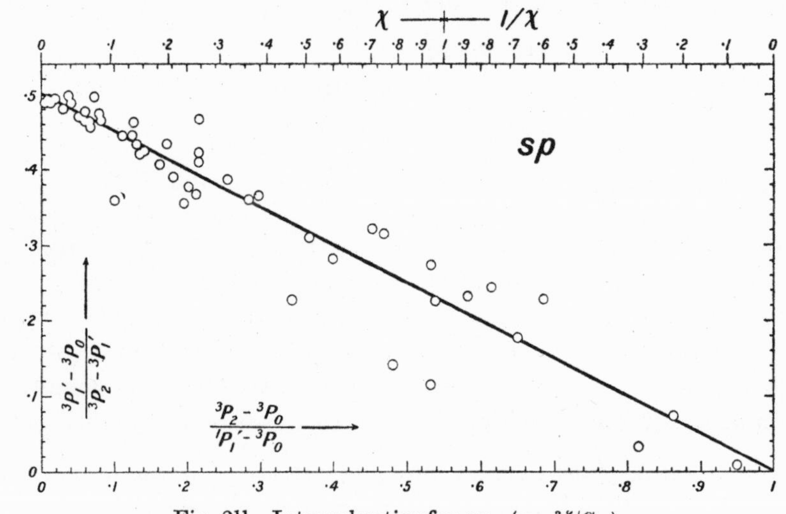

Fig. 2[11]. Interval ratios for sp. ($\chi = \frac{3}{4}\zeta/G_1$.)

In order to see how well the observed sl configurations fit these formulas, we have drawn in Figs. 2[11] and 3[11] for sp and sd respectively curves which show the theoretical value of $(^3L_l' - {}^3L_{l-1})/(^3L_{l+1} - {}^3L_l')$ plotted as ordinate against $(^3L_{l+1} - {}^3L_{l-1})/(^1L_l' - {}^3L_{l-1})$ as abscissa. These curves are in fact straight lines: if we represent the ordinate by o and the abscissa by a,

$$o = -\frac{l}{l+1}a + \frac{l}{l+1}.$$

The plotted points represent these ratios for the observed *sl* configurations. It is seen that in general they follow the theoretical formulas rather well. While there is no essential requirement that G_l be positive, and there are many *sd* configurations in which the singlet is below the triplet as if G_2 were

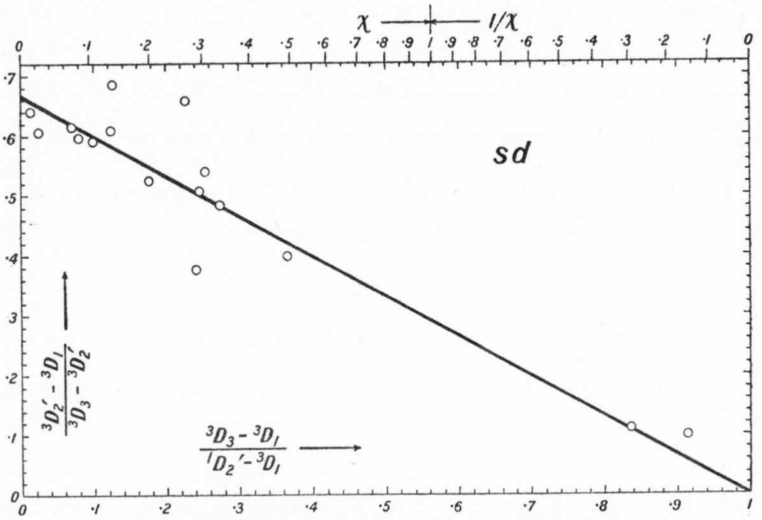

Fig. 3^{11}. Interval ratios for *s d*. $(\chi = \frac{5}{4}\zeta/G_2.)$

negative, the interval ratios for these cases do not obey the above formulas at all accurately; they are probably all to be explained as in § 1^{15} by a strong perturbation of the singlet by another configuration. No plot is made for *sf* configurations, since there are just a few known, all close to the Russell-Saunders limit.*

* Formulas which make an approximate allowance for the terms in the Hamiltonian which express the interaction of the spin of one electron with the orbit of the other have been calculated for *s l* by use of group-theoretical methods by WOLFE [Phys. Rev. **41**, 443 (1932)]. We may derive Wolfe's formulas very simply in the following way. For a two-electron configuration this interaction is assumed to have the form (cf. $7^7 1$)

$$H^N = \pi(\boldsymbol{r}_1, \boldsymbol{r}_2)\,[\boldsymbol{L}_1\cdot\boldsymbol{S}_2 + \boldsymbol{L}_2\cdot\boldsymbol{S}_1]. \tag{1}$$

The second factor here has no elements connecting different configurations. By writing

$$\boldsymbol{L}_1\cdot\boldsymbol{S}_2 + \boldsymbol{L}_2\cdot\boldsymbol{S}_1 = \boldsymbol{L}\cdot\boldsymbol{S} - \boldsymbol{L}_1\cdot\boldsymbol{S}_1 - \boldsymbol{L}_2\cdot\boldsymbol{S}_2 = \boldsymbol{L}\cdot\boldsymbol{S} - \sum_i \boldsymbol{L}_i\cdot\boldsymbol{S}_i, \tag{2}$$

we see that its matrix is that of $\boldsymbol{L}\cdot\boldsymbol{S}\,[=\frac{1}{2}(\boldsymbol{J}^2 - \boldsymbol{L}^2 - \boldsymbol{S}^2)]$ minus that of the spin-orbit interaction with $\xi(r_i)$ and hence ζ_i set equal to unity: both of these are known in *LS* coupling. This matrix must be multiplied by that of $\pi(\boldsymbol{r}_1, \boldsymbol{r}_2)$ for the configuration under consideration. If in $\pi(\boldsymbol{r}_1, \boldsymbol{r}_2)$ we replace the mutual distance r_{12} by $r_>$, we have a function $\rho(r_1, r_2)$ merely of the magnitudes of the radii. The matrix of this for a configuration in which the l values of the two

p^{2*}

$$\left.\begin{matrix}{}^1D_2' \\ {}^3P_2'\end{matrix}\right\} = F_0 - 2F_2 + \tfrac{1}{4}\zeta \pm \sqrt{9F_2^2 - \tfrac{3}{2}F_2\zeta + \tfrac{9}{16}\zeta^2}$$

$${}^3P_1 = F_0 - 5F_2 - \tfrac{1}{2}\zeta$$

$$\left.\begin{matrix}{}^1S_0' \\ {}^3P_0'\end{matrix}\right\} = F_0 + \tfrac{5}{2}F_2 - \tfrac{1}{2}\zeta \pm \sqrt{\tfrac{225}{4}F_2^2 + \tfrac{15}{2}F_2\zeta + \tfrac{9}{4}\zeta^2}.$$

If ϵ/F_2 is expanded as a power series in ζ/F_2, keeping just the first power, these results agree with those obtained in Chapter VII for LS coupling. If ϵ/ζ is expanded in terms of F_2/ζ, the results agree with those of Chapter X for jj coupling. See Fig. 1[10]. The complete transition is plotted in Fig. 4[11] as a function of $\chi = \frac{1}{5}\zeta/F_2$. We superpose on this figure the observed levels of Ge I $4p^2$, Sn I $5p^2$, and Pb I $6p^2$, placed at such values of χ that ${}^3P_1$, the mean of ${}^1D_2'$ and ${}^3P_2'$, and the mean of ${}^1S_0'$ and ${}^3P_0'$ fit the theory exactly. This corresponds to the parameter values

	Ge I	Sn I	Pb I
F_2	1016·9	918·6	921·5
ζ	880·1	2097·3	7294
χ	0·173	0·457	1·583

electrons are different is diagonal in the zero-order scheme, the diagonal elements having the constant value

$$\eta = (nl\,n'l'|\rho(r_1, r_2)|nl\,n'l').$$

Hence in any scheme this matrix is η times the unit matrix, and the whole matrix of H^N is just η times the matrix of (2). To this approximation, the Landé interval rule still holds in the Russell-Saunders case.

For $s\,l$, the matrix of H^N is η times the matrix of $\boldsymbol{L}\cdot\boldsymbol{S}$ (which is diagonal with elements $0, l, -1, -l-1$, for ${}^1L_l$, ${}^3L_{l+1}$, ${}^3L_l$, ${}^3L_{l-1}$) minus the matrix of spin-orbit interaction of Table 1[11] with η written for ζ_l. When we include this matrix, we obtain the energies

$${}^3L_{l+1} = F_0 - G_0 + \tfrac{1}{2}l(\zeta + \eta)$$

$$\left.\begin{matrix}{}^1L_l' \\ {}^3L_l'\end{matrix}\right\} = F_0 \qquad - \tfrac{1}{4}(\zeta + \eta) \pm \sqrt{[G_0 + \tfrac{1}{4}(\zeta + \eta)]^2 + \tfrac{1}{4}l(l+1)(\zeta - \eta)^2}$$

$${}^3L_{l-1} = F_0 - G_0 - \tfrac{1}{2}(l+1)(\zeta + \eta).$$

We cannot compare these formulas satisfactorily with experiment because we have now as many parameters as energy levels.

These considerations can readily be extended to other configurations.

* The secular equations for this configuration were first calculated by Goudsmit, Phys. Rev. **35**, 1325 (1930), and those for p^3 later by Inglis, Phys. Rev. **38**, 862 (1931), by the following procedure: Since all the matrix elements of the Hamiltonian are linear functions of the F's, G's, and ζ's, the secular equation for a J value occurring n times will be homogeneous of nth degree in the F's, G's, ζ's, and ϵ, the energy variable. For small ζ's, the roots of these equations are linear functions of the F's, G's, and ζ's, known from the electrostatic energies and Landé splittings of Chapter VII. For small electrostatic interaction the roots are linear functions of these same parameters known from the considerations of Chapter X. Knowing these roots for limiting values of the parameters serves in simple cases to determine all the coefficients in the secular equations. Since this procedure does not give the energy matrix, it is not possible in this way to determine the eigenstates in intermediate coupling in terms of those for pure coupling for use in Zeeman effect or intensity calculations. In simple cases, the weak-field Zeeman effect is given by addition of terms in $\mathscr{H}$ to the secular equations.

It is seen that the coupling gets progressively closer to jj as the atom gets heavier. The separation between the two levels of $J=2$ and the separation between the two levels of $J=0$, which are absolutely predicted by the theory, are seen to agree well with the observed data.

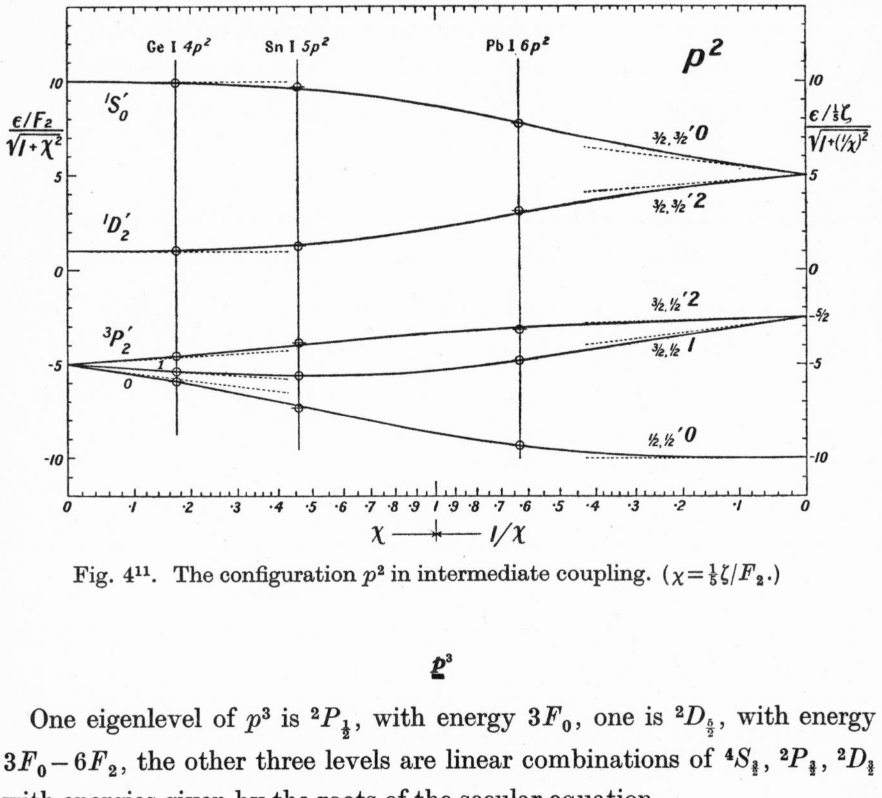

Fig. 4^{11}. The configuration p^2 in intermediate coupling. ($\chi=\frac{1}{5}\zeta/F_2$.)

p^3

One eigenlevel of p^3 is $^2P_{\frac{1}{2}}$, with energy $3F_0$, one is $^2D_{\frac{5}{2}}$, with energy $3F_0-6F_2$, the other three levels are linear combinations of $^4S_{\frac{3}{2}}$, $^2P_{\frac{3}{2}}$, $^2D_{\frac{3}{2}}$ with energies given by the roots of the secular equation

$$\epsilon^3+21F_2\epsilon^2+(90F_2^2-\tfrac{9}{4}\zeta^2)\epsilon-\tfrac{99}{4}F_2\zeta^2=0.$$

These energies are plotted in Fig. 5^{11} as a function of $\chi=\frac{1}{5}\zeta/F_2$, together with the observed levels for As I $4p^3$, Sb I $5p^3$, and Bi I $6p^3$, which are seen to depart progressively from Russell-Saunders coupling. The values of χ for these configurations were determined by making all the levels fit as well as possible. The parameter values are

	As I	Sb I	Bi I
F_2	1210	1050	990
ζ	1500	3400	10100
χ	0·25	0·64	2·05

The configuration p^4 is similarly discussed in Chapter XIII.

For more complex configurations the formulas become more complicated and the number of parameters necessary to determine the intervals greater than two, so that it is no longer possible to plot the complete transition in two dimensions. No comparisons with experiment have been made for such configurations. The secular equations are readily obtained from the electrostatic energies of Chapter VII and the spin-orbit matrices of Table 1[11].*

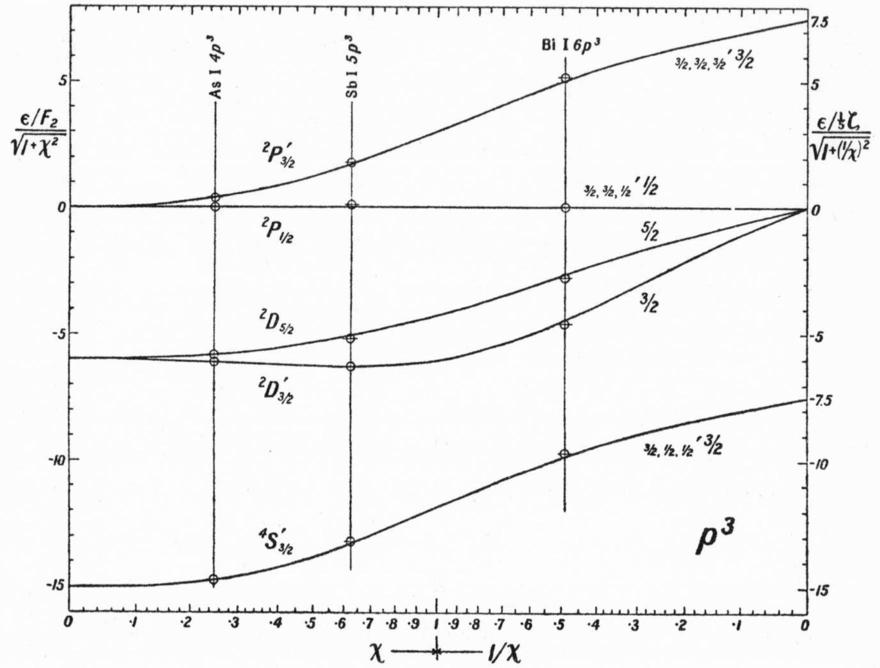

Fig. 5[11]. The configuration p^3 in intermediate coupling. ($\chi = \frac{1}{5}\zeta/F_2$.)

At this point we can merely call attention to a recent paper by Bacher and Goudsmit,† which makes a good start on the problem of calculating the *absolute* energy of a state of an atom from the experimentally known energy states of its ions. It is unfortunately too late for us to treat this important paper in detail.

* VAN VLECK [Phys. Rev. **45**, 412 (1934)] has shown that in the addition of an s electron to an ion, the electrostatic energies for the atom may be easily calculated from those of the ion by the Dirac vector model, and that *in this case the spin of the parent term is always an exact quantum number.* By the procedure of § 1[11], the spin-orbit interaction for such an atom is easily obtained in terms of that of the parent ion. MERRILL [Phys. Rev. **46**, 487 (1934)], combining these calculations, has made an interesting comparison with experiment of the energy levels based on certain definite terms of the d^k ion in a number of configurations of the type $d^k\,ns$ and $d^k\,ns\,n's$.

† BACHER and GOUDSMIT, Phys. Rev. **46**, 948 (1934).

4. Line strengths in intermediate coupling.

The strength of a line from a level aJ to a level bJ' is given in any coupling by formula $7^4 5$ in terms of the quantity $(aJ\vdots P\vdots bJ')$. If the expansion of the state aJ is known in terms of a set of states $\Psi(\alpha J)$ and the expansion of bJ' is known in terms of the set $\Psi(\beta J')$, we need only to know the values of $(\alpha J\vdots P\vdots \beta J')$ to obtain the strength of the line in question. Since the factors $(\ \vdots P\vdots\ )$ transform like the components of a matrix of an observable ($9^3 12$),

$$(aJ\vdots P\vdots bJ') = \sum_{\alpha\beta} (aJ|\alpha J)(\alpha J\vdots P\vdots \beta J')(\beta J'|bJ'). \tag{1}$$

The matrix components on the right of this equation must be taken between states having the same phases as those for which the transformation coefficients are calculated. Hence we cannot immediately replace these matrix components by the square root of the line strengths as given by Chapter IX. However it is convenient to replace these components by quantities which we shall write as $\mathbf{S}^{\frac{1}{2}}(\alpha J, \beta J')$, and which are understood to be *the square root of the line strength taken + or − according to the sign of the matrix component* $(\alpha J\vdots P\vdots \beta J')$. Hence we write

$$\mathbf{S}^{\frac{1}{2}}(aJ, bJ') = \sum_{\alpha\beta} (aJ|\alpha J)\, \mathbf{S}^{\frac{1}{2}}(\alpha J, \beta J')(\beta J'|bJ'). \tag{2}$$

In this way we can use directly the results obtained in Chapter IX for the magnitude of the $\mathbf{S}$'s, and need only learn to choose the proper phases.

Phases in the matrix $\mathbf{S}^{\frac{1}{2}}$.

Let us consider, as in § 3^9, the transitions from configuration I + II to configuration I + IV and ask the phase of

$$(\gamma^{\mathrm{I}} S^{\mathrm{I}} L^{\mathrm{I}}\, \gamma^{\mathrm{IV}} S^{\mathrm{IV}} L^{\mathrm{IV}}\, SLJ\vdots P\vdots \gamma^{\mathrm{I}} S^{\mathrm{I}} L^{\mathrm{I}}\, \gamma^{\mathrm{II}} S^{\mathrm{II}} L^{\mathrm{II}}\, S'L'J').$$

This element is expressed by $11^3 8$ with $\boldsymbol{S}+\boldsymbol{L}=\boldsymbol{J}$ in terms of

$$(\gamma^{\mathrm{I}} S^{\mathrm{I}} L^{\mathrm{I}}\, \gamma^{\mathrm{IV}} S^{\mathrm{IV}} L^{\mathrm{IV}}\, L\vdots P\vdots \gamma^{\mathrm{I}} S^{\mathrm{I}} L^{\mathrm{I}}\, \gamma^{\mathrm{II}} S^{\mathrm{II}} L^{\mathrm{II}}\, L'),$$

which is in turn expressed by $11^3 8$ with $\boldsymbol{L}^{\mathrm{I}}+\boldsymbol{L}^{\mathrm{IV}}$ or $\boldsymbol{L}^{\mathrm{I}}+\boldsymbol{L}^{\mathrm{II}}=\boldsymbol{L}$ in terms of $(\gamma^{\mathrm{IV}} L^{\mathrm{IV}}\vdots P\vdots \gamma^{\mathrm{II}} L^{\mathrm{II}})$. In each case the proper sign of the coefficient is given by $11^3 8$. In the same way, we find the relative strengths of the lines in a multiplet from $2^9 2$, and then the relative strengths of the multiplets in a group having the same parent terms by a reapplication of $2^9 2$ with SLJ, $SL'J'$ replaced by $L^{\mathrm{I}} L^{\mathrm{IV}} L$, $L^{\mathrm{I}} L^{\mathrm{II}} L'$. This expresses the strengths of all lines in the group in terms of $\mathbf{S}(\gamma^{\mathrm{IV}} S^{\mathrm{IV}} L^{\mathrm{IV}}, \gamma^{\mathrm{II}} S^{\mathrm{II}} L^{\mathrm{II}})$. We get the proper phases for the $\mathbf{S}^{\frac{1}{2}}$'s if on each application of $2^9 2$ we carry the factor (-1) where $(-1)^2$ is written in these formulas and also in the second formula of $2^9 2$b, when $J(J+1)-S(S+1)+L(L+1)$ is negative. This amounts to carrying *in Table* 1^9 *a minus sign for second order satellites when* $L\to L\pm 1$, *and for all satellites and those principal lines for which* $S(S+1) > J(J+1)+L(L+1)$ *when* $L\to L$.

In this way we find the phase of $\mathbf{S}^{\frac{1}{2}}$ for a given line in terms of the phase of $\mathbf{S}^{\frac{1}{2}}(\gamma^{\text{IV}}S^{\text{IV}}L^{\text{IV}}, \gamma^{\text{II}}S^{\text{II}}L^{\text{II}})$ correctly for eigenstates with the phases we agreed to use in § 1^{11} in calculating the energy matrix.

For transitions in which the relative multiplet intensities cannot be obtained by the method sketched above, we must calculate the matrix $(\gamma SL\vdots P\vdots\gamma' SL')$ directly from the eigenfunctions used for the expression of the energy matrix. We assume that the states of a term have the proper relative phases as given by the calculation of § 5^8. Then $(\gamma SL\vdots P\vdots\gamma' SL')$ may be obtained by finding the matrix element $(\gamma SLM_SM_L|\boldsymbol{P}|\gamma' SL'M_SM'_L)$ or $(\gamma SLJM|\boldsymbol{P}|\gamma' SL'J'M')$ for any *one* component of the multiplet, since the values for all components may by the formulas of Chapter III be readily expressed in terms of $(\gamma SL\vdots P\vdots\gamma' SL')$. This matrix element is obtained from the zero-order eigenfunctions by the formulas of § 6^6 and the one-electron components (cf. 4^91 and 3^96)

$$(n\,l\,m_s\,m_l|e\boldsymbol{r}|n'\,l-1\,m'_s\,m'_l) = \mathsf{s}\,\delta(m_s, m'_s)\,\{\delta(m_l, m'_l)\,\sqrt{l^2-m_l^2}\,\boldsymbol{k} \pm \delta(m'_l, m_l \pm 1)\,\tfrac{1}{2}\,\sqrt{(l\mp m_l)(l\mp m_l-1)}\,(\boldsymbol{i}\pm i\boldsymbol{j})\}. \quad (3)$$

The procedure for finding intensities in intermediate coupling is thus the following: Put the calculated eigenvalues of the energies into the energy matrix to find the transformations from the LS-coupling states in the usual way. Find the matrix of $\mathbf{S}^{\frac{1}{2}}$ for states with the same phases as those used in obtaining the energy matrix. Multiply the two transformation matrices and the $\mathbf{S}^{\frac{1}{2}}$ matrix according to (2).

There are at present practically no data suitable for comparison with the theory, so we shall not carry through a detailed calculation.

Sum rules.

Concerning the line strengths in a transition array, we can make certain general predictions in the form of sum rules which are independent of the coupling of the configurations.

The most general of these is the *J-group sum rule.** We call the array of lines connecting all levels of a given J in one configuration with all levels of a given J' in another configuration a J group. If neither of the configurations is perturbed by outside configurations, it follows immediately from $4^{11}2$ and the principle of spectroscopic stability (2^225) that *in any transition array the sum of the strengths of the lines in a J group is independent of coupling.* If several configurations are perturbing, we must enlarge the J group to include all perturbing initial levels and all perturbing final levels before the above statement becomes true.

* Harrison and Johnson, Phys. Rev. **38**, 757 (1931).

The next sum rule of importance is the *J-file sum rule*,* which is an extension of that proved in § 3[9] for *LS* coupling. In a transition array, the set of lines connecting a single given level of one configuration with all levels of the other configuration we shall call the *J file* referring to that level. We saw in § 3[9] that for a transition array like $d^4p \leftrightarrows d^4d$, the strengths of the different files referring to the levels of d^4p were proportional to $2J+1$, and the strengths of the different files referring to d^4d were proportional to $2J+1$, in *LS* coupling. We shall here show that this is true for any intermediate coupling. In this case the jumping electron is not equivalent to any of the electrons in the ion in either configuration, and the *J*-file sum rule holds for both rows and columns. For such an array as $d^4p \leftrightarrows d^5$, in which the jumping electron is equivalent to other electrons in one of the configurations, the strengths of the different files referring to d^5 are, in any coupling, proportional to $2J+1$, but this is not true of the files referring to d^4p.

In general, we shall prove that *for any coupling, the strengths of the J files referring to the level of the* $\left\{\begin{matrix}\textit{initial}\\ \textit{final}\end{matrix}\right\}$ *configuration are proportional to* $2J+1$ *provided that the jumping electron is not equivalent to any other in the* $\left\{\begin{matrix}\textit{final}\\ \textit{initial}\end{matrix}\right\}$ *configuration.* In this statement, the jumping electron may be equivalent to others in the $\left\{\begin{matrix}\text{initial}\\ \text{final}\end{matrix}\right\}$ configuration. In case it is not equivalent to another in *either* configuration, the sum rule holds for the files referring to *both* configurations.

That this sum rule holds in any coupling for the files referring to ps and to pp in $ps \leftrightarrows pp$ and for the files referring to p^2 in $ps \leftrightarrows p^2$ follows from an examination of Table 2[9] or Table 3[10]. In these tables, we see that the sums of the $\mathbf{S}$'s in all columns headed by the same J value are the same, while the sum of the products of the $\mathbf{S}^{\frac{1}{2}}$'s in any two columns headed by the same J value is zero. Therefore, since the $\mathbf{S}^{\frac{1}{2}}$'s transform like the matrix elements of an observable, it follows from 2[2]26 that these sums are invariant under any transformations of initial and final states which do not mix up the J values. A similar statement applies to the rows of $ps \leftrightarrows pp$ in these tables, but not to the rows of $ps \leftrightarrows p^2$.

In order to prove in general the sum rule as we have stated it, we shall apply considerations similar to these to the whole electric-moment matrix for the transition array in the zero-order nlm_sm_l scheme. Let us consider the transition array connecting the configurations λ and μ. We suppose that in the nl and $n'l'$ shells (where $l'-l=\pm 1$ or 0), λ contains nl^k and

* SHORTLEY, Phys. Rev. **47**, 295, 419 (1935).

μ contains $nl^{k-1}n'l'$. In addition λ and μ have the same numbers of electrons in all other shells. For a non-vanishing matrix element of $\boldsymbol{P}$ in the zero-order scheme, the quantum numbers in these other shells cannot differ in λ and μ. When they do not differ the element has the same value as in the arrry $nl^k \leftrightarrows nl^{k-1}n'l'$, which we now consider.

First we calculate the sum of the squares of the elements of $\boldsymbol{P}$ connecting the state $\Phi(m_s^1 m_l^1, m_s^2 m_l^2, \ldots, m_s^k m_l^k)$ of nl^k to all states of $nl^{k-1}n'l'$. The k sets of quantum numbers in Φ must be all different if the state is to be allowed. This state will combine with the states

$$^{(1)}\Psi(m_s^2 m_l^2, \ldots, m_s^k m_l^k, m_s' m_l'), \quad ^{(2)}\Psi(m_s^1 m_l^1, m_s^3 m_l^3, \ldots, m_s^k m_l^k, m_s' m_l'), \quad \ldots,$$
$$^{(k)}\Psi(m_s^1 m_l^1, \ldots, m_s^{k-1} m_l^{k-1}, m_s' m_l') \tag{4}$$

of $nl^{k-1}n'l'$. For combinations between Φ and $^{(i)}\Psi$ we must have $m_s' = m_s^i$ and $m_l' = m_l^i$ or $m_l^i \pm 1$. From $6^6 8$ we see that the square of the matrix element of $\boldsymbol{P}$ connecting Φ and $^{(i)}\Psi$ is just $|(nl\, m_s^i m_l^i|\boldsymbol{P}|n'l'\, m_s' m_l')|^2$, calculated for the one electron which changes its quantum numbers. The total strength of transitions from Φ to $^{(i)}\Psi$ for all $m_s' m_l'$ is given by $13^3 1'$ as

$$\sum_{m_s' m_l'} |(nl\, m_s^i m_l^i|\boldsymbol{P}|n'l'\, m_s' m_l')|^2 = |(nl\vdots P\vdots n'l')|^2\, \Xi(l, l'). \tag{5}$$

The total strength of all transitions from Φ to all the Ψ's is just k times this, and is independent of the quantum numbers in Φ.

We shall now show that for two states Φ^a and Φ^b of nl^k, the sum of the scalar products of the elements of $\boldsymbol{P}$ in the transition array is zero, i.e., that the two files of the matrix of $\boldsymbol{P}$ referring to Φ^a and Φ^b are orthogonal. The states Φ^a and Φ^b do not combine with the same states of $nl^{k-1}n'l'$ unless they agree in $k-1$ of their sets of quantum numbers—say that Φ^a has $m_s^{ia} m_l^{ia}$ where Φ^b has $m_s^{ib} m_l^{ib}$. If Φ^a and Φ^b do not agree in $k-1$ sets, the files referring to them are obviously orthogonal. If they do agree in all but the ith set, they combine simultaneously only with the states $^{(i)}\Psi$ of (4). The sum of the scalar products of the elements connecting them with the states of $^{(i)}\Psi$ is

$$\sum_{m_s' m_l'} (nl\, m_s^{ia} m_l^{ia}|\boldsymbol{P}|n'l'\, m_s' m_l') \cdot (n'l'\, m_s' m_l'|\boldsymbol{P}|nl\, m_s^{ib} m_l^{ib}),$$

and this, by $13^3 1$, is seen to vanish unless $m_s^{ib} = m_s^{ia}$, $m_l^{ib} = m_l^{ia}$, i.e., unless $\Phi^a \equiv \Phi^b$.

Hence in the $nlm_s m_l$ scheme, the sum of the absolute squares of the matrix elements of $\boldsymbol{P}$ having a common state of nl^k is independent of that state, and the files of matrix elements from two states of nl^k have vanishing scalar product. A simple extension of $2^2 26$ shows these properties to be invariant under unitary transformations. Hence if we return to the complete configurations λ and μ in a scheme characterized by J and M

values, we may say that the lines having the initial state λJM have the strength sum $k\,|(nl\vdots P\vdots n'l')|^2\,\Xi(l,l')$ and hence that the J file referring to the level λJ has the strength

$$k(2J+1)\,|(nl\vdots P\vdots n'l')|^2\,\Xi(l,l'). \qquad (6)$$

This proves the J-file sum rule as stated, and gives the absolute values, in terms of the one-electron component $(nl\vdots P\vdots n'l')$, of the strengths of the files.

With regard to cases of configuration interaction, we may say that if a pure configuration λ combines with two perturbing configurations μ and ν, then if the sum rule holds with respect to the J files referring to the levels of λ in the transitions to μ and ν separately, it holds with respect to these files in the perturbed case.

In the still more special case of transitions $\alpha s \leftrightarrows \alpha p$ in which the valence electron jumps from s to p or p to s and is not equivalent to any electron in the ion in either configuration, we can obtain yet another rule which may be called the *J-group-file sum rule*. This rule says that for these arrays the individual J groups may be broken up in one direction into invariant files. The direction is such that all the lines in the file have a level of αs in common. These J-group files in Tables 2⁹ and 3¹⁰ are set off by solid and broken lines, and their strengths are shown in parentheses. The strength of each allowed J-group file is the same as that of the J files referring to the levels of αp which cross it in the array. These statements are proved (using 2²26), in jj coupling, by noting that two files of the same J group are orthogonal since they refer to different quantum numbers of the ion (the valence electron must have quantum numbers $ns_{\frac{1}{2}}$), and by showing directly from 11³8 that

$$\sum_{j'=\frac{1}{2}}^{\frac{3}{2}} \mathsf{S}(J^1 sj\,J,\, J^1 pj'\,J') = (2J'+1)\,|(p\vdots P\vdots s)|^2\,\Xi(p,s), \qquad (7)$$

for $J'=J+1$, J, and $J-1$.

The best experimental data with which to compare these sum rules are the anomalous-dispersion determinations by Ladenburg and Levy* of the strengths in the neon array $2p^5\,3p \rightarrow 2p^5\,3s$. (We shall see in § 8¹³ that the strengths in this array should be the same as those in $pp \rightarrow ps$.) Such a comparison has been made by Shortley (*loc. cit.*) with agreement within the experimental error. In the same paper is reported an attempt to calculate the detailed strength pattern for this array by the method we have outlined, using the parameters of Table 1¹³ for $2p^5\,3p$. This calculation was not satisfactory because a small change in the parameters has a very large effect on the calculated strengths, and the parameters used were necessarily inaccurate.

* LADENBURG and LEVY, Zeits. für Phys. **88**, 461 (1934).

5. The forbidden lines of astrophysical interest.

For a long time several prominent lines in the spectra of some nebulae remained unclassified. They were often called 'nebulium' lines because it was thought that they might be due to a new element not yet known on the earth—a repetition of the history of helium's discovery on the sun before it was found terrestrially. Bowen* finally showed that they are due to transitions in oxygen and nitrogen in various stages of ionization. The atoms involved have p^2, p^3 and p^4 for their normal configuration and the lines arise from transitions between different levels belonging to the normal configuration. They cannot, therefore, be electric-dipole radiation. We shall see that they are partly quadrupole and partly magnetic dipole in character.

The most systematic study of astrophysical data for identifications of such lines following Bowen is that of Boyce, Menzel and Payne.† Calculations of the transition probabilities have been made by Stevenson and by Condon.‡ An earlier calculation by Bartlett§ is based on approximations not sufficiently exact to be of interest. We shall limit ourselves to the p^2 configuration.

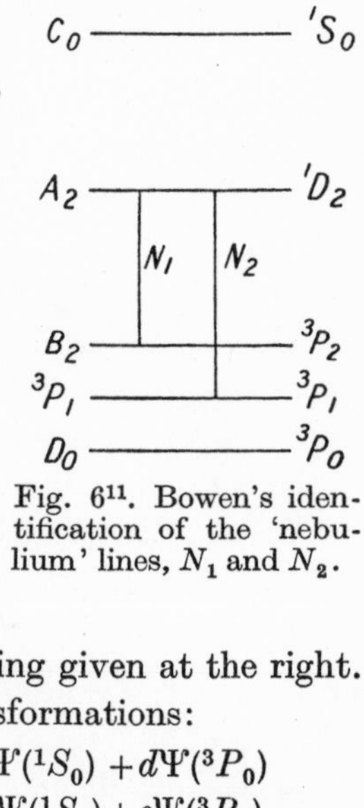

Fig. 6^{11}. Bowen's identification of the 'nebulium' lines, N_1 and N_2.

The most famous of the nebulium lines, called N_1 and N_2, were identified by Bowen as the transitions indicated in Fig. 6^{11} in O III. These intersystem transitions are made possible by the slight departures from Russell-Saunders coupling. As we have seen in the preceding sections, the eigenfunction of the level usually called ${}^1S_0$ is a linear combination of those for ${}^1S_0$ and ${}^3P_0$. The actual levels will be designated by the letters on the left in the figure, the approximate Russell-Saunders labels being given at the right. The breakdown of coupling is expressed by the transformations:

$$\Psi(A_2) = a\Psi({}^1D_2) + b\Psi({}^3P_2) \qquad \Psi(C_0) = c\Psi({}^1S_0) + d\Psi({}^3P_0)$$
$$\Psi(B_2) = -b\Psi({}^1D_2) + a\Psi({}^3P_2) \qquad \Psi(D_0) = -d\Psi({}^1S_0) + c\Psi({}^3P_0).$$

The matrix components of the spin-orbit interaction are given in Table 1^{11}. We evaluate the parameter ζ from the intervals in the ${}^3P$ term, obtaining $\zeta = 210\,\text{cm}^{-1}$ for O III. To the first order of the perturbation theory b and d are given by

$$b = \frac{({}^1D_2|H^I|{}^3P_2)}{{}^1D - {}^3P} = 0{\cdot}0074; \quad d = \frac{({}^1S_0|H^I|{}^3P_0)}{{}^1S - {}^3P} = -0{\cdot}0069.$$

These values, though small, are what make the nebular lines possible.

* Bowen, Astrophys. J. **67**, 1 (1928).
† Boyce, Menzel and Payne, Proc. Nat. Acad. Sci. **19**, 581 (1933).
‡ Stevenson, Proc. Roy. Soc. **A137**, 298 (1932);
Condon, Astrophys. J. **79**, 217 (1934).
§ Bartlett, Phys. Rev. **34**, 1247 (1929).

By means of the results of Chapter VIII the Russell-Saunders eigenfunctions are expressible in terms of those of the zero-order scheme. From the latter the matrix components of quadrupole moment and magnetic-dipole moment may be computed. In this way the following results were obtained for the line strengths:

Line	Magnetic dipole	Quadrupole
$\mathsf{S}(A_2, B_2)$	$\frac{15}{2}a^2b^2$	$\frac{28}{15}a^2b^2\mathsf{s}_2^2$
$\mathsf{S}(A_2, {}^3P_1)$	$\frac{5}{2}b^2$	$\frac{9}{15}b^2\mathsf{s}_2^2$

Here s_2 is the integral $$\mathsf{s}_2 = -e\int_0^\infty r^2\,R^2(2p)\,dr.$$

The value of this in atomic units, using the eigenfunction of Hartree and Black,* is 1·24; so using the formulas of Chapter IV connecting line strength with transition probability we find for the absolute value of the transition probability for the two lines

$$N_1:\quad \mathsf{A}(A_2,\ B_2) = 0{\cdot}018\,\text{sec}^{-1}$$

$$N_2:\quad \mathsf{A}(A_2, {}^3P_1) = 0{\cdot}006\,\text{sec}^{-1}.$$

The lines are almost entirely due to magnetic-dipole radiation, the quadrupole term in the transition probability giving only 0·1 per cent. of the whole amount.

The line $C_0 \rightarrow A_2$ is called an auroral line since in O I this transition gives rise to the green line of the aurora and the night sky. Being a line for which $\Delta J = 2$ it is entirely of quadrupole character, but not being an intersystem combination ($\Delta S = 0$) it does not depend on the partial breakdown of coupling. The calculated transition probability is

$$\mathsf{A}(C_0, A_2) = 1{\cdot}8\,\text{sec}^{-1}.$$

The triplet $C_0 \rightarrow {}^3P$ is interesting. The line $C_0 \rightarrow D_0$ is forbidden in all approximations by the general exclusion of $0 \rightarrow 0$ transitions in J. Likewise $C_0 \rightarrow {}^3P_1$ cannot be a quadrupole line as $0 \rightarrow 1$ change in J is of vanishing strength for this type of radiation. On the other hand $C_0 \rightarrow B_2$ cannot occur with magnetic-dipole radiation because of the dipole selection rule on J. We therefore have two lines close together, one of which is due to purely quadrupole and the other to purely magnetic-dipole radiation. The calculations give

$$\mathsf{A}(C_0,\ B_2) = 1{\cdot}5 \times 10^{-4}\,\text{sec}^{-1}, \qquad \text{(Quadrupole)}$$

$$\mathsf{A}(C_0, {}^3P_1) = 0{\cdot}102\,\text{sec}^{-1}, \qquad \text{(Magnetic dipole)}$$

indicating that the quadrupole line is only about 10^{-3} as strong as the magnetic-dipole line.

* Hartree and Black, Proc. Roy. Soc. **A139**, 311 (1933).

CHAPTER XII

TRANSFORMATIONS IN THE THEORY OF COMPLEX SPECTRA*

We wish in this chapter to consider, more in detail than heretofore, the transformations between the various schemes of states of interest in the theory of atomic spectra. There are five representations of importance, the two zero-order schemes, which we shall designate as the $m_s m_l$ scheme and the jm scheme, the two LS-coupling schemes characterized by SLM_SM_L and $SLJM$ respectively, and the jj-coupling scheme of § 1[10], characterized by the electronic j values and resultant JM. We shall consider in particular the four transformations

$$\begin{array}{ccc} nljm & \rightleftharpoons & nlm_sm_l \\ \upharpoonleft\downharpoonright & & \upharpoonleft\downharpoonright \\ jjJM & \rightleftharpoons & SLJM \end{array}$$

and incidentally the transformation $nlm_sm_l \rightleftharpoons SLM_SM_L$. In preparation for Chapter XIII on configurations containing almost closed shells we shall establish a correlation in each scheme between states of a given configuration containing an almost closed shell and those of a 'corresponding' simpler configuration, and determine the relation between the 'corresponding' transformation matrices.†

1. Configurations containing almost closed shells.

Let us consider a configuration which contains, outside of closed shells, a shell '$\mathscr{S}$' which is complete except for ϵ 'missing' electrons, and in addition η other electrons. Here ϵ and η are considered to be small integers, although formally there is no restriction placed on their magnitude. This configuration we shall designate as 'configuration $\mathscr{R}$,' since such configurations occur mainly for elements near the *right* of the periodic table. To this configuration we shall correlate a simpler configuration '$\mathscr{L}$' which contains the same closed shells, ϵ electrons *present* in shell $\mathscr{S}$, and the same η other electrons.

In the nlm_sm_l scheme we shall correlate to a given state of $\mathscr{L}$ with quantum numbers listed in the standard order of § 5[6], that state of $\mathscr{R}$ whose ϵ missing electrons have the *negatives* of the m_s and m_l values of the ϵ electrons in shell $\mathscr{S}$ of $\mathscr{L}$, and whose η other electrons have the same quantum numbers as

* Shortley, Phys. Rev. **40**, 185 (1932), § 5; *ibid.* **43**, 451 (1933).

† The considerations of this chapter can all be readily extended to configurations containing more than one 'almost closed shell.' This case is, however, of little interest.

those of $\mathscr{L}$—taken when the quantum numbers are listed in standard order with the phase $(-1)^w$, where w is the sum of the m_l values of the electrons in shell $\mathscr{S}$. It is clear that this gives a one-to-one correlation between all states of the two configurations, the correlated states having the same M_S and M_L values. With this correlation we shall show in § 2^{12} that the matrices of $\boldsymbol{S}^2$, $\boldsymbol{L}^2$, and $\boldsymbol{L}\cdot\boldsymbol{S}$ are the same for the two configurations, and hence that the same matrix will transform them to LS coupling. *The allowed states in LS coupling thus have the same SLM_SM_L or $SLJM$ values for the two configurations,* and corresponding states in LS coupling will be the same linear combinations of the correlated m_sm_l states.

In the $nljm$ scheme we shall correlate to a given state of $\mathscr{L}$ with quantum numbers listed in the standard order of § 5^6, that state of $\mathscr{R}$ whose ϵ missing electrons have the same j values but the negatives of the m values of the ϵ electrons in shell $\mathscr{S}$ of $\mathscr{L}$, and whose η other electrons have the same quantum numbers as those of $\mathscr{L}$—taken when in standard order with the phase $(-1)^{(\epsilon+1)(q+l)}$. Here l is the azimuthal quantum number for the shell $\mathscr{S}$, which in $\mathscr{L}$ contains p electrons with $j=l+\frac{1}{2}$, and q electrons with $j=l-\frac{1}{2}$ $(p+q=\epsilon)$. These correlated states have the same M values. With this correlation we shall show in § 3^{12} that the matrix of $\boldsymbol{J}^2$ is the same for the two configurations. Hence the allowed states in jj coupling have the same JM values, but the electronic j values of $\mathscr{L}$ are those missing in the correlated states of $\mathscr{R}$. Corresponding states in jj coupling will be the same linear combinations of the correlated jm states.

We shall show in § 5^{12} that with these correlations between zero-order states of $\mathscr{L}$ and $\mathscr{R}$, the transformation from the m_sm_l scheme to the jm scheme is the same for the two configurations. Hence the transformation from LS coupling to jj coupling is the same for the correlated states of $\mathscr{L}$ and $\mathscr{R}$. This transformation will be given explicitly for a number of two-electron configurations in § 6^{12}.

2. The transformation to LS coupling.

In order to show that the same matrix will transform the correlated nlm_sm_l states of $\mathscr{L}$ and $\mathscr{R}$ to LS coupling, it is only necessary to show that the matrices of $\boldsymbol{L}^2$, $\boldsymbol{S}^2$, and $\boldsymbol{L}\cdot\boldsymbol{S}$ are the same for the two configurations. Formulas for the elements of these matrices were given in § 4^8. Let us denote by $A_{\mathscr{L}}$, $A'_{\mathscr{L}}$, ... zero-order states of $\mathscr{L}$ and by $A_{\mathscr{R}}$, $A'_{\mathscr{R}}$, ... the correlated states of $\mathscr{R}$.

Matrix of $\boldsymbol{L}^2$. The general non-diagonal element of $\boldsymbol{L}^2$ is given by 4^83a. If neither a nor b is in shell $\mathscr{S}$, the value is obviously the same for $\mathscr{L}$ and $\mathscr{R}$. If a, but not b, is in shell $\mathscr{S}$, the second term vanishes. In order to obtain a value for the first term a and a' must be of the form $(m_l)^{\pm}$,

$(m_l-1)^{\pm}$.* Equation $4^8 2$ shows that in this case the elements for $\mathscr{L}$ and $\mathscr{R}$ are essentially equal—we must consider the phases in greater detail. If the individual set lying between $(m_l)^{\pm}$ and $(m_l-1)^{\pm}$ is present in $\mathscr{L}$, so far as shell $\mathscr{S}$ is concerned $A_{\mathscr{L}}$ and $A'_{\mathscr{L}}$ differ by an odd permutation, while $A_{\mathscr{R}}$ and $A'_{\mathscr{R}}$ do not differ in order. The converse is true if this element is missing. This introduces one difference of sign between $(A_{\mathscr{L}}|\boldsymbol{L}^2|A'_{\mathscr{L}})$ and $(A_{\mathscr{R}}|\boldsymbol{L}^2|A'_{\mathscr{R}})$. However, the sum of the m_l values in shell $\mathscr{S}$ differs by one in $A_{\mathscr{R}}$ and $A'_{\mathscr{R}}$. This results in a second difference in sign from the different choices of phase for $\Psi(A_{\mathscr{R}})$ and $\Psi(A'_{\mathscr{R}})$, which makes these two elements just equal. In a similar way one demonstrates the equality in case a and b are both in shell $\mathscr{L}$.

The diagonal element of $\boldsymbol{L}^2$ is given by $4^8 3$b. The first term is the same for $A_{\mathscr{R}}$ and $A_{\mathscr{L}}$, as also are the parts of the second and third terms arising from shells other than $\mathscr{S}$. The parts of those terms which arise from shell $\mathscr{S}$ are the same if calculated for $A_{\mathscr{L}}$ as for the group of *missing* electrons of $A_{\mathscr{R}}$. The equivalence of the calculation for the missing electrons to the calculation for the electrons present in $A_{\mathscr{R}}$ is a direct consequence of the following interesting relation: If one takes the integers (or half-odd integers) $-l, \ldots, l$ and arranges them into two groups α and β, then the following sum has the same value for groups α and β (the individual integers are denoted by m_l):

$$\sum_{m_l} [l(l+1) - m_l^2] + \sum [-l(l+1) + m_l^2 - m_l], \tag{1}$$

where the second sum is taken over only those m_l's for which $m_l - 1$ is also in the group.†

This completes the proof that the matrix of $\boldsymbol{L}^2$ is the same for $\mathscr{L}$ and $\mathscr{R}$. The calculation for $\boldsymbol{S}^2$ is very similar and will not be discussed in detail.

Matrix of $\boldsymbol{L}\cdot\boldsymbol{S}$. If A' differs from A in two electrons, the calculation is much the same as for $\boldsymbol{L}^2$ and $\boldsymbol{S}^2$. If A' and A differ in one electron, the matrix element is given by $4^8 5$b. If a and a' are in shell $\mathscr{S}$, and for $\mathscr{L}$ have the

* For example shell $\mathscr{S}$ of $A_{\mathscr{L}}$ and $A'_{\mathscr{L}}$ might have the following electrons:

	...	$(m_l)^+$	$(m_l)^-$	$(m_l-1)^+$	...
$A_{\mathscr{L}}$	...	×	$\underline{\times}$	–	...
$A'_{\mathscr{L}}$	...	–	$\underline{\times}$	×	...

Here × indicates the presence, – the absence, of a given individual set; all sets except those noted explicitly are the same for $A_{\mathscr{L}}$ and $A'_{\mathscr{L}}$. In this case $A_{\mathscr{R}}$ and $A'_{\mathscr{R}}$ will have the forms:

	...	$(-m_l+1)^-$	$(-m_l)^+$	$(-m_l)^-$	...
$A_{\mathscr{R}}$	...	×	$\overline{\times}$	–	...
$A'_{\mathscr{R}}$	...	–	$\overline{\times}$	×	...

† This may be proved as follows: The division into groups α and β may be made by splitting the series $-l, \ldots, l$ into r sections, putting the first section into group α, the second into β, the third into α, etc. Consider the sections as defined by section points $p_1, p_2, \ldots, p_{r+1}$, such that the tth section contains the numbers $p_t+\frac{1}{2}, p_t+\frac{3}{2}, \ldots, p_{t+1}-\frac{1}{2}$. The value of the sum (1) when m_l runs over the numbers of this section is $f(p_t)+f(p_{t+1})$, where $f(p)=\frac{1}{2}(l+\frac{1}{2})^2-\frac{1}{2}p^2$. Then the sum for group α has the value $f(p_1)+f(p_2)+f(p_3)+f(p_4)+\ldots$, while the sum for β has the value $f(p_2)+f(p_3)+f(p_4)+\ldots$. One of these expressions ends with the term $f(p_r)$, the other with $f(p_{r+1})$. But $f(p_1)=f(-l-\frac{1}{2})=0$ and $f(p_{r+1})=f(l+\frac{1}{2})=0$. Hence the sums for α and β are equal. This proof was suggested to us by Professor Bennett of Brown University.

form $(m_l)^-$, $(m_l-1)^+$, for $\mathscr{R}$ they will have the form $(-m_l+1)^-$, $(-m_l)^+$. One may easily verify that $A_{\mathscr{L}}$ and $A'_{\mathscr{L}}$, $A_{\mathscr{R}}$ and $A'_{\mathscr{R}}$ then differ by the same permutation; however, one difference in sign of the matrix elements is introduced by the different m_l values for $A_{\mathscr{R}}$ and $A'_{\mathscr{R}}$. For $\mathscr{L}$ the second, third terms vanish unless $b=(m_l-1)^-$, $c=(m_l)^+$ respectively are present. For $\mathscr{R}$ the second, third terms vanish unless $b=(-m_l)^-$, $c=(-m_l+1)^+$ are present. Now if neither b nor c is in $\mathscr{L}$, they must both be in $\mathscr{R}$. Hence in this case

$$(A_{\mathscr{L}}|\boldsymbol{L}\cdot\boldsymbol{S}|A'_{\mathscr{L}})=(a|\boldsymbol{L}\cdot\boldsymbol{S}|a')$$
$$(A_{\mathscr{R}}|\boldsymbol{L}\cdot\boldsymbol{S}|A'_{\mathscr{R}})=-\{(a|\boldsymbol{L}\cdot\boldsymbol{S}|a')-2(a|\boldsymbol{L}\cdot\boldsymbol{S}|a')\}=(A_{\mathscr{L}}|\boldsymbol{L}\cdot\boldsymbol{S}|A'_{\mathscr{L}})$$

(cf. $4^{8}2$ and the discussion under $4^{8}5$b). Similar considerations hold for the other two possibilities.

The diagonal element ($2^{8}5$c) of $\boldsymbol{L}\cdot\boldsymbol{S}$ is obviously the same for $\mathscr{L}$ and $\mathscr{R}$ with our correlation. This completes the proof of the equality of the *LS*-coupling transformations for $\mathscr{L}$ and $\mathscr{R}$.

3. The transformation to *jj* coupling.

The transformation from the *nljm* scheme to the *jjJM* scheme may be obtained by any of the methods sketched for *LS* coupling in Chapter VIII; the diagonalization method is however much simpler than for *LS* coupling, since only one matrix, that of $\boldsymbol{J}^2$, need be diagonalized, and the transformation is diagonal with respect to the j values of all electrons in addition to M. The elements of the matrix of $\boldsymbol{J}^2$ in the *jm* scheme may be written down in exact analogy to those of $\boldsymbol{L}^2$ in the $m_s m_l$ scheme (cf. $4^{8}3$). It has non-diagonal elements only between states which differ in regard to two individual sets of quantum numbers; say that A contains a and b where A' contains a' and b'. This element has the value

$$(A|\boldsymbol{J}^2|A')=\pm 4\{(a|J_x|a')(b|J_x|b')-(a|J_x|b')(b|J_x|a')\}\,\delta(M,M'),$$

where $(a|J_x|b)=\delta(n^a l^a, n^b l^b)(j^a m^a|J_x|j^b m^b)$ is given by $3^{3}4$. The diagonal element of $\boldsymbol{J}^2$ is

$$(A|\boldsymbol{J}^2|A)=\hbar^2M^2+\hbar^2\sum_a\{j^a(j^a+1)-(m^a)^2\}-4\sum_{a<b}(a|J_x|b)^2.$$

Now with the correlation given in § 1^{12} it is easily seen, by an argument which is the direct analogue of that given above for $\boldsymbol{L}^2$ in the $m_s m_l$ scheme (noting that $\boldsymbol{J}^2$ is diagonal with respect to the number of electrons in shell $\mathscr{S}$ with $j=l-\frac{1}{2}$, i.e., to the q occurring in the correlation factor), that the matrix of $\boldsymbol{J}^2$ is the same for the two configurations and hence that the transformation to *jj* coupling is the same.

4. The transformation between zero-order states.

We shall in this section consider in general the transformation connecting two systems of zero-order states, and the relation between these transformations for $\mathscr{L}$ and $\mathscr{R}$ with certain general correlations. In the next section we shall specialize to the particular transformation of interest to us.

Let us consider an N-electron configuration for which there are n possible sets of one-electron quantum numbers.* In the first zero-order scheme let us denote these individual sets, in some 'standard' order, by the numbers $1, 2, \ldots, n$, and the corresponding states by $\psi(1), \psi(2), \ldots, \psi(n)$. In the second scheme let us use the notation ${}_{,}1, {}_{,}2, \ldots, {}_{,}n$ for the quantum numbers, and $\phi({}_{,}1), \phi({}_{,}2), \ldots, \phi({}_{,}n)$ for the states. The transformation between these two one-electron systems is given by

$$\psi(\alpha) = \sum_{\beta=1}^{n} \phi({}_{,}\beta)({}_{,}\beta|\alpha). \tag{1}$$

The zero-order antisymmetric function belonging to the complete set $\alpha^1, \alpha^2, \ldots, \alpha^N$ of quantum numbers is defined by

$$\Psi(\alpha^1\alpha^2\ldots\alpha^N) = \mathscr{A}\,\psi_1(\alpha^1)\,\psi_2(\alpha^2)\ldots\psi_N(\alpha^N) \tag{2}$$

(cf. § 3^6). Here $\alpha^1, \alpha^2, \ldots, \alpha^N$ are N numbers of the set $1, 2, \ldots, n$, arranged in the order $\alpha^1 < \alpha^2 < \ldots < \alpha^N$.

The transformation of this state to the primed scheme is given by the following calculation. From (2) and (1)

$$\begin{aligned}\Psi(\alpha^1\alpha^2\ldots\alpha^N) &= \mathscr{A} \sum_{\beta^1=1}^{n} \phi_1({}_{,}\beta^1)\,({}_{,}\beta^1|\alpha^1)\ldots \sum_{\beta^N=1}^{n} \phi_N({}_{,}\beta^N)({}_{,}\beta^N|\alpha^N)\\ &= \sum_{\beta^1\ldots\beta^N} ({}_{,}\beta^1|\alpha^1)\ldots({}_{,}\beta^N|\alpha^N)\,\mathscr{A}\,\phi_1({}_{,}\beta^1)\ldots\phi_N({}_{,}\beta^N).\end{aligned}$$

If two of the β's are equal the antisymmetric factor $\mathscr{A}\ldots$ vanishes, while for a given set of β's this factor is the same, irrespective of the order of the β's, to within a sign which is just correct to give us a determinant of the transformation coefficients; i.e.

$$\Psi(\alpha^1\alpha^2\ldots\alpha^N) = \sum_{\beta^1<\beta^2<\ldots<\beta^N} \mathscr{D}\begin{pmatrix}{}_{,}\beta^1\ldots{}_{,}\beta^N\\ \alpha^1\ldots\alpha^N\end{pmatrix}\Phi({}_{,}\beta^1\,{}_{,}\beta^2\ldots{}_{,}\beta^N), \tag{3}$$

where we use the notation

$$\mathscr{D}\begin{pmatrix}{}_{,}\beta^1\ldots{}_{,}\beta^N\\ \alpha^1\ldots\alpha^N\end{pmatrix} = \begin{vmatrix}({}_{,}\beta^1|\alpha^1) & \ldots & ({}_{,}\beta^1|\alpha^N)\\ \ldots & \ldots & \ldots\\ \ldots & \ldots & \ldots\\ ({}_{,}\beta^N|\alpha^1) & \ldots & ({}_{,}\beta^N|\alpha^N)\end{vmatrix}. \tag{4}$$

The transformation coefficient connecting the state ${}_{,}\beta^1\,{}_{,}\beta^2\ldots{}_{,}\beta^N$ with the state $\alpha^1\alpha^2\ldots\alpha^N$ (quantum numbers in standard order) is given, then, by just this determinant:

$$\begin{aligned}({}_{,}\beta^1\,{}_{,}\beta^2\ldots{}_{,}\beta^N|\alpha^1\alpha^2\ldots\alpha^N) &= \int\bar{\Phi}({}_{,}\beta^1\,{}_{,}\beta^2\ldots{}_{,}\beta^N)\,\Psi(\alpha^1\alpha^2\ldots\alpha^N)\\ &= \mathscr{D}\begin{pmatrix}{}_{,}\beta^1\ldots{}_{,}\beta^N\\ \alpha^1\ldots\alpha^N\end{pmatrix}.\end{aligned} \tag{5}$$

* For example, for the configuration $2s\,3p^2$, $N=3$ and $n=8$, the possible sets of quantum numbers being given, for the $m_s m_l$ scheme, by ${}_{,}1=2s\,0^+$, ${}_{,}2=2s\,0^-$, ${}_{,}3=3p\,1^+$, ${}_{,}4=3p\,1^-$, ${}_{,}5=3p\,0^+$, ${}_{,}6=3p\,0^-$, ${}_{,}7=3p\,-1^+$, ${}_{,}8=3p\,-1^-$; and for the jm scheme by $1=2s\,\frac{1}{2}\,\frac{1}{2}$, $2=2s\,\frac{1}{2}\,-\frac{1}{2}$, $3=3p\,\frac{3}{2}\,\frac{3}{2}$, $4=3p\,\frac{3}{2}\,\frac{1}{2}$, $5=3p\,\frac{3}{2}\,-\frac{1}{2}$, $6=3p\,\frac{3}{2}\,-\frac{3}{2}$, $7=3p\,\frac{1}{2}\,\frac{1}{2}$, $8=3p\,\frac{1}{2}\,-\frac{1}{2}$.

Since the transformation (1) has no components connecting sets belonging to different shells, the determinant (4) and hence the transformation coefficient (5) splits up into a product of factors each referring to a single shell. For example, if the configuration consists of M electrons $(1 \dots M)$ in shell $\mathscr{S}$ and $N-M$ in other shells, we may split off a factor giving the dependence on shell $\mathscr{S}$:

$$({}_{\prime}\beta^1 \dots {}_{\prime}\beta^M\, {}_{\prime}\beta^{M+1} \dots {}_{\prime}\beta^N | \alpha^1 \dots \alpha^M\, \alpha^{M+1} \dots \alpha^N) = \mathscr{D}\begin{pmatrix} {}_{\prime}\beta^1 \dots {}_{\prime}\beta^M \\ \alpha^1 \dots \alpha^M \end{pmatrix} \mathscr{D}\begin{pmatrix} {}_{\prime}\beta^{M+1} \dots {}_{\prime}\beta^N \\ \alpha^{M+1} \dots \alpha^N \end{pmatrix}. \tag{6}$$

Because of this property, in considering the relation between these coefficients for $\mathscr{L}$ and $\mathscr{R}$ we may restrict ourselves purely to the shell $\mathscr{S}$ so long as the other quantum numbers are the same for the correlated states of $\mathscr{L}$ and $\mathscr{R}$. We shall suppose that there are altogether m states in the shell $\mathscr{S}$, and consider the relation of these components for the configuration consisting of ϵ electrons to that consisting of $m-\epsilon$ electrons in this shell.

Let us seek an invariant correlation between an ϵ-electron state and the $(m-\epsilon)$-electron state whose ϵ missing electrons have the same quantum numbers. This correlation is to be independent of the system of zero-order states used in the description. We shall denote a given ϵ-electron state by $\Psi(\alpha^1 \alpha^2 \dots \alpha^\epsilon)$ and the corresponding $(m-\epsilon)$-electron state by $\Psi(a^1 a^2 \dots a^{m-\epsilon})$, where the α's and a's together make up the set $1, 2, \dots, m$.

Let us define a linear operator Z by the relations

$$Z\Psi(\alpha^1 \alpha^2 \dots \alpha^\epsilon) = (-1)^{\alpha^1+\alpha^2+\dots+\alpha^\epsilon}\, \bar{\Psi}(a^1 a^2 \dots a^{m-\epsilon}). \tag{7}$$

This operator is seen to be unitary, since the states on the right form a normalized orthogonal system. In the primed scheme we shall define a corresponding operator ${}_{\prime}Z$ by

$${}_{\prime}Z\Phi({}_{\prime}\beta^1\, {}_{\prime}\beta^2 \dots {}_{\prime}\beta^\epsilon) = (-1)^{\beta^1+\beta^2+\dots+\beta^\epsilon}\, \bar{\Phi}({}_{\prime}b^1\, {}_{\prime}b^2 \dots {}_{\prime}b^{m-\epsilon}). \tag{8}$$

The relation between ${}_{\prime}Z$ and Z is given by the following calculation:

$$\begin{aligned} {}_{\prime}Z\Psi(\alpha^1 \alpha^2 \dots \alpha^\epsilon) &= {}_{\prime}Z \sum_{\beta^1<\dots<\beta^\epsilon} \mathscr{D}\begin{pmatrix} {}_{\prime}\beta^1 \dots {}_{\prime}\beta^\epsilon \\ \alpha^1 \dots \alpha^\epsilon \end{pmatrix} \Phi({}_{\prime}\beta^1\, {}_{\prime}\beta^2 \dots {}_{\prime}\beta^\epsilon) \\ &= \sum_{\beta^1<\dots<\beta^\epsilon} \mathscr{D}\begin{pmatrix} {}_{\prime}\beta^1 \dots {}_{\prime}\beta^\epsilon \\ \alpha^1 \dots \alpha^\epsilon \end{pmatrix} (-1)^{\Sigma\beta^\nu}\, \bar{\Phi}({}_{\prime}b^1\, {}_{\prime}b^2 \dots {}_{\prime}b^{m-\epsilon}) \\ &= \sum_{\beta^1<\dots<\beta^\epsilon} (-1)^{\Sigma\beta^\nu} \mathscr{D}\begin{pmatrix} {}_{\prime}\beta^1 \dots {}_{\prime}\beta^\epsilon \\ \alpha^1 \dots \alpha^\epsilon \end{pmatrix} \sum_{c^1<\dots<c^{m-\epsilon}} \mathscr{D}\begin{pmatrix} {}_{\prime}b^1 \dots {}_{\prime}b^{m-\epsilon} \\ c^1 \dots c^{m-\epsilon} \end{pmatrix} \bar{\Psi}(c^1 c^2 \dots c^{m-\epsilon}) \\ &= (-1)^{\Sigma\alpha^\nu} \mathscr{D}\begin{pmatrix} {}_{\prime}1 \dots {}_{\prime}m \\ 1 \dots m \end{pmatrix} \bar{\Psi}(a^1 a^2 \dots a^{m-\epsilon}), \end{aligned}$$

since the terms for $c^1 = a^1, \dots, c^{m-\epsilon} = a^{m-\epsilon}$ furnish just the Laplace expansion of $\mathscr{D}\begin{pmatrix} ,1 \dots ,m \\ 1 \dots m \end{pmatrix}$, while the other terms vanish. Since any state may be expressed in terms of the $\Psi(\alpha^1 \alpha^2 \dots \alpha^\epsilon)$, we see by comparison with (7) that

$$,Z = \mathscr{D}\begin{pmatrix} ,1 \dots ,m \\ 1 \dots m \end{pmatrix} Z. \tag{9}$$

This shows the operator Z to be invariant to within a constant factor. If we apply this operator to an ϵ-electron state and then take the complex conjugate, we obtain an invariantively correlated $(m-\epsilon)$-electron state. We take the complex conjugate because by (7) Z acting on a state gives a state in the dual space, whereas we want a state in the same space.*

We may now calculate the relation between the transformation coefficients:

$$\begin{aligned}
({}_,b^1\, {}_,b^2 \dots {}_,b^{m-\epsilon} | a^1 a^2 \dots a^{m-\epsilon}) &= \int \bar{\Phi}({}_,b^1\, {}_,b^2 \dots {}_,b^{m-\epsilon})\, \Psi(a^1 a^2 \dots a^{m-\epsilon}) \\
&= (-1)^{\Sigma\alpha^\nu + \Sigma\beta^\nu} \int {}_,Z\Phi({}_,\beta^1\, {}_,\beta^2 \dots {}_,\beta^\epsilon)\, \overline{Z\Psi(\alpha^1 \alpha^2 \dots \alpha^\epsilon)} \\
&= (-1)^{\Sigma\alpha^\nu + \Sigma\beta^\nu} \mathscr{D}\begin{pmatrix} ,1 \dots ,m \\ 1 \dots m \end{pmatrix} \int \overline{Z\Psi(\alpha^1 \alpha^2 \dots \alpha^\epsilon)}\, Z\Phi({}_,\beta^1\, {}_,\beta^2 \dots {}_,\beta^\epsilon) \\
&= (-1)^{\Sigma\alpha^\nu + \Sigma\beta^\nu} \mathscr{D}\begin{pmatrix} ,1 \dots ,m \\ 1 \dots m \end{pmatrix} (\alpha^1 \alpha^2 \dots \alpha^\epsilon | {}_,\beta^1\, {}_,\beta^2 \dots {}_,\beta^\epsilon),
\end{aligned} \tag{10}$$

since a transformation coefficient is invariant under a unitary transformation. This shows that the transformation coefficient connecting two ϵ-electron states is essentially the complex conjugate of that connecting the $(m-\epsilon)$-electron states whose ϵ missing electrons have the same quantum numbers; and gives the exact phase relation between these coefficients.

5. The transformation $nlm_s m_l \leftrightarrows nljm$.

Of transformations of the type considered in the previous section, we are most interested in that from the $m_s m_l$ scheme to the jm scheme. We shall now show that this transformation is the same for configurations $\mathscr{L}$ and $\mathscr{R}$, with the correlations of § 1^{12}. Because the transformation factors according to the shells (cf. $4^{12}6$), and since in our correlations all electrons except those in shell $\mathscr{S}$ have the same quantum numbers for $\mathscr{L}$ and $\mathscr{R}$, we need only show that this transformation is the same for the ϵ electrons in shell $\mathscr{S}$ in $\mathscr{L}$ and the $m-\epsilon$ electrons in shell $\mathscr{S}$ in $\mathscr{R}$.

* This is not a linear correlation, since taking the complex conjugate is an invariant, but not linear ($\overline{c\Psi} \neq c\overline{\Psi}$ unless c is real), operation. A linear correlation such as is obtained directly by omitting the bar on the right side of (7) is invariant only under real orthogonal transformations of the zero-order states.

If we denote the m_sm_l states by $\phi(nlm_sm_l)$ and the jm states by $\psi(nljm)$, correlating these schemes respectively with the primed and unprimed schemes above, the transformation $4^{12}1$ is given by $4^{5}8$, and is entirely real. The determinant of this transformation for an l shell is easily shown to be $(-1)^l$. This value is to be inserted in $4^{12}10$. Moreover, from the standard order of listing quantum numbers (§ 5^6) it is seen that $(-1)^{\Sigma\beta^\nu}=(-1)^{M_S+\frac{1}{2}\epsilon}$, and $(-1)^{\Sigma\alpha^\nu}=(-1)^{-M+q+\epsilon(l-\frac{1}{2})}$, where q is the number of electrons in $\Psi'(\alpha^1\alpha^2\ldots\alpha^\epsilon)$ with $j=l-\frac{1}{2}$. Since $M=M_S+M_L$, $4^{12}10$ becomes

$$({}_{,}b^1\,{}_{,}b^2\ldots{}_{,}b^{m-\epsilon}|a^1a^2\ldots a^{m-\epsilon})=(-1)^{-M_L+q+l(\epsilon+1)}({}_{,}\beta^1\,{}_{,}\beta^2\ldots{}_{,}\beta^\epsilon|\alpha^1\alpha^2\ldots\alpha^\epsilon). \quad (1)$$

The states on the right side of this equation have just the quantum numbers which are missing on the left. Our correlation requires that they have the negatives of the m_s, m_l, and m values missing on the left. But this is easily accomplished. Let the coefficient on the right of (1) be

$$(m_s^1m_l^1, m_s^2m_l^2, \ldots, m_s^\epsilon m_l^\epsilon|l+\tfrac{1}{2}:m^1, m^2, \ldots, m^p; l-\tfrac{1}{2}:m'^1, m'^2, \ldots, m'^q), \quad (2)$$

where $p+q=\epsilon$. The coefficient we would like to compare with this is

$$(-m_s^\epsilon-m_l^\epsilon, \ldots, -m_s^1-m_l^1|l+\tfrac{1}{2}:-m^p, \ldots, -m^1; l-\tfrac{1}{2}:-m'^q, \ldots, -m'^1). \quad (3)$$

If (2) is in standard order, (3) is seen to be also. Now from a comparison of the two determinants of type $4^{12}5$ which give the values of these two coefficients, and by using the relations (cf. $4^{5}8$)

$$(m_s\,m_l|l+\tfrac{1}{2}\,m)=(-m_s-m_l|l+\tfrac{1}{2}-m),$$
$$(m_s\,m_l|l-\tfrac{1}{2}\,m)=-(-m_s-m_l|l-\tfrac{1}{2}-m),$$

it is seen that the element (3) is $(-1)^{\epsilon q}$ times the element (2).

Hence, returning to (1), it is seen that the ratio of the transformation coefficient for the almost closed shell to that of the ϵ-electron configuration whose electrons have the negatives of the m_s, m_l, m values of those missing from the closed shell is

$$(-1)^{M_L+(\epsilon+1)(q+l)}. \quad (4)$$

When we make the correlations of § 1^{12} which include a phase factor $(-1)^{M_L}$ for the m_sm_l scheme and a factor $(-1)^{(\epsilon+1)(q+l)}$ for the jm scheme, we may say that the m_sm_l-jm transformation is the same for the two configurations.

6. The transformation $jjJM \rightleftharpoons SLJM$.

We have found a correlation between the zero-order states of $\mathscr{L}$ and $\mathscr{R}$ such that the transformations

$$jjJM \rightleftharpoons nljm, \quad nljm \rightleftharpoons nlm_sm_l, \quad \text{and} \quad nlm_sm_l \rightleftharpoons SLJM$$

are the same for the two configurations. This is seen to imply that the transformation $jjJM \rightleftharpoons SLJM$ is the same. More explicitly, the situation is as follows: A given state of the configuration $\mathscr{L}$ characterized by the quantum numbers $SLJM$ (this state is not uniquely determined for two reasons—first, there is an arbitrary phase; second, the set $SLJM$ of quantum numbers is not in general complete for more than two electrons so there may be two independent states characterized by the same $SLJM$) is a certain linear combination of the $m_s m_l$ states; the same linear combination of the correlated $m_s m_l$ states of $\mathscr{R}$ is a state of $\mathscr{R}$ characterized by this same $SLJM$. A similar statement holds for a $jjJM$ state except that the *missing* j's of $\mathscr{R}$ are those of $\mathscr{L}$. The transformation between the LS- and jj-coupling states of $\mathscr{R}$ obtained in this way is the same as the corresponding transformation for $\mathscr{L}$.

Now to obtain this SL-jj transformation by the combination of the other three is in general a very tedious process involving the multiplication of three matrices of high order, for no two of the three transformations are diagonal in common with respect to more than M. On the other hand the resulting transformation is diagonal with respect to both J and M and splits in general into steps of a relatively low order. For example, no two-electron configuration or its equivalent can have more than four states of the same J and M, while pp or p^5p has altogether 10 states of $M=0$, dd or d^9d has 19.

It seems, then, to be desirable to find a direct method which will enable this transformation to be calculated with greater ease. The following accomplishes this result for two-electron configurations, this being in general the only case in which the quantum numbers $SLJM$ or $jjJM$ define a unique state (to within a phase).

The procedure we shall follow is that of calculating the matrices of $\boldsymbol{L}^2$ and $\boldsymbol{S}^2$ in the jj-coupling scheme and then diagonalizing them to obtain the transformation to LS coupling. Since these matrices will split into steps of at most fourth order, and since their eigenvalues are known, this diagonalization is a very simple procedure.

The matrices of $\boldsymbol{L}^2$ and $\boldsymbol{S}^2$ in a non-antisymmetric scheme in which the first electron has the quantum numbers n, l, j, and the second electron the quantum numbers n', l', j', with resultant J, M, are given by formulas 12^32 and 10^32. We shall define our phases by correlating, in the process of vector addition, s, l, j; s', l', j'; j, j', J; with the j_1, j_2, j respectively of § 10^3. Now $\boldsymbol{L}^2 = \boldsymbol{L}_1^2 + \boldsymbol{L}_2^2 + 2\boldsymbol{L}_1 \cdot \boldsymbol{L}_2$. The first two terms have known diagonal matrices. The matrix of the third term is diagonal with respect to nl, $n'l'$, JM. If we omit these quantum numbers, writing simply the values of j and j', the elements of this matrix are given with the abbreviations

$$w = l + \tfrac{1}{2}, \quad w' = l' + \tfrac{1}{2},$$

by
$$\begin{aligned}
&(l\pm\tfrac{1}{2}, l'\dot{\pm}\tfrac{1}{2}|2\boldsymbol{L}_1\cdot\boldsymbol{L}_2|l\pm\tfrac{1}{2}, l'\dot{\pm}\tfrac{1}{2}) \\
&\quad =\{J(J+1)-w(w\pm 1)-w'(w'\dot{\pm}1)\}\left(\frac{2w\mp 1}{2w}\right)\left(\frac{2w'\dot{\mp}1}{2w'}\right) \\
&(l\pm\tfrac{1}{2}, l'+\tfrac{1}{2}|2\boldsymbol{L}_1\cdot\boldsymbol{L}_2|l\pm\tfrac{1}{2}, l'-\tfrac{1}{2}) \\
&\quad =\sqrt{(J\mp w+w')(J\pm w-w'+1)(J\pm w+w'+1)(w'\pm w-J)}\left(\frac{2w\mp 1}{2w}\right)\left(-\frac{1}{2w'}\right) \\
&(l+\tfrac{1}{2}, l'\pm\tfrac{1}{2}|2\boldsymbol{L}_1\cdot\boldsymbol{L}_2|l-\tfrac{1}{2}, l'\pm\tfrac{1}{2}) \\
&\quad =-\sqrt{(J+w\mp w')(J-w\pm w'+1)(J+w\pm w'+1)(w\pm w'-J)}\left(-\frac{1}{2w}\right)\left(\frac{2w'\mp 1}{2w'}\right) \\
&(l+\tfrac{1}{2}, l'+\tfrac{1}{2}|2\boldsymbol{L}_1\cdot\boldsymbol{L}_2|l-\tfrac{1}{2}, l'-\tfrac{1}{2}) \\
&\quad =-\sqrt{(J+w+w'+1)(J+w+w')(w+w'-J)(w+w'-J-1)}\left(-\frac{1}{2w}\right)\left(-\frac{1}{2w'}\right) \\
&(l-\tfrac{1}{2}, l'+\tfrac{1}{2}|2\boldsymbol{L}_1\cdot\boldsymbol{L}_2|l+\tfrac{1}{2}, l'-\tfrac{1}{2}) \\
&\quad =\sqrt{(J-w+w')(J-w+w'+1)(J+w-w')(J+w-w'+1)}\left(-\frac{1}{2w}\right)\left(-\frac{1}{2w'}\right). \qquad (1)
\end{aligned}$$

(In the first formula either the upper or lower undotted sign is to be taken with either the upper or lower dotted sign.)

The matrices of $2\boldsymbol{S}_1\cdot\boldsymbol{S}_2$—and of $2\boldsymbol{L}_1\cdot\boldsymbol{S}_2$ and $2\boldsymbol{S}_1\cdot\boldsymbol{L}_2$—are given by formulas closely related to the above: In (1) we may replace $\boldsymbol{L}_1$ by $\boldsymbol{S}_1$ if we replace $(2w\mp 1)/2w$ by $\pm 1/2w$ and $-1/2w$ by $+1/2w$; similarly we may replace $\boldsymbol{L}_2$ by $\boldsymbol{S}_2$ if we replace $(2w'\mp 1)/2w'$ by $\pm 1/2w'$ and $-1/2w'$ by $+1/2w'$.

In terms of these non-antisymmetric matrix components, the components between antisymmetric states may be obtained by considerations such as those of § 3^{10}. Let $\alpha=nl$, $\beta=n'l'$ and let all differences in quantum numbers be indicated by primes, subscripts merely denoting the electrons. Then unless $\alpha=\beta$ and $j=j'$, the antisymmetric

$$\Psi(\alpha j\,\beta j'\,JM)=2^{-\frac{1}{2}}[\psi(\alpha_1 j_1\,\beta_2 j_2'\,JM)-\psi(\alpha_2 j_2\,\beta_1 j_1'\,JM)]. \qquad (2)$$

Now if v is a symmetric observable, diagonal with respect to n_1l_1 and n_2l_2, it is easily seen that for *non-equivalent electrons*

$$(\alpha j\,\beta j'\,JM|v|\alpha j''\,\beta j'''\,JM)=(\alpha_1 j_1\,\beta_2 j_2'\,JM|v|\alpha_1 j_1''\,\beta_2 j_2'''\,JM). \quad (\alpha\not\equiv\beta) \qquad (3)$$

Consideration of the matrix element

$$(\alpha j\,\alpha j'\,JM|v|\alpha j''\,\alpha j'''\,JM)\equiv(jj'|v|j''j''') \qquad (4)$$

for *equivalent electrons* must be divided into three cases. First, if $j\neq j'$, $j''\neq j'''$, (2) applies to both states, and

$$\begin{aligned}
(jj'|v|j''j''') &= \tfrac{1}{2}[(j_1j_2'|v|j_1''j_2''')+(j_2j_1'|v|j_2''j_1''')-(j_1j_2'|v|j_2''j_1''')-(j_2j_1'|v|j_1''j_2''')] \\
&= (j_1j_2'|v|j_1''j_2''')-(j_1j_2'|v|j_2''j_1''') \\
&= (j_1j_2'|v|j_1''j_2''')-(-1)^{j''+j'''-J}(j_1j_2'|v|j_1'''j_2''). \quad (j\neq j', j''\neq j''') \qquad (5a)
\end{aligned}$$

Second, if $j\neq j'$, $j''=j'''$, $\psi(j_1''j_2'')$ is already either antisymmetric or symmetric, corresponding to allowed and excluded states. For an allowed state we have (cf. $3^{10}2$)

$$(jj'|v|j''j'')=\sqrt{2}\,(j_1j_2'|v|j_1''j_2''). \qquad (j\neq j') \qquad (5b)$$

Third, if $j=j'$, $j''=j'''$, and both states are allowed, we have simply

$$(jj|v|j''j'')=(j_1j_2|v|j_1''j_2''). \qquad (5c)$$

These formulas, then, enable us to calculate the matrices of $\boldsymbol{L}^2$ and $\boldsymbol{S}^2$ for antisymmetric states in jj coupling using the non-antisymmetric values (1).

The diagonalization of these matrices for all two-electron configurations up to dd gives the set of transformation matrices in Table 1^{12}. On the left of the matrices are given the values of j and j' for the jj-coupling states—the J value is shown at the top in the Russell-Saunders notation. The transformations are independent of M.

TABLE 1^{12}.

sl

$$\begin{array}{c|c} & {}^3L_{l+1} \\ \hline \frac12\; l+\frac12 & 1 \end{array}$$

$$\begin{array}{c|cc} & {}^1L_l & {}^3L_l \\ \hline \frac12\; l+\frac12 & \sqrt{l+1} & \sqrt{l} \\ \frac12\; l-\frac12 & -\sqrt{l} & \sqrt{l+1} \end{array}\;\frac{1}{\sqrt{2l+1}}$$

$$\begin{array}{c|c} & {}^3L_{l-1} \\ \hline \frac12\; l-\frac12 & 1 \end{array}$$

d²

$$\begin{array}{c|cc} & {}^1G_4 & {}^3F_4 \\ \hline \frac52\frac52 & 1 & 2 \\ \frac52\frac32 & -2 & 1 \end{array}\;\frac{1}{\sqrt5}$$

$$\begin{array}{c|c} & {}^3F_3 \\ \hline \frac52\frac32 & 1 \end{array}$$

$$\begin{array}{c|ccc} & {}^3P_2 & {}^1D_2 & {}^3F_2 \\ \hline \frac52\frac52 & 2\sqrt{14} & 2\sqrt{15} & -3 \\ \frac52\frac32 & 3\sqrt7 & -\sqrt{30} & 4\sqrt2 \\ \frac32\frac32 & -\sqrt6 & \sqrt{35} & 2\sqrt{21} \end{array}\;\frac{1}{5\sqrt5}$$

$$\begin{array}{c|c} & {}^3P_1 \\ \hline \frac52\frac32 & 1 \end{array}$$

$$\begin{array}{c|cc} & {}^3P_0 & {}^1S_0 \\ \hline \frac52\frac52 & -\sqrt2 & \sqrt3 \\ \frac32\frac32 & \sqrt3 & \sqrt2 \end{array}\;\frac{1}{\sqrt5}$$

p²

$$\begin{array}{c|cc} & {}^3P_2 & {}^1D_2 \\ \hline \frac32\frac32 & \sqrt2 & 1 \\ \frac32\frac12 & 1 & -\sqrt2 \end{array}\;\frac{1}{\sqrt3}$$

$$\begin{array}{c|c} & {}^3P_1 \\ \hline \frac32\frac12 & 1 \end{array}$$

$$\begin{array}{c|cc} & {}^1S_0 & {}^3P_0 \\ \hline \frac32\frac32 & \sqrt2 & 1 \\ \frac12\frac12 & 1 & -\sqrt2 \end{array}\;\frac{1}{\sqrt3}$$

pp

$$\begin{array}{c|c} & {}^3D_3 \\ \hline \frac32\frac32 & 1 \end{array}$$

$$\begin{array}{c|ccc} & {}^1D_2 & {}^3D_2 & {}^3P_2 \\ \hline \frac32\frac32 & \sqrt2 & 0 & -2 \\ \frac32\frac12 & -\sqrt2 & \sqrt3 & -1 \\ \frac12\frac32 & \sqrt2 & \sqrt3 & 1 \end{array}\;\frac{1}{\sqrt6}$$

$$\begin{array}{c|cccc} & {}^3S_1 & {}^3P_1 & {}^1P_1 & {}^3D_1 \\ \hline \frac32\frac32 & 2\sqrt5 & 0 & -\sqrt{30} & -2 \\ \frac32\frac12 & 4 & 3\sqrt3 & \sqrt6 & \sqrt5 \\ \frac12\frac32 & -4 & 3\sqrt3 & -\sqrt6 & -\sqrt5 \\ \frac12\frac12 & -\sqrt2 & 0 & -2\sqrt3 & 2\sqrt{10} \end{array}\;\frac{1}{\sqrt{54}}$$

$$\begin{array}{c|cc} & {}^1S_0 & {}^3P_0 \\ \hline \frac32\frac32 & \sqrt2 & 1 \\ \frac12\frac12 & 1 & -\sqrt2 \end{array}\;\frac{1}{\sqrt3}$$

pd

$$\begin{array}{c|c} & {}^3F_4 \\ \hline \frac32\frac52 & 1 \end{array}$$

$$\begin{array}{c|ccc} & {}^1F_3 & {}^3F_3 & {}^3D_3 \\ \hline \frac32\frac52 & -2\sqrt3 & 1 & 4\sqrt2 \\ \frac32\frac32 & 3\sqrt2 & 2\sqrt6 & \sqrt3 \\ \frac12\frac52 & -\sqrt{15} & 2\sqrt5 & -\sqrt{10} \end{array}\;\frac{1}{\sqrt{45}}$$

$$\begin{array}{c|cccc} & {}^3P_2 & {}^3D_2 & {}^3F_2 & {}^1D_2 \\ \hline \frac32\frac52 & -3\sqrt{21} & \sqrt{35} & -4 & -\sqrt{210} \\ \frac32\frac32 & -6 & 4\sqrt{15} & 2\sqrt{21} & 3\sqrt{10} \\ \frac12\frac52 & 6\sqrt6 & 4\sqrt{10} & -\sqrt{14} & -2\sqrt{15} \\ \frac12\frac32 & 3 & -\sqrt{15} & 4\sqrt{21} & -3\sqrt{10} \end{array}\;\frac{1}{15\sqrt2}$$

$$\begin{array}{c|ccc} & {}^3D_1 & {}^3P_1 & {}^1P_1 \\ \hline \frac32\frac52 & -\sqrt3 & 3 & 3\sqrt2 \\ \frac32\frac32 & 2\sqrt3 & 4 & -\sqrt2 \\ \frac12\frac32 & \sqrt{15} & -\sqrt5 & \sqrt{10} \end{array}\;\frac{1}{\sqrt{30}}$$

$$\begin{array}{c|c} & {}^3P_0 \\ \hline \frac32\frac32 & 1 \end{array}$$

dd

$$\begin{array}{c|c} & {}^3G_5 \\ \hline \frac52\frac52 & 1 \end{array}$$

$$\begin{array}{c|ccc} & {}^3G_4 & {}^3F_4 & {}^1G_4 \\ \hline \frac52\frac52 & 0 & 2\sqrt2 & -\sqrt2 \\ \frac52\frac32 & \sqrt5 & 1 & 2 \\ \frac32\frac52 & \sqrt5 & -1 & -2 \end{array}\;\frac{1}{\sqrt{10}}$$

$$\begin{array}{c|cccc} & {}^3F_3 & {}^1F_3 & {}^3G_3 & {}^3D_3 \\ \hline \frac52\frac52 & 0 & -3\sqrt{14} & -2\sqrt2 & 6\sqrt6 \\ \frac52\frac32 & 5\sqrt7 & 2\sqrt{21} & 3\sqrt3 & 8 \\ \frac32\frac52 & 5\sqrt7 & -2\sqrt{21} & -3\sqrt3 & -8 \\ \frac32\frac32 & 0 & -2\sqrt{14} & 12\sqrt2 & -\sqrt6 \end{array}\;\frac{1}{5\sqrt{14}}$$

$$\begin{array}{c|cccc} & {}^3D_2 & {}^1D_2 & {}^3F_2 & {}^3P_2 \\ \hline \frac52\frac52 & 0 & 2\sqrt{30} & 3\sqrt2 & -4\sqrt7 \\ \frac52\frac32 & 5\sqrt5 & -\sqrt{30} & -4\sqrt2 & -3\sqrt7 \\ \frac32\frac52 & 5\sqrt5 & \sqrt{30} & 4\sqrt2 & 3\sqrt7 \\ \frac32\frac32 & 0 & \sqrt{70} & -2\sqrt{42} & 2\sqrt3 \end{array}\;\frac{1}{5\sqrt{10}}$$

$$\begin{array}{c|cccc} & {}^3P_1 & {}^1P_1 & {}^3D_1 & {}^3S_1 \\ \hline \frac52\frac52 & 0 & 2\sqrt7 & 2\sqrt2 & -\sqrt{14} \\ \frac52\frac32 & 5 & -\sqrt2 & -\sqrt7 & -4 \\ \frac32\frac52 & 5 & \sqrt2 & \sqrt7 & 4 \\ \frac32\frac32 & 0 & 3\sqrt2 & -2\sqrt7 & 2 \end{array}\;\frac{1}{\sqrt{50}}$$

$$\begin{array}{c|cc} & {}^3P_0 & {}^1S_0 \\ \hline \frac52\frac52 & -\sqrt2 & \sqrt3 \\ \frac32\frac32 & \sqrt3 & \sqrt2 \end{array}\;\frac{1}{\sqrt5}$$

The Transformation from jj Coupling to LS Coupling

The jj-coupling states here have phases in accord with those used in deriving (1). The phases in LS coupling are arbitrarily chosen. These transformations could be used to obtain the matrix of spin-orbit interaction in LS coupling, since this matrix has a simple diagonal form in jj coupling. The transformation to jj coupling can alternatively be obtained by diagonalizing the matrix (Table 1^{11}) of spin-orbit interaction in LS coupling; in this case the phases of the jj-coupling states would not be determined. This would possibly be useful for a configuration containing more than two equivalent electrons if the matrix of spin-orbit interaction is non-degenerate.

CHAPTER XIII

CONFIGURATIONS CONTAINING ALMOST CLOSED SHELLS. X-RAYS

While most of the considerations up to this point have held in general for configurations containing any number of electrons, actual calculations by the methods given have usually become very laborious for configurations containing more than a very few electrons. There is little hope of improving this situation for multi-electron configurations which actually are very complicated, such as those which occur for atoms near the middle of the periodic table. But there is a class of multi-electron configurations occurring near the right side of the table which are given unexpectedly simple properties by the Pauli principle. These are the configurations containing almost closed shells which were shown in § 1^{12} to have essentially the same allowed states as the simpler configurations obtained by replacing the 'holes' in the almost closed shell by electrons. The transformations for such configurations were shown in Chapter XII to be immediately obtainable from the transformations for the correlated simpler configurations. In this chapter we shall give an arrangement of the energy-level and intensity calculations which will enable the matrices to be either obtained directly from those for the corresponding simpler configurations, or calculated with no more labour than for those configurations.

1. The electrostatic energy in *LS* coupling.*

The electrostatic energy matrix for configuration $\mathscr{R}$ is in general quite different from that for $\mathscr{L}$ and must be separately calculated. Configuration $\mathscr{R}$ consists of a closed shell minus ϵ electrons, η other electrons, and any number of completely closed shells. For a given state A in a zero-order scheme, denote the individual sets of the closed shell $\mathscr{S}$ by $a^1, a^2, \ldots, a^\epsilon, a^{\epsilon+1}, \ldots, a^m$, where $a^{\epsilon+1}, \ldots, a^m$ are the sets occurring in A and $a^1, \ldots, a^\epsilon$ are the 'missing' sets. Denote the rest of the sets occurring in A outside of closed shells by $\alpha^1, \alpha^2, \ldots, \alpha^\eta$. It will usually be most convenient to characterize such a state by giving the ϵ missing sets and the η others. The missing sets must satisfy the Pauli principle—no two of them can be identical. By writing down all possible combinations of missing sets satisfying the exclusion principle, we obtain all possible zero-order states of the configuration.

* SHORTLEY, Phys. Rev. **40**, 185 (1932); JOHNSON, Phys. Rev. **43**, 632 (1933).

The electrostatic energy matrix has non-diagonal elements between states of the same configuration only when these differ in regard to two individual sets. Since these elements have the simple form given by $7^6 4$, involving no summation over the electrons of shell $\mathscr{S}$, they cause no particular difficulty for configuration $\mathscr{R}$. Since most of the electrostatic energies of interest are given by the diagonal-sum rule in terms of the zero-order *diagonal* elements, we proceed directly to a discussion of the diagonal element for the state A. This diagonal element is given by $9^6 2$. The summation in $9^6 2$ may conveniently be divided into several parts.

First, there are the terms in the summation in which both of the individual sets are contained in completely closed shells; the sums of these terms are given by $9^6 12$ and $9^6 13$. Second, there are terms in which one set is in a closed shell and the other is one of the η electrons of group α. These terms are given in simplest form by $9^6 11$. Third, there are terms in which one set is in a closed shell, the other in shell $\mathscr{S}$. These terms are given, for each closed shell, by $(m-\epsilon)$ times $9^6 11$. The terms already considered are the same for all diagonal elements of the configuration. Finally, the remainder of the diagonal element consists of terms in which both sets are outside of closed shells, and has the value

$$\left.\begin{aligned} &\sum_{i>j=\epsilon+1}^{m} [J(a^i a^j) - K(a^i a^j)] && (\lambda) \\ +&\sum_{i=\epsilon+1}^{m}\sum_{j=1}^{\eta} [J(a^i \alpha^j) - K(a^i \alpha^j)] && (\mu) \\ +&\sum_{i>j=1}^{\eta} [J(\alpha^i \alpha^j) - K(\alpha^i \alpha^j)] && (\nu) \end{aligned}\right\}. \tag{1}$$

The inconvenience here is caused by the long summations over the range $\epsilon+1$ to m; these we wish to reduce to summations over the more convenient range 1 to ϵ. [For example, if shell $\mathscr{S}$ contains nine d electrons, the sum in (λ) runs over 36 terms, that in (μ) over 9η terms, while if the range were 1 to ϵ the first sum would contain no, the second η, terms.]

Consider first the sum (λ). We may write

$$\begin{aligned} (\lambda) &= \left(\sum_{i>j=1}^{m} - \sum_{i=\epsilon+1}^{m}\sum_{j=1}^{\epsilon} - \sum_{i>j=1}^{\epsilon}\right)[J(a^i a^j) - K(a^i a^j)] \\ &= \left(\sum_{i>j=1}^{m} - \sum_{i=1}^{m}\sum_{j=1}^{\epsilon} + \sum_{i=1}^{\epsilon}\sum_{j=1}^{\epsilon} - \sum_{i>j=1}^{\epsilon}\right)[J(a^i a^j) - K(a^i a^j)] \\ &= \sum_{i>j=1}^{m} [J(a^i a^j) - K(a^i a^j)] - \sum_{i=1}^{m}\sum_{j=1}^{\epsilon} [J(a^i a^j) - K(a^i a^j)] \\ &\quad + \sum_{i>j=1}^{\epsilon} [J(a^i a^j) - K(a^i a^j)], \end{aligned} \tag{2}$$

where we have used the relations $J(a^i a^i) \equiv K(a^i a^i)$, $J(a^i a^j) = J(a^j a^i)$, and $K(a^i a^j) = K(a^j a^i)$. Of the three sums to which we have reduced (λ), the first is a sum over the closed shell $\mathscr{S}$ and is given by 9[6]12. The second is given by ϵ times an expression of the form 9[6]11 with $nl = n'l'$. Only the third term depends on the particular state A of the configuration, and this has a simple sum over pairs of electrons chosen from ϵ.

The term (μ) of (1) becomes similarly

$$(\mu) = \sum_{i=1}^{m} \sum_{j=1}^{\eta} [J(a^i \alpha^j) - K(a^i \alpha^j)] - \sum_{i=1}^{\epsilon} \sum_{j=1}^{\eta} [J(a^i \alpha^j) - K(a^i \alpha^j)], \tag{3}$$

of which the first term is given by 9[6]11 and depends only on the configuration.

Since the term (ν) is already in its simplest form, we find for *that part of* $(A|Q|A)$ *which may vary within the configuration* the value

$$\left.\begin{array}{lr} \displaystyle\sum_{i>j=1}^{\epsilon} [J(a^i a^j) - K(a^i a^j)] & (\lambda) \\ \displaystyle -\sum_{i=1}^{\epsilon} \sum_{j=1}^{\eta} [J(a^i \alpha^j) - K(a^i \alpha^j)] & (\mu) \\ \displaystyle +\sum_{i>j=1}^{\eta} [J(\alpha^i \alpha^j) - K(\alpha^i \alpha^j)] & (\nu) \end{array}\right\}. \tag{4}$$

Hence, aside from common terms, we may calculate the diagonal elements using the quantum numbers of the *missing* electrons of shell $\mathscr{S}$ exactly as for a simple configuration except that we reverse the sign of the interaction of a missing electron with an electron in another shell (i.e. in group α). In using the diagonal-sum rule to obtain from these the LS-coupling energies we must remember that the M_L and M_S values of the state A are given by the *negatives* of the values for the missing electrons plus the values for group α.

As an illustration we shall sketch the calculation of the electrostatic energies for $np^5 n'p$. We shall use a notation of the type $(0^+ 1^-)$, in which the first entry gives the quantum numbers of the electron missing from the np shell, the second the quantum numbers of the electron $n'p$. The double entry table giving the M_L and M_S values of the zero-order states then has the form (cf. 1[7]2)

$p^5 p$		M_S = 1	M_S = 0
M_L	2	$(-1^- 1^+)$	$(-1^- 1^-)(-1^+ 1^+)$
	1	$(-1^- 0^+)(0^- 1^+)$	$(-1^- 0^-)(0^- 1^-)(-1^+ 0^+)(0^+ 1^+)$
	0	$(-1^- -1^+)$ $(0^- 0^+)(1^- 1^+)$	$(-1^- -1^-)(0^- 0^-)(1^- 1^-)$ $(-1^+ -1^+)(0^+ 0^+)(1^+ 1^+)$

. (5)

From this we obtain, using the above results, by the usual diagonal-sum procedure

	$p^5 p$
$^3D = -F_0 - F_2$	$^3D = -F_0 - F_2$
$^1D + {}^3D = -2F_0 - 2F_2 + 12G_2$	$^1D = -F_0 - F_2 + 12G_2$
$^3D + {}^3P = -2F_0 + 4F_2$	$^3P = -F_0 + 5F_2$
$^3P + {}^3D + {}^1D + {}^1P = -4F_0 + 8F_2 + 12G_2$	$^1P = -F_0 + 5F_2$
$^3S + {}^3P + {}^3D = -3F_0 - 6F_2$	$^3S = -F_0 - 10F_2$
$^3S + {}^3P + {}^3D + {}^1S + {}^1P + {}^1D = -6F_0 - 12F_2 + 12G_2 + 6G_0$	$^1S = -F_0 - 10F_2 + 6G_0.$

This is one of the important almost-closed-shell configurations which is correlated to a (non-equivalent) two-electron configuration. With regard to such two-electron-like configurations we may make several general statements.

In the first place, *the configuration $l^{m-1}s$ has the same singlet-triplet separation as ls*; this means, e.g., that the separations for p^5s and d^9s are given by the same integral as for ps and ds. This may be inferred, if we like, from a generalization of the above scheme to include two or more almost-closed shells. The procedure is essentially the same as the above with the additional direction that one assigns the usual (not the reversed) sign to the integral connecting electrons missing from two different shells. If we consider the s shell of $l^{m-1}s$ as having one missing electron, the result follows immediately from 8^616.

Second, *the electrostatic energies for the two configurations, $l^{m-1}l'$, and $l'^{m'-1}l$, correlated to ll' are the same*. This follows from 8^616 if we note that the double entry table of the type (5) for $l^{m-1}l'$ may be obtained from that for $l'^{m'-1}l$ by replacing the state $(m_s'm_l', m_sm_l)$ by the state $(-m_s - m_l, -m_s' - m_l')$. Hence, e.g., p^5d and d^9p have the same electrostatic energies.

Third, since $J(i,j)$ depends on only the absolute values of $m_s^im_l^i, m_s^jm_l^j$, the terms in the electrostatic energies involving F_k's for $l^{m-1}l'$ will be just the negatives of those for ll'. Furthermore $K(i,j)$ vanishes unless $m_s^i = m_s^j$, hence in a table such as (5) we obtain G_k's only for states having $M_S = 0$; none of the triplets have G's in their electrostatic energy expressions. These facts reduce our calculation for such cases to a determination of the coefficients of the G_k's for the singlets. The following formulas show certain striking characteristics with reference to these G_k's. For each singlet, at most only one G_k has a non-vanishing coefficient, namely that with $k = L$. Since for even $l + l'$ coefficients of G_k's with odd k all vanish, and for odd $l + l'$ coefficients of G_k's with even k vanish, every other singlet has the same energy as the corresponding triplet. This peculiarity was not noticed experimentally because for such configurations the spin-orbit interaction is usually as large or larger than the electrostatic.

The electrostatic energies for several configurations correlated to two-non-equivalent-electron configurations follow:

$p^5 d$ or $d^9 p$

$^3F = -F_0 - 2F_2$
$^1F = -F_0 - 2F_2 + 90G_3$
$^3D = -F_0 + 7F_2$
$^1D = -F_0 + 7F_2$
$^3P = -F_0 - 7F_2$
$^1P = -F_0 - 7F_2 + 20G_1$

$p^5 f$ or $f^{13} p$

$^3G = -F_0 - 5F_2$
$^1G = -F_0 - 5F_2 + 56G_4$
$^3F = -F_0 + 15F_2$
$^1F = -F_0 + 15F_2$
$^3D = -F_0 - 12F_2$
$^1D = -F_0 - 12F_2 + 126G_2$

$d^9 d$

$^3G = -F_0 - 4F_2 - F_4$
$^1G = -F_0 - 4F_2 - F_4 + 140G_4$
$^3F = -F_0 + 8F_2 + 9F_4$
$^1F = -F_0 + 8F_2 + 9F_4$
$^3D = -F_0 + 3F_2 - 36F_4$
$^1D = -F_0 + 3F_2 - 36F_4 + 28G_2$
$^3P = -F_0 - 7F_2 + 84F_4$
$^1P = -F_0 - 7F_2 + 84F_4$
$^3S = -F_0 - 14F_2 - 126F_4$
$^1S = -F_0 - 14F_2 - 126F_4 + 10G_0$

$d^9 f$ or $f^{13} d$

$^3H = -F_0 - 10F_2 - 3F_4$
$^1H = -F_0 - 10F_2 - 3F_4 + 420G_5$
$^3G = -F_0 + 15F_2 + 22F_4$
$^1G = -F_0 + 15F_2 + 22F_4$
$^3F = -F_0 + 11F_2 - 66F_4$
$^1F = -F_0 + 11F_2 - 66F_4 + 120G_3$
$^3D = -F_0 - 6F_2 + 99F_4$
$^1D = -F_0 - 6F_2 + 99F_4$
$^3P = -F_0 - 24F_2 - 66F_4$
$^1P = -F_0 - 24F_2 - 66F_4 + 70G_1$

For configurations of the type $l^{m-\epsilon}$ containing purely an almost closed shell we may say at once that the electrostatic energies are the same as for the correlated configuration l^ϵ. This follows from $8^6 16$ if we note that the double entry table of the type (5) for $l^{m-\epsilon}$ may be obtained from that for l^ϵ merely by reversing the signs of all m_l's and m_s's. Hence d^7, p^4, and f^7 have exactly the electrostatic energies of d^3, p^2, and f^7 respectively. We may also show that $l^{m-\epsilon} s$ has the same electrostatic energies as $l^\epsilon s$, for any ϵ.

2. The spin-orbit interaction.

We shall show here that *the matrix of spin-orbit interaction in any coupling scheme with the correlations of Chapter XII is the same for configurations $\mathscr{L}$ and $\mathscr{R}$, except for reversal of the sign of the spin-orbit coupling parameter ζ_{nl} referring to the almost closed shell $\mathscr{S}$.* Since with the correlations of Chapter XII, all transformations are the same for $\mathscr{L}$ and $\mathscr{R}$, it will suffice to show that this relation is satisfied for any one scheme. The simplest scheme to consider is the *nljm* scheme in which the spin-orbit interaction is diagonal and independent of the m's. Its value is given by $1^{10}2$, in which the coefficient of ζ_{nl} is given as a sum over the $(m-\epsilon)$ electrons of shell $\mathscr{S}$. If this is written as a sum over the whole shell minus a sum over the ϵ missing electrons our theorem is proved, since the sum over the whole shell vanishes. (This vanishing may be inferred, e.g., from the fact that $M_L = M_S = 0$ in $4^7 2$ for a whole closed shell.)

Hence if the LS-coupling states of $\mathscr{R}$ and $\mathscr{L}$ are taken as the same linear combinations of the correlated zero-order states, the spin-orbit matrices in LS coupling are the same except for the sign of the spin-orbit parameter referring to shell $\mathscr{S}$.

3. Pure almost-closed-shell configurations.

We have shown in § 1^{13} that the electrostatic energies of the two corresponding configurations l^ϵ and $l^{m-\epsilon}$, which contain electrons in only one shell outside of closed shells, are given by the same formulas. This was first shown by Heisenberg,* who gave two examples of related configurations of this type which show an almost constant ratio between the term separations. These are repeated here, with the positions of the centres of gravity of the terms given in cm^{-1} above the lowest term of the configuration:

	Ti I $3d^2$	Ni I $3d^8$	Ni/Ti
^3F	0	0	—
^3P	8324	14724	1·769
^1S	14944	†	—
^1D	7032	12549	1·785
^1G	11895	21130	1·776

	O III $2p^2$	O I $2p^4$	O III/O I
^3P	0	0	—
^1D	20066	15791	1·271
^1S	42979	33716	1·275

The constancy of the ratios is striking in view of the fact that the individual configurations do not fit the first-order theory particularly well (see § 5^7). This constancy would imply that the electrostatic parameters all increased in constant ratio. Such, for example, is the case for the F's if the two fields are both hydrogen-like with different effective nuclear charges.

The matrix of spin-orbit interaction for $l^{m-\epsilon}$ is just the negative of that for the correlated states of l^ϵ. This means that in Russell-Saunders coupling, all the multiplets which are normal in l^ϵ will be inverted in $l^{m-\epsilon}$, while those inverted in l^ϵ will be normal in $l^{m-\epsilon}$. This is a fact well known to empirical spectroscopists (cf. Fig. 6^7). For example, the lowest configuration (p^5) of the rare-gas ions Ne II, A II, and Kr II consists of an inverted doublet, with the level of lowest j value highest. Since there is only one spin-orbit parameter if only one shell is involved, the ratios of the term splittings are the same for the configurations l^ϵ and $l^{m-\epsilon}$, except for this inversion.

From this argument we see that to the first order the terms of the configuration $l^{m/2}$ have no splitting. This is illustrated by Fig. 5^{11} for the case of p^3. We cannot infer from this argument that the whole matrix of spin-orbit interaction is zero because for example in LS coupling, while the

* Heisenberg, Ann. der Phys. **10**, 888 (1931).
† Not observed.

correlated states have the same $SLJM$ or SLM_SM_L values, they have different phases since they are the same linear combinations of different correlated zero-order states.

The secular equations for the configuration p^4 are immediately obtained from those given in § 3[11] for p^2 by reversing the sign of ζ. The energies are plotted in Fig. 1[13] as a function of $\chi = \frac{1}{5}\zeta/F_2$ in the same manner as for p^2 in Fig. 4[11]. The only available experimental data are for Te I $5p^4$, which is plotted at $\chi = 0{\cdot}685$ ($F_2 = 1227$, $\zeta = 4203$), determined by making ${}^3P_1$, the mean of ${}^1D_2'$ and ${}^3P_2'$, and the mean of ${}^1S_0'$ and ${}^3P_0'$ fit exactly.* The separations between the levels of $J = 0$ and between those of $J = 2$ are absolutely predicted by the theory. This configuration shows slightly the partial, instead of complete, inversion of the triplet which occurs for $\chi > \frac{2}{3}$.

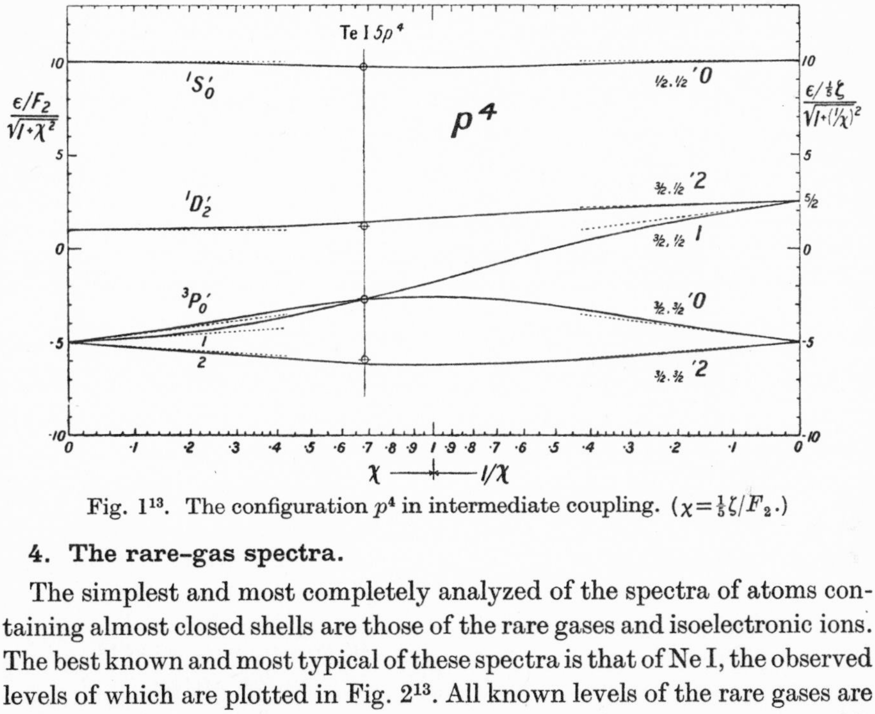

Fig. 1[13]. The configuration p^4 in intermediate coupling. ($\chi = \frac{1}{5}\zeta/F_2$.)

4. The rare-gas spectra.

The simplest and most completely analyzed of the spectra of atoms containing almost closed shells are those of the rare gases and isoelectronic ions. The best known and most typical of these spectra is that of Ne I, the observed levels of which are plotted in Fig. 2[13]. All known levels of the rare gases are included in the systems $n'p^5\,nl$, where $n' = 2, 3, \ldots$ for neon, argon, ...; this makes these spectra greatly resemble one-electron spectra so far as the distribution of configurations is concerned.

The configurations of each series rapidly divide, with increasing n, into two groups of levels. That the upper group approaches as limit the ${}^2P_{\frac{1}{2}}$ level of the ion and the lower the ${}^2P_{\frac{3}{2}}$ level shows that the interaction which causes

* This comparison was originally made by GOUDSMIT, Phys. Rev. 35, 1325 (1930). We have corrected a slight numerical error in his comparison.

this splitting is the spin-orbit interaction of the almost closed shell. This is a large and practically constant interaction for all configurations, while the spin-orbit interaction of the valence electron and the electrostatic interaction between this electron and the p^5 core rapidly diminish with increasing n. This gives the higher configuration members to the resolution of Fig. 2^{13} the appearance of doublets.

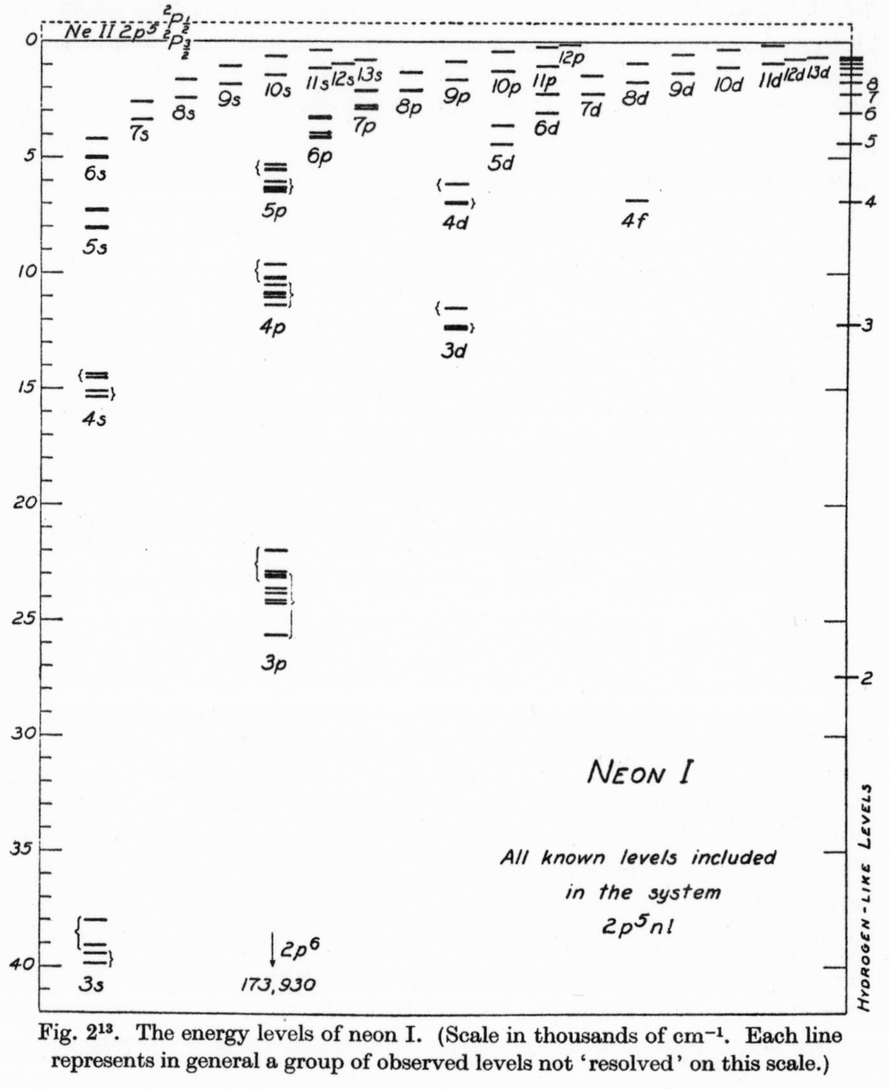

Fig. 2^{13}. The energy levels of neon I. (Scale in thousands of cm^{-1}. Each line represents in general a group of observed levels not 'resolved' on this scale.)

The quantum number which distinguishes levels belonging rigorously to one parent level from those belonging to the other is the J value of the p^5 group, which is also the j value of the missing electron. Adding an s electron (of $j=\frac{1}{2}$) to the $J=\frac{1}{2}$ level of p^5 gives two states, of $J=0$, 1; adding it to the $J=\frac{3}{2}$ level gives two states, $J=1$, 2. This parentage may be shown schematically as follows:

p^5		$p^5 s_{\frac{1}{2}}$
$\frac{1}{2}$	$\rightarrow$	1, 0
$\frac{3}{2}$	$\rightarrow$	2, 1.

The numbers here indicate J values. Similarly the jj-coupling parentage for p^5p is given by

p^5		$p^5 p_{\frac{1}{2}}$	$p^5 p_{\frac{3}{2}}$
$\frac{1}{2}$	$\rightarrow$	1, 0	2, 1
$\frac{3}{2}$	$\rightarrow$	2, 1	3, 2, 1, 0,

while for p^5l with $l>1$, we have

p^5		$p^5 l_{l-\frac{1}{2}}$	$p^5 l_{l+\frac{1}{2}}$
$\frac{1}{2}$	$\rightarrow$	$l, l-1$	$l+1, l$
$\frac{3}{2}$	$\rightarrow$	$l+1, l, l-1, l-2$	$l+2, l+1, l, l-1.$

Hence in Fig. 2[13], the upper group of levels (those having the limit $p^5_{\frac{1}{2}}$) contains two levels for p^5s and four levels for p^5p, p^5d, ...; while the lower group contains two levels for p^5s, six for p^5p, and eight for p^5d, ... For Ne I, the configuration p^5s is completely known up to $11s$, the p^5p's are complete up to $7p$, while the p^5d's are completely analysed up to $10d$. Only two levels of the lowest p^5f have been found.

For those configurations which appear in Fig. 2[13] as doublets without structure, the characterization by parentage is almost exact, i.e. the J value of the p^5 group is a very good quantum number. Since this is the case, a glance at the transformations of Table 1[12] will show that assignment of L and S values is impossible. The states are not, however, in pure jj coupling, for the j value of the valence electron is in general not a good quantum number.* We shall consider the exact characterization of these states in the succeeding sections.

A plot for the heavier rare gases would present essentially the same appearance as to the absolute location of configurations as Fig. 2[13], but with all n values increased by 1 for argon, 2 for krypton, 3 for xenon, etc. However, the parent doublet splitting increases rapidly. It is about 800 cm^{-1} for neon, 1400 for argon, 5400 for krypton, and 10,000 for xenon. The fine structure of the parts of the doublet increases less rapidly than this, so for corresponding configurations the division into two groups becomes sharper as we go down the series of gases. For krypton and xenon even the lowest

* This is typical of the coupling near the limit of any series in any atom. The quantum numbers of the parent levels, in particular the J values of these levels, become asymptotically exact quantum numbers, because the interactions of the valence electron eventually become very small compared with the separations of the levels of the ion.

p^5p's occur in two well-separated groups, so that the parentage may be quite accurately assigned.

5. The configurations p^5s and d^9s.*

The secular equations for $l^{m-1}s$ are the same as those of § 3^{11} for ls if we reverse the sign of ζ. Hence for $G_l > 0$—and all observed cases come under this category—we have in LS coupling a singlet above an inverted triplet. The transition from LS to jj coupling for p^5s is plotted in Fig. 3^{13} as a function of $\chi = \frac{3}{4}\zeta/G_1$.

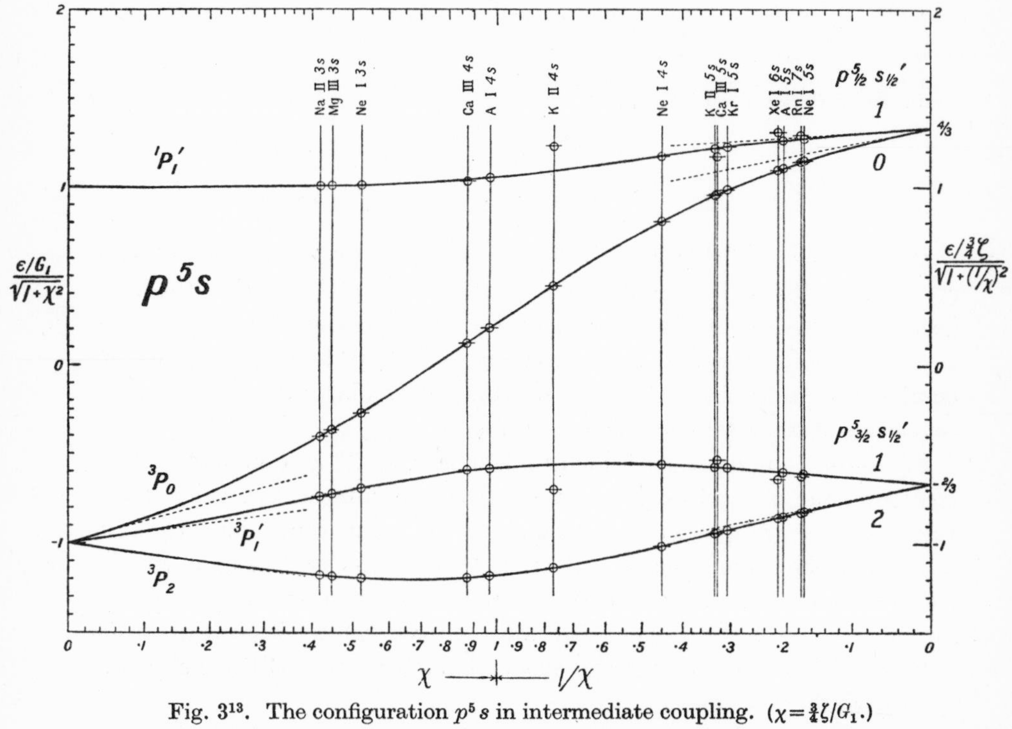

Fig. 3^{13}. The configuration p^5s in intermediate coupling. ($\chi = \frac{3}{4}\zeta/G_1$.)

We see in Fig. 2^{13} that the p^5s's of neon are all close to the jj-coupling limit, where they appear as two double levels. This is true of all the rare gases and rare-gas-like ions. For configurations very close to jj coupling, the theory predicts that the electrostatic splitting of $p^5_{\frac{3}{2}}s_{\frac{1}{2}}'$ be twice that of $p^5_{\frac{1}{2}}s_{\frac{1}{2}}'$, as seen at the right of Fig. 3^{13}. The experimental ratios for the higher series members of neon are

$2p^5\,6s$	$7s$	$8s$	$9s$	$10s$	$11s$
1·97	2·18	2·16	1·95	0	1·43

* Laporte and Inglis, Phys. Rev. **35**, 1337 (1930); Condon and Shortley, *ibid.* **35**, 1342 (1930).

The agreement is good up to $10s$, which has a zero splitting for $2p^5_{\frac{3}{2}}\,10s_{\frac{1}{2}}$ in place of an expected splitting of about 8 cm^{-1}. This term lies very close to $2p^5_{\frac{1}{2}}\,7d$, on which we shall see in § 7^{13} that there is indication of perturbation—but a glance at Fig. 8^{13} shows a perturbation as large as 8 cm^{-1} to be quite unexpected. For argon, the agreement of the higher series members is not good; the observed ratios are

$2p^5\,6s$	$7s$	$8s$	$9s$
1·19	4·24	−1·13	1·29

All other observed $p^5 s$ configurations have larger values of $1/\chi$ and are plotted in Fig. 3^{13} at that χ for which the levels $J=0$, 2, and the mean of the two levels of $J=1$ fit the theory exactly. The agreement is in general good.

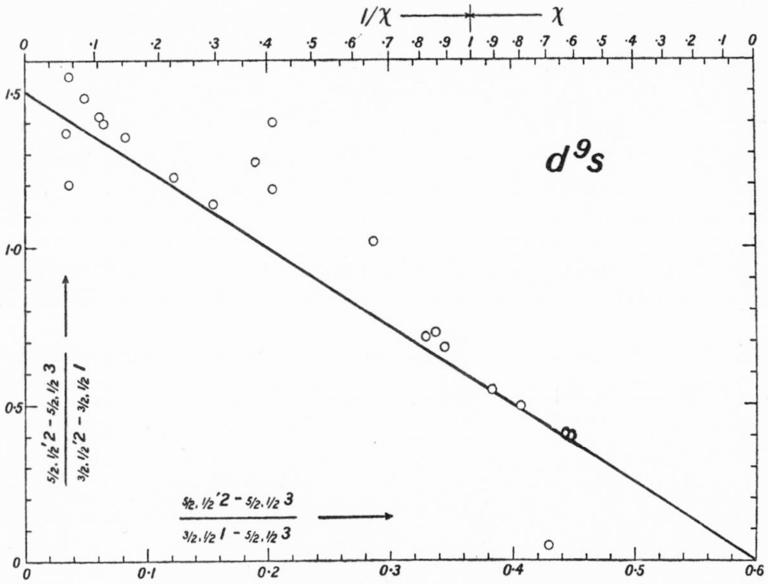

Fig. 4^{13}. Interval ratios for $d^9 s$. ($\chi = \frac{5}{4}\zeta/G_2$.)

Just as for ls (cf. § 3^{11}), we can find for $l^{m-1}s$ a pair of linearly related interval ratios to compare with the theory. If we take as ordinate $({}^3L'_l - {}^3L_{l+1})/({}^1L'_l - {}^3L_{l-1})$ and as abscissa $({}^3L'_l - {}^3L_{l+1})/({}^3L_{l-1} - {}^3L_{l+1})$, we find that

$$o = -\frac{2l+1}{l}a + \frac{l+1}{l}.$$

The ordinate here is the ratio of the splitting of $l^{m-1}_{l+\frac{1}{2}}\,s_{\frac{1}{2}}$ to the splitting of $l^{m-1}_{l-\frac{1}{2}}\,s_{\frac{1}{2}}$, which is $(l+1)/l$ for jj coupling (abscissa 0). The intervals in $d^9 s$ are compared with the theory in this way in Fig. 4^{13}. These configurations tend to lie closer to jj coupling, which is on the left in this plot, than to LS coupling. The agreement is again seen to be in general good.

Unfortunately the spectra of the latter rare earths is not known, so we have no observed instances of $f^{13}s$.

6. The configuration $p^5\,p$ in the rare gases.*

The configuration $p^5\,p$ contains 1 level with $J=3$, 3 with $J=2$, 4 with $J=1$, and 2 with $J=0$. Hence the diagonalization of the Hamiltonian for this configuration involves the solution of a linear, a quadratic, a cubic, and a quartic equation. These equations contain as parameters six radial integrals. Given the values of these parameters, the solution of these equations is rather complicated; to obtain the values of these parameters from the observed energies and then check the self-consistency of the equations, as we propose to do, is much more complicated. A simplification valid in many cases can, however, be made. Just as our consideration of one configuration at a time is an approximation in which we neglect the interaction between configurations, so we may use an approximation in which we also neglect the interaction between the groups of levels of the same configuration which have different levels of the ionic doublet as parents. This latter approximation is expected (cf. Fig. 2^{13}) to be increasingly good as we go up the series of configurations in the rare gases.

Since the quantum number which distinguishes levels belonging to one parent from those belonging to the other is the J value of the p^5 group (i.e., the j value of the missing electron), in order to split the secular equation according to parentage, it is necessary to obtain the matrix of the Hamiltonian in the jj-coupling scheme.

The matrix of spin-orbit interaction is seen from §§ 1^{10} and 2^{13} to be diagonal in jj coupling, with the elements

$$\begin{array}{llll@{\qquad}llll}
 & n'p^5\,np & & & & n'p^5\,np & & \\
a & (\tfrac{3}{2},\tfrac{3}{2}): & -\tfrac{1}{2}\zeta'+\tfrac{1}{2}\zeta & & c & (\tfrac{1}{2},\tfrac{3}{2}): & \zeta'+\tfrac{1}{2}\zeta & \\
b & (\tfrac{3}{2},\tfrac{1}{2}): & -\tfrac{1}{2}\zeta'-\zeta & & d & (\tfrac{1}{2},\tfrac{1}{2}): & \zeta'-\zeta. &
\end{array} \qquad (1)$$

In the parentheses are given the j values first of the p^5 core and then of the added p electron. The letters a, b, c, d are introduced as a convenient abbreviation for the four possible combinations of j values. States labelled by a and b belong to the lower doublet level, states labelled by c and d to the higher doublet level of the ion. These elements are independent of the J and M values of the states.

The matrix of electrostatic interaction is diagonal in LS coupling; the diagonal elements are given in § 1^{13}. With the transformation of Table 1^{12} this matrix is found in jj coupling to have the value

* SHORTLEY, Phys. Rev. **44**, 666 (1933).

$$
-F_0 +
\begin{array}{c|c|}
 & 3a \\ \hline
3a & -F_2 \\ \hline
\end{array}
$$

$$
\begin{array}{c|cc:c|}
 & 2a & 2b & 2c \\ \hline
2a & 3F_2+4G_2 & 2F_2-4G_2 & -2F_2+4G_2 \\
2b & 2F_2-4G_2 & 4G_2 & -F_2-4G_2 \\ \hdashline
2c & -2F_2+4G_2 & -F_2-4G_2 & 4G_2 \\ \hline
\end{array}
$$

$$
\begin{array}{c|cc:cc|}
 & 1a & 1b & 1c & 1d \\ \hline
1a & -F_2 & -2\sqrt{5}F_2 & 2\sqrt{5}F_2 & \sqrt{10}F_2 \\
1b & -2\sqrt{5}F_2 & 0 & 5F_2 & 0 \\ \hdashline
1c & 2\sqrt{5}F_2 & 5F_2 & 0 & 0 \\
1d & \sqrt{10}F_2 & 0 & 0 & 0 \\ \hline
\end{array}
\qquad (2)
$$

$$
\begin{array}{c|c:c|}
 & 0a & 0d \\ \hline
0a & -5F_2+4G_0 & -5\sqrt{2}F_2+2\sqrt{2}G_0 \\ \hdashline
0d & -5\sqrt{2}F_2+2\sqrt{2}G_0 & 2G_0 \\ \hline
\end{array}
$$

In the notation here used the J value is given as an arabic numeral, followed by a letter which specifies the electronic j values according to the scheme (1). The matrices are independent of the value of M.

Adding the spin-orbit interaction (1) and splitting these matrices according to the parent j values as indicated by the broken lines, we obtain the following equations for the energy levels:

Upper levels

$$
\begin{aligned}
(2c) &= -F_0 + \zeta' + \tfrac{1}{2}\zeta \qquad + 4G_2 \\
(1c) &= -F_0 + \zeta' + \tfrac{1}{2}\zeta \\
(1d) &= -F_0 + \zeta' - \zeta \\
(0d) &= -F_0 + \zeta' - \zeta \qquad + 2G_0
\end{aligned}
$$

Lower levels (3)

$$
\begin{aligned}
(3a) &= -F_0 - \tfrac{1}{2}\zeta' + \tfrac{1}{2}\zeta - F_2 \\
\left.\begin{matrix}2a' \\ 2b'\end{matrix}\right\} &= -F_0 - \tfrac{1}{2}\zeta' - \tfrac{1}{4}\zeta + \tfrac{3}{2}F_2 + 4G_2 \pm \sqrt{(\tfrac{3}{4}\zeta + \tfrac{3}{2}F_2)^2 + (2F_2 - 4G_2)^2} \\
\left.\begin{matrix}1a' \\ 1b'\end{matrix}\right\} &= -F_0 - \tfrac{1}{2}\zeta' - \tfrac{1}{4}\zeta - \tfrac{1}{2}F_2 \qquad \pm \sqrt{(\tfrac{3}{4}\zeta - \tfrac{1}{2}F_2)^2 + 20F_2^2} \\
(0a) &= -F_0 - \tfrac{1}{2}\zeta' + \tfrac{1}{2}\zeta - 5F_2 + 4G_0.
\end{aligned}
$$

The six unprimed levels of this group belong in our approximation rigorously to the quantum numbers assigned to them. The eigenstates for the primed levels are linear combinations of those for the corresponding unprimed levels. We arbitrarily denote the higher of the two levels by a' and the lower by b' since the level a would lie above the level b if the electrostatic interaction vanished.

Now the value of one of the parameters, ζ', should in all cases in which our approximation is good be obtainable from the splitting of the parent doublet, which is just $\frac{3}{2}\zeta'$. But this splitting is known from the spectra of the ions only to an accuracy of one cm^{-1}. Hence we prefer to determine this parameter, along with the rest, from the data for each configuration, and then to check it against the doublet splitting of the ion later.

The six linear expressions in (3) should determine the six parameters. From the values of the parameters we could then predict absolute positions

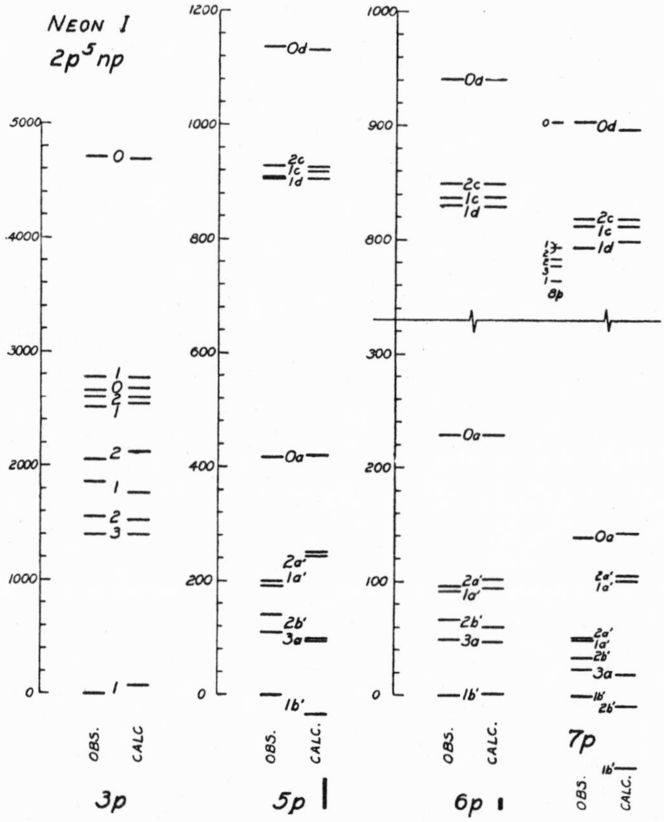

Fig. 5[13]. The configuration p^5p in neon.

of the other four levels. However, in order to make the values of the parameters less sensitive to the small perturbations which must exist both between the two parts of one configuration and between neighbouring configurations, we prefer to fit by least squares the eight quantities obtained by adding to the above six levels the means of $2a'$ and $2b'$ and of $1a'$ and $1b'$. *This leaves the separations* $2a'-2b'$ *and* $1a'-1b'$ *to be absolutely predicted.*

The results of such a calculation are shown in Fig. 5[13] for neon, Fig. 6[13] for argon, and Fig. 7[13] for krypton and xenon. In these figures the empirical value for $1b'$ has been taken as the zero for each configuration; a break in the wave-number scale indicates the separation between the two groups of levels. The scales are in cm^{-1}.

The first question one asks concerns the validity of the neglect of interaction between the two groups of levels. If the parameters we have obtained

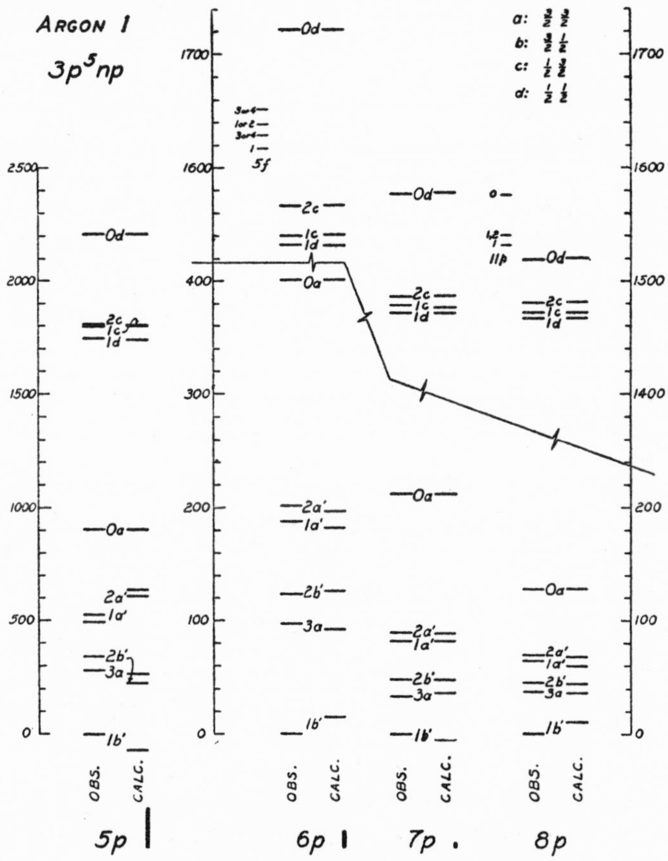

Fig. 6[13]. The configuration p^5p in argon.

are approximately correct, the magnitude of this interaction may be found by using the interaction elements of the matrices (2) according to the second-order perturbation theory. In this way the *largest* interaction was found to have approximately the value represented by the length of the short black bar drawn underneath each configuration. This gives an idea of the agreement to be expected in the comparison.

Having obtained the values of the parameters, one can readily find the coefficients in the expansion

$$\begin{aligned}\Psi(Ja') &= \Psi(Ja)(Ja|Ja') + \Psi(Jb)(Jb|Ja')\\ \Psi(Jb') &= \Psi(Ja)(Ja|Jb') + \Psi(Jb)(Jb|Jb').\end{aligned} \qquad (4)$$

We may call $|(Ja|Ja')|^2$ the purity of the levels Ja' and Jb'. It is, so far as this work is concerned, purely accidental that the electrostatic interaction

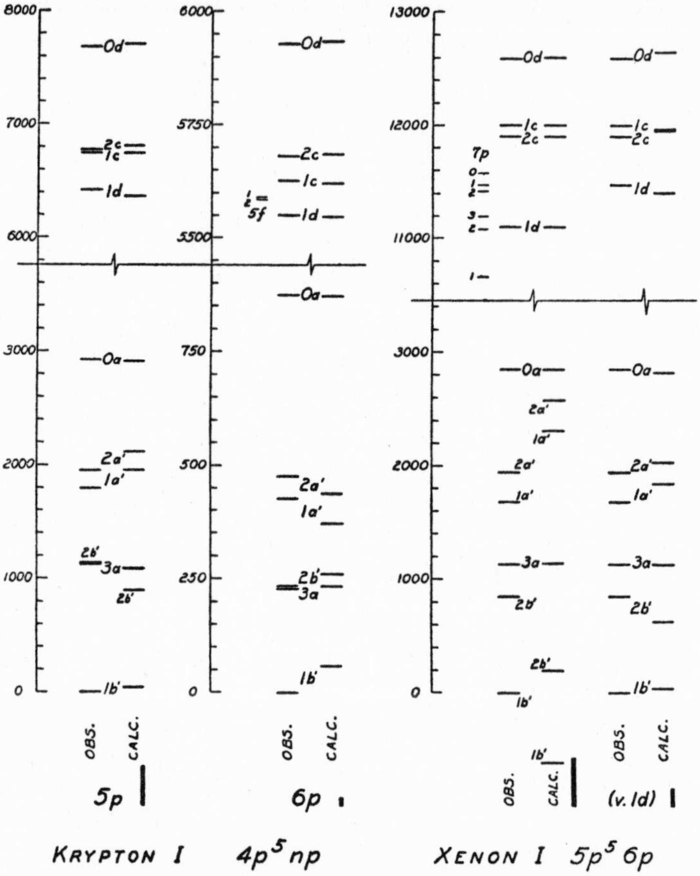

Fig. 7^{13}. The configuration p^5p in krypton and xenon.

between 1c and 1d vanishes identically and hence that these levels are to our approximation 100 per cent. pure. To the same approximation the separation between these two levels represents just the doublet splitting of the np electron in the central field due to the core.

Neon I (*cf. Fig.* 5^{13}).

The $2p^5\,3p$ is taken from the work of Inglis and Ginsburg,* who solved the complete secular equation in LS coupling using the levels of $J=0$, 2, 3, and

* INGLIS and GINSBURG, Phys. Rev. **43**, 194 (1933); corrected by SHORTLEY, *ibid.* **47**, 295 (1935). In regard to this configuration see also ROZENTAL, Zeits. für Phys. **83**, 534 (1933).

the mean of the four levels with $J=1$ to obtain the parameters, leaving otherwise the $J=1$ levels to be absolutely predicted.

The $2p^5\,5p$ is the first of the series to which our approximation is at all applicable. These levels agree to within the rather large maximum interaction between the groups.

For $2p^5\,6p$, the interaction is much smaller and the agreement correspondingly better.

TABLE 1[13]. *Neon I $2p^5\,np$.*

	3p	5p	6p	7p
F_0	−1737·0	−385·8	−314·7	−291·9
F_2	157·7	31·0	10·4	18·3
G_0	750·5	112·6	55·8	49·0
G_2	44·8	1·9	3·0	−1·7
ζ'	403·0	528·6	520·2	521·0
ζ	40·0	8·1	5·4	13·2
$(2a\|2a')$	—	0·921	0·977	
$(2b\|2a')$	—	0·390	0·208	
$(1a\|1a')$	—	0·683	0·697	
$(1b\|1a')$	—	−0·730	−0·716	
% Purity $2a'b'$	—	84·8	95·5	
% Purity $1a'b'$	—	46·6	48·6	

$2p^5\,7p$ is found to agree not at all; the reason is that the lower $8p$ group exactly overlaps the upper $7p$ and is therefore expected to interact strongly with it.

The values of the parameters (in cm^{-1}) and the coefficients in (4) are given by Table 1[13]. The significance to be attached to these values must be judged by comparison with Fig. 5[13]. The value of ζ' is to be compared with the value 521 obtained from the parent-doublet splitting of 782 cm^{-1}.

Argon I (cf. Fig. 6[13]).

The p^5p configurations of argon remain rather well separated from each other up to $3p^5\,8p$, the last of the series which is completely known. $8p\,(0a)$ falls about 200 cm^{-1} below $7p\,(1d)$ and the lower group of $11p$ falls close to the upper group of $8p$ as indicated in the figure. There is definite evidence of perturbation on $8p$. The $5f$ group overlapping $6p$ does not have a pronounced effect.

TABLE 2[13]. *Argon I $3p^5\,np$.*

	5p	6p	7p	8p
F_0	−805·8	−585·6	−521·4	−515·7
F_2	75·7	18·7	9·8	5·5
G_0	235·6	95·5	53·6	28·3
G_2	2·0	6·4	2·5	2·5
ζ'	977·2	952·3	953·5	953·3
ζ	44·1	7·1	3·6	5·0
$(2a\|2a')$	0·927	0·986	0·968	0·999
$(2b\|2a')$	0·378	0·168	0·254	0·045
$(1a\|1a')$	0·702	0·690	0·690	0·723
$(1b\|1a')$	−0·712	−0·724	−0·725	−0·692
% Purity $2a'b'$	85·9	97·2	93·7	99·8
% Purity $1a'b'$	49·3	47·6	47·6	52·3

The constants are given in Table 2^{13}. The value of ζ' is to be compared with the 954 obtained from the ionic doublet splitting of 1431 cm^{-1}.

Krypton I and Xenon I (*cf. Fig.* 7^{13}).

Even the lowest p^5p's of krypton and xenon are amenable to treatment with our approximation. For krypton the $2p^5\,5p$ and $6p$ are completely known. The $2p^5\,5p$ agrees to within the error of the approximation. The $6p$ shows definite evidence of outside perturbation which is undoubtedly mainly due to the lower group of $7p$ which lies at only 6200 on the $6p$ scale, rather than to the $5f$ levels plotted.

In xenon only the $5p^5\,6p$ is complete. This in the first plot agrees very poorly with the calculations. It is definitely perturbed by the lower $7p$ group which overlaps as indicated. Although one cannot make a rigorous assignment of these overlapping levels to configurations, one suspects on comparing this figure with the others that the highest $J=1$ of $7p$ might belong more exactly to $6p$ than the $1d$ assigned to it. If we make this rearrangement, we obtain the plot on the right, which shows a decidedly better agreement.

TABLE 3^{13}.

	Krypton I $4p^5\,np$		*Xenon I* $5p^5\,6p$	
	$5p$	$6p$	Given	Rearranged
F_0	−2984	−2028·4	−4764	−4682
F_2	213	34·8	321	198
G_0	671	194·1	748	622
G_2	20	16·3	−25	−2
ζ'	3632	3567·6	7028	7094
ζ	246	49·0	599	374
$(2a\|2a')$	0·955	0·999	—	0·954
$(2b\|2a')$	0·297	0·022	—	0·299
$(1a\|1a')$	0·736	0·750	—	0·775
$(1b\|1a')$	−0·678	−0·662	—	−0·632
% Purity $2a'b'$	91·2	99·0	—	91·0
% Purity $1a'b'$	54·2	56·2	—	60·1

The constants for these spectra have the values given in Table 3^{13}. ζ' for krypton is to be compared with the 3581 obtained from the ion. The doublet splitting for the xenon ion is not known.

7. The configuration $p^5\,d$ in the rare gases.

The diagonal elements of spin-orbit interaction for p^5d in jj coupling are

$$\begin{array}{llll} & n'p^5\,nd & & n'p^5\,nd \\ a\ (\tfrac{3}{2},\tfrac{5}{2}): & -\tfrac{1}{2}\zeta_p+\zeta_d & c\ (\tfrac{1}{2},\tfrac{5}{2}): & \zeta_p+\zeta_d \\ b\ (\tfrac{3}{2},\tfrac{3}{2}): & -\tfrac{1}{2}\zeta_p-\tfrac{3}{2}\zeta_d & d\ (\tfrac{1}{2},\tfrac{3}{2}): & \zeta_p-\tfrac{3}{2}\zeta_d, \end{array} \qquad (1)$$

where we have introduced a notation a, b, c, d similar to that used for p^5p. Here the parameter ζ_p, which is the same as the former ζ', accomplishes the splitting of the configuration into two groups, states characterized by a and b lying in the lower group, those characterized by c and d in the upper.

Transforming the electrostatic energies of § 1[13] to jj coupling gives the matrices

$-F_0+$

	$4a$
$4a$	$-2F_2$

	$3a$	$3b$	$3c$
$3a$	$\frac{22}{5}F_2+24G_3$	$\frac{4}{5}\sqrt{6}F_2-12\sqrt{6}G_3$	$-\frac{8}{5}\sqrt{5}F_2+12\sqrt{5}G_3$
$3b$	$\frac{4}{5}\sqrt{6}F_2-12\sqrt{6}G_3$	$-\frac{7}{5}F_2+36G_3$	$-\frac{1}{5}\sqrt{30}F_2-6\sqrt{30}G_3$
$3c$	$-\frac{8}{5}\sqrt{5}F_2+12\sqrt{5}G_3$	$-\frac{1}{5}\sqrt{30}F_2-6\sqrt{30}G_3$	$30G_3$

	$2a$	$2b$	$2c$	$2d$
$2a$	$\frac{4}{5}F_2$	$-\frac{2}{5}\sqrt{21}F_2$	$\frac{8}{5}\sqrt{14}F_2$	$\frac{3}{5}\sqrt{21}F_2$
$2b$	$-\frac{2}{5}\sqrt{21}F_2$	$\frac{21}{5}F_2$	$\frac{7}{5}\sqrt{6}F_2$	$-\frac{14}{5}F_2$
$2c$	$\frac{8}{5}\sqrt{14}F_2$	$\frac{7}{5}\sqrt{6}F_2$	0	0
$2d$	$\frac{3}{5}\sqrt{21}F_2$	$-\frac{14}{5}F_2$	0	0

(2)

	$1a$	$1b$	$1d$
$1a$	$-\frac{28}{5}F_2+12G_1$	$-\frac{14}{5}F_2-4G_1$	$-\frac{7}{5}\sqrt{5}F_2+4\sqrt{5}G_1$
$1b$	$-\frac{14}{5}F_2-4G_1$	$-\frac{7}{5}F_2+\frac{4}{3}G_1$	$\frac{14}{5}\sqrt{5}F_2-\frac{4}{3}\sqrt{5}G_1$
$1d$	$-\frac{7}{5}\sqrt{5}F_2+4\sqrt{5}G_1$	$\frac{14}{5}\sqrt{5}F_2-\frac{4}{3}\sqrt{5}G_1$	$\frac{20}{3}G_1$

	$0b$
$0b$	$-7F_2$

Just as in the case of p^5p we find, if we neglect the interaction between levels belonging to different parents, the following formulas to express the twelve levels in terms of six parameters:

Upper levels

$$(3c) = -F_0 + \zeta_p + \zeta_d + 30G_3$$
$$(2c) = -F_0 + \zeta_p + \zeta_d$$
$$(2d) = -F_0 + \zeta_p - \tfrac{3}{2}\zeta_d$$
$$(1d) = -F_0 + \zeta_p - \tfrac{3}{2}\zeta_d + \tfrac{20}{3}G_1$$

Lower levels

$$(4a) = -F_0 - \tfrac{1}{2}\zeta_p + \zeta_d - 2F_2 \tag{3}$$
$$\left.\begin{matrix}3a'\\3b'\end{matrix}\right\} = -F_0 - \tfrac{1}{2}\zeta_p - \tfrac{1}{4}\zeta_d + \tfrac{3}{2}F_2 + 30G_3 \pm \sqrt{(\tfrac{29}{10}F_2 - 6G_3 + \tfrac{5}{4}\zeta_d)^2 + 96(3G_3 - \tfrac{1}{5}F_2)^2}$$
$$\left.\begin{matrix}2a'\\2b'\end{matrix}\right\} = -F_0 - \tfrac{1}{2}\zeta_p - \tfrac{1}{4}\zeta_d + \tfrac{5}{2}F_2 \pm \sqrt{(\tfrac{5}{4}\zeta_d - \tfrac{17}{10}F_2)^2 + \tfrac{84}{25}F_2^2}$$
$$\left.\begin{matrix}1a'\\1b'\end{matrix}\right\} = -F_0 - \tfrac{1}{2}\zeta_p - \tfrac{1}{4}\zeta_d - \tfrac{7}{2}F_2 + \tfrac{20}{3}G_1 \pm \sqrt{(\tfrac{5}{4}\zeta_d - \tfrac{21}{10}F_2 + \tfrac{16}{3}G_1)^2 + (4G_1 + \tfrac{14}{5}F_2)^2}$$
$$(0b) = -F_0 - \tfrac{1}{2}\zeta_p - \tfrac{3}{2}\zeta_d - 7F_2.$$

We take the six levels given by linear expressions and the means of $3a'$, $3b'$; $2a'$, $2b'$; and $1a'$, $1b'$; fitting the six parameters to these nine quantities by least squares. *This leaves the $3a'$-$3b'$, $2a'$-$2b'$, $1a'$-$1b'$ separations to be absolutely predicted.*

From Fig. 2[13] we see that our approximation should be applicable to all the $p^5 d$'s of neon I. Since these are known completely from $3d$ through $10d$, we obtain the long series of configurations plotted in Fig. 8[13]. These all agree within the accuracy of the calculation except $6d$ and $7d$, which are clearly perturbed. The most of this perturbation is due to the fact that the upper group of $6d$ lies only 26 cm^{-1} below the lower of $7d$. That there is further perturbation on the *upper group* of $7d$ is shown by the abnormally large $2c$-$2d$ separation, which should be just the $7d$-electron doublet splitting. This further perturbation may be attributed to the fact that the lower part of $9d$ lies 100 cm^{-1} above the upper of $7d$ and perhaps to the $2p^5\,10s$ which lies about 20 cm^{-1} above $7d$.

The constants used in these calculations are given by Table 4[13]. The value of ζ_p is to be compared with the 521 obtained from the ionic doublet splitting. We should from the values here given predict a doublet splitting of $780{\cdot}4 \pm 0{\cdot}2$ cm^{-1} in comparison with the observed 782 cm^{-1}.

Thus the present theory accounts very satisfactorily for the observed structure of these rare-gas configurations. The perturbations which occur seem to be small except when two configurations of the same series overlap. Since there can be no spin-orbit perturbation corresponding to the interaction which splits the configuration into two groups, in estimating the mutual perturbation of two configurations we should compare the distance of the nearest levels not to the overall configuration size but to the much smaller spread of each group of levels. This requires that two configurations lie very close in order appreciably to affect each other.

TABLE 4[13]. *Neon I $2p^5\,nd$.*

	$3d$	$4d$	$5d$	$6d$	$7d$	$8d$	$9d$	$10d$
F_0	−374·23	−306·59	−285·78	−273·97	−269·61	−266·52	−264·38	−263·55
F_2	15·507	6·595	3·620	2·008	1·433	0·900	0·615	0·467
G_1	2·98	1·70	1·04	0·64	0·47	0·29	0·20	0·16
G_3	0·002	0·026	0·016	0·020	−0·004	0·004	0·001	0·002
ζ_p	530·11	521·62	520·40	519·10	519·49	520·24	520·20	520·35
ζ_d	1·178	0·002	0·141	0·476	0·064	0·043	0·000	0·044
$(3a\vert 3a')$	0·957	0·959	0·961	0·970	0·955	0·962	0·96	0·96
$(3b\vert 3a')$	0·285	0·281	0·276	0·241	0·297	0·276	0·29	0·27
$(2a\vert 2a')$	0·413	0·399	0·406	0·444	0·41	0·41	0·40	0·41
$(2b\vert 2a')$	−0·911	−0·917	−0·914	−0·898	−0·91	−0·92	−0·91	−0·91
$(1a\vert 1a')$	0·615	0·638	0·659	0·697	0·682	0·677	0·678	0·67
$(1b\vert 1a')$	−0·789	−0·770	−0·751	−0·718	−0·733	−0·736	−0·735	−0·74
% Purity $3a'b'$	91·6	92·0	92·4	94·1	91·2	92·5	92	92
% Purity $2a'b'$	17·1	15·9	16·5	19·7	17	16	16	17
% Purity $1a'b'$	37·8	40·7	43·4	48·6	46·5	45·8	46·0	45

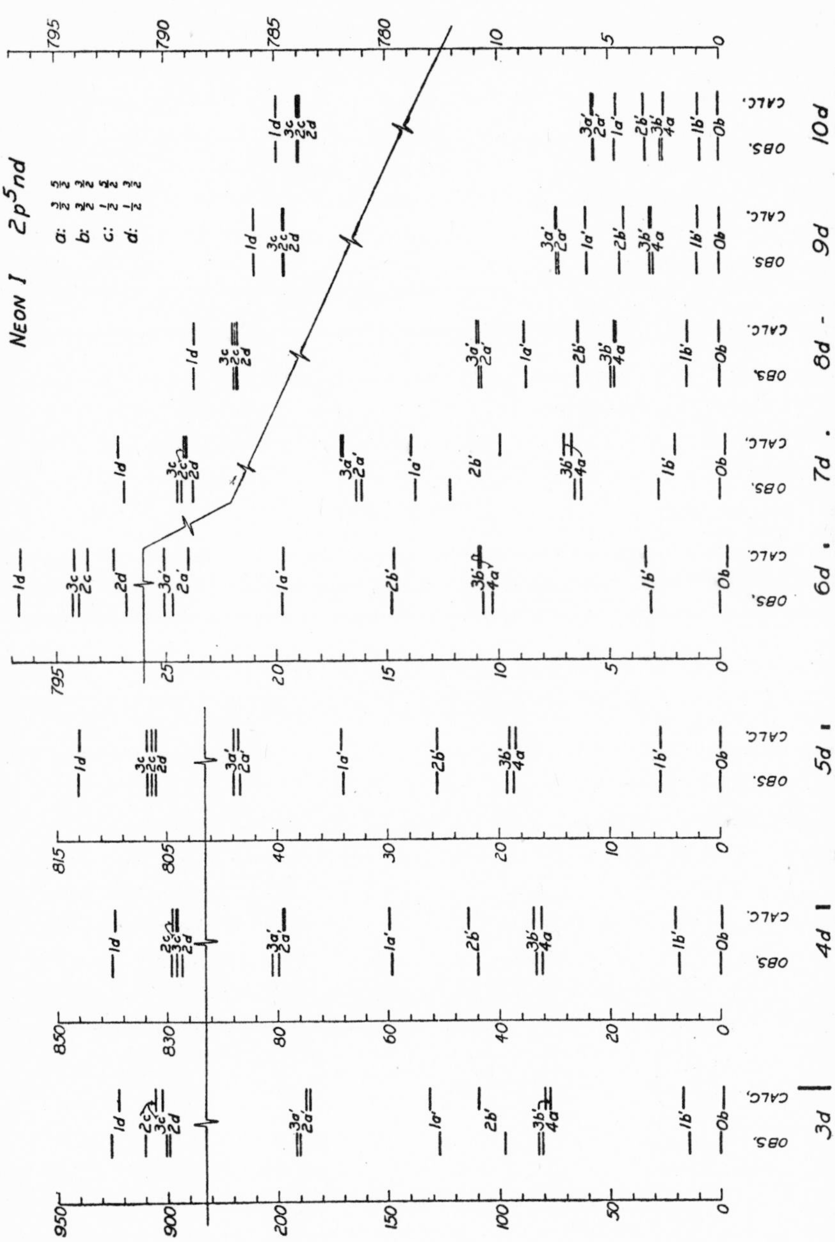

Fig. 8[13]. The configuration p^5d in neon. (The lengths of the black bars represent the maximum interaction between the two groups of levels in each configuration. Note the extreme coarseness of the cm^{-1} scale for the right half of the plot.)

8. Line strengths.

We may say immediately that the line strengths in any coupling scheme, for transitions between two configurations $\mathscr{R}$ and $\mathscr{R}'$ having the same almost closed shell, are the same as for the corresponding configurations $\mathscr{L}$ and $\mathscr{L}'$. To see this we note that in the $nlm_s m_l$ scheme the quantum numbers in the almost closed shell cannot change since one of the other electrons definitely jumps. Since all quantum numbers except those of the almost closed shell are the same for the correlated states of $\mathscr{L}$ and $\mathscr{R}$, the strengths in this scheme must be the same. The transformation to any other scheme is the same for the two cases, hence the transition arrays have the same strengths in any coupling scheme.

Thus, e.g., the array $p^5 s \to p^5 p$ has the relative strengths given by Table 2[9] or by Table 3[10] for $ps \to pp$. We may obtain the relative strengths in any of the rare-gas transition arrays by using the matrices of $\mathbf{S}^{\frac{1}{2}}$ in jj coupling and the transformations of the eigenfunctions to the jj-coupling scheme as calculated from the observed energy levels. The neon array $2p^5\,3p \to 2p^5\,3s$ is discussed briefly at the end of § 4[11].

9. X-ray spectra.

The characteristic X-ray spectrum of an element is obtained usually by making that element, or a compound containing it, the target in an electron tube so that it is struck by a beam of electrons which have energies corresponding to some tens of kilovolts. Under this excitation the atoms emit radiation of wave-lengths in the general range one to ten Ångström units. Owing to their great wave-number, the spectroscopy of these radiations requires a special experimental technique. This is fully treated for example in Siegbahn's *Spektroskopie der Röntgenstrahlen.** We confine ourselves to the theoretical problems involved.

These high-frequency radiations arise from transitions between highly excited states of an atom, corresponding to configurations in which an electron is missing from one of the inner closed shells of the normal atom. Practically all X-ray spectroscopy is done with the emitting matter in the solid state. It turns out that the interaction energy between the atoms in a solid is of the order of a Rydberg unit or less, whereas in ordinary X-ray spectra the energies are a thousand times greater. Thus to a first approximation such interactions may be neglected and the spectra interpreted as due to isolated atoms. Of course the term 'X-ray spectra' is a practical category arising from a classification by experimental technique. From the theoretical standpoint there is no sharp distinction between optical spectra and X-ray spectra, the two merge in a natural way. But experimentally they have been quite effectively separated because of the great experimental

* Second edition, Julius Springer, 1931.

difficulties of working in the soft X-ray region and the extreme ultra-violet, where the merging takes place.

Let us consider one of the heavier elements. Its normal configuration will be $1s^2\,2s^2\,2p^6\,3s^2\,3p^6\,3d^{10}\,4s^2\ldots$. Here the shells have been written down in order of the firmness with which their electrons are bound to the nucleus, that is, in the order of the energy levels of the central field best suited for the starting point of the perturbation theory. If we remove an electron from the $1s$ shell and put it on the outside (just where, on the outside, does not matter in the first approximation), the energy of the resulting configuration will be very high. This level (really a group of levels quite close together if the structure due to the outer electrons not in closed shells could be observed) is called the K level. Similarly if an electron is removed from the $2s$ shell we obtain a level not as high as the K level which is called the L_{I} level.

If an electron is removed from the $2p$ shell we are left with a p^5 configuration plus closed shells. This gives rise to an inverted ^2P term. Evidently removal of just one electron in this way from any one closed shell will give rise to a doublet term, the nature of the term being the same as in alkali spectra but with the doublets inverted. The relation of the conventional X-ray level notation to the usual spectroscopic notation is given here:

n	s	p^5		d^9		f^{13}	
	$^2S_{\frac{1}{2}}$	$^2P_{\frac{1}{2}}$	$^2P_{\frac{3}{2}}$	$^2D_{\frac{3}{2}}$	$^2D_{\frac{5}{2}}$	$^2F_{\frac{5}{2}}$	$^2F_{\frac{7}{2}}$
1	K						
2	L_{I}	L_{II}	L_{III}				
3	M_{I}	M_{II}	M_{III}	M_{IV}	M_{V}		
4	N_{I}	N_{II}	N_{III}	N_{IV}	N_{V}	N_{VI}	N_{VII}
5	O_{I}	O_{II}	O_{III}	O_{IV}	O_{V}		
6	P_{I}	P_{II}	P_{III}				

The table ends with O_{V} and P_{III} since the $5f$ and $6d$ shells do not get filled in the known elements.

Of course the full scheme of levels as here presented is only present in those atoms which contain normally all of these closed shells. If we start with uranium, where this scheme of levels is fully developed, and consider in turn elements of lower atomic number, the scheme of X-ray levels becomes simpler through the absence of the later entries in the table, for one cannot remove an electron from a closed shell that does not exist in the atom. For example in Mo (42) the scheme of the table extends to N_{III} as the $4d$ shell is only half-filled with this element (see § 1[14]).

To ignore the structure of the X-ray levels due to the outer electrons not in closed shells is formally the same as supposing that all electrons present are in closed shells except for the one shell whose openness is indicated by the above terminology.

The X-ray lines are produced by transitions between these levels. This level scheme, together with the selection rules $\Delta l = \pm 1$, $\Delta j = 0, \pm 1$, gives an adequate description of the main features of the observed spectra. In addition there are observed weaker lines some of which receive their interpretation as quadrupole transitions in this level scheme. Other lines, sometimes called 'non-diagram lines' because there is no place for them in this simple level scheme, are supposed to arise from transitions in a level scheme associated with removal of two electrons from the inner closed shells.

The K series of lines in an X-ray spectrum is produced by transitions from the K level to various lower levels. Thus the $K\alpha_1$ line is produced by the transition from the K level to the L_{III} level. In this transition an electron jumps from the $2p$ shell to the $1s$ shell, that is, in a transition from the K level to the L level an electron jumps from the L shell to the K shell. It is sometimes convenient to focus attention on the missing electron or hole in the shell and to say that in the transition from the K level to the L_{III} level, the hole jumps from the K shell to the L shell. The behaviour of a hole is just opposite to that of an electron: the normal state of the atom corresponds to the hole being in the outermost shell, that is, to all of the electrons being as far as possible in inner shells.

The shift of emphasis from electron to hole is also useful in discussing X-ray absorption spectra. We cannot have an absorption line in normal atoms corresponding to a jump of the hole from the L shell to the K shell since in normal atoms there is no hole in the L shell. In normal atoms the holes are on the outside so the X-ray absorption spectrum is produced by the transition of one of these outer holes to an inside shell. The absorption spectrum is thus directly affected by the external structure of the atoms and the way in which this is influenced by the physical and chemical state in the absorbing substance. Such details we do not treat as being outside the scope of the theory of atomic spectra.

The theory of the X-ray energy levels, neglecting structure due to outer electrons, is thus that of a one-electron spectrum. Suppose our central field approximation is based on the potential function $U(r)$ and that we define the effective nuclear charge $Z(r)$ at distance r by the equation

$$U(r) = -\frac{Z(r)\,e^2}{r}. \tag{1}$$

The force on an electron at distance r is then

$$-\frac{\partial U}{\partial r} = -\frac{[Z(r) - rZ'(r)]e^2}{r^2}. \tag{2}$$

The quantity in brackets is the effective nuclear charge for the force on the electron at distance r, which we denote by $Z_f(r)$:

$$Z_f(r) = Z(r) - rZ'(r). \tag{3}$$

We know that $Z_f(r) \to Z$ as $r \to 0$ and $Z_f(r) \to 1$ for r greater than the 'radius of the atom.' If we are dealing with a state of the atom in which the radial probability distribution has a strong maximum at $r = b$, it is natural to suppose, as a rough approximation, that the energy of the electron will be close to that of an electron in the Coulomb field

$$U_b(r) = -\frac{Z_f(b)\,e^2}{r} - e^2 Z'(b) \tag{4}$$

as this has the same value and the same slope at $r = b$ as has the effective field $U(r)$. For such a field the energy levels are given by $5^5 11$ including the relativity and spin effects.

The development of such a theory of the X-ray levels is largely due to Sommerfeld.* It fits the facts surprisingly well. The account of it we have just given is incomplete in that it does not say how b is to be chosen. Let us approach this question through an examination of the empirical data. If we subtract the rest-energy μc^2 from $5^5 11$ and measure energy with the Rydberg unit, then each electron has an energy given by that formula on writing in the proper quantum numbers, writing $Z_f(b)$ for Z and subtracting the constant term $e^2 Z'(b)$. (Note that as $Z'(b)$ is negative this adds to the energy and hence decreases the absolute value of the negative numerical value of the energy of each electron. It represents the fact that the inner electrons move in a region where their average potential energy is increased by the presence of the outer electrons; this is sometimes called 'external screening.') Consequently the atom's energy is increased by the amount of one of these energy levels when the corresponding electron is removed. Accordingly the X-ray levels should be given by this formula with the opposite sign, which puts the K level highest as it corresponds to removal of the most tightly bound electron.

Let us consider first the K levels. Their values are given for most of the elements in Table 5^{13}, as summarized in Siegbahn's book. The energies increase roughly as the square of Z, corresponding to the first known empirical regularity in X-ray spectra, as discovered by Moseley. Moseley found that the square roots of the line frequencies are approximately a linear function of Z. If the square root of the frequencies in Rydberg units be plotted against Z, the curve is strikingly linear, with slope close to 1. The departures from the Moseley law are put strongly in evidence by plotting $\sqrt{\nu} - Z$ against Z to remove the main linear trend. Such a plot is given in Fig. 9^{13}.

The formula for the K level is, in Rydberg units,

$$\nu(K) = \frac{e^2 Z'(b)}{\mathsf{R}hc} - \frac{\mu c^2}{\mathsf{R}hc}\left\{\left(1 + \frac{\alpha^2 Z_f^2}{1 - \alpha^2 Z_f^2}\right)^{-\frac{1}{2}} - 1\right\}. \tag{5}$$

* SOMMERFELD, *Atombau und Spektrallinien*, 5th edition.

Expanding the second term in powers of α and taking the square root, we find

$$\sqrt{\nu(K)} - Z = Z_f - Z + \frac{e^2 Z'(b)}{2\mathsf{R} Z_f hc} + \tfrac{1}{8}\alpha^2 Z_f^3 + \ldots. \quad (6)$$

TABLE 5^{13}. *The K energy levels of the atoms in Rydberg units.*

Element	$\nu(K)$	$\sqrt{\nu(K)}$	Element	$\nu(K)$	$\sqrt{\nu(K)}$
92 U	8477·0	92·07	47 Ag	1878·9	43·35
90 Th	8073·5	89·85	45 Rh	1709·1	41·34
83 Bi	6646·7	81·53	42 Mo	1473·1	38·38
82 Pb	6463·0	80·39	41 Cb	1401·3	37·43
81 Tl	6289·0	79·31	40 Zr	1325·8	36·41
80 Hg	6115·9	78·20	29 Cu	661·1	25·71
79 Au	5940·4	77·07	28 Ni	612·0	24·74
78 Pt	5764·0	75·92	27 Co	568·9	23·85
74 W	5113·8	71·51	26 Fe	523·8	22·89
67 Ho	4115·9	64·16	25 Mn	482·4	21·96
66 Dy	3972·5	63·03	24 Cr	441·1	21·00
64 Gd	3711·9	60·93	23 V	402·3	20·06
63 Eu	3583·4	59·86	22 Ti	365·4	19·11
62 Sa	3457·0	58·80	21 Sc	331·2	18·20
60 Nd	3214·2	56·69	20 Ca	297·5	17·25
59 Pr	3093·3	55·62	19 K	265·3	16·29
58 Ce	2972·2	54·52	17 Cl	207·8	14·42
56 Ba	2756·4	52·50	16 S	181·8	13·48
55 Cs	2649·1	51·47	15 P	158·3	12·58
53 I	2448·3	49·48	13 Al	114·7	10·71
52 Te	2345·0	48·43	12 Mg	95·8	9·79
51 Sb	2241·7	47·35			

The quantity $(Z - Z_f)$ is often called the internal screening constant. It might be expected to be equal to one or two: it is some kind of average of the amount of charge of the other electrons in the atoms that is closer to the nucleus than the $1s$ electron. It is harder to say how the term $e^2 Z'(b)/2\mathsf{R}Z_f hc$ will depend on Z. If we had n electrons on a sphere of radius ρ the potential energy of an electron inside the sphere would be ne^2/ρ and this term measures such a potential. As Z increases we have more shells so this term becomes $\Sigma ne^2/\rho$ for the shells present. On the whole it seems reasonable that this increases more rapidly with Z than the first power; hence in spite of the Z_f occurring in the denominator this is perhaps a slowly increasing function of Z. As it appears here with reversed sign it will contribute a slowly decreasing function of Z to the K level. The third term is quite definite as soon as we know the screening constant $Z - Z_f$. Using $Z - Z_f = 2$, the dotted curve in Fig. 9^{13} is obtained by subtracting off the third term from the experimental values of $\sqrt{\nu_K} - Z$. The dotted curve therefore shows the combined effect of the external and internal screening. We see that it decreases rapidly except at the large values of Z, where higher terms in the expansion become appreciable so that the curve becomes meaningless. The value of $Z - Z_f$ obtained is between one and two, so we see that the effect of the internal screening is as surmised above.

In view of the crudeness of the model we do not believe there is much to be gained by attempting to choose Z_f and $Z'(b)$ in such a way as to represent the data best. The formula evidently is in accord with the main facts and that is all that can be expected.

This same model has also been used to discuss the doublet intervals in X-ray spectra. Let us consider the L levels. L_{I} corresponds to removal of a $2s$ electron while L_{II} and L_{III} correspond to the inverted ^2P given by the $2p^5$ configuration. For the same n the hydrogenic s states penetrate more closely to the nucleus than the p states. Therefore the $2s$ electron is more tightly bound than the $2p$ so the L_{I} level will be higher than L_{II} or L_{III}. The difference will be partly due to the difference in screening and partly to the spin-relativity effects. For a reason that has since become meaningless,

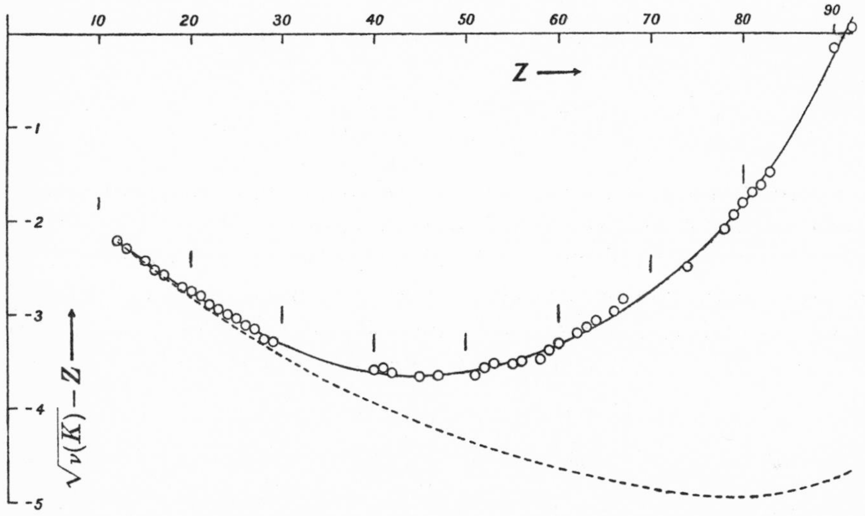

Fig. 9^{13}. Plot of $\sqrt{\nu(K)} - Z$ against Z to show departure from the Moseley law for the K levels of the elements.

the L_{I} and L_{II} levels, and in general the pairs of levels of the same n and J but different L, are commonly called 'irregular doublets.'

If we calculate the doublet interval between L_{II} and L_{III} by means of the hydrogenic formula, assuming the same values of $Z'(b)$ and $Z_f(b)$ for the two levels, we find to the first order that the interval should vary as Z_f^4. Hence we can conveniently test the formula by calculating empirical values of $Z - Z_f$ from the doublet intervals and seeing if they have reasonable magnitudes. The quantity calculated in this way is sometimes called the screening constant for doublet separation. Such comparisons have been made by

Sommerfeld and others,* and have shown that the simple screened hydrogenic model gives a good account of the main facts.

10. Line strengths in X-ray spectra.

As emphasized in the preceding section, the X-ray energy-level scheme of the atoms has the structure of a one-electron spectrum. From § 8^{13} we see that the theory of the relative line strengths in a doublet spectrum is applicable here. For example, the relative strengths of $K\alpha_2$ $(1\,^2S_{\frac{1}{2}}\rightarrow 2\,^2P_{\frac{1}{2}})$ and $K\alpha_1$ $(1\,^2S_{\frac{1}{2}}\rightarrow 2\,^2P_{\frac{3}{2}})$ will be as 1:2, according to the results of § 6^5. This is in accord with the experimental data:

Element	Intensity ratio	Observer
W (74)	2·0	(1)
Mo (42)	1·93	(2)
Zn (30)	2·00	(3)
Cu (29)	1·96	(3)
Fe (26)	2·00	(3)

(1) Duane and Stenstrom, Proc. Nat. Acad. Sci. **6**, 477 (1920).
(2) Duane and Patterson, *ibid.* **8**, 85 (1922).
(3) Siegbahn and Žáček, Ann. der Phys. **71**, 187 (1923).

Likewise in the K series the lines $K\beta_1$ and $K\beta_3$ form the same kind of doublet. Allison and Armstrong† found intensity ratios of 2·1 and 2·0 in Mo and Cu respectively for this doublet. They also measured the relative intensities of the doublets arising in the hole transitions $1s\rightarrow 3p$ $(K\beta_1, K\beta_3)$ and $1s\rightarrow 4p$ $(K\beta_2)$. The ratio of the intensity of the first of these doublets to that of the second increases with decreasing atomic number:

Element	Intensity ratio	Strength ratio
W (74)	2·3	2·6
Mo (42)	7·7	8·3
Cu (29)	41·0	42·0

The third column gives the ratio of the strengths as obtained by dividing the intensity ratio by the fourth power of the frequency ratio. The great ratio obtained in the lighter elements is due to the greater difference in the relative effect of screening on the $3p$ and $4p$ radial function for small Z. No calculations of these relative strengths have been published.

In the L series we have the possibility of studying the relative intensities in a $^2P\rightarrow{}^2D$ multiplet. For the hole transition $2p\rightarrow 3d$ the lines and measured relative intensities are

* Sommerfeld, *Atombau und Spektrallinien*, 5th edition, p. 297;
Pauling and Goudsmit, *The Structure of Line Spectra*, Chapter x;
Pauling and Sherman, Zeits. für Kristallographie, **81**, 1 (1932).

† Allison and Armstrong, Proc. Nat. Acad. Sci. **11**, 563 (1925); Phys. Rev. **26**, 701, 714 (1925).

Element	$L\alpha_1$ $^2P_{\frac{3}{2}}\to{}^2D_{\frac{5}{2}}$	$L\alpha_2$ $^2P_{\frac{3}{2}}\to{}^2D_{\frac{3}{2}}$	$L\beta_1$ $^2P_{\frac{1}{2}}\to{}^2D_{\frac{3}{2}}$
Mo (42)	1	0·130	0·33 (0·28)
Rh (45)	1	0·133	0·24 (0·20)
Pd (46)	1	0·119	0·20 (0·16)
Ag (47)	1	0·122	0·17 (0·14)

As the first two lines have the same initial state, their relative intensities should be the ratio of the relative strengths multiplied by the fourth power of the frequency ratio. The relative strengths from § 2[9] are as 9:1 or 1:0·111. The ^2D interval in the M levels is so small here that the frequency correction is negligible, so the relative intensities are the same as relative strengths and are seen to be in good agreement with the theoretical ratio. Assuming natural excitation the relative strengths of the first and third line are obtained from the relative intensities by dividing by the fourth power of the frequency ratio. The results are given in parentheses. All the experimental values are less than half of the theoretical value $\frac{5}{9}$; this is probably due to the lack of natural excitation, the higher level $^2P_{\frac{1}{2}}$ being considerably less excited than the lower $^2P_{\frac{3}{2}}$ level. Jönsson* has estimated the correction for excitation and has found values which check much better with the theoretical 9:1:5 strength ratio.

Wentzel† has given a theoretical calculation of the relative strengths of the different multiplets originating in the L levels based on use of approximate eigenfunctions. Experimental measurements bearing on this question have been made by Allison.‡

11. X-ray satellites.

In addition to the lines which are accounted for by transitions in the doublet energy-level diagram of the preceding sections there are observed a number of other lines known as 'satellites' or 'non-diagram lines.' These are close to the more prominent lines which are accounted for by the discussion just given. Such satellites were first studied by Siegbahn § when observing the $L\alpha$ lines of Sn, Ag, and Mo. Since then they have been the object of considerable study.

The doublet structure is so simple that it is evidently necessary to consider the energy diagram corresponding to more than one non-closed shell to get any more detail. Several possibilities are open. Wentzel || suggested double ionization (two holes) of the $1s$ shell as an initial state of the satellites of the K lines. Then he supposed the satellites to be emitted when one of the

* Jönsson, Zeits. für Phys. **46**, 383 (1928).
† Wentzel, Naturwiss. **14**, 621 (1926).
‡ Allison, Phys. Rev. **34**, 7, 176 (1929).
§ Siegbahn, Ark. f. Mat. Astron. Fys. **14**, No. 9 (1920).
|| Wentzel, Ann. der Phys. **66**, 437 (1921); **73**, 647 (1924).

holes jumped to an outer shell. For initial states of some of the lines he suggested triple ionization with two holes in the K shell and one in the L shell. The relations proposed by Wentzel are best considered from Fig. 10^{13}, which is a schematic level diagram.

The first column shows the ordinary level diagram. In the second column we have this diagram extended for the case of two holes. The energy of removal, K', of the second $1s$ electron will be greater than K, that for the first, but owing to screening not as much as the K of the element whose atomic number is one greater. Similar remarks hold with regard to L' as compared with L. Now L' will be greater than L'' because $2p$ is screened

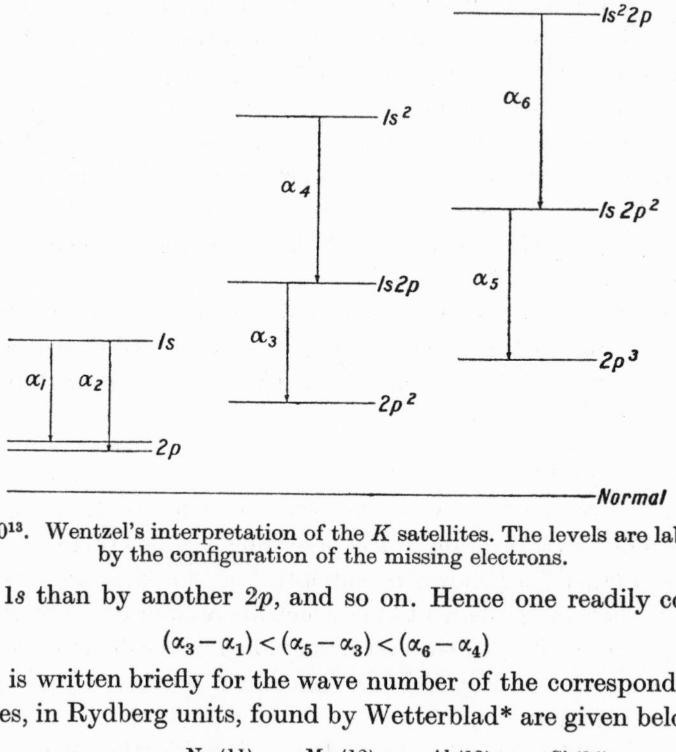

Fig. 10^{13}. Wentzel's interpretation of the K satellites. The levels are labelled by the configuration of the missing electrons.

more by $1s$ than by another $2p$, and so on. Hence one readily concludes that

$$(\alpha_3-\alpha_1)<(\alpha_5-\alpha_3)<(\alpha_6-\alpha_4)$$

where α_n is written briefly for the wave number of the corresponding line. The values, in Rydberg units, found by Wetterblad* are given below:

	Na (11)	Mg (12)	Al (13)	Si (14)
$\alpha_3-\alpha_1$	0·52	0·64	0·71	0·83
$\alpha_5-\alpha_3$	0·57	0·67	0·76	0·91
$\alpha_6-\alpha_4$	0·65	0·76	0·83	0·94

Moreover, since $1s$ is so much closer to the nucleus than $2p$, absence of a $1s$ electron is almost like a full unit increase in the nuclear charge in its effect on the others, so $(\alpha_6-\alpha_4)$ of one element should be nearly the $(\alpha_3-\alpha_1)$ of the next element. This is the case here.

* WETTERBLAD, Zeits. für Phys. **42**, 611 (1927).

The main objection to Wentzel's interpretation concerns the question of producing these multiply-ionized states. The mean life is so short that excitation by two successive singly-ionizing impacts cannot be effective; therefore the electron impacts must produce double excitation. This requires more than twice the usual K critical potential to produce the line $K\alpha_4$. Bäcklin* studied this point in Al and found that $K\alpha_4$ appeared at definitely lower voltages than expected on this hypothesis. It does appear however that the satellites require higher voltages (by about 25 per cent.) on the tube for their appearance than the related main lines.

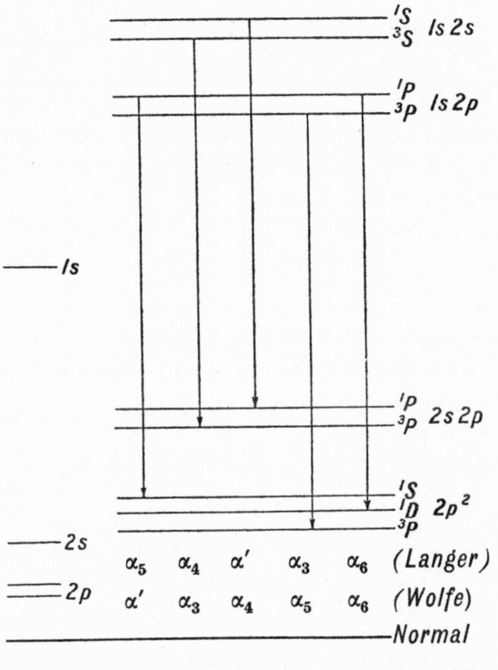

Fig. 11^{13}. The Langer-Wolfe interpretation of the K satellites. (Configuration labels are for the missing electrons.)

In Wentzel's work no account was taken of the complex structure arising from electrostatic interaction of the holes, although he mentioned it as a possibility in later papers.† The importance of this structure was emphasized by Langer.‡ On account of the existence of such structure the K satellites can be provided for without using the triple ionization process or the double ionization of the K shell. Langer's level scheme and assignment of the satellites is shown in Fig. 11^{13}. This scheme is consistent with the small

* Bäcklin, Zeits. für Phys. **27**, 30 (1924).
† Wentzel, Zeits. für Phys. **31**, 445 (1925); **34**, 730 (1925).
‡ Langer (abstract only), Phys. Rev. **37**, 457 (1931).

excess in the critical potentials of the satellites over those of the main lines. Similar views have been put forward by Druyvesteyn.* Some calculations of the terms arising from these configurations have been made by Wolfe† using a Hartree field for potassium. His assignments of the lines differs from Langer's and is shown in the lower row of the figure. The calculations show that the structural energies are of the right order of magnitude to account for the satellites but the detailed assignments are quite uncertain. It seems quite likely that the X-ray satellites are to be accounted for by double excitation levels, but there still remains much to be done before the interpretations are satisfactory.

Richtmyer‡ has developed a somewhat different view of the origin of the satellites in which two-electron jumps as well as double-hole levels are employed. As, generally speaking, two-electron jumps are weaker than single-electron transitions, it seems that such processes should not be considered unless it is found that the one-electron jump picture is inadequate.

Another possible source of more detailed structure is the open shells in the outer structure of the atom, as was noted by Coster and Druyvesteyn.§ The relation of the filling of electron shells to the periodic system of the elements is discussed in § 1^{14}. As mentioned in § 9^{13} the doublet structure of the X-ray levels depends on a model in which there is but one hole in the shells. For elements in which an outer shell is being filled the outer shell is incomplete, so there should be a complex structure due to the interaction of the outer electrons with the open inner shell. Such interactions are presumably somewhat smaller than one Rydberg unit and in case of ordinary resolution give rise to broadened lines more than to observable structure. Structure of this kind, if appreciable, we should expect to be sensitive to the state of chemical combination of the element in the source. These ideas have had practically no detailed development thus far.

* Druyvesteyn, Diss. Groningen, 1928; Zeits. für Phys. **43**, 707 (1927).

† Wolfe, Phys. Rev. **43**, 221 (1933); see also Kennard and Ramberg, Phys. Rev. **46**, 1040 (1934).

‡ Richtmyer, Phil. Mag. **6**, 64 (1928); Phys. Rev. **34**, 574 (1929); J. Franklin Inst. **208**, 325 (1929).

§ Coster and Druyvesteyn, Zeits. für Phys. **40**, 765 (1927); Coster, *ibid.* **45**, 797 (1927).

CHAPTER XIV

CENTRAL FIELDS

1. The periodic system.

We have now to consider some of the relations between the spectra of different atoms. Chemists long ago discovered the great power of Mendelejeff's periodic system of elements as a coordinator of their empirical knowledge of the chemical properties of atoms. As these properties are all, in the last analysis, related to the energy states of the atoms, we shall expect the periodic system to play an important part in coordinating the different atomic spectra. More than that, as Bohr showed, the theory of atomic structure through its picture of the arrangement of electrons in closed shells provides a clear understanding of how it is that elements with similar properties recur periodically in the list of elements. In this section we shall, then, consider the empirical data on the gross structure of the atomic spectra in its relation to the chemist's periodic table.

In Fig. 1^{14} are shown the energies of the highest and lowest observed levels of the principal configurations of the first eighteen elements, plotted up from the lowest level of the atom, the even configurations on the left and the odd on the right. This takes us past the initial hydrogen and helium pair through the second Mendelejeff period of eight. We have included, in addition to all the low-lying configurations, at least the lowest configuration of each observed series. The lowest level of the ion of each element is indicated by cross-hatching, and configurations of the ion which may have importance as parents are shown in broken lines.

We see that in normal helium the second electron has gone into a $1s$ state and that the next excited state is very high—much higher than the whole ionization energy of hydrogen. Moreover we see that $1s\,2s$ is definitely lower in energy than $1s\,2p$. This we shall connect with the fact that the $2s$ state comes nearer to the nucleus than the $2p$. At any rate, the fact of $2s$ being lower than $2p$ in helium leds us to expect that $2s$ will be the normal state of lithium. In lithium we see that $2p$ is considerably higher than $2s$; this points to $2s^2$ as the normal state of beryllium.* Now the $2s$ shell is closed. The fact that in beryllium the $2p$ energy is considerably less than the $3s$ leads us to expect boron to have $2p$ for its lowest configuration.† In the next six elements the normal configuration is in each case obtained by

* That He is, and Be is not, a rare gas, although both have closed s shells for their ground states, is connected with the extreme difference in energy necessary to excite one of the electrons in this closed shell.

† For brevity we omit mention of the closed shells already present and speak of $2p$ when of course the complete designation is $1s^2\,2s^2\,2p$.

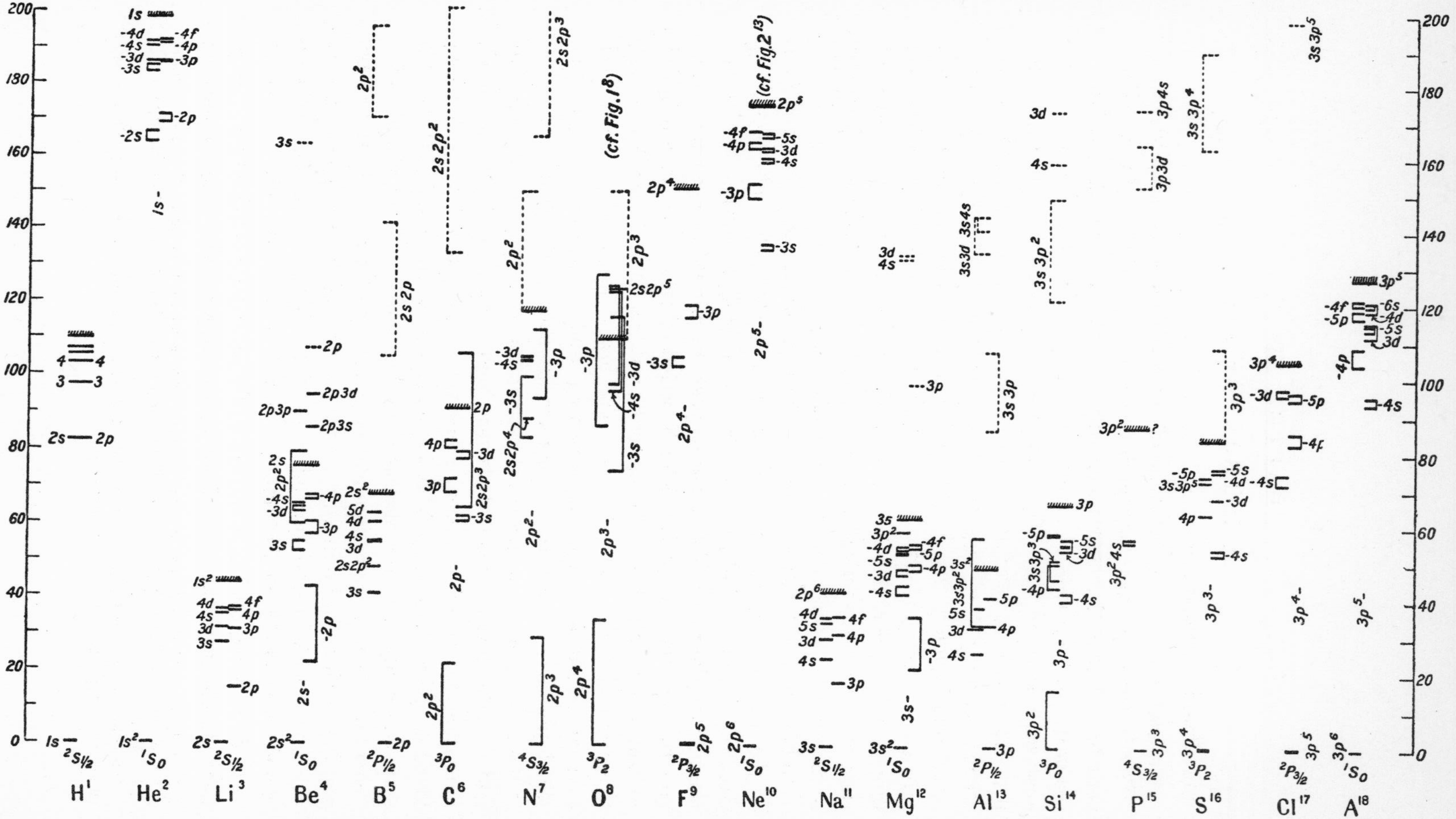

Fig. 1[14]. Main features of the energy-level scheme of the elements from hydrogen to argon. (Scale in thousands of cm^{-1}.)

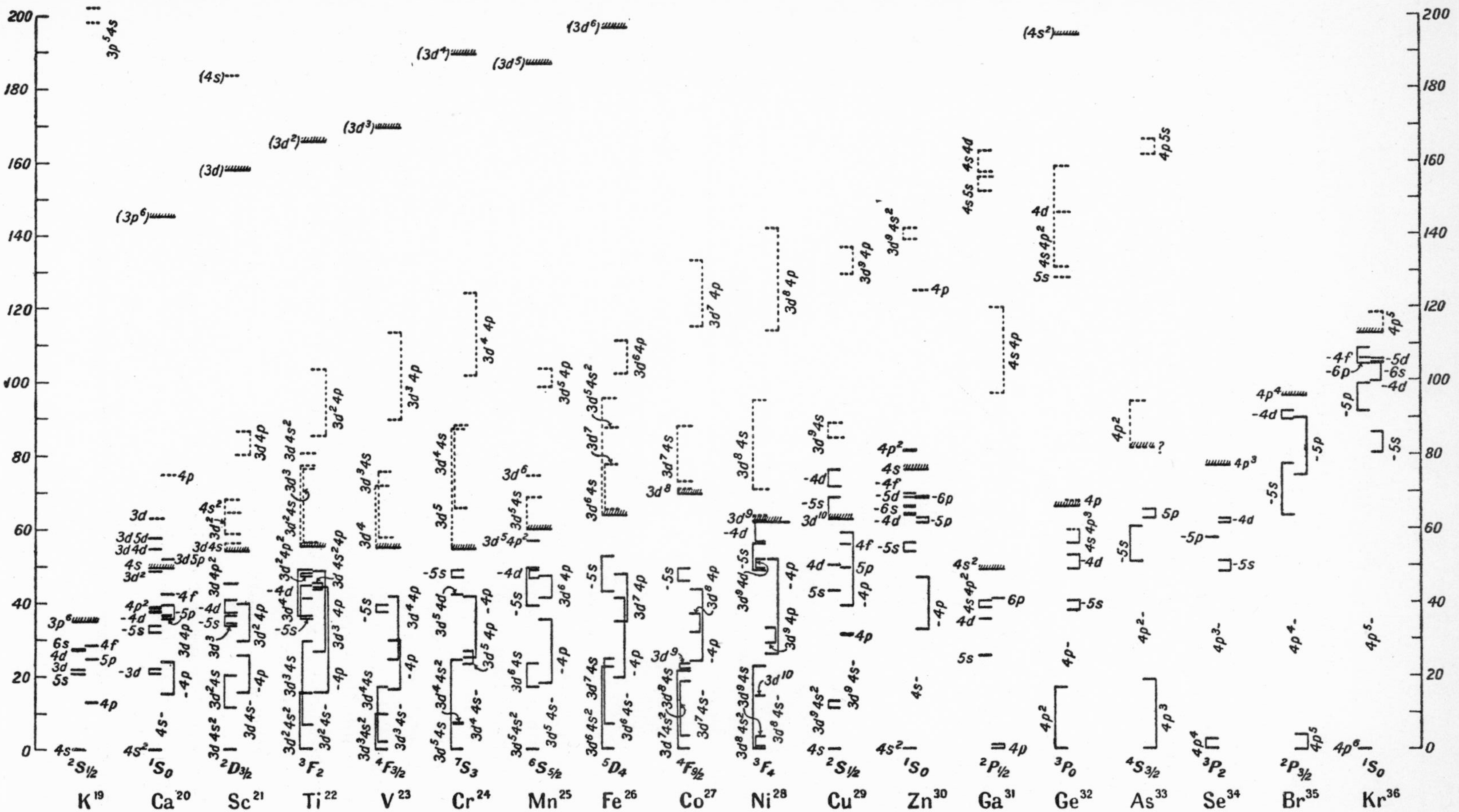

Fig. 2[14]. Main features of the energy-level scheme of the elements from potassium to krypton. (Scale in thousands of cm^{-1}.)

successive addition of a $2p$ electron to the normal configuration of the preceding atom. At neon the process is brought to a stop by the operation of the Pauli principle, as six is the maximum number of electrons in any p shell. As we go from boron to neon we observe that the interval between the normal configuration and the lowest excited configuration progressively increases. There is thus little ground for ambiguity here as to prediction of the order of the low configurations in the element $(Z+1)$ from a knowledge of the spectrum of element Z.

Thus we see that the first period of eight is connected with the sequence of normal configurations obtained by successive addition of two $2s$ electrons followed by six $2p$ electrons. Looking at the spectrum of neon we see that $2p^5\,3s$ lies below $2p^5\,3p$, with the other configurations definitely higher. This makes us expect a repetition of the previous development in the next eight elements by successive addition of two $3s$ electrons followed by six $3p$ electrons. This is in fact what happens. The close parallelism of the period from Li to Ne and that from Na to A is very striking, although our knowledge of the second period is far from complete, especially in regard to phosphorus and sulphur.

What next? We notice in argon that $4s$ is lower than $3d$, the departure of the effective field from the Coulomb form being sufficiently great to upset the order of n values as it occurs in hydrogen. Looking back we see that the inversion occurred at least as far back as neon where $p^5\,4s$ is definitely below $p^5\,3d$. Even in nitrogen the known terms of $p^2\,4s$ are below $p^2\,3d$, but not so much but that the complete configurations probably overlap. This inversion has not occurred in boron; but by the time we reach sulphur, chlorine, and argon it is quite definite. Therefore we expect the next element, potassium, to have $4s$ for its lowest configuration.

The eighteen elements from potassium to krypton are shown in Fig. 2^{14}. Here we have a situation which is quite different from that in the first two periods. There is a close competition going on between configurations of the type $d^n s^2$, $d^{n+1} s$ and d^{n+2} for the distinction of contributing the normal state of the atom. As all of these are even configurations their proximity undoubtedly gives rise to strong interactions so that the assignment of a definite term to a definite configuration has much less meaning than in the cases heretofore considered. In scandium, the configurations $d\,s^2$, $d^2\,s$ and d^3 are rather widely separated, but in titanium the configurations $d^2\,s^2$ and $d^3\,s$ overlap considerably; similarly for succeeding elements. The competition is especially close in nickel where, although the lowest level is $d^8\,s^2\,{}^3F_4$, the term intervals are such that $d^8\,s^2\,{}^3F_3$ is considerably higher than $d^9\,s\,{}^3D_3$ which lies only $205\,\mathrm{cm}^{-1}$ above the normal level. Our knowledge of these spectra, while extensive owing to their great complexity, is by no means complete, but generally it may be said that the configurations $d^n s^2$ and $d^{n+1} s$ overlap completely so that the question of which configuration gives the normal state, as far as it has significance, is as much dependent on the internal interactions in a configuration as on the central field. A striking property of these atoms is their almost constant ionization energy as contrasted with the upward trend in the ionization energy during the filling of a p shell. The lowest levels of the doubly ionized atom are indicated in parentheses for some of these elements.

Copper is somewhat akin to the alkalis in that its level system is mainly due to various states occupied by one electron outside the closed $3d$ shell. But configurations like $d^9 s^2$ and $d^9 sp$ are still in evidence relatively low, whereas the corresponding $p^5 s^2$ and $p^5 sp$ are not observed in alkali spectra. In copper and zinc we note that $4p$ is definitely below $5s$; this leads us to expect that the $4p$ shell will be filled in the next six elements. This is the case, although our knowledge of some of these spectra is quite fragmentary. We thus arrive at another inert gas, krypton, the eighteenth element after the last preceding inert gas, argon. In these eighteen elements the $4s$, $3d$ and $4p$ shells have been built in. This is known as the first long period in the periodic table.

There are many variants of the periodic system employed by chemists: a fairly common form is reproduced in Table 1^{14}. Let us note how their treatment of the first long period differs from that of the short periods and the relation of this to the facts of spectroscopy. In Groups I and II, potassium and calcium are clearly alkali and alkaline earth, but from Group III to Group VII the elements do not have such a close relationship to the elements of the first two periods: the metal, manganese, has certainly little in common with the halogens, fluorine and chlorine. That is correlated here with the fact that a d shell rather than a p shell is being filled. As the d shell is longer than a p shell, they have to introduce a Group VIII and lump three elements into it. Then they omit an entry from Group 0 and put the remaining elements into the second or 'odd' series of the long period. With these an s and a p shell are being filled, so we expect them to show more chemical similarity with elements of the first and second short periods than did the first part of the long period. This is certainly the case for the elements bromine and krypton at the end, but not for copper and zinc at the beginning of this group. In the table the rare gases are counted in at the beginning of new periods, whereas we prefer to count them in at the end: this is just one example of the kind of arbitrariness present in the empirical system, which has led some to suggest that the elements should be distributed on a spiral curve drawn on a cylinder to emphasize the arbitrariness in the choice of beginning and end of a period.

TABLE 1[14]. *The periodic table.*

Group	0	I	II	III	IV	V	VI	VII	VIII
First short period	**He^{2}**	**H^{1}** **Li^{3}**	**Be^{4}**	**B^{5}**	**C^{6}**	**N^{7}**	**O^{8}**	**F^{9}**	
Second short period	**Ne^{10}**	**Na^{11}**	**Mg^{12}**	**Al^{13}**	**Si^{14}**	**P^{15}**	**S^{16}**	**Cl^{17}**	
First long period {Even series Odd series	**A^{18}**	**K^{19}** **Cu^{29}**	**Ca^{20}** **Zn^{30}**	**Sc^{21}** **Ga^{31}**	**Ti^{22}** **Ge^{32}**	**V^{23}** **As^{33}**	**Cr^{24}** **Se^{34}**	**Mn^{25}** **Br^{35}**	**Fe^{26} Co^{27} Ni^{28}**
Second long period {Even series Odd series	**Kr^{36}**	**Rb^{37}** **Ag^{47}**	**Sr^{38}** **Cd^{48}**	**Y^{39}** **In^{49}**	**Zr^{40}** **Sn^{50}**	**Cb^{41}** **Sb^{51}**	**Mo^{42}** **Te^{52}**	**Ma^{43}** **I^{53}**	**Ru^{44} Rh^{45} Pd^{46}**
Third long period {Even series Odd series	**Xe^{54}**	**Cs^{55}**	**Ba^{56}**	**La^{57}** The rare earth elements **Ce^{58}** to **Lu^{71}**					
Fourth long period {Even series Odd series		**Au^{79}**	**Hg^{80}**	**Tl^{81}**	**Hf^{72}** **Pb^{82}**	**Ta^{73}** **Bi^{83}**	**W^{74}** **Po^{84}**	**Re^{75}** —^{85}	**Os^{76} Ir^{77} Pt^{78}**
Fifth long period {Even series	**Rn^{86}**	—^{87}	**Ra^{88}**	**Ac^{89}**	**Th^{90}**	**Pa^{91}**	**U^{92}**		

Table 2[14]. *The normal electron configurations of the elements.*

An atom contains all the closed shells occurring above and to the left of its position in the table. Since d shells are filled in competition with an s shell, the d columns are subdivided to show the number of s electrons in the normal electron configuration. It should be remembered that the configurations d^n, $d^{n-1}\,s$ and $d^{n-2}\,s^2$ usually overlap considerably (see Figs. 2^{14}, 3^{14}, 4^{14}). The rare earth elements are built in at the position of the star: their normal configurations are probably of the type $6s^2\,5d\,4f^n$ or $6s^2\,4f^{n+1}$, with $n=1$ to 14, but their spectra have not been analysed thus far.

	s	s^2	p	p^2	p^3	p^4	p^5	p^6		d	d^2	d^3	d^4	d^5	d^6	d^7	d^8	d^9	d^{10}	f to f^{14}
$1s$	H^{1}	He^{2}																		
$2s$	Li^{3}	Be^{4}																		
$2p$			B^{5}	C^{6}	N^{7}	O^{8}	F^{9}	Ne^{10}												
$3s$	Na^{11}	Mg^{12}																		
$3p$			Al^{13}	Si^{14}	P^{15}	S^{16}	Cl^{17}	A^{18}												
									$4s^0$											
$4s$, $3d$	K^{19}	Ca^{20}							$4s$					Cr^{24}					Cu^{29}	
									$4s^2$	Sc^{21}	Ti^{22}	V^{23}		Mn^{25}	Fe^{26}	Co^{27}	Ni^{28}		Zn^{30}	
$4p$			Ga^{31}	Ge^{32}	As^{33}	Se^{34}	Br^{35}	Kr^{36}												
									$5s^0$										Pd^{46}	
$5s$, $4d$	Rb^{37}	Sr^{38}							$5s$				Cb^{41}	Mo^{42}	Ma^{43}	Ru^{44}	Rh^{45}		Ag^{47}	
									$5s^2$	Y^{39}	Zr^{40}								Cd^{48}	
$5p$			In^{49}	Sn^{50}	Sb^{51}	Te^{52}	I^{53}	Xe^{54}												
									$6s^0$									Ir^{77}		
$6s$, $5d$, $4f$	Cs^{55}	Ba^{56}							$6s$									Pt^{78}	Au^{79}	*Ce^{58} Pr^{59} Nd^{60} Il^{61} Sa^{62} Eu^{63} Gd^{64} Tb^{65} Dy^{66} Ho^{67} Er^{68} Tu^{69} Yb^{70} Lu^{71}
									$6s^2$	La^{57}*	Hf^{72}	Ta^{73}	W^{74}	Re^{75}	Os^{76}				Hg^{80}	
$6p$			Tl^{81}	Pb^{82}	Bi^{83}	Po^{84}	—^{85}	Rn^{86}												
									$7s^0$											
$7s$, $6d$	—^{87}	Ra^{88}							$7s$											
									$7s^2$	Ac^{89}	Th^{90}	Pa^{91}	U^{92}							

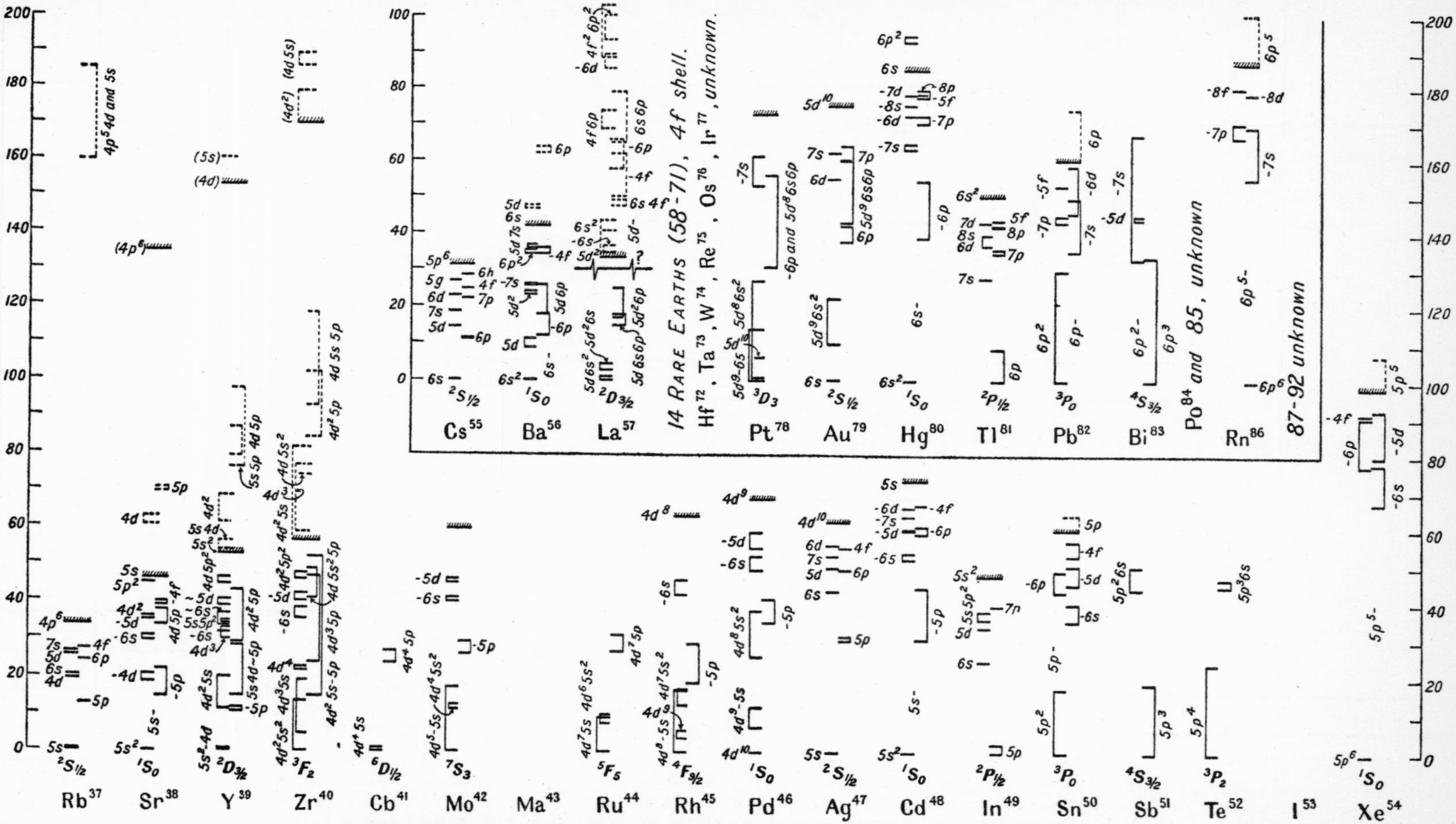

Fig. 3[14]. Main features of the energy-level scheme of the elements from rubidium to xenon.

Fig. 4[14]. (Inset.) Main features of the energy-level scheme of the elements from caesium to uranium.

It is not necessary to discuss the remainder of the system in such detail. The second long period builds in the $5s$, $4d$, and $5p$ shells, ending with xenon. This is illustrated in Fig. 3[14] where the similarities with the first long period are evident.

With the third long period greater complications set in (Fig. 4[14]). It starts off like the others with the alkali, caesium, and the alkaline earth, barium, closing the $6s$ shell. After this the $6s$ and $5d$ electrons start a long-period competition as before, but now the $4f$ states enter the competition. The complexity that presents itself is so great that thus far very little progress has been made in disentangling the spectra of the elements which have f electrons in their normal configuration.* These are the rare earths—the chemists do not have such an easy time with them either. There are fourteen places, from cerium ($Z=58$) to lutecium ($Z=71$), where the $4f$ shell is being filled. From Hf 72 to Pt 78 the filling of the $5d$ shell, interrupted by the rare earths, continues. In the process the $6s$ electrons have been displaced so that they must come in again at gold (79) and mercury (80) to give elements chemically resembling copper and zinc, having respectively one and two s electrons built on just completed d shells. Finally in the elements thallium to radon the $6p$ shell is built in. In these spectra, due to large spin-orbit interactions, there is a pronounced tendency to jj coupling. The radioactive metals follow: almost nothing is known about their energy-level schemes.

For many purposes it is convenient to have a periodic table of the elements which shows the electron configuration of the normal state of the atom. Table 2[14] is arranged in this way, and provides a convenient summary of the main facts which have been discussed in this section.

2. The statistical method of Fermi-Thomas.†

The first method of getting an approximate central field which we shall consider is that of Fermi and Thomas, who treated the cluster of electrons around a nucleus by the methods of statistical mechanics as modified to include the Pauli exclusion principle, that is, by the Fermi-Dirac statistics. In this method the phase space associated with the positional coordinates and momenta of each electron is divided into elements of volume h^3 and the exclusion principle is brought in by the restriction of the number of electrons in each cell to two (corresponding to the two possible spin orientations).

If p is the resultant momentum of an electron, the volume of phase space corresponding to electrons having less than this resultant momentum and located in volume dv of physical space is $\frac{4}{3}\pi p^3\,dv$. Let us suppose the number

* See the recent paper by Albertson, Phys. Rev. **47**, 370 (1935).

† Fermi, Rend. Lincei **6**, 602 (1927); **7**, 342, 726 (1928); Zeits. für Phys. **48**, 73; **49**, 550 (1928); Thomas, Proc. Camb. Phil. Soc. **23**, 542 (1927).

of electrons in unit volume is n. We assume that the kinetic energy of electrons at each point is as small as possible consistent with the density at each point. This means that p_0, the maximum value of p at each point, is connected with n through the relation $2\frac{4\pi p_0^3}{3h^3}=n$, so that

$$\frac{1}{2\mu}p_0^2=\frac{h^2}{2\mu}\left(\frac{3n}{8\pi}\right)^{\frac{2}{3}} \tag{1}$$

gives the maximum value of the kinetic energy of the electrons occurring at a place where the density is n. If φ is the electrostatic potential at the point in question, counted from zero at infinity, the potential energy of the electron is $-e\varphi$. If $p_0^2/2\mu<e\varphi$, the kinetic energy of the electron is insufficient to permit the escape of that electron to infinity. As this must be the case in the atom, we write

$$\frac{1}{2\mu}p_0^2=e(\varphi-\varphi_0), \tag{2}$$

where φ_0 is a positive constant.* Since for a neutral atom φ tends to zero at infinity we also assume that $n=0$ for regions of space where (2) leads to a negative maximum kinetic energy. Using (1) and (2) we see that the electron density at a point is related to the potential φ by the equation

$$n=[2\mu e(\varphi-\varphi_0)]^{\frac{3}{2}}\frac{8\pi}{3h^3}. \tag{3}$$

Treating the charge distribution as approximately continuous we may use the Poisson equation of electrostatics, $\Delta\varphi=-4\pi\rho$, to obtain the basic equation of the statistical method:

$$\Delta\eta=\frac{32\pi^2e}{3h^3}[2\mu e\eta]^{\frac{3}{2}}, \tag{4}$$

where $$\eta=\varphi-\varphi_0.$$

This equation has been used for an extension of the statistical method to molecules† and to metals.‡ We confine our attention to its application to the atomic problem. For the atom we need a solution of (4) which is such that $\varphi\to Ze/r$ as $r\to 0$, and $r\varphi\to 0$ as $r\to\infty$, and which has spherical symmetry. It turns out that for the neutral atom we may put $\varphi_0=0$. Using the polar coordinate form of the Laplacian and writing

$$\left.\begin{aligned}\varphi(r)&=\frac{Ze}{r}\chi(r),\\ r=xb,\quad b&=\frac{1}{2}\left(\frac{3\pi}{4}\right)^{\frac{2}{3}}\frac{\mathrm{a}}{Z^{\frac{1}{3}}}=\frac{0{\cdot}88534\mathrm{a}}{Z^{\frac{1}{3}}},\end{aligned}\right\} \tag{5}$$

* The introduction of φ_0 is due to GUTH and PEIERLS, Phys. Rev. **37**, 217 (1931).
† HUND, Zeits. für Phys. **77**, 12 (1932).
‡ LENNARD-JONES and WOODS, Proc. Roy. Soc. **A 120**, 727 (1928).

we find that $\chi(x)$ satisfies a differential equation free from universal constants:

$$x^{\frac{1}{2}}\frac{d^2\chi}{dx^2}=\chi^{\frac{3}{2}}. \tag{6}$$

We need a solution of this such that $\chi=1$ at $r=0$ and $\chi=0$ at $r=\infty$. Such a solution has been computed numerically* and is given in Table 3[14].

TABLE 3[14]. *The function* $\chi(x)$.

x	χ	x	χ	x	χ	x	χ
0·000	1·000	0·417	0·651	0·917	0·449	7·083	0·0450
0·010	0·985	0·458	0·627	0·958	0·436	8·125	0·0355
0·030	0·959	0·500	0·607	1·000	0·425	9·167	0·0287
0·060	0·924	0·542	0·582	1·250	0·364	10·000	0·0244
0·080	0·902	0·584	0·569	1·667	0·287	11·16	0·0198
0·100	0·882	0·625	0·552	2·083	0·234	12·01	0·0171
0·150	0·835	0·667	0·535	2·500	0·193	13·33	0·0139
0·200	0·793	0·709	0·518	3·125	0·150	15·01	0·0109
0·250	0·755	0·750	0·502	3·960	0·110	20·00	0·0058
0·292	0·727	0·792	0·488	4·375	0·0956	24·00	0·0038
0·333	0·700	0·833	0·475	5·000	0·0788	30·00	0·0022
0·375	0·675	0·875	0·461	6·042	0·0587	36·92	0·0011

With this approximate potential function Fermi was able to make a striking calculation of the number of s, p, d electrons in the atom as a function of atomic number. At a point where the maximum value of p is p_0 the actual momentum vectors are equally likely to be anywhere inside the centred momentum-space sphere of radius p_0. Hence the number of electrons in unit volume for which the component of p perpendicular to the radius vector has a value between $p_\perp$ and $p_\perp+dp_\perp$ is

$$\frac{8\pi p_\perp\sqrt{p_0^2-p_\perp^2}}{h^3}\,dp_\perp=\frac{8\pi\sqrt{p_0^2-L^2/r^2}}{h^3r^2}\,L\,dL$$

if we introduce the orbital angular momentum $L=p_\perp r$. Hence the whole number for which L lies between L and $L+dL$ at all places in the shell between r and $r+dr$ is

$$\frac{32\pi^2}{h^3}\sqrt{p_0^2-L^2/r^2}\,dr\,L\,dL.$$

Replacing p_0^2 by its value $2\mu e\varphi$ and integrating over those values of r for which the radical has real values, we obtain for the number of electrons for which L lies between L and $L+dL$

$$N(L)\,dL=\frac{32\pi^2}{h^3}\left(\int\sqrt{2\mu e\varphi-L^2/r^2}\,dr\right)L\,dL.$$

* A short table was given by Fermi. The values given here are taken from the table calculated with the differential analyser by BUSH and CALDWELL, Phys. Rev. **38**, 1898 (1931).

We shall suppose in this semi-classical argument that we have to put $L=(l+\frac{1}{2})\hbar$ and $dL=\hbar$ in order to find

$$N_l(Z)=2\,(2l+1)\frac{1}{\hbar}\oint\sqrt{2\mu\left(e\varphi-\frac{\hbar^2(l+\frac{1}{2})^2}{2\mu r^2}\right)}\,dr, \tag{7}$$

where $N_l(Z)$ is the number of electrons whose orbital angular momentum quantum number is l in the atom of atomic number Z. Written in this form the expression has a simple interpretation: $\hbar^{-1}$ times the integral is the value of the radial quantum number, calculated classically, for an electron of zero energy and orbital angular momentum $(l+\frac{1}{2})\hbar$ moving in a central field of potential energy, $-e\varphi(r)$. The factor $2\,(2l+1)$ is the number of electrons in a closed shell and the other factor is the number of closed shells occurring. The integral in this expression has to be evaluated numerically. Fermi has done this: the results are compared with the empirical data in Fig. 5^{14} where the smooth curves give the values of the theoretical $N_l(Z)$ and the

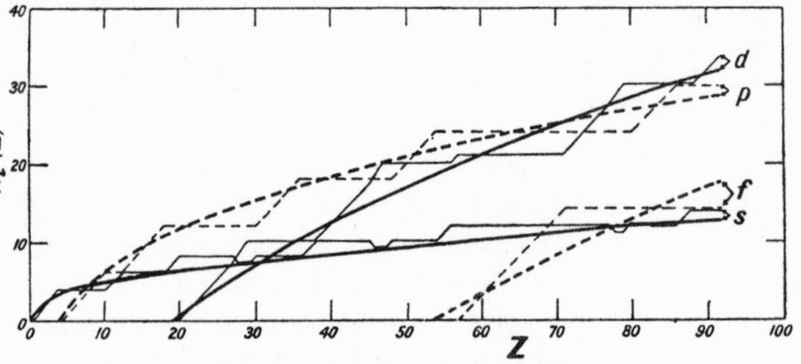

Fig. 5^{14}. Numbers of *s*, *p*, *d*, and *f* electrons in the normal state as a function of Z, as given by the statistical-field method.

irregular curves give the empirical values as discussed in the preceding section. The agreement is extremely good.

Another question which has been studied by the statistical method is that of the total ionization energy, that is the energy necessary to remove all of the electrons from a neutral atom. This was calculated by Milne and Baker.* The method is to calculate the electrostatic energy of the charge distribution. The total energy is half of this since by the virial theorem the average kinetic energy is minus half of the average potential energy in a system of particles interacting by the Coulomb law. The result is that the total energy of the normal state of the neutral atom of atomic number Z is

$$W=20{\cdot}83\,Z^{\frac{7}{3}}\text{ electron volts.}$$

An examination of the experimental data has been made by Young† for the

* Milne, Proc. Camb. Phil. Soc. **23**, 794 (1927);
Baker, Phys. Rev. **36**, 630 (1930).
† Young, Phys. Rev. **34**, 1226 (1929).

elements up to neon. He finds good agreement with the exponent of Z but that the data are better represented by the formula

$$W = 15{\cdot}6\, Z^{\frac{7}{3}} \text{ electron volts.}$$

Milne gave the theoretical coefficient the value 17 because of numerical inaccuracies; this accidentally gave a much better agreement.

3. The Wentzel-Brillouin-Kramers approximation.*

An important method of finding approximately the eigenvalues and eigenfunctions of one-dimensional Schrödinger equations was developed by Wentzel, Brillouin, and Kramers† independently. It is of interest because it exhibits the connection with the older quantization rules of Bohr and Sommerfeld. Let us first consider motion along a single coordinate under the potential energy $U(x)$. For convenience we suppose this has but one minimum and increases monotonically to the right and left of it.

We enter the wave equation

$$\psi'' + \frac{2\mu}{\hbar^2}(E-U)\psi = 0 \tag{1}$$

with the assumption $$\psi = e^{i\varphi/\hbar}, \tag{2}$$

so that φ has to satisfy $$\varphi'^2 + 2\mu(U-E) = i\hbar\varphi''. \tag{3}$$

If $\hbar$ were zero this would be the same as the Hamilton-Jacobi equation of classical mechanics for the action function $S(x)$. This suggests writing

$$\varphi = S + i\hbar\varphi_1 + \ldots + (i\hbar)^n\varphi_n + \ldots.$$

Equating successive powers of $\hbar$ to zero in the equation for φ, we find

$$S'^2 + 2\mu(U-E) = 0,$$

$$2\varphi_1' S' = S''$$

for the first two approximations. These lead to

$$S(x) = \int^x \sqrt{2\mu(E-U)}\,dx,$$

$$\varphi_1(x) = \tfrac{1}{4}\log|2\mu(E-U)|.$$

Hence in this approximation the wave function is given by

$$\psi_a = |\sigma|^{-\frac{1}{2}} \exp\left[\pm i \int^x \sigma\, dx\right], \tag{4}$$

where $$\sigma = \frac{1}{\hbar}\sqrt{2\mu(E-U)}. \tag{5}$$

The approximation is evidently not valid where $E-U=0$ as ψ_a becomes infinite there. In this approximation $\bar{\psi}_a\psi_a \sim |2\mu(E-U)|^{-\frac{1}{2}}$; inside the

* A good discussion of this topic has been given by HARTREE, Proc. Manchester Lit. and Phil. Soc. **77**, 91 (1933).

† WENTZEL, Zeits. für Phys. **38**, 518 (1926);
BRILLOUIN, Jour. de Physique, **7**, 353 (1926);
KRAMERS, Zeits. für Phys. **39**, 828 (1926).

classical range of motion, where $E > U$, this is equal to the classical probability of being between x and $x+dx$, the latter being taken proportional to the fraction of the whole period spent in going through this interval.

Next let us investigate the behaviour near the classical turning points of the motion, that is near the roots x_1 and x_2 $(x_1 < x_2)$ of the equation $E - U(x) = 0$. In the neighbourhood of these two points

$$\sigma = \sqrt{\alpha_1 (x - x_1)}, \quad \sigma = \sqrt{\alpha_2 (x_2 - x)},$$

where α_1 and α_2 are both positive. Writing

$$\xi = \alpha_1^{\frac{1}{3}} (x - x_1) \quad \text{or} \quad \alpha_2^{\frac{1}{3}} (x_2 - x),$$

the wave equation near either of the turning points assumes the canonical form

$$\psi'' + \xi\psi = 0. \tag{6}$$

The solution of this which is finite everywhere is an Airy integral* which Kramers took in the form

$$\psi = \omega(\xi) = \frac{-i}{2\sqrt{\pi}} \int_C \exp(\xi t + \tfrac{1}{3}t^3)\, dt, \tag{7}$$

where the path C comes from infinity along the ray $t = \tau e^{-i\pi/3}$ and goes to infinity along the ray $t = \tau e^{+i\pi/3}$. The function may be expressed in terms of Bessel functions of order one-third, so that the numerical tables and functional properties given in Watson are applicable:

$$\left.\begin{aligned} \omega(\xi) &= \frac{\sqrt{\pi}}{3} \xi^{\frac{1}{2}} [J_{\frac{1}{3}}(\tfrac{2}{3}\xi^{\frac{3}{2}}) + J_{-\frac{1}{3}}(\tfrac{2}{3}\xi^{\frac{3}{2}})], && (\xi > 0) \\ \omega(\xi) &= \frac{1}{\sqrt{3\pi}} |\xi|^{\frac{1}{2}} K_{\frac{1}{3}}(\tfrac{2}{3}|\xi|^{\frac{3}{2}}). && (\xi < 0) \end{aligned}\right\} \tag{8}$$

Applying the asymptotic formulas given in Watson (Chapter VII), we find that for large ξ

$$\omega(\xi) \sim \xi^{-\frac{1}{4}} \cos(\tfrac{2}{3}\xi^{\frac{3}{2}} - \tfrac{1}{4}\pi), \qquad (\xi > 0)$$

$$\omega(\xi) \sim \tfrac{1}{2}|\xi|^{-\frac{1}{4}} \exp(-\tfrac{2}{3}|\xi|^{\frac{3}{2}}). \qquad (\xi < 0)$$

Comparing this with the approximation valid away from the turning points we see that

$$\omega\big(\alpha_1^{\frac{1}{3}}(x - x_1)\big) \sim \frac{\alpha_1^{\frac{1}{6}}}{\sigma^{\frac{1}{2}}} \cos\left(\int_{x_1}^{x} \sigma\, dx - \frac{\pi}{4}\right) \qquad (x > x_1)$$

$$\sim \frac{1}{2} \frac{\alpha_1^{\frac{1}{6}}}{|\sigma|^{\frac{1}{2}}} \exp\left(-\int_{x}^{x_1} |\sigma|\, dx\right); \qquad (x < x_1)$$

$$\omega\big(\alpha_2^{\frac{1}{3}}(x_2 - x)\big) \sim \frac{\alpha_2^{\frac{1}{6}}}{\sigma^{\frac{1}{2}}} \cos\left(-\int_{x}^{x_2} \sigma\, dx + \frac{\pi}{4}\right) \qquad (x < x_2)$$

$$\sim \frac{1}{2} \frac{\alpha_2^{\frac{1}{6}}}{|\sigma|^{\frac{1}{2}}} \exp\left(-\int_{x_2}^{x} |\sigma|\, dx\right). \qquad (x > x_2)$$

* WATSON, *Theory of Bessel Functions*, p. 188.

These forms will join up smoothly in the region $x_1 < x < x_2$ with the solution (4) only if the arguments of the two cosine functions differ by an integral multiple of π. This implies that

$$2\int_{x_1}^{x_2} \sigma\, dx = (2n+1)\pi,$$

or, in terms of the classical momentum at the point x, that

$$\oint p\, dx = (n+\tfrac{1}{2})h. \tag{9}$$

This is exactly of the form of the Bohr-Sommerfeld quantization rule except for the occurrence of $(n+\frac{1}{2})$ in place of n. It thus appears that the use of quantum numbers increased by $\frac{1}{2}$ in the old rule gives an approximation to the energy levels as determined by quantum mechanics. With the energy level so determined the approximate eigenfunction (not normalized) is given by

$$\psi_a(x) = \begin{cases} \frac{1}{2}|\sigma|^{-\frac{1}{2}} \exp\left(i\int_x^{x_1} \sigma\, dx\right), & (x < x_1) \\ \alpha_1^{-\frac{1}{6}}\, \omega\left(\alpha_1^{\frac{1}{3}}(x - x_1)\right), & (x \sim x_1) \\ \sigma^{-\frac{1}{2}} \cos\left(\int_{x_1}^{x} \sigma\, dx - \frac{\pi}{4}\right), & (x_1 < x < x_2) \\ \alpha_2^{-\frac{1}{6}}\, \omega\left(\alpha_2^{\frac{1}{3}}(x_2 - x)\right), & (x \sim x_2) \\ \frac{1}{2}|\sigma|^{-\frac{1}{2}} \exp\left(i\int_{x_2}^{x} \sigma\, dx\right). & (x > x_2) \end{cases} \tag{10}$$

This completes the discussion for a single rectilinear coordinate where the boundary conditions require that ψ be finite in the range $-\infty \leqslant x \leqslant +\infty$. For atomic theory we are more interested in the equation for the radial eigenfunctions, where the coordinate range is 0 to $+\infty$ and the boundary conditions $R(0) = R(\infty) = 0$. This was treated by Kramers as follows:

The equation is

$$R'' + \frac{2\mu}{\hbar^2}\left(E - U(r) - \frac{\hbar^2}{2\mu}\frac{l(l+1)}{r^2}\right)R = 0. \tag{11}$$

Near the nucleus $U(r)$ is $-Ze^2/r + C$, so the equation has the form

$$R'' + \frac{2\mu}{\hbar^2}\left[(E - C) + \frac{Ze^2}{r} - \frac{\hbar^2}{2\mu}\frac{l(l+1)}{r^2}\right]R = 0.$$

If we neglect the constant term in the coefficient of R, this equation may be solved in terms of Bessel functions to give

$$R(r) \sim A\sqrt{\frac{r}{\mathrm{a}}}\, J_{2l+1}(\sqrt{8Zr/\mathrm{a}}), \tag{12}$$

the asymptotic expression for which is

$$\frac{A}{\sqrt{\pi}}\left(\frac{r}{2Z\mathrm{a}}\right)^{\frac{1}{4}}\cos\left[\sqrt{8Zr/\mathrm{a}}+\frac{2l(l+1)+\frac{3}{8}}{\sqrt{8Zr/\mathrm{a}}}-\pi(l+\tfrac{1}{2})-\frac{\pi}{4}\right].$$

Let us suppose that we have to take

$$\int^r \sqrt{\frac{2Z\mathrm{a}}{r}-\frac{\mathrm{a}^2k^2}{r^2}}\,d\left(\frac{r}{\mathrm{a}}\right)$$

as the argument of the complex exponential in (2). To terms of order r^{-1} this becomes

$$\sqrt{8Zr/\mathrm{a}}+\frac{2k^2}{\sqrt{8Zr/\mathrm{a}}}-\pi k,$$

so with this accuracy the approximation in the region of classical motion becomes

$$R\sim\left(\frac{r}{2Z\mathrm{a}}\right)^{\frac{1}{4}}\cos\left[\sqrt{8Zr/\mathrm{a}}+\frac{2k^2}{\sqrt{8Zr/\mathrm{a}}}-\pi k-\beta\right].$$

Comparing this with the asymptotic expression for the Bessel function we see that they will agree if

$$A=\sqrt{\pi},\quad \beta=\pi/4,\quad k=l+\tfrac{1}{2}.$$

At the outer bound of the classical motion in r, one may join the approximation smoothly by means of the ω function as before. Thus Kramers is led to the conclusion that one should use the classical quantum condition in the form

$$\oint p_r\,dr=(n_r+\tfrac{1}{2})h=(n-l-\tfrac{1}{2})h, \tag{13}$$

where

$$\frac{1}{2\mu}p_r^2+U(r)+\frac{\hbar^2}{2\mu}\frac{(l+\frac{1}{2})^2}{r^2}=E.$$

That is, one uses half-integral values of the radial phase integral and uses $(l+\frac{1}{2})^2$ instead of $l(l+1)$ in the term representing the effect of the orbital angular momentum on the radial motion.

This result is the basis of a method for finding an effective central field $U(r)$ which was developed before quantum mechanics by Fues and Hartree.* This early work is characterized by the use of integral quantum numbers. This made the work easier because for the circular orbits the radial integral was zero, so their empirical energies gave several points on the curve directly. The first calculations with the half-quantum numbers demanded by the new theory were made by Sugiura and Urey.† They worked by a graphical process. Prokofjew‡ made calculations for sodium by a numerical method.

* Fues, Zeits. für Phys. **11**, 364; **12**, 1; **13**, 211 (1922); **21**, 265 (1924); Hartree, Proc. Camb. Phil. Soc. **21**, 625 (1923).

† Sugiura and Urey, Kgl. Danske Vid. Selskab, Math. fys. **7**, No. 13 (1926); Sugiura, Phil. Mag. **4**, 495 (1927).

‡ Prokofjew, Zeits. für Phys. **48**, 255 (1929).

If we write $E=-\epsilon\dfrac{e^2}{2\mathrm{a}}$, $U(\rho)=-\dfrac{e^2}{\mathrm{a}}Q(\rho)/\rho^2$, $\rho=r/\mathrm{a}$,

the quantum condition is

$$(n-l-\tfrac{1}{2})=\frac{\sqrt{2}}{\pi}\int\sqrt{Q(\rho)-\tfrac{1}{2}[\epsilon\rho^2+(l+\tfrac{1}{2})^2]}\,\frac{d\rho}{\rho},$$

where the limits of integration bound the range in which the integrand is real. For large ρ we have $Q(\rho)\doteq\rho$ and for all values $Q(\rho)>\rho$. Since the f terms of sodium are essentially hydrogenic, to a good approximation $Q(\rho)=\rho$ for $\rho>6{\cdot}125$. For smaller values Prokofjew assumes $Q(\rho)$ in the form $\alpha\rho^2+\beta\rho+\gamma$, choosing two coefficients in such a way as to join smoothly to

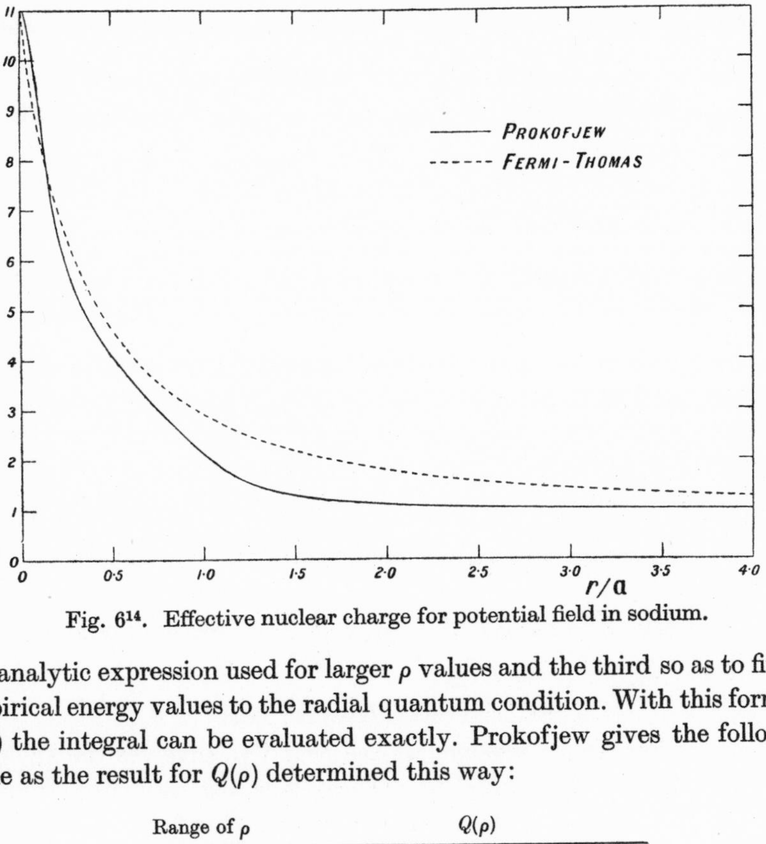

Fig. 6[14]. Effective nuclear charge for potential field in sodium.

the analytic expression used for larger ρ values and the third so as to fit the empirical energy values to the radial quantum condition. With this form for $Q(\rho)$ the integral can be evaluated exactly. Prokofjew gives the following table as the result for $Q(\rho)$ determined this way:

Range of ρ		$Q(\rho)$
0	to 0·01	11ρ
0·01	0·15	$-24{\cdot}4\,\rho^2+11{\cdot}53\,\rho-0{\cdot}00264$
0·15	1·00	$-2{\cdot}84\,\rho^2+4{\cdot}46\,\rho+0{\cdot}5275$
1·00	1·55	$+1{\cdot}508\,\rho^2-4{\cdot}236\,\rho+4{\cdot}876$
1·55	3·30	$0{\cdot}1196\rho^2+0{\cdot}2072\rho+1{\cdot}319$
3·30	6·74	$0{\cdot}0005\rho^2+0{\cdot}9933\rho+0{\cdot}0222$
6·74	∞	ρ

In Fig. 6^{14} is plotted the effective nuclear charge for potential

$$U(\rho) = -\frac{e^2}{a\rho} Z(\rho),$$

as a function of ρ for this field as determined by Prokofjew and also for the Fermi-Thomas potential for an electron in sodium. We see that the latter does not approach the asymptotic value sufficiently rapidly in this case.

4. Numerical integration of the radial equation.

Given an approximate potential function, the eigenvalues and eigenfunctions may be found by a numerical integration of the radial equation. We shall not attempt a thorough discussion of this matter as it involves varied questions of technique. A sketch of the method follows.

A trial eigenvalue is chosen; the differential equation to be solved is then of the form

$$D^2 y(x) = g(x), \quad \text{where} \quad g(x) = f(x)\,y$$

and $f(x)$ is a known function ($D = d/dx$). By assuming a power series for $y(x)$ near the origin one may find the numerical values of the coefficients and compute a table of values of $y(x)$ for small values of x. In terms of the backward difference operator ∇ the finite difference tables

$$\begin{array}{ccccccccc}
y_0 & & & & & g_0 & & & \\
 & \nabla y_1 & & & & & \nabla g_1 & & \\
y_1 & & \nabla^2 y_2 & & & g_1 & & \nabla^2 g_2 & \\
 & \nabla y_2 & & \nabla^3 y_3 & & & \nabla g_2 & & \nabla^3 g_3 \\
y_2 & & \nabla^2 y_3 & & & g_2 & & \nabla^2 g_3 & \\
 & \nabla y_3 & & & & & \nabla g_3 & & \\
y_3 & & & & & g_3 & & &
\end{array}$$

may be set up for interval h in x. We now try to calculate the next value of $\nabla^2 y$ in terms of quantities either known or almost known. We have $D^2 y = g(x)$ so $D^2 \nabla^2 y = \nabla^2 g$; hence

$$\nabla^2 y = D^{-2} \nabla^2 g.$$

Since $1 - \nabla = e^{-hD}$ we may express the operator $D^{-2}\nabla^2$ in powers of ∇ by the series

$$\nabla^2 \log^{-2}(1-\nabla) = 1 - \nabla + \tfrac{1}{12}\nabla^2 - \tfrac{1}{240}\nabla^4(1+\nabla) + \text{terms of order } \nabla^6,$$

from which follows

$$\nabla^2 y_n = h^2 [g_{n-1} + \tfrac{1}{12}\nabla^2 g_n - \tfrac{1}{240}\nabla^4 g_{n+1} + \ldots].$$

Referring to the above scheme, we see that $\nabla^2 y_4$ involves the known g_3 and the unknown $\nabla^2 g_4$ and $\nabla^4 g_5$. These must be estimated from the trend in the table and a trial value of $\nabla^2 y_4$ calculated. From this we get ∇y_4 and y_4 which gives us g_4 and so permits the extension of the table of differences of g by one more row. The computed value of $\nabla^2 g_4$ serves to verify the estimate already made. If they do not agree the computed $\nabla^2 g_4$ is used to get a new value of $\nabla^2 y_n$ and the process repeated. With experience the first estimate will be right more than half the time and it ought never to be necessary to make more than one correction to the estimate of $\nabla^2 g_4$. In this way the table is successively extended.

The process is carried out until the behaviour at large values of x is discernible. In general the trial value of the parameter will not be an eigenvalue so the process must be repeated with a different trial. By comparing the behaviour of the two trial functions one attempts to interpolate or extrapolate to a considerably improved trial value. No general rules for this have been given.

5. Normal state of helium.

We wish now to consider another type of approximate method of obtaining atomic energy levels, namely the variation method, and to illustrate its use with reference to the very exact calculations of the normal state of helium made by Hylleraas.* The key idea consists in recognizing that the Schrödinger equation is the differential equation corresponding to a minimum problem in the calculus of variations. From the general principles of Chapter II we know that $\bar{\psi}H\psi$ will be greater than the least allowed value of H for all ψ except the eigen-ψ belonging to this least allowed value. Moreover it is easy to see that $\bar{\psi}H\psi$ is stationary for ψ in the neighbourhood of each of the eigen-ψ's of H and that the stationary value is equal to the corresponding allowed value. In a general way it is a consequence of this stationary property that one can often calculate a fairly accurate value of the allowed energy from an evaluation of $\bar{\psi}H\psi$ from a relatively inaccurate ψ.

From the standpoint of the calculus of variations the Schrödinger equation is simply the Euler equation which expresses the fact that the normalized ψ minimizes $\bar{\psi}H\psi$. Instead of seeking ψ through its Schrödinger equation, we may use the Ritz method.† In this one tries to represent the unknown function ψ as belonging to a class of functions, say $\phi(a, b, c, \ldots)$ depending on several parameters $a, b, c, \ldots$. Then one chooses the parameters in such a way that $\bar{\phi}H\phi$ is a minimum when ϕ is normalized. This implies a set of equations

$$\frac{\partial}{\partial a}(\bar{\phi}H\phi) = 0, \quad \frac{\partial}{\partial b}(\bar{\phi}H\phi) = 0, \quad \ldots \tag{1}$$

which determine the minimizing values of the parameters. Evidently if the family $\phi(a, b, c, \ldots)$ happens to contain the exact eigenfunction for some set of values of the parameters, the exact energy and eigenfunction will be found in this way. That would be a lucky accident. Generally the method furnishes that member of the family which gives the closest fit to the true eigenfunction ψ. Moreover the value of $\bar{\phi}H\phi$ found this way will surely be greater than

* Hylleraas, Zeits. für Phys. **48**, 469 (1928); **54**, 347 (1929); Norske Vid. Akads. Skrifter, Mat. Kl. 1932, No. 6.

† Ritz, *Gesammelte Werke*, Paris, 1911, p. 192; or Jour. für reine und angew. Math. **135**, 1 (1907).

or equal to the least allowed value of H. This one-sided approach of the successive approximations to the true value is a useful feature of this method. It gives us a semi-test of the theory in comparison with experiment, since if we ever find a value lower than the observed lowest energy level the theory must be wrong.

This method of dealing approximately with variation problems was developed for classical problems, especially in the theory of elasticity, and has proved to be a valuable tool in that field. It was first used in quantum mechanics by Kellner* to calculate the normal state of helium; the problem has been later treated with much greater precision by Hylleraas. It is evident that the success of the method depends principally on a fortunate choice of the family of functions $\phi(a, b, c, \ldots)$ on which the approximation is based. In the work on helium it has been found convenient to use the distances r_1, r_2, and r_{12} and the three Euler angles specifying the orientation of the plane determined by the two electrons and the nucleus as coordinates. For the normal state, ψ does not depend on these angles.

The simplest possible choice of eigenfunction is obtained by normalizing

$$e^{-(Z-b)(r_1+r_2)},$$

the product of two hydrogenic 1s eigenfunctions (r is in atomic units). Here the integrations are simply performed and give, as the lowest energy level, $-(Z-\frac{5}{16})^2$ atomic units, and the eigenfunction (not normalized)

$$e^{-(Z-\frac{5}{16})(r_1+r_2)}.$$

This gives 1·695 Rydberg units instead of the observed value 1·810 for the ionization potential of helium. For the ionization potential as a function of Z we find (in Rydberg units)

	H^- (1)	He (2)	Li^+ (3)	Be^{++} (4)
Calculated	−0·055	1·695	5·445	11·195
Observed	—	1·810	5·560	11·307
% error	—	6·4	2·1	1·0

This shows that the simple screened hydrogenic eigenfunction gives quite good values as Z increases, corresponding to the fact that the interaction of each electron with the nucleus increases as Z^2 while the electron interaction energy increases as Z.

Using atomic units for length, let us write

$$s = r_1 + r_2, \quad t = r_1 - r_2, \quad \text{and} \quad u = r_{12}.$$

Then one parameter may be handled at once, in any trial function, namely the one which merely changes the scale of lengths in the trial function. For the energy of a function $\phi(ks, kt, ku)$ we find

$$E = \frac{1}{N}(k^2 M - kL),$$

* Kellner, Zeits. für Phys. **44**, 91 (1927).

where, if we let $\varphi = \phi(s, t, u)$,

$$L = \int_0^\infty ds \int_0^s du \int_0^u dt\,(4Zsu - s^2 + t^2)\,\varphi^2,$$

$$M = \int_0^\infty ds \int_0^s du \int_0^u dt \left\{ u\,(s^2 - t^2)\left[\left(\frac{\partial\varphi}{\partial s}\right)^2 + \left(\frac{\partial\varphi}{\partial t}\right)^2 + \left(\frac{\partial\varphi}{\partial u}\right)^2\right] + 2s\,(u^2 - t^2)\frac{\partial\varphi}{\partial s}\frac{\partial\varphi}{\partial u} + 2t\,(s^2 - u^2)\frac{\partial\varphi}{\partial t}\frac{\partial\varphi}{\partial u}\right\},$$

$$N = \int_0^\infty ds \int_0^s du \int_0^u dt\, u\,(s^2 - t^2)\,\varphi^2.$$

The condition $\partial E/\partial k = 0$ leads to $k = L/2M$, so

$$E = -L^2/4MN.$$

This fixes the value of k in terms of the other parameters and gives us the minimum value of E with regard to variation of the scale of lengths in ϕ. For ϕ Hylleraas assumes

$$\phi(s, t, u) = e^{-\frac{1}{2}s} \sum_{n, l, m=0}^{\infty} C_{n, 2l, m}\, s^n t^{2l} u^m.$$

Odd powers of t cannot occur since ϕ must be symmetric in the two electrons.

The results of such calculations are extremely interesting and show that quantum mechanics is undoubtedly able to give the correct ionization potential to within quantities of the order of the neglected spin-relativity and nuclear kinetic energy terms. In the third approximation Hylleraas finds

$$\phi(s, t, u) = e^{-\frac{1}{2}s}(1 + 0{\cdot}08u + 0{\cdot}01t^2), \qquad k = 3{\cdot}63$$

$$E = -1{\cdot}80488\,\mathsf{R}hc,$$

while in the sixth approximation

$$\phi(s, t, u) = e^{-\frac{1}{2}s}(1 + 0{\cdot}0972u + 0{\cdot}0097t^2 - 0{\cdot}0277s + 0{\cdot}0025s^2 - 0{\cdot}0024u^2),$$

$$E = -1{\cdot}80648\,\mathsf{R}hc.$$

An eighth approximation led to $1{\cdot}80749\,\mathsf{R}hc = 198322\,\text{cm}^{-1}$ for the ionization potential as compared with an experimental value of $198298 \pm 6\ \text{cm}^{-1}$. It will be noticed that the theoretical value exceeds the experimental value by $24\ \text{cm}^{-1}$, which appears to contradict our statement that the Ritz method always gives too high an energy value. The discrepancy is due to the neglect of the finite mass of the helium nucleus and to relativity effects. When these are included the theoretical value becomes $198307\ \text{cm}^{-1}$, which is in agreement with the experimental value. This is an important accomplishment of quantum mechanics since it is known that the older quantized-orbit theories led definitely to the wrong value.

In order to calculate the ionization potentials of the ions iso-electronic with He, Hylleraas* modified the details of the use of the Ritz method so as

* Hylleraas, Zeits. für Phys. **65**, 209 (1930).

to obtain the ionization potential as a series in Z^{-1}. The result, in Rydberg units, neglecting the finite mass of the nucleus and the relativity correction is given by the formula

$$E = Z^2 - 1{\cdot}25Z + 0{\cdot}31488 - 0{\cdot}01752Z^{-1} + 0{\cdot}00548Z^{-2}.$$

The data for comparison of theory and experiment are given in Table 4^{14}, from which it is seen that there is agreement within the experimental error of the known ionization potentials.

TABLE 4^{14}. *Ionization potentials of two-electron atoms.*

	H^-	He	Li II	Be III	B IV	C V
Calculated in Rhc	0·05284	1·80749	5·55965	11·31084	19·06160	28·81211
Mass correction	−20	−29	−47	−67	−94	−130
Relativity correction	−2	−9	−29	−46	−77	−105
Total	0·05262	1·80711	5·55859	11·30971	19·06143	28·81586
Calculated in cm^{-1}	5 774	198 308	609 985	1 241 222	2 091 770	3 161 770
Measured in cm^{-1}	*	198 298	610 090	1 241 350	2 092 000	3 161 900
Probable error of measurement	—	8	100	200	300	800

6. Excited levels in helium.

The variation method is peculiarly adapted to calculation of the lowest energy level of a system, for this is the level which corresponds to an absolute minimum of $\bar{\psi}H\psi$. The next higher level is characterized as the minimum value of $\bar{\psi}H\psi$ under the auxiliary condition that ψ be orthogonal to the normal state, and the third level is that which makes $\bar{\psi}H\psi$ a minimum under the two auxiliary conditions that ψ be orthogonal to each of the two states of lower energy. In approximate work these auxiliary conditions make trouble, since in the absence of exact knowledge of the ψ's for the lower states all one can do is require orthogonality to the approximately known lower states. This inaccurately applied auxiliary condition in general introduces a large error in the calculations by permitting the trial ψ to contain a component along the true ψ of the normal state, with a resultant 'sagging' of the minimum value of $\bar{\psi}H\psi$ below the correct value.

Nevertheless the existence of quantum numbers that are exact permits the exact fulfilment of the auxiliary conditions in some cases. This is exemplified by the calculations of Hylleraas and Undheim† on the $1s\,2s\,^3S$ level of helium. In so far as He has exact Russell-Saunders coupling this ψ must be an antisymmetric function of the position of the two electrons. This condition alone serves to make it exactly orthogonal to the normal state, which is a symmetrical function of the position coordinates. Using hydro-

* LOZIER, Phys. Rev. **36**, 1417 (1930), finds that the electron affinity of the hydrogen atom is roughly 0·6 electron volt (5000 cm^{-1}).

† HYLLERAAS and UNDHEIM, Zeits. für Phys. **65**, 759 (1930).

genic eigenfunctions with $Z=2$ for the $1s$ state and $Z=1$ for the $2s$ state, the eigenfunction (not normalized) for $2\,^3S$ becomes

$$\psi \sim e^{-\frac{5}{4}s}\left[\left(\tfrac{1}{2}s-2\right)\sinh\tfrac{3}{4}t-\tfrac{1}{2}t\cosh\tfrac{3}{4}t\right], \tag{1}$$

where s and t are the elliptic coordinates of the preceding section, measured in atomic units. The value of $\int\bar{\psi}H\psi$ for this, relative to the normal state of the He ion, is $-0{\cdot}2469\,\mathsf{R}hc$ which is considerably higher than the experimental value, $-0{\cdot}35048\,\mathsf{R}hc$. Preserving the requirement of antisymmetry, one may generalize (1) to

$$\psi \sim e^{-ks}\left[(C_1+C_2s+C_4u+C_5us)\sinh ct+t\,(C_3+C_6u)\cosh ct\right] \tag{2}$$

and choose the parameters to minimize $\int\bar{\psi}H\psi$. In this way Hylleraas and Undheim found $-0{\cdot}35044\,\mathsf{R}hc$ for the energy of this level, agreeing with the experimental value to about 0·01 per cent.

To illustrate the error involved in failing to apply the auxiliary condition properly, they calculated $\int\bar{\psi}H\psi$ for the symmetric function analogous to (1) for the $2\,^1S$ level. The value, $-0{\cdot}3422\,\mathsf{R}hc$, lies considerably below the observed $-0{\cdot}29196$ because of the 'impurity' of the normal state ψ which is contained in the trial wave function. They tried a more general form

$$\psi \sim e^{-ks}\left[(C_1+C_2s+C_4u+C_5us)\cosh ct+t\,(C_3+C_6u)\sinh ct\right]$$

and applied the orthogonality condition by a special device. The minimizing conditions expressed by $5^{14}1$ are satisfied by several values of the energy. Instead of taking the least root, they chose the parameters in such a way as to minimize the next to the least value of the energy. This gave a value $-0{\cdot}28980\,\mathsf{R}hc$, that is, 0·7 per cent. higher than the observed value. A more significant way of estimating the accuracy is to note that it gives the departure from the hydrogenic value $-0{\cdot}25\,\mathsf{R}hc$ to within 5·4 per cent. of the actual departure from this value.

Orthogonality of the $2\,P$ terms relative to the normal state and the $2\,S$ levels may be accurately obtained by the proper use of the dependence on the angular coordinates of these eigenfunctions. Calculation of the $2\,^3P$ level was made in this way by Breit,* by using the hydrogenic functions for $1s$ and $2p$ in the proper combination, with the effective nuclear charges Z_{1s} and Z_{2p} as the parameters. The value obtained was $-0{\cdot}2616\,\mathsf{R}hc$, the experimental value being $-0{\cdot}2664\,\mathsf{R}hc$, an error of 1·8 per cent. on the term value or 29 per cent. on the departure from the hydrogenic value. Eckart† made the corresponding calculation for the $2\,^1P$ term, obtaining $-0{\cdot}245$ instead of the observed value $-0{\cdot}2475$. He made calculations with hydro-

* Breit, Phys. Rev. **35**, 569 (1930); **36**, 383 (1930).
† Eckart, Phys. Rev. **36**, 878 (1930).

genic functions and variable Z for the $2\,^3S$, $2\,^1P$ and $2\,^3P$ levels in Li II as well as He I. The results were, in Rydberg units:

	Li II		He I	
	Calc.	Obs.	Calc.	Obs.
$2\,^3S$	1·21	1·22	0·334	0·350
$2\,^3P$	1·04	1·05	0·262	0·266
$2\,^1P$	0·99	1·00	0·245	0·247

Evidently the method could be applied to find the lowest levels in any series of configurations because of the orthogonality of the trial functions in the angle variables.

Historically the first helium calculations are contained in a paper by Heisenberg.* It was in this paper that he laid the foundations of the theory of atomic spectra by showing the importance of the symmetry of the eigenfunctions. Heisenberg's calculations were confined to the $1s\,nl$ configurations with $l \neq 0$. He assumed that the $1s$ eigenfunction was almost entirely confined to smaller values of the radius than the nl eigenfunction, so he could take for the potential energy function of each electron

$$U(r) = \left[-\frac{Z}{r} + v(r)\right]e^2 \quad \text{with} \quad v(r) = \begin{cases} 1/r_0 & (r < r_0) \\ 1/r. & (r > r_0) \end{cases}$$

An electron inside r_0 moves in the full field while one outside r_0 moves in a screened field of effective nuclear charge, $Z-1$.

If r_0 is taken rather larger than a/Z and at the same time smaller than the values of r for which $R(nl)$ becomes appreciable, one may use as a good approximation the hydrogenic eigenfunction with full Z for the $1s$ electron and the corresponding function with nuclear charge $(Z-1)$ for the nl electron. This implies a correction to the energy of

$$(1s|v(r)|1s) + (nl|v(r) - e^2/r|nl)$$

to allow for the fact that they do not correspond to $U(r)$ but to purely Coulomb fields instead. In calculating the first order perturbation one needs the direct and exchange integrals of the Coulomb interaction as well as the perturbation energy corresponding to the fact that the true central field is the field, $-Ze^2/r$. This provides the additional terms

$$(1s\,nl|e^2/r_{12}|1s\,nl) \pm (1s\,nl|e^2/r_{12}|nl\,1s) - (1s|v(r)|1s) - (nl|v(r)|nl)$$

with the result that the whole energy becomes independent of the choice of r_0, to the first approximation. It is

$$-\frac{\mathsf{R}hZ^2}{1^2} - \frac{\mathsf{R}h(Z-1)^2}{n^2} + (1s\,nl|e^2/r_{12}|1s\,nl) - (nl|e^2/r|nl) \pm (1s\,nl|e^2/r_{12}|nl\,1s).$$

The third and fourth terms are nearly equal and opposite: on account of the spherical symmetry of the $1s$ eigenfunction its field is like that of a charge

* HEISENBERG, Zeits. für Phys. 39, 499 (1927).

concentrated at the centre. These two terms fail to compensate exactly because some of the $R(nl)$ charge distribution penetrates to smaller radii than some of the $1s$ charge.

The numerical value of the third and fourth terms together (in Rydberg units) as calculated by Heisenberg is

	$2p$	$3p$	$3d$
He (2)	$-0{\cdot}0020$	$-0{\cdot}00070$	$6{\cdot}7\times10^{-6}$
Li^+ (3)	$-0{\cdot}0098$	$-0{\cdot}0032$	

These values do not give a good approximation to the location of the mean of the singlet and triplet terms because we have neglected polarization. When the outer electron is present its field acts on the inner electron to distort its wave function and give an induced dipole moment which gives rise to an altered interaction between the two electrons. This polarization effect can, in principle, be calculated with the aid of the second order perturbation theory. The most careful calculations of the polarization effect are those given by Bethe.*

The separation of the singlet and triplet states is given by calculating the exchange integral $(1s\,nl|e^2/r_{12}|nl\,1s)$. The values given by Heisenberg are (Rydberg units):

	$2p,\ 3p$	$3d,\ 4d$	$4f,\ 5f$
He (2)	$0{\cdot}00765$	$2{\cdot}57\times10^{-5}$	$5{\cdot}25\times10^{-8}$
	$0{\cdot}00246$	$1{\cdot}50\times10^{-5}$	$4{\cdot}31\times10^{-8}$
Li^+ (3)	$0{\cdot}0307$	$0{\cdot}000189$	$6{\cdot}95\times10^{-7}$
	$0{\cdot}00935$	$0{\cdot}000108$	$5{\cdot}72\times10^{-7}$

7. Normal states of first-row atoms.

In this section we shall review the work that has been done on the use of the variation method for finding the energies of the normal states of the atoms in the first row of the periodic table. For the Li iso-electronic sequence the most complete work is that of Wilson.† He works with an antisymmetric combination of one-electron functions as introduced in 3[6]6. For the radial functions he uses the forms

$$R(1s)/r \sim e^{-\xi r},$$
$$R(2s)/r \sim \xi\alpha r\, e^{-\xi\eta r} - e^{-\xi\zeta r} \tag{1}$$

so that the trial function contains four variable parameters, ξ, η, ζ and α. The minimizing with regard to the scale factor ξ can be carried out directly since ξ here plays the same role as k in § 5[14]. Because of the complicated way in which α, η and ζ appear in the expression for $\bar{\psi}H\psi$ it was not possible to solve analytically for the best values of the parameters so a graphical-numerical method was used.

* Bethe, Handbuch der Physik **24/1**, 2[d] ed., 339 (1933).
† Wilson, J. Chem. Phys. **1**, 210 (1933).

To compare with earlier work by Eckart and by Guillemin and Zener* it should be noted that Eckart's $R(2s)$ is obtained from Wilson's by the specialization, $\zeta=\eta=\xi\alpha$, so he has just two parameters to vary, while Guillemin and Zener's is equivalent to writing $\zeta=\eta$ in Wilson's, leaving them with three parameters to vary. Comparing with Hylleraas' precise calculations on helium we note the absence of any dependence of the trial functions on the mutual electronic distances, r_{ij}; this would make the numerical work extremely difficult. The results show that the first term in $R(2s)$ is much larger than the second, so Zener and Slater† have pointed out that good approximations can be obtained by simply omitting the second term in $R(2s)$.

The results of all such calculations as summarized by Wilson are given in Table 5[14] where the energies are expressed in atomic units (twice the Rydberg unit). The calculated values of the ionization potential are the difference between the calculated normal state of the three-electron problem and that of the corresponding two-electron problem.

TABLE 5[14]. *Energies of the Li iso-electronic sequence.*

	Total energy	% difference	Ion. potential	% difference
Li I				
Experimental	− 7·4837	—	0·1983	
Wilson	− 7·4192	0·86	0·1965	0·91
Guillemin-Zener	− 7·4183	0·87	0·1956	1·36
Slater	− 7·4179	0·88	0·1953	1·51
Eckart	− 7·43922	1·22	0·1696	14·5
Be II				
Experimental	− 14·3422	—	0·6704	
Wilson	− 14·2639	0·55	0·6663	0·61
Slater	− 14·2584	0·58	0·6607	1·45
B III				
Experimental	− 23·476	—	1·395	
Wilson	− 23·363	0·48	1·390	0·36
Slater	− 23·350	0·54	1·378	1·22
C IV				
Experimental	− 34·778	—	2·3722	
Wilson	− 34·713	0·19	2·3650	0·30
Guillemin-Zener	− 34·698	0·23	2·3496	0·95
Slater	− 34·690	0·25	2·3422	1·26

The $R(1s)$ functions used in all this work are hydrogenic in character except for the variable scale factor. The $R(2s)$ functions as used in the variation problem are not directly comparable as they stand, since they are not

* ECKART, Phys. Rev. **36**, 878 (1930);
GUILLEMIN and ZENER, Zeits. für Phys. **61**, 199 (1930).

† ZENER, Phys. Rev. **36**, 51 (1930);
SLATER, *ibid.* **36**, 57 (1930).

orthogonal to the $R(1s)$ functions and therefore cannot be regarded as belonging to the same effective central field. One can correct this by considering $R^{\circ}(2s) = R(2s) + \beta R(1s)$ as the proper $2s$ radial function, choosing β so that $R^{\circ}(2s)$ is orthogonal to $R(1s)$. This does not alter the value of the Ψ in determinant form, since this change in $R(2s)$ corresponds to adding to one row of the determinant a constant multiple of another row.

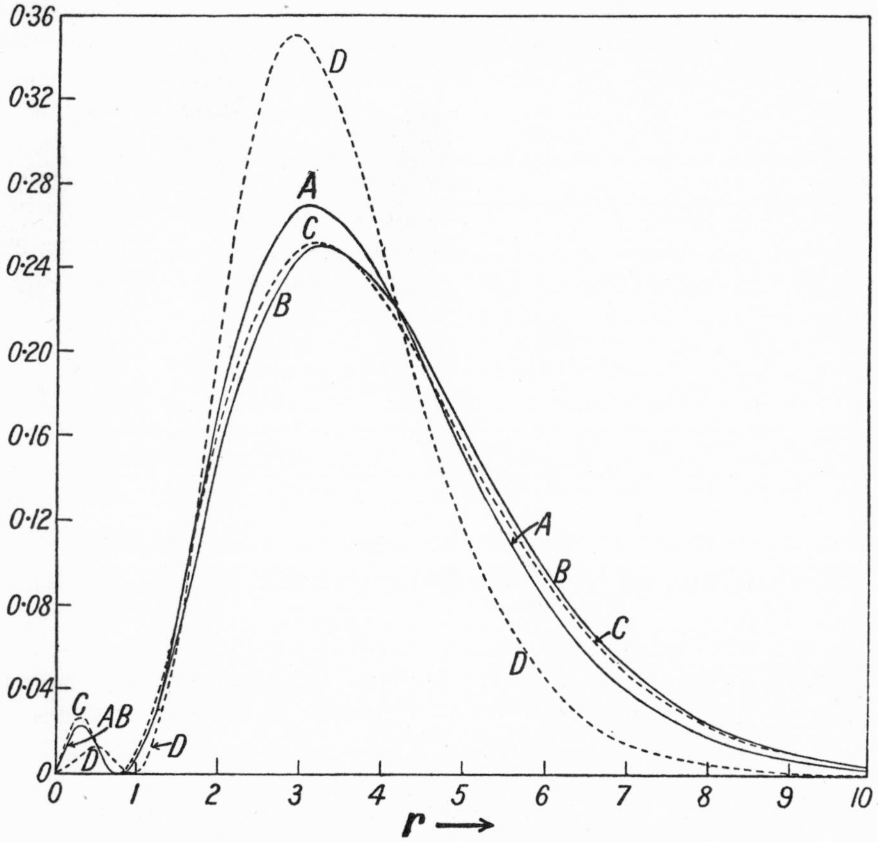

Fig. 7^{14}. Comparison of different approximations to $R^2(2s)$ for Li I.
A, Wilson; *B*, Guillemin-Zener; *C*, Slater; *D*, hydrogen-like.

The four different $R(2s)$ functions obtained by the four processes are compared by Wilson in Fig. 7^{14}, where $R^{\circ 2}(2s)$ is plotted after the $R(2s)$ has been made orthogonal to the $R(1s)$. It will be noticed that this process has introduced a node into Slater's form for $R(2s)$ although it was node-less in the original form. The figure shows a striking similarity of the $R(2s)$ for the three forms which gave best agreement with experiment.

Zener (*loc. cit.*) has applied the variation method to the normal states of the other atoms in the first row from Be to F.

8. Hartree's self-consistent fields.

The different applications of the variation method, that of seeking the ψ's which make $\bar{\psi}H\psi$ stationary for normalized ψ, may be classified according to the type of trial function admitted for ψ. In the Ritz method a trial function depending on several parameters is used. This makes the value of $\bar{\psi}H\psi$ depend on these parameters and the stationary values are sought by ordinary calculus methods. At the other extreme we have the case in which the variation of ψ is wholly unrestricted; then the Euler variation equation is just the Schrödinger equation of the problem. In between these extremes we may admit trial functions of special forms and determine their detailed character by the variation principle. The most useful of these is a method devised by Hartree* on physical considerations; recognition of its connection with the variation principle is due to Slater and to Fock.†

Although we know from Chapter VI that ψ must be antisymmetric in all the electrons, let us neglect that requirement for simplicity and write for the ψ belonging to the complete set A the simple form (2^64)

$$\psi(A) = u_1(a^1)\,u_2(a^2)\,u_3(a^3)\dots u_N(a^N)$$

instead of the properly antisymmetrized function (3^66). Let us suppose each of the u's normalized, although as yet they are not specified to be solutions of a particular central-field problem as they were in §§ 1^6 and 2^6. If we use the approximate Hamiltonian 1^61, neglecting spin-orbit interaction, the value of $\bar{\psi}H\psi$ is

$$E = \bar{\psi}H\psi = \sum_{i=1}^{N}\left(a^i\left|\frac{1}{2\mu}\boldsymbol{p}^2 - \frac{Ze^2}{r}\right|a^i\right) + \sum_{i>j=1}^{N}(a^i a^j|e^2/r_{ij}|a^i a^j), \qquad (1)$$

the calculation of the matrix components being exactly as in §§ 6^6 and 7^6 only simpler because there are no permutations and hence no exchange integrals. The dependence of E on any particular factor of ψ, say $u_i(a^i)$, is shown more explicitly by writing

$$E = \int \bar{u}_i(a^i)\left[\frac{1}{2\mu}\boldsymbol{p}_i^2 - \frac{Ze^2}{r_i}\right]u_i(a^i)\,dv_i + \int\!\!\int \bar{u}_i(a^i)\,u_i(a^i)\sum_j^{(\neq i)}\frac{e^2}{r_{ij}}\bar{u}_j(a^j)\,u_j(a^j)\,dv_j\,dv_i$$
$$+\,\text{terms independent of } u_i(a^i). \quad (1')$$

This dependence on $u_i(a^i)$ is exactly of the form of the dependence for a one-electron problem in which the electron having the quantum numbers a^i moves in an effective field for which the potential energy function is

$$V(a^i, \boldsymbol{r}_i) = -\frac{Ze^2}{r_i} + \sum_j^{(\neq i)}\int\frac{e^2}{r_{ij}}\bar{u}_j(a^j)\,u_j(a^j)\,dv_j. \qquad (2)$$

This is true for each of the u's, since they occur symmetrically in ψ. Hence if

* HARTREE, Proc. Camb. Phil. Soc. **24**, 89 (1928).

† SLATER, Phys. Rev. **35**, 210 (1930);
FOCK, Zeits. für Phys. **61**, 126 (1930).

we make E stationary by varying each $u_i(a^i)$ independently, each must satisfy a one-electron Schrödinger equation in which the effective field is that of the nucleus plus the classical potential energy field due to the other electrons calculated according to their quantum-mechanical probability distributions.

In this way the variation principle leads us to a set of simultaneous equations for the u's. But they are very difficult, being non-linear and integro-differential through the appearance of $\bar{u}(a^j)\,u(a^j)$ under the integral sign in (2). Hartree's procedure is to solve this system of equations by numerical integration using a successive approximation process.

Physically the set of equations appears to be very reasonable. Each electron actually does move in a field that is due to the fixed nucleus and to the action of the other electrons. It was by such an argument that Hartree set up his equations rather than by way of the variation principle. This field, or set of fields, Hartree calls *self-consistent* in the sense that their own eigenfunctions are consistent with the potential field from which they are determined. The method is evidently applicable in principle to the calculation of other problems where several electrons are involved. For example, Brillouin* has developed it for use in the theory of metals.

Generally the potential in (2) due to the other electrons will not be spherically symmetric owing to a departure from spherical symmetry of the charge distribution $\sum_j \bar{u}(a^j)\,u(a^j)$. To consider such departures from spherical symmetry in the atomic problem would be very difficult and probably would not correspond to an improvement in the final result. For that reason Hartree does not use the actual non-spherically-symmetric potential field defined by (2) but the result of averaging this field over all directions. This results in each electron's moving in an effective central field. To keep this symmetrizing process within the form of a variation problem we may assume for each $u(a^i)$ that it is of the form of the wave function of a central-field problem,

$$u(a^i) = \frac{1}{r}\,R(n^i l^i)\,S(l^i m_l^i), \tag{3}$$

where $S(l^i m_l^i) = \Theta(l^i m_l^i)\,\Phi(m_l^i)$ is the normalized spherical harmonic appropriate to the set of quantum numbers a^i. This means in the variation problem that only the radial function R/r is subject to variation. In (1′) the integral representing the interaction of the electrons with quantum numbers a^i and a^j is now of the form

$$J(a^i, a^j) = \iint \bar{u}_i(a^i)\,u_i(a^i)\,\frac{e^2}{r_{ij}}\,\bar{u}_j(a^j)\,u_j(a^j)\,dv_j dv_i,$$

* BRILLOUIN, Jour. de Physique, 3, 373 (1932).

whose evaluation we have considered in § 8^6. There we have seen that

$$J(a^i, a^j) = \sum_k a^k(l^i m_l^i, l^j m_l^j)\, F^k(n^i l^i, n^j l^j),$$

where the a's are the coefficients in Table 2^6 and the F's are the radial integrals defined in $8^6$15a. From this expression it is clear that the effective central field for the electron with quantum numbers a^i becomes

$$V(a^i, r_i) = -\frac{Ze^2}{r_i} + \sum_j^{(\neq i)} \sum_k a^k(l^i m_l^i, l^j m_l^j) \int \frac{e^2 r_<^k}{r_>^{k+1}} R^2(n^j l^j)\, dr_j. \tag{4}$$

This is a central field, to be sure, but its value depends on the m_l of the electron in question. This is an undesirable feature since we have to use the various complete sets A belonging to a configuration to calculate the details of the level structure of that configuration, as in Chapter VII and later chapters. As that structure involves integrals of the type F^k anyway, it is not likely that we gain much in accuracy by dealing with different central fields for each m_l value. For any fixed value of $l^j m_l^j$, the a's of Table 2^6 have the property that the average over m_l^i vanishes for $k \neq 0$ and is equal to unity for $k = 0$. Therefore if we use for each value of m_l^i not the field given by (4) but the value on averaging over the values of m_l^i, we have the same result as if we had taken the non-central field (2) and averaged it over all directions of $\boldsymbol{r}_i$, that is,

$$V(a^i, r_i) = -\frac{Ze^2}{r_i} + \sum_j^{(\neq i)} \int \frac{e^2}{r_>} R^2(n^j l^j)\, dr_j.$$

This is the field with which the Hartree method actually works.

By this operation of introducing central symmetry, the system of equations becomes a system of ordinary integro-differential equations for the N radial functions, $R(a^i)$, instead of a system of N partial integro-differential equations, each in three independent variables for the N functions, $u(a^i)$.

Let us next see what value this method gives for the energy of the atom. The characteristic value $\epsilon(a^i)$ of the equation for the ith electron is not the work necessary to remove it to infinity with no kinetic energy, for in the actual process of removal its contribution to the effective field in which the others move is removed which in turn alters the characteristic values of all the other electrons. Nevertheless it turns out empirically, as we shall see, that these characteristic values for the deep-lying electrons do provide good approximations to the X-ray term values, so it must be that the correction terms are quite small.

The value of E from (1) is given now by using our results for the self-consistent wave functions. It is evident that we shall have

$$E = \sum_i \epsilon(a^i) - \sum_{i>j=1}^{N} (a^i a^j | e^2/r_{ij} | a^i a^j), \tag{5}$$

since the interaction of each other electron with the ith electron is counted once in the equations by which each $\epsilon(a^i)$ is determined. On summing over i in the first term therefore the interaction of the electrons is counted twice whereas in (1) it should be counted but once, so the interaction energy has to be subtracted from $\Sigma\epsilon(a^i)$ to allow for this.* Suppose we compare this with the energy E' for the $N-1$-electron problem in which the electron a^i is removed:

$$E' = \sum_j^{(\neq i)} \epsilon'(a^j) - \sum_{j>k}^{(\neq i)} (a^j a^k|e^2/r_{jk}|a^j a^k)',$$

where, however, the values of the terms occurring are calculated from the altered self-consistent field (indicated by $'$). The increase in energy for removal of this electron is

$$(E'-E) = -\epsilon(a^i) + \sum_j^{(\neq i)} [\epsilon'(a^j) - \epsilon(a^j)] + \sum_j^{(\neq i)} (a^i a^j|e^2/r_{ij}|a^i a^j)$$
$$- \sum_{j>k}^{(\neq i)} [(a^j a^k|e^2/r_{jk}|a^j a^k)' - (a^j a^k|e^2/r_{jk}|a^j a^k)],$$

which exhibits the terms responsible for the difference between $\epsilon(a^i)$ and the ionization energy of the ith electron. Roughly, the removal from the equation for each other electron of the positive potential due to the ith electron will make each $\epsilon'(a^j) < \epsilon(a^j)$, so the first sum is negative. In so far as we can calculate $\epsilon'(a^j) - \epsilon(a^j)$ by taking the average over the state $u(a^j)$ of the change in potential energy without allowing for the change in $u(a^j)$ itself, the term $\epsilon'(a^j) - \epsilon(a^j) = -(a^i a^j|e^2/r_{ij}|a^i a^j)$ so that the first two sums cancel each other approximately. In this same approximation the third sum vanishes, for its whole value arises from the changes in the functions $u(a^j)$ and $u(a^k)$. This shows us roughly why the quantity $\epsilon(a^i)$ is equal to the energy of removal of the ith electron.

Before proceeding to discussion of the numerical results of the Hartree method we observe that each electron's field depends on the particular configuration being considered. As a consequence the ψ's obtained for the different configurations are not orthogonal and so cannot be made the basis for a calculation in which configuration interaction is taken into account by the perturbation theory. On the other hand the various complete sets belonging to the same configuration do have orthogonal eigenfunctions since their orthogonality depends simply on the spherical harmonics and the spin functions. Therefore we can use the radial functions found by the self-consistent field method to calculate the term structure of any one configuration by the methods developed in previous chapters on the basis of a more strict central-field approximation as set up in Chapter VI.

* Gaunt, Proc. Camb. Phil. Soc. **24**, 89, 111, 426 (1928); **25**, 225, 310 (1929).

9. Survey of consistent-field results.

In this section we shall give an account of the numerical results which have been obtained by the application of Hartree's method to particular atoms. The computing labour involved for accurate results is considerable but the results so far obtained indicate that the method is an excellent one for obtaining good radial functions for a first approximation to the atomic structure problem and as the starting point for more exact calculations. Therefore a number of workers are now engaged in a programme of these calculations, the chief activity being that of Hartree and his students at Manchester. He is constructing a differential analyser of the Bush* type which will be largely employed on this work and will soon add considerably to our knowledge of this type of atomic wave function. Owing to the rapidity with which these calculations are being made this section of our book will probably be out of date soon after publication, so the reader who wants to keep abreast of the field will need to follow the current literature.

According to a summary prepared by Hartree for the summer spectroscopic conference at the Massachusetts Institute of Technology in 1933, the work may be divided into two classes according to the standard of accuracy with which the self-consistency condition is fulfilled. In class A the maximum error is two or three units in the fourth decimal place. Class B includes calculations of a decidedly lower standard of accuracy. Of class A the following were complete at that time:

Oxygen I, II, III, IV	Hartree and Black, Proc. Roy. Soc. **A139**, 311 (1933).
Neon I	McDougall, unpublished.
Sodium II	McDougall, unpublished.
Copper II and chlorine negative ion	Hartree, Proc. Roy. Soc. **A141**, 282 (1933).
Potassium II and caesium II	Hartree, *ibid.* **A143**, 506 (1934).
Silicon V	McDougall, *ibid.* **A138**, 550 (1932).

In addition to these he reported that various persons associated with him were at work on Be, Al IV, A, and Rb II and that he would shortly undertake the negative fluorine ion and Ca III.

In class B the following were complete:

Helium I (and two-electron ions)	Caldwell (Mass. Tech.) and unpublished work by Hartree.
Lithium I	Hargreaves, Proc. Camb. Phil. Soc. **25**, 75 (1928).
Be II and Be I, also B III	Hartree, unpublished.
Ne I, F I, and fluorine negative ion	Brown, Phys. Rev. **44**, 214 (1933).
Boron I	Brown, Bartlett, and Dunn, *ibid.* **44**, 296 (1933).

and in progress were carbon by Torrance (Princeton)†, silicon by Lindsay (Brown), and titanium V, silver II, and mercury III by Hartree.‡

It would take too much space to make a full presentation of all details. Therefore we shall tell only of the programme of Hartree, in collaboration

* Bush, J. Franklin Inst. **212**, 447 (1931).

† Now completed: Torrance, Phys. Rev. **46**, 388 (1934).

‡ For mercury see Hartree, Phys. Rev. **46**, 738 (1934).

with Comrie and Sadler, and the work of Hartree and Black on oxygen, as representing some of the most interesting possibilities of the work.* After standardizing the computing method, Hartree arranged with Comrie and Sadler to carry out the detailed computations. The procedure, as Hartree† explains, was as follows:

"From the earlier work, I make revised estimates of the contributions to the field from the various electron groups, and the computing work carried out professionally is concerned with the calculation of wave functions in the field so constructed, regarded as *given*, and of charge distributions from these wave functions. For reference I will call these calculations the 'standard calculations.' Unless estimates of the contributions to the field have been unusually fortunate, the results of these standard calculations are not yet near enough to the self-consistent field to be quite satisfactory, but they should be near enough for the effect of any variation of the estimates to be treated as a first order variation from the results of the standard calculations.

"A further revision of the estimate is made if necessary and the *variations* in the wave functions, etc. due to the *variations* in the estimates from those used in the standard calculations are calculated, and the variations of wave functions, etc. added to the results of the standard calculations; the variations are so small that this variation calculation is very much shorter and easier than the main calculation. If necessary, further revisions of the estimates are made and corresponding variations from the results of the standard calculations are worked out, until a thoroughly satisfactory approximation to the self-consistent field is obtained."

We now turn to the results for Cl^- and Cu^+. If $R(nl)$ is the normalized radial function of the self-consistent field, then the total charge due to an electron in this state lying within r is given by

$$Z_0(nl, nl; r) = \int_0^r R^2(nl)\, dr,$$

this being a special case of a more general set of radial functions

$$Z_k(n^\alpha l^\alpha, n^\beta l^\beta; r) = r^{-k} \int_{r_1=0}^{r} r_1^k\, R(n^\alpha l^\alpha)\, R(n^\beta l^\beta)\, dr_1, \qquad (1)‡$$

which arise in other parts of the work. The effective nuclear charge for field strength (which we called Z_f in § 9[13]) is then, at radius r,

$$Z(r) = Z - \Sigma\, Z_0(nl, nl; r),$$

the sum being over the nl values of the configuration under consideration. The quantity $[1 - Z_0(nl, nl; r)]$ is the contribution to Z at radius r from unit

* For an extended discussion of the calculation of the terms of the one-electron spectrum, Si IV, starting from the Hartree field for Si V, see McDougall, Proc. Roy. Soc. **A138**, 550 (1932).

† Hartree, Proc. Roy. Soc. **A141**, 282 (1933).

‡ Hartree's notation for this function is $Z_k(n^\alpha l^\alpha, n^\beta l^\beta | r)$.

charge on the nucleus as screened by an electron in an nl state. "It is in terms of these contributions to Z that work on the self-consistent field is usually done, and the extent of the agreement between estimated and calculated contributions usually expressed."

In the standard calculations the contributions to Z were evaluated to three decimals with the last uncertain by one or two units. The standard of self-consistency was such that the difference between the estimated and calculated contributions to Z for each whole group of electrons of the same value did not exceed 0·02 at any radius. Hartree says:

"The contributions to Z may be called 'stable' in the sense that if the estimated contributions from any group is increased over a range of r, the effect of this is to decrease the calculated contribution from this group (and from others also); for an increase of Z means that the attractive field on an electron towards the nucleus is increased, the wave functions of electrons in the field become more compact, and the proportion of the electron distribution lying inside any given radius is increased....

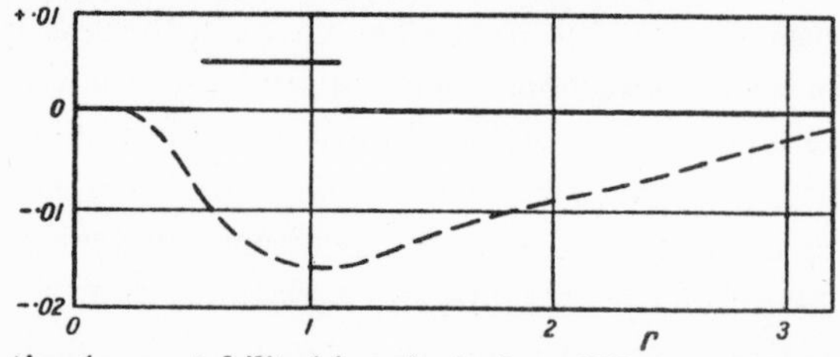

Fig. 8[14]. Illustrating 'over-stability' in calculation of Hartree field for Cu II. Full line shows change of estimated contribution to Z; broken curve shows the consequent change in the calculated contribution to Z.

"For all but the groups of the outermost shell, it is usually if not always the case that the change in the calculated contributions is smaller than the change in the estimated contributions. If this were so for all groups, an iterative process, taking the calculated contributions from one approximation as the estimates for the next, would give a series of calculations with results converging to those for the self-consistent field; though this process would be unnecessarily lengthy, as with experience it is usually possible to make revised estimates better than those obtained by simply taking the calculated contributions of the previous approximation. But for the groups of the outer shell, and particularly the most loosely bound group, there sometimes occurs a phenomenon which may be termed over-stability, in which a change of estimated contribution to Z causes a change in the calculated contribution larger than the change in the estimate. When a group is over-stable in this sense, an iterative process would not converge, but, for small variations, would oscillate and diverge, and it is then quite necessary to choose, as revised estimates of contributions to Z, values better than the calculated contributions of the previous approximation; it is also unusually difficult to make satisfactory estimates and adjustments to them, so the process of approximation to the self-consistent field is most troublesome in such cases."

An example (Fig. 8[14]) of this kind of over-stability is given in the calculations for the $3d^{10}$ group in Cu II. The tables of the radial functions obtained are given in full in Hartree's paper. To exemplify the point discussed in the preceding section that the individual electronic eigenvalues, $\epsilon(a^i)$, agree fairly well with the X-ray terms, Hartree gives this table:

nl	ϵ	ν/R from X-ray data	Per cent. difference
$1s$	658·0	661·6	0·54
$2s$	78·45	81·0	3·1
$2p$	69·86	68·9	1·5
$3s$	8·968	8·9	0·8
$3p$	6·078	5·7	6·7
$3d$	1·195	0·4	200

From the radial functions the total charge distribution in electrons per atomic unit of the radius can be calculated. Using the results for K II and Rb II, the alkali ions preceding and following Cu II in the periodic system, an interesting comparison is made in Fig. 9[14] from Hartree's paper. This shows clearly that the charge distribution for Cu II is much more compact than for either of the alkali ions, a fact which is reflected in the interatomic distances in the crystal lattices of the three metals:

	Half interatomic distance
K II	2·309
Cu II	1·275
Rb II	2·43

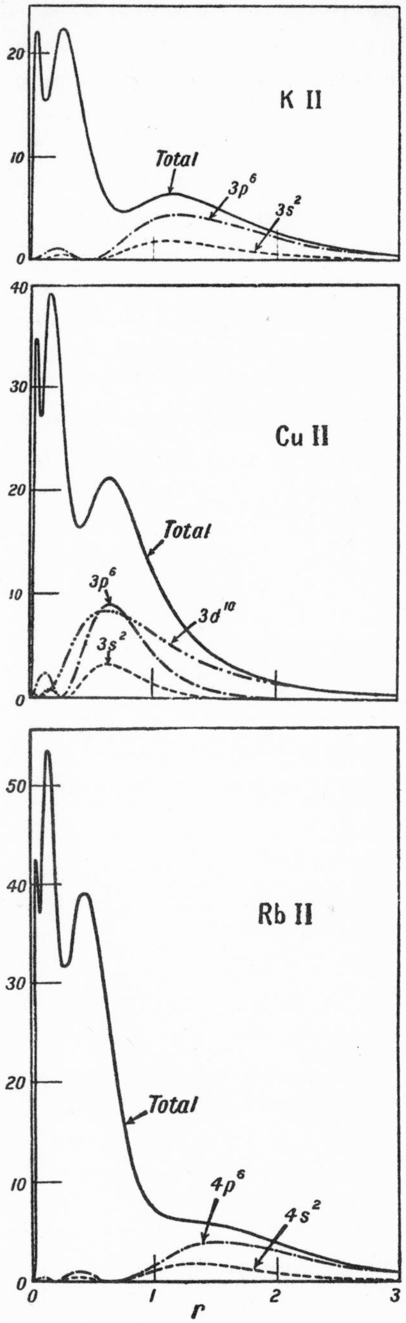

Fig. 9[14]. Charge distributions in K II, Cu II, and Rb II as obtained by the Hartree method.

Comparison of these values with the charge distribution curves shows that in each case half the interatomic distance is just a little greater than the abscissa of the last inflection point in the total charge distribution curve.

10. Self-consistent fields for oxygen.

While most of the calculations on self-consistent fields are for atoms or ions in which all the electrons are in closed shells, this is not the case in oxygen for which an interesting study has been made by Hartree and Black.* This affords a more severe test of the method than closed shell structures in which there is no approximation involved in averaging the field over all directions in space. The results obtained for the total charge

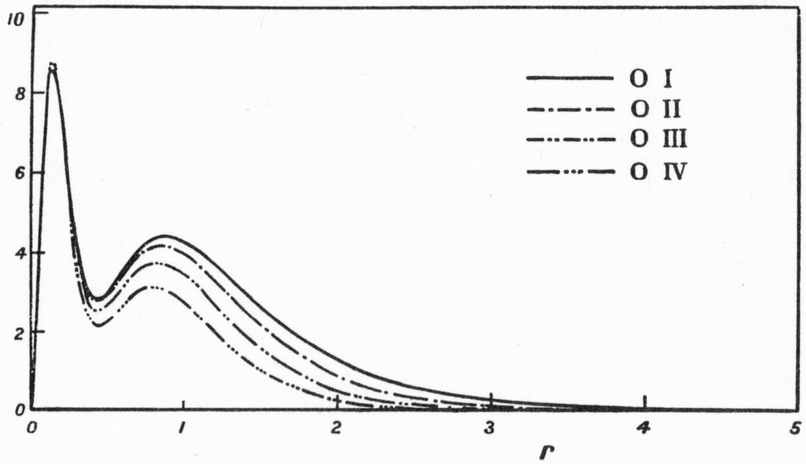

Fig. 10^{14}. Charge distributions in O I, O II, O III, and O IV, as obtained by the Hartree method.

density distribution are shown in Fig. 10^{14} which indicates clearly the increasing compactness of the atom as the degree of ionization is increased.

This closing-in is mostly due to the change in the $2p$ radial function as is brought out nicely in Fig. 11^{14} which shows the radial charge distribution of the $2p$ electron for several stages of ionization. The curves are not accurately representable as a single function plotted to different scales of abscissas, however, which is the approximation implied when the wave functions are treated as hydrogenic with appropriate screening constants.

In using these radial functions to calculate the energy levels of the normal configurations of the oxygen ions, the quantity $\bar{\Psi} H \Psi$ is evaluated, where Ψ is the properly antisymmetrized combination of the one-electron functions, as in $3^{6}6$. This integral can be evaluated for each state of the zero-order

* Hartree and Black, Proc. Roy. Soc. **A139**, 311 (1933).

approximation in terms of the F and G integrals as in Chapter VI and then these may be used as in Chapter VII to find the energies of the Russell-Saunders terms. The present calculation goes beyond that of Chapter VII, for now we have the radial functions and so can get approximate theoretical values of these integrals instead of treating them as adjustable parameters.

There are two points to be noted in extension of the developments of Chapter VI:

First, the results of Chapter VI, especially of §§ 6^6 and 7^6, depend essentially on the fact that the one-electron eigenfunctions, $u(a^i)$, are orthogonal. If the functions are not orthogonal, analogous results can be developed but they are much more complicated. The self-consistent functions are orthogonal

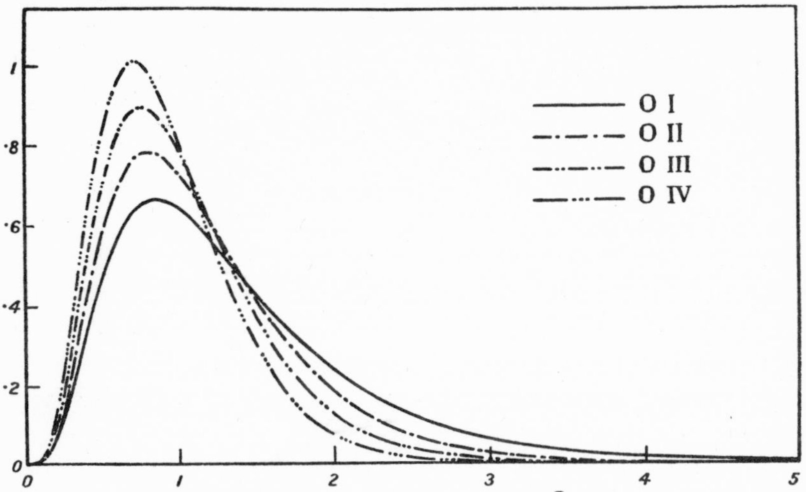

Fig. 11^{14}. Change in $R^2(2p)$ with increasing ionization in O I, O II, O III, and O IV.

with respect to l, m_s, and m_l because the spherical harmonic and spin factors have been retained, but they are not orthogonal with regard to n since different central fields are used for $R(nl)$ and $R(n'l)$. This is most simply remedied by using for $R(nl)$ and $R(n'l)$ not the functions given directly by the Hartree method but linear combinations of them that are orthogonal. This does not affect the value of Ψ because it amounts to performing the same linear transformation on all the functions in the same rows or columns of a determinant. (Compare § 7^{14}.) We pause to note that the fact that this may be done in a large variety of ways without affecting the result shows that the exact distribution of the charge density among different values of n, for the same l, is without significance. Thus for the $1s$ and $2s$ electrons any two orthogonal functions for which $R^2(1s) + R^2(2s)$ is the same function of r as is given by the Hartree field will do equally well. In the actual numerical work it is found convenient to take the following combinations as the orthogonal functions (denoted by °),

$$R^\circ(1s) = R(1s),$$
$$R^\circ(2s) = (1-S^2)^{-\frac{1}{2}}[R(2s) - SR(1s)],$$

where
$$S = \int_0^\infty R(1s)\, R(2s)\, dr.$$

Second, in the Hartree method the radial functions are not, as in Chapter VI, eigenfunctions of the same central-field problem. As a consequence the calculation of the quantities

$$\int \bar{u}(a^i) \left(\frac{1}{2\mu} \boldsymbol{p}^2 - \frac{Ze^2}{r}\right) u(a^i)\, dr$$

becomes an actual calculation, whereas before we could replace

$$\int \bar{u}(a^i) \left(\frac{1}{2\mu} \boldsymbol{p}^2 + U(r)\right) u(a^i)\, dv$$

at once by $E(a^i)$ since the $u(a^i)$ were characteristic functions of this central-field problem. This makes necessary a fair amount of additional computing.

The results of the calculations are very good, with regard to both the absolute values of the terms and the inter-term separations. Measuring each term down from the normal level of the ion with one less electron, the values in atomic units are:

		Calculated		Observed	
O III $2p^2$	^3P	1·988		2·025	
			0·099		0·091
	^1D	1·889		1·934	
			0·148		0·105
	^1S	1·741		1·829	
O II $2p^3$	^4S	1·258		1·290	
			0·129		0·123
	^2D	1·129		1·167	
			0·087		0·062
	^2P	1·042		1·105	
O I $2p^4$	^3P	0·416		0·500	
			0·073		0·072
	^1D	0·343		0·428	
			0·109		0·081
	^1S	0·234		0·347	

The agreement here is as good as can be expected. On the standard first approximation the interval ratio $(^1D - {}^1S)/(^3P - {}^1D)$ should be 1·5 while the observed ratio is only 1·14 in O I and O III (cf. § 5^7), so evidently large perturbations of the second order are important here.

CHAPTER XV

CONFIGURATION INTERACTION

Hitherto we have neglected matrix components of the electrostatic and spin-orbit interactions which connect different configurations when applying the perturbation theory. Now we have to consider what properties of atomic spectra are distinctively associated with these matrix components. So long as the inter-configuration components are neglected, the resulting eigenfunctions are precisely associated with definite configurations; this has been the standpoint of the preceding chapters. If they are no longer neglected, their effect may be treated as a perturbation which causes interacting energy levels to be pushed apart and results in an intermingling of character through linear combination of the Ψ's of the interacting levels.

Such interactions are of quite general importance in atomic spectra. Generally the Ψ of a level cannot be accurately approximated by a single definite function of the type $3^6 6$ or of combinations of them belonging to one configuration such as we have considered exclusively except in the last chapter. In Chapter XIV we have seen that more general forms were needed by Hylleraas in order to obtain an accurate calculation of the normal state of helium. When the variation method is used, all attempts to recognize the trial functions as the wave functions of a central-field approximation are given up. The reader may readily satisfy himself, however, that any wave function depending explicitly on the distance r_{12} of the two electrons in helium corresponds to no definite configuration assignment in any central-field problem. In such calculations of energy levels the central-field terminology is more in the background, however, so we must look elsewhere for really distinctive manifestations of the interaction of configurations.

These are of several kinds. So far very little theoretical work has been done in the way of definite and detailed calculations of these effects, so this chapter must be in the nature of an outline sketch.

First, it may happen that two particular configurations have a large electrostatic interaction which is sufficient to make the order of the terms be other than that given by the ordinary first-order theory of Chapter VII. The best example is the interaction of *sd* and p^2 configurations in Mg I, which we consider in § 1.

Second, one term of a configuration may interact with an entire series of terms with a consequent departure of the series from the simple Rydberg or Ritz formulas which usually hold for series. This we shall discuss in § 2.

Third, terms lying higher than the ionization potential of an atom are in a position to interact strongly with states in the continuous spectrum corresponding to unclosed electron orbits of one electron relative to the ion. As a result such levels lose their sharpness and acquire something of the characteristics of the continuous spectrum. They give rise to broad fuzzy lines, their lifetime is altered and their intensity relationships become quite sensitive to the partial pressure of free electrons in the source. This phenomenon, which is analogous to predissociation in molecules, has not been treated very exactly theoretically in any special case. We shall report the known facts in § 3.

Fourth, as already mentioned in § 1^9, the possibility of so-called 'two-electron jumps,' that is, line emission in which the apparent configuration change involves two of the *nl* values, is connected with breakdown in configuration assignments. This is discussed in § 4.

Finally, in § 5 we discuss the intensity anomalies in the alkali spectra which are associated with spin-orbit interaction between configurations.

1. Interaction of *sd* and p^2 in magnesium.

The general theory of the Russell-Saunders term energies (Chapter VII) gives ^1D and ^3D for the simple $s\,d$ configuration and for the interval $^1D - {}^3D$ the formula

$$^1D - {}^3D = \tfrac{2}{5}G^2(ns, n'd),$$

where the G is defined by $8^6 15$b. From their definition the G's do not need to be positive like the F's, and so at first sight it appears that the singlet might on occasion lie below the triplet on the energy scale. This is actually the case in the $3s\,3d$ configuration of Mg I. However, Bacher* noticed that no reasonable approximation to the radial functions $R(3s)$ and $R(3d)$ would make the G integral negative. Using radial functions determined by a method devised by Slater he calculated $^1D - {}^3D \sim +4000$ cm^{-1}, whereas the actual experimental value is ~ -1600 cm^{-1}.

Thus the ^1D is actually about 5600 cm^{-1} below its position calculated by the ordinary method. This Bacher ascribes to interaction with the $3p^2$ configuration, which gives ^1D, ^3P and ^1S terms. Since $\boldsymbol{L}^2$ and $\boldsymbol{S}^2$ commute with $\Sigma e^2/r_{ij}$, the electrostatic interaction will have no matrix components connecting terms differing in regard to L and S. Likewise since the electrostatic interaction commutes with the parity operator $\mathscr{P}$ of § 11^6 there will only be interaction between configurations of like parity. The parity condition is satisfied, so we expect an interaction of p^2 and *sd*, but only of the ^1D of p^2 with the ^1D of *sd*. The former is above the latter and as all such interactions make the levels move apart the situation is right for an explanation of the effect. What is surprising is that the interaction is large enough, for the ^1D of p^2 is above the ionization level of Mg I, about 15,000 cm^{-1} above the mean of ^1D and ^3D of the *sd* configuration.

By setting up the electrostatic interaction in terms of the zero-order wave

* Bacher, Phys. Rev. **43**, 264 (1933).

functions and transforming to LS eigenfunctions, by methods explained in Chapter VIII, Bacher finds that the non-diagonal matrix component connecting $sd\,{}^1D$ and $p^2\,{}^1D$ amounts to 13,200 cm^{-1}. Their interaction energy is thus comparable with their unperturbed separation. Calculating the effect in detail he finds that the interaction pushes the $sd\,{}^1D$ down to 4000 cm^{-1} below the triplet, whereas it is only observed 1600 cm^{-1} below. The conclusion is that the configuration interaction is certainly adequate to account for the inversion. As to the inaccuracy of the final result, perhaps no better agreement can be expected because of the approximate character of the radial functions used.

With such a large interaction, of course, the exact assignment of the ${}^1D$ below $sd\,{}^3D$ to the configuration sd is quite meaningless. Its true wave function will be a linear combination of those for $sd\,{}^1D$ and $p^2\,{}^3D$ in which there is a large component of the latter. Naturally this alteration in the character of the wave function will bring with it other special features of the spectrum, e.g. altered intensities.

Another detailed study of configuration interaction has been made by Ufford,* who calculated the interaction between the configurations $nd^2n's$, nd^3, $nd\,n's^2$ and $nd^2n''s$. He compares the results with observed data in Ti II and Zr II, and finds definite evidence that the configuration interaction has altered the intervals between terms. We have seen in § 1^{14} that in all the elements where a d-shell is being filled, the energy of binding of an s electron of one higher n value is about the same as that of a d electron, so that there is a large amount of overlapping of the terms arising from such configurations as d^x, $d^{x-1}s$ and $d^{x-2}s^2$. Hence quite generally we may expect large effects due to configuration interaction here, although as yet the only detailed calculations we have are those due to Ufford.

2. Perturbed series.

Perhaps the most interesting effect of configuration interaction is that of producing strong departures from the usual Rydberg-Ritz formulas. Such 'irregular' series have been known for a long time and various explanations for them have been advanced, but Shenstone and Russell,† following a suggestion of Langer,‡ have given very convincing evidence that they are due to configuration interaction.

The Ritz formula is of the form

$$\sigma_n = \frac{\mathsf{R}}{(n+\mu+\alpha\sigma_n)^2} = \frac{\mathsf{R}}{n^{\star 2}}, \tag{1}$$

where σ_n is the absolute energy value of the nth term measured from the series limit, R the Rydberg constant and μ and α are constants, α being small.

* UFFORD, Phys. Rev. **44**, 732 (1933).

† SHENSTONE and RUSSELL, Phys. Rev. **39**, 415 (1932).

‡ LANGER, Phys. Rev. **35**, 649 (1930).

The effective quantum number $n^\star$ is defined by the relation given in (1). It is convenient to study the behaviour of series by plotting $(n^\star - n)$ against σ_n. In that case the Rydberg formula ($\alpha = 0$) is represented by a straight line $n^\star - n = \mu$, parallel to the axis of σ, while the Ritz formula is a straight line with intercept μ on the axis of ordinates and slope α.

Shenstone and Russell find that they can represent many series which depart from (1) by the formula

$$n^\star = n + \mu + \alpha\sigma_n + \frac{\beta}{\sigma_n - \sigma_0}, \tag{2}$$

the added term in the formula for the effective quantum number representing the effect of perturbation of the series members by a foreign level at the term value σ_0. Such a formula plotted with $n^\star - n$ against σ_n represents a hyperbola with a vertical asymptote at $\sigma = \sigma_0$. Before the perturbation of series levels by an extraneous term was recognized, the extraneous term was often counted in as a member of the series. As a result the value of $n^\star - n$ would change rather rapidly by a whole unit in going past the value of the perturbing term because by counting it in the series all higher terms would be erroneously assigned an n one unit too high.

This is illustrated in Fig. 1^{15} which shows the series of the ${}^3D_1$ levels of Ca I. Curve (1) shows the $n^\star - n$ values as usually given and curve 1a shows how this becomes a hyperbola with asymptotes at $\sigma = \sigma_0$ when the higher series members are given the altered n assignments. To indicate how accurately the hyperbolic law is obeyed, curve (2) shows a plot of $(n^\star - n)(\sigma_n - \sigma_0)$, which should not contain the singularity in the curve of $(n^\star - n)$, against σ. In curve (3) the values of $(n^\star - n) - \beta/(\sigma_n - \sigma_0)$ are plotted to an exaggerated scale which shows in another way how accurately the formula is satisfied.

Related to the perturbation of the ${}^3D$ levels of the series is the perturbation in the intervals between the levels of each term. In this series the intervals increase to a maximum up to the perturbing term and then rapidly sink to zero as one goes up the series. This is due to the fact that there are three series, ${}^3D_3$, ${}^3D_2$ and ${}^3D_1$, each of which are separately perturbed by the corresponding level of the extraneous term. As the perturbing term has larger intervals than the series would have if unperturbed, this tends to increase the other intervals by the differential effect of the perturbation.

Numerous examples of other perturbed series are given by Shenstone and Russell. The subject has recently attracted the attention of a number of spectroscopists.* Pincherle has given a theoretical calculation of the perturbation of the $3s\,nd\,{}^1D$ series in Al II by the $3p^2\,{}^1D$ term and finds order-of-magnitude agreement. Beutler has extended the work of Shenstone and

* Pincherle, Nuovo Cimento **10**, 37 (1933);
Beutler, Zeits. für Phys. **83**, 404 (1933);
Rasmussen, *ibid.* **83**, 404 (1933);
Langstroth, Proc. Roy. Soc. **A142**, 286 (1933).

Russell on the perturbed ^1P and ^3P series in Hg I. Rasmussen has obtained new experimental data on the F series in Ba II showing perturbations. Langstroth has made quite a detailed study of perturbations in Ba I, including the effect on intensities. Some general theoretical work on the way in which configuration interaction affects line strengths has been done by Harrison and Johnson.*

The most thorough calculation by quantum-mechanical methods of series irregularities are those given by Whitelaw and Van Vleck for Al II.†

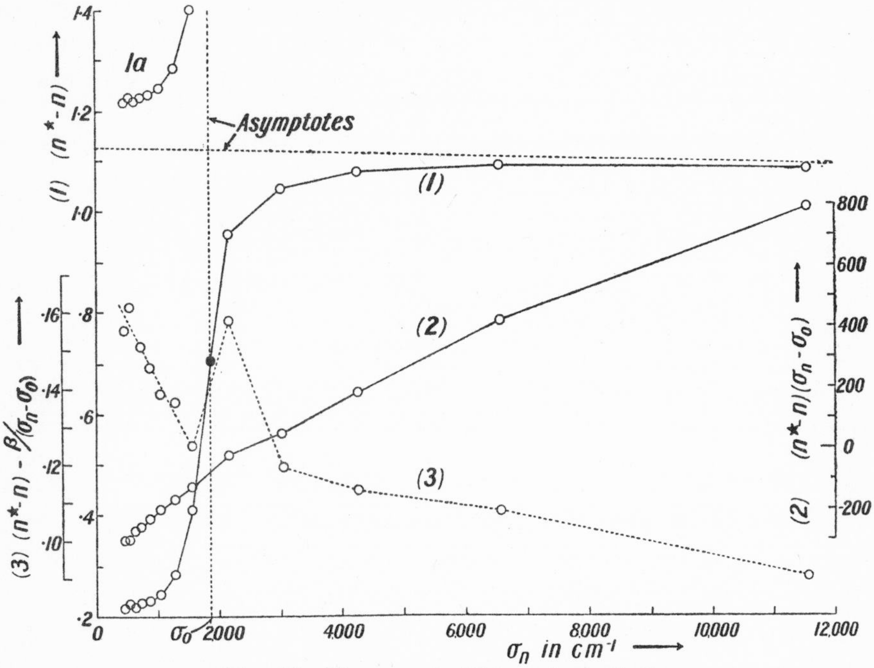

Fig. 1[15]. The perturbed $^3D_1$ series in Ca I.

An explanation involving configuration interaction for the occurrence of inverted doublets in the alkali spectra (§ 8[5]) has been proposed by White and considered in detail by Phillips.‡ This inversion is shown to be capable of production by interaction with configurations of extremely high energy in which one of the electrons of the p^6 shell in the core is excited.

3. Auto-ionization.

At all energies higher than the minimum necessary to remove one electron of an atom to infinity, the spectrum of allowed energy levels is continuous. In the one-electron problem the continuous range of positive energies has

* Harrison and Johnson, Phys. Rev. **38**, 757 (1931).
† Whitelaw, Phys. Rev. **44**, 544 (1933);
Van Vleck and Whitelaw, *ibid.* **44**, 551 (1933).
‡ White, Phys. Rev. **40**, 316 (1932);
Phillips, *ibid.* **44**, 644 (1933).

associated with it a continuum of states for each value of the orbital angular momentum. To label these states the integral total quantum number n of the discrete states is replaced by the continuous variable E which gives the limiting value of the kinetic energy of the electron when at large distances from the nucleus. It may happen that a configuration in which two electrons are excited, but in which both are in discrete levels of the basic one-electron problem, gives rise to energy levels lying above the least ionization energy of the atom, i.e. in the midst of the continuous spectrum. If there is no interaction between these states and the configurations involved in the continuous spectrum, these levels do not exhibit any special properties on account of their location in the continuous spectrum. However, if there is interaction the Ψ of the quasi-discrete level becomes coupled with the Ψ's of the neighbouring levels in the continuum. As a consequence the state assumes something of the character of the states of the continuous spectrum. The most important feature of the states of the continuum is that they are unstable in the sense that one of the electrons moves in an orbit which extends to infinity. Hence as a result of interaction with them the discrete level acquires to some extent the property of spontaneous ionization through one of the electrons moving off to infinity. This property is called *auto-ionization*.

The direct spectroscopic effect associated with auto-ionization is a broadening of lines whose initial levels are subject to the effect and an alteration in the intensity of these lines with variation of the concentration of free electrons in the source. There are two mathematically equivalent ways in which we may regard the broadening of the lines.

We may say that the discrete level, as a level, remains sharp, but that owing to coupling with the continuous spectrum there is a probability per unit time that the atom will make a radiationless change of state over into a state of the continuum of equal energy. Owing to this possibility of leaving the discrete state rather quickly, the interaction of the atom with the radiation field is limited to the production of short wave trains of mean duration equal to the mean life of the atom in the discrete level. In the spectroscope these short wave trains give rise to a broadened line because the Fourier integral representation of the short wave train involves a band of frequencies in the neighbourhood of the mean frequency.

Another way of regarding the matter is to treat the discrete level as completely assimilated into the continuum. At each energy E the Ψ of the corresponding state in the continuum will contain a certain component of the eigenfunction of the assimilated discrete level. This will be larger the nearer E is to the original position of the discrete level. Assuming that the assimilated discrete level, A, is capable of strong radiative combination with a lower discrete level, B, while the ordinary continuum combines weakly or not at all with B, it is clear that the intensity of radiation from the continuum in combination with the discrete level B will be dependent mainly on the amount of $\Psi(A)$ contained in the eigenfunction of the continuum at each energy. As this is a maximum at the original location of the discrete level A, we obtain also by this argument a broadened line at this place.

Auto-ionization is thus a consequence of the presence of matrix components of the Hamiltonian connecting a discrete level above the ionization energy with the states of the continuum at the same energy. If these are not to vanish, these states must be of the same parity and the same J value; moreover in case of Russell-Saunders coupling they must have the same L and S values. The perturbation theory for this case requires some modification to take into account the fact that the interaction is with a continuous spectrum of states. This has been discussed by Wentzel.* Apart from the general formulation and some qualitative discussion based on the application of the selection rules, very little has been done so far in the way of attempted quantitative calculations of the amount of auto-ionization. The theoretical problem is related to that underlying the theory of predissociation, the analogous phenomenon in molecules, whereby a molecule may spontaneously dissociate if put into a quasi-discrete state in which its energy exceeds the energy necessary for dissociation.

In view of the lack of an accurate detailed theory we shall have to be content with a review of the main experimental facts. The first evidence for auto-ionization was non-spectroscopic in character. Auger† showed that when a gas in a Wilson cloud chamber absorbs X-rays there are frequently several electron tracks diverging from the same point. One trail is long and is interpreted as due to the primary photo-electron ejected from the K shell by the light quantum. Another of the trails was found to be of the correct length to correspond to a kinetic energy of about $K-2L$, where K and L are written for the corresponding excitation energies of the atom. Other trails, when present, correspond to electrons of considerably less energy. The interpretation of the trail of energy $K-2L$ is clear from the diagram (Fig. 2[15]). The atom is left in the K level by removal of an electron from the K shell. Of equal energy

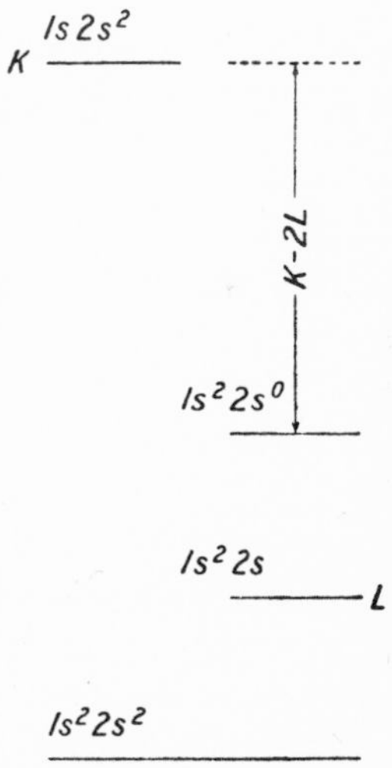

Fig. 2[15]. Auto-ionization or Auger effect in a K level.

* Wentzel, Zeits. für Phys. **43**, 524 (1927); Phys. Zeits. **29**, 321 (1928).

† Auger, Comptes Rendus, **180**, 65 (1925); **182**, 773, 1215 (1926); Jour. de Physique **6**, 205 (1925); Ann. de Physique **6**, 183 (1926). See also Locher, Phys. Rev. **40**, 484 (1932).

with the K level is the configuration $1s^2\,2s^0\,Es$, where $E=K-2L$. Hence the atom may pass over into this configuration by the auto-ionization process, which implies expulsion of an electron with energy $K-2L$. The argument is correct in principle, though rather rough in detail in that we have supposed the energy of removal of the second $2s$ electron also equal to L, but such roughnesses are acceptable in view of the lack of precision with which the electron energies are measured. The short electron trails are supposed to be due to further auto-ionization processes occurring in the ion in the $1s^2\,2s^0$ configuration. Such spontaneous ionization

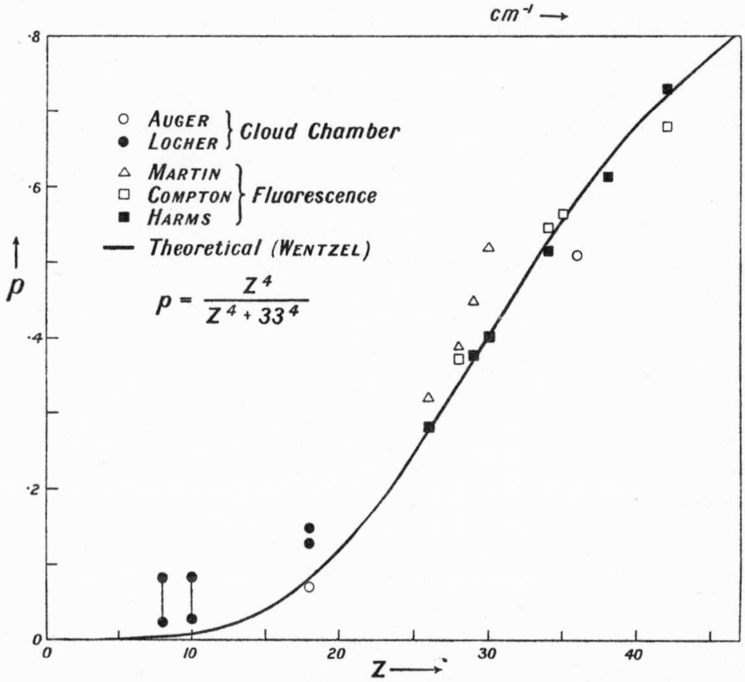

Fig. 3^{15}. Fraction, p, of atoms emitting K radiation as a function of the atomic number. The probability of auto-ionization of the K level is given by $1-p$.

processes are by no means rare. An atom in the K level may get out either by radiation of a K line or by auto-ionization. For light elements the auto-ionization processes occur much more frequently than the radiation, while the reverse is true for heavy elements. This is shown in Fig. 3^{15}, where the ordinate, p, gives the fraction of all atoms in the K level which emit K radiation, so that $1-p$ gives the fraction which pass from the K level by auto-ionization. The abscissas are the atomic numbers Z. The heavy curve is the theoretical value as estimated by Wentzel. According to him the probability of auto-ionization is nearly independent of Z, so the whole variation of p

with Z is due to the increase of the radiative transition probability, approximately as Z^4.

Recognition of the importance of auto-ionization in atomic spectra in the optical region is due to Shenstone and to Majorana.* Shenstone discussed the effect in general and especially in connection with the spectrum of Hg I. Among other things he showed how it provides an explanation of the ultra-ionization potentials discovered by Lawrence.† Lawrence found that the probability of ionization of Hg atoms by impact of electrons having energies close to the ionization potential shows discontinuities of slope as if new modes of producing ionization are becoming effective with increase in the voltage. These were interpreted by Shenstone as due to excitation to the levels which are subject to auto-ionization. This view seems adequate to explain the observed facts, although at present there is no detailed correlation between the values of the ultra-ionization potentials and known spectroscopic levels. Shenstone thought the levels responsible for the ultra-ionization potentials were probably connected with the d^9s^2p configuration in Hg, but later work by Beutler‡ on the far-ultraviolet absorption spectrum shows that this particular identification cannot be the correct one.

Majorana's paper deals with the ^3P terms due to $4p^2$, $5p^2$ and $6p^2$ in the spectra of Zn, Cd and Hg respectively. Here the striking thing is that in all three cases the $^3P_0$ and $^3P_1$ levels are known but the $^3P_2$ cannot be found. The absence of $^3P_2$ is interpreted as due to strong auto-ionization probability, and this raises the problem as to why this level should be so much more affected than the other levels of the same term. The auto-ionization arises by interaction with the continua associated with the $4s\infty l$ limit in Zn (correspondingly the $5s\infty l$ in Cd and the $6s\infty l$ in Hg). The $4s\,Ep$ continuum is due to an odd configuration and hence cannot give rise to auto-ionization of the even $4p^2$ configuration, which can only be unstable through interaction with even parts of the continuum, that is, with $4s\,Es$ and $4s\,Ed$. These give rise directly to interactions making the $4p^2\,{}^1S$ and ^1D terms subject to auto-ionization. Through partial breakdown of the LS coupling in the p^2 configuration, $\Psi(^3P_2)$ acquires a component of $\Psi(^1D_2)$ and thus becomes subject to auto-ionization through interaction with $4s\,Ed$. As $^3P_1$ is the only level in the configuration with $J=1$, it is not subject to breakdown of LS coupling and so escapes auto-ionization. However, $^3P_0$ should become mixed with $^1S_0$ and hence unstable; the fact is that the interaction here seems to be much weaker than for $^3P_2$ but Majorana's theoretical discussion does not show clearly why this should be so.

* SHENSTONE, Phys. Rev. **38**, 873 (1931);
MAJORANA, Nuovo Cimento **8**, 107 (1931).
† LAWRENCE, Phys. Rev. **28**, 947 (1926). See also SMITH, Phys. Rev. **37**, 808 (1931).
‡ BEUTLER, Zeits. für Phys. **86**, 710 (1933).

The arc spectrum of copper is known to have a large number of levels above the ionization limit, some of which give rise to broad lines. Results of the experiments of Allen* find satisfactory explanation in terms of the auto-ionization process as was pointed out by Shenstone in a note appended to Allen's papers. Particularly interesting is the behaviour of lines in multiplets arising from a ^4D term which lies in the range from 95 to 2164 cm^{-1} above the $d^{10}\infty l$ ionization limit of Cu. This ^4D arises from an even configuration ($3d^9\,4s\,5s$) and so can show auto-ionization through interaction with ^2D of $d^{10}\,Ed$ if there is enough breakdown of LS coupling to permit violation of the selection rule on S. But according to the J selection rule only the levels $^4D_{\frac{3}{2}}$ and $^4D_{\frac{5}{2}}$ can show auto-ionization. Therefore in a multiplet, lines arising from the $^4D_{\frac{1}{2}}$ and $^4D_{\frac{7}{2}}$ levels should be sharp while those originating from the other two levels of the quartet should be broadened. This is in fact the case. Allen measured the line breadths in a Cu arc running in air at various pressures up to 80 atmospheres. The breadth of all lines originating from the quartet levels was found to increase linearly with the pressure, with the same rate of increase for unit pressure increment. But lines from $^4D_{\frac{3}{2}}$ and $^4D_{\frac{5}{2}}$ were found to approach finite width at zero pressure whereas those from the other two levels approached vanishing width at zero pressure.

Allen also studied the variation of relative intensity of the lines arising in a ^4D term with change in the arc current. The arc in these experiments ran at atmospheric pressure. It was found that the lines from the unstable levels of the quartet term were very sensitive to arc current, approaching zero relative intensity at currents below one ampere and approaching constant values for currents above twelve amperes. These intensities are relative to other lines of the spectrum, auxiliary experiments having shown that lines not subject to auto-ionization retain constant relative intensity with variation of arc current. It seems likely that this effect is due to the fact that the atoms in the unstable states normally fall apart by auto-ionization before they can radiate. But if there is a high concentration of free electrons around the ions the inverse process, in which free electrons are caught up by an ion, comes into play. This could nullify the effect of auto-ionization and build up the strength of the lines from the unstable levels. This view has not been subjected to a quantitative discussion.

Auto-ionization effects observed in spectra of the alkaline earths and in rare-gas spectra, as shown by White,† provide further examples of the operation of the selection rules.

Beutler‡ has found a long absorption series in Hg vapour corresponding

* Allen, Phys. Rev. **39**, 42, 55 (1932).
† White, Phys. Rev. **38**, 2016 (1931).
‡ Beutler, Zeits. für Phys. **86**, 710 (1933); **87**, 19, 176 (1933).

to the transitions $5d^{10}\,6s^2\,{}^1S_0 \rightarrow 5d^9\,6s^2\,np\,{}^1P_1$. These are all subject to broadening by the auto-ionization process and the experiments show that the amount of the broadening decreases as one goes up the series, showing that terms near the ionization limit are more unstable than those of similar character at higher energies.

4. Many-electron jumps.

As we observed in § 1[9], it is a consequence of the fact that the various moments—electric-dipole, electric-quadrupole, magnetic-dipole, etc.—are quantities of type F that the only non-vanishing matrix components connect states differing in regard to one individual set of quantum numbers. Therefore in the approximation in which an energy level is assigned definitely to one configuration of a central-field problem, radiative transitions occur only between configurations differing in regard to one of the nl values. Such transitions are called one-electron jumps. However, in many spectra, especially in the elements of the iron group, lines are observed corresponding to transitions in which two of the nl values change. These are known as two-electron jumps.

This is clearly due to the breakdown of precise configuration assignments.* If we know accurately the eigenfunctions $\Psi(A)$ and $\Psi(B)$ corresponding to two levels A and B, the existence of radiative transitions between them arises from the non-vanishing of the matrix components $(A|\alpha|B)$ connecting the states of A with those of B, where α stands for any type of electric or magnetic moment of the atom. If we try to describe the atom in terms of a central-field approximation, $\Psi(A)$ and $\Psi(B)$ will appear as expansions in terms of the eigenfunctions of the states built on that central field. In general this expansion will involve several different configurations for $\Psi(A)$ and for $\Psi(B)$.

If the configuration interaction is not too great, one configuration in each expansion will appear with a considerably larger coefficient than the others, and the experimental spectroscopist will assign the level to that configuration. The one-electron jump rule will not then apply to this approximate label, for A may combine with configurations excluded by this rule in virtue of the other configurations involved in its eigenfunction. In this way apparent two- or many-electron jumps may be permitted.

This view appears to be fully adequate for the interpretation of the apparent two-electron jumps. However, there have not been as yet any quantitative investigations of this point. What is needed are estimates of the amount of configuration interaction either from a precise calculation from the fundamental wave equation, or inferentially from the observed

* Condon, Phys. Rev. **36**, 1121 (1930);
Goudsmit and Gropper, *ibid.* **38**, 225 (1931).

perturbations of the energy levels. With this one could calculate the relative transition probabilities for the different configuration transitions to compare with measurements of these transition probabilities.

5. Spin-orbit perturbation of doublet intensities.

We have seen in § 2^9 that one of the simplest conclusions from the sum rules for line strengths is that the two components of the doublet $^2P \to {}^2S$ should have strengths in the ratio 2:1. This result involves the assumption that the radial factor of the wave function is the same for the two levels of the ^2P term. The result is generally in agreement with experimental results, as mentioned in § 9^5, but in the case of the higher members of the principal series of Cs the ratio is found to be larger than two. The different experimental results are not in complete accord, but agree in indicating that the ratio for the second member of the principal series, $7\,^2P \to 6\,^2S$, is between 3·5 and 4·5, although the ratio seems to have the normal value for the first member, $6\,^2P \to 6\,^2S$. This was explained by Fermi* as an effect of the matrix components of the spin-orbit interaction connecting different terms of the ^2P series.

The spin-orbit interaction, from § 4^5, is given by $\xi(r)\,\boldsymbol{L}\cdot\boldsymbol{S}$. For $^2P_{\frac{3}{2}}$ the value of $\boldsymbol{L}\cdot\boldsymbol{S}$ is $\frac{1}{2}\hbar^2$, while for $^2P_{\frac{1}{2}}$ it is $-\hbar^2$. The non-diagonal matrix components connecting different members of the ^2P series are then

$$(n\,^2P_J\,M|\xi(r)\,\boldsymbol{L}\cdot\boldsymbol{S}|n'\,^2P_J M) = (\boldsymbol{L}\cdot\boldsymbol{S})'\int_0^\infty \xi(r)\,R(np)\,R(n'p)\,dr,$$

where $(\boldsymbol{L}\cdot\boldsymbol{S})'$ is the appropriate value of $\boldsymbol{L}\cdot\boldsymbol{S}$. Because of this perturbation, the first order eigenfunctions are altered and by different amounts for the two levels of ^2P. As the perturbation is diagonal in l, J and M, it is only the radial factor of the eigenfunction which is altered. If we write $R(np\,J)$ for the radial factor in ^2P_J in the first approximation, and $R_0(np)$ for the radial factor before the spin-orbit interaction is considered, we have

$$R(np\,\tfrac{3}{2}) = R_0(np) + \tfrac{1}{2}\hbar^2 \sum_{n'} R_0(n'p) \frac{\int_0^\infty \xi(r)\,R_0(np)\,R_0(n'p)\,dr}{E_n - E_{n'}},$$

$$R(np\,\tfrac{1}{2}) = R_0(np) - \hbar^2 \sum_{n'} R_0(n'p) \frac{\int_0^\infty \xi(r)\,R_0(np)\,R_0(n'p)\,dr}{E_n - E_{n'}}.$$

When we calculate the strengths of the two lines $^2P_{\frac{3}{2}} \to {}^2S$ and $^2P_{\frac{1}{2}} \to {}^2S$ the calculations go through as before except that we must use these altered radial functions in calculating the integral $\int r\,R(np)\,R(n''s)\,dr$ whose square enters as a factor in the strength of the line. The ratio of the line strengths

* Fermi, Zeits. für Phys. **59**, 680 (1929).

for ${}^2P_{\frac{3}{2}} \to {}^2S$ and ${}^2P_{\frac{1}{2}} \to {}^2S$ is thus 2, from the angle and spin factors, multiplied by the square of the ratio of the two radial integrals,

$$\left[\int r\, R(np\,\tfrac{3}{2})\, R(n''s)\, dr\right] : \left[\int r\, R(np\,\tfrac{1}{2})\, R(n''s)\, dr\right].$$

This is what makes possible the departure of the line strength ratio from the value given by the simpler considerations of § 2[9].

The numerical values of the integrals involved have been estimated by Fermi, who finds that in the case of Cs the values are probably large enough to account for the observed intensity ratio. The quantities

$$\hbar^2 \int_0^\infty \xi(r)\, R_0(np)\, R_0(n'p)\, dr$$

which measure the change in the radial function are small compared with the energy differences $E_n - E_{n'}$, so the actual change in the radial functions is quite small. For the first member of the principal series the small change does not produce an observable effect because its strength is so great relative to the higher members of the series (cf. Tables 7[5] and 8[5]) that the introduction into it of small components of the radial functions of the higher members produces no appreciable effect. For the higher series members, however, the situation is reversed. Introduction of a small component of the radial function of the lowest ${}^2P$ term into their radial functions produces a relatively large effect because of the fact that the first series member combines so much more strongly with the normal ${}^2S$ level.

CHAPTER XVI

THE ZEEMAN EFFECT

In § 10^5 we have treated the Zeeman effect for one-electron spectra. This serves as a simple pattern for the present chapter, which is devoted to the Zeeman effect for the general case.

1. The 'normal' Zeeman effect.

The argument of § 10^5 which leads to $10^5 5$ is valid for each of the N electrons in the atom, so that a magnetic field of strength $\mathscr{H}$ in the direction of the z-axis contributes to the Hamiltonian the term

$$H^M = o \sum_{i=1}^{N} (L_{zi} + 2S_{zi}) = o\,(L_z + 2S_z), \tag{1}$$

with $o = e\mathscr{H}/2\mu c$ as in $10^5 6$. The whole theory of the effect of a magnetic field on the energy levels of an atom is therefore given by a study of this perturbation term.

Before developing the theory from this standpoint it will be instructive to consider a little of the history of the Zeeman effect. Prior to the introduction of the electron-spin hypothesis in 1925 physicists had attempted to give a formal description of atomic spectra in terms of a purely orbital scheme of electronic states, so that the entire angular momentum of the atom was given by the sum of the $\boldsymbol{L}$ vectors for the individual electrons. In such a scheme L_z, the sum of the z-components of the orbital angular momenta, is a constant of the motion which is quantized to integer values $M_L\hbar$.* Likewise the magnetic perturbation energy is given by (1) without the S_z term. Therefore the magnetic energy is simply o times a constant of the motion, so the effect on an energy level characterized by the quantum number L would be to split it into $2L+1$ equally-spaced levels by addition of the quantity $o\hbar M_L$ to the unperturbed value, where $-L \leqslant M_L \leqslant L$.

The ordinary dipole radiation involves transitions between states for which $\Delta M_L = 0$, in which case the radiation is linearly polarized with the electric vector in the plane determined by the z-axis and the direction of propagation, and $\Delta M_L = \pm 1$, in which case the radiation is circularly polarized when viewed along the z-axis. Since, according to the foregoing, all the energy levels are split by the magnetic field in the same way, the observed splitting of all lines is the same on this view of the matter. The transitions for which $\Delta M_L = 0$ will all have the same frequency and will coincide with the unperturbed line. The transitions for which $M_L \to M_L + 1$

* This is actually the case for singlet levels ($S = 0$) in Russell-Saunders coupling.

on emission will all have the same frequency which will be in fact shifted toward lower frequency from the unperturbed line by an amount $o/2\pi$ sec^{-1} or $o/2\pi c$ cm^{-1}. The transitions for which $M_L \rightarrow M_L - 1$ will be shifted toward higher frequency by the same amount. For longitudinal observation, that is, observation in the direction of the magnetic field, the radiation is circularly polarized: for the high frequency component it is polarized in the direction of the positive current which in a solenoid would produce the applied magnetic field, for the low frequency component it is oppositely polarized.

The conventional way of exhibiting this result, which we shall adopt, is shown in Fig. 1[16]. With longitudinal observation one sees simply the two components which are circularly polarized in the directions indicated if one imagines the magnetic field to be up from the paper. Observation in a direction transverse to the field shows all three components, the undisplaced one showing linear polarization parallel (π) to the field, the others showing linear polarization perpendicular (σ, for *senkrecht*) to the field.

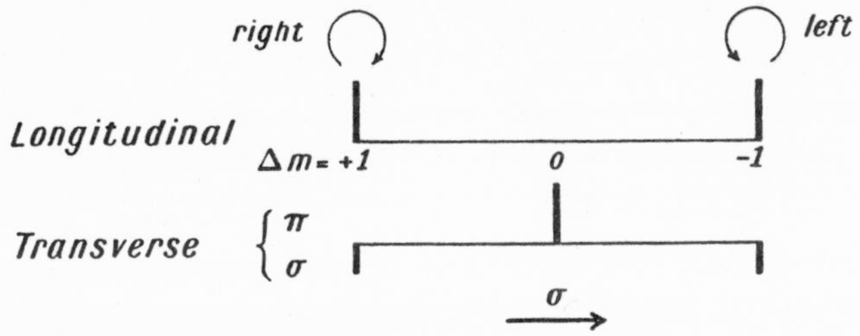

Fig. 1[16]. The normal Zeeman triplet.

The influence of a magnetic field on spectral lines was discovered by Zeeman in 1896. Soon after, a simple electron theory of the effect predicting the normal triplet was given by Lorentz. Although we have described the theory of the normal triplet in terms of quantum mechanics, this is by no means necessary. The same result follows from a consideration of the action of a magnetic field on a vibrating electron moving according to classical mechanics. By comparison of the observed behaviour of some of the zinc and cadmium lines with the Lorentz theory, it was found that the displacement corresponded to negatively charged particles having the same value of e/μ as had been found for electrons by deflection of cathode ray beams. Thus, in the very beginning of modern atomic theory, the Zeeman effect provided very strong evidence that the emission of light is connected with the motions of electrons in the atoms.

This agreement between theory and experiment for the lines of zinc and cadmium gave a great impetus to the infant electron theory. Almost a year elapsed before further studies of the new effect on other lines showed that the normal Lorentz triplet is by no means the general behaviour of a spectral line under the influence of the field. Generally speaking the patterns are much more complicated, so that it was something of a happy accident that the first lines studied were those to which the simple Lorentz theory was applicable. With characteristic love of the simple, physicists fell into the habit of referring to the triplet pattern as 'normal'; all other forms of the effect were called 'anomalous.' This characterization remains in use even to-day although we are now in possession of a complete theory which gives a rational account of all the observed effects. In terms of the complete theory the so-called normal effect is simply a special case in which the effects of electron spin are absent.

Nevertheless the 'anomalous' effect remained a great puzzle until the electron-spin hypothesis was introduced—and that was a quarter of a century later. Perhaps the continued use of the adjective anomalous is appropriate in view of the long time during which the general Zeeman effect resisted the attempts of physicists to understand it. Before leaving this brief review of the historical setting it may be remarked that the essential feature of the spin theory is the occurrence of (L_z+2S_z) in (1) rather than (L_z+S_z). That is, that the ratio of magnetic moment to angular momentum is twice as great for spin angular momentum as for orbital angular momentum. If this were not the case introduction of electron spin would make no alteration in the theory of the normal triplet as just sketched, for J_z would simply be written for L_z everywhere with no change in the observable results.

2. The weak-field case: Russell–Saunders terms.

Most of the work on the Zeeman effect applies to spectra in which Russell-Saunders coupling holds quite accurately, so it will be convenient to begin the study with this case. As we have already seen in § 10^5, special effects arise if the magnetic field is strong enough to produce energy changes comparable with the separation of the levels of a term, so it is also convenient at first to consider the case of weak fields, meaning by this fields whose effects are small compared with the unperturbed intervals—the fields commonly used are weak in this sense. The name Paschen-Back effect is given to the special features of the Zeeman effect which arise when the field is not weak.

To find the weak-field perturbation of a Russell-Saunders level characterized by SLJ, we must calculate that part of the matrix of H^M ($1^{16}1$) which refers to this level. Since L_z+2S_z can be written as J_z+S_z, and since S_z commutes with J_z, this part of the matrix will be diagonal with respect to M. The diagonal element of J_z for the state $SLJM$ is $M\hbar$; the diagonal element of S_z for this state is obtained as a special case of the results given in § 10^3. If we identify $\boldsymbol{J}_1$ with $\boldsymbol{S}$ and $\boldsymbol{J}_2$ with $\boldsymbol{L}$, the diagonal matrix element of S_z is given by a combination of $10^3 2a$ and $9^3 11$. Altogether we find for the diagonal element of the perturbation energy the value

$$(\gamma SLJM|H^M|\gamma SLJM) = o\hbar gM, \tag{1}$$

in which
$$g(SLJ) = 1 + \frac{J(J+1) - L(L+1) + S(S+1)}{2J(J+1)}. \tag{2}$$

If g were equal to unity, (1) would give the splitting which corresponds to the normal Lorentz triplet.

We see that the energies of the perturbed states are distributed symmetrically around that of the unperturbed level. These states are $2J+1$ in number, as before, but the scale of the splitting differs from the simple theory by the factor g. The factor g differs from unity by a term which arises from the matrix component of S_z. For singlet levels $S=0$ and $L=J$, so the additional term vanishes. In other words, the theory of the normal Lorentz triplet applies to the Zeeman effect of lines which are combinations of singlet levels.

These results which we have so easily obtained are in good accord with the empirical data. The winning of the result (2) as a generalization from the empirical data was not so easy and represents a great amount of study by spectroscopists. This formula expresses implicitly Preston's rule* which says that all the lines in a spectral series have exactly the same Zeeman pattern. This is due to the fact that the perturbation energy (1) is independent of γ, which stands for all quantum numbers other than those explicitly written. More generally, the Zeeman pattern depends only on the S, L, and J of the initial and final levels, provided Russell-Saunders coupling obtains for the atom in question.

Runge's rule† says that the displacements of the Zeeman components from the unperturbed line are rational multiples of the Lorentz splitting $o/2\pi c$. Since the displacement in a line is actually the difference of the displacements of the initial and final states, and since the g's are rational fractions, it follows that this rule is contained in our results.

The empirical fact that the Zeeman effect of Russell-Saunders terms is given by the formulas (1) and (2) was worked out by Landé, and g is usually known as the Landé factor. Landé's formulation was based on the modern experimental measurements by Back. This work is admirably summarized in the book by Back and Landé, *Zeemaneffekt und Multiplettstruktur der Spektrallinien*, which furnishes a good account of the subject as it stood just before the electron spin and quantum mechanics gave the theoretical basis for the experimental material.

In Table 1^{16} are given the g values for the terms of interest. The table exhibits some interesting properties. For $J=0$ the formula for g gives $\frac{0}{0}$, but

* Preston, Trans. Roy. Soc. Dublin **7**, 7 (1899).

† Runge, Phys. Zeits. **8**, 232 (1907).

the splitting is of course zero since $J=0$ implies $M=0$. As already remarked, singlets give the normal separation. For $L=S$ the g factor equals $\frac{3}{2}$ for all values of J. A somewhat surprising feature is the fact that certain terms such as ${}^4D_{\frac{1}{2}}$, ${}^5F_1$ and ${}^6G_{\frac{3}{2}}$ have $g=0$ and so are not split by the magnetic field; others such as ${}^6F_{\frac{1}{2}}$, ${}^7G_1$ and ${}^8G_{\frac{1}{2}}$ show negative g values, which means that the sense of the effect is turned around, as it would be if the electron were a positive charge and the behaviour 'normal.'

TABLE 1^{16}. *Landé g factors for Russell-Saunders terms.*

S	J	L: $S\,0$	$P\,1$	$D\,2$	$F\,3$	$G\,4$	$H\,5$
$\frac{1}{2}$	$L-\frac{1}{2}$	—	$\frac{2}{3}=0.667$	$\frac{4}{5}=0.800$	$\frac{6}{7}=0.857$	$\frac{8}{9}=0.889$	$\frac{10}{11}=0.909$
	$L+\frac{1}{2}$	2	$\frac{4}{3}=1.333$	$\frac{6}{5}=1.200$	$\frac{8}{7}=1.143$	$\frac{10}{9}=1.111$	$\frac{12}{11}=1.091$
1	$L-1$	—	$\frac{0}{0}$	$\frac{1}{2}=0.500$	$\frac{2}{3}=0.667$	$\frac{3}{4}=0.750$	$\frac{4}{5}=0.800$
	L	—	$\frac{3}{2}=1.500$	$\frac{7}{6}=1.167$	$\frac{13}{12}=1.083$	$\frac{21}{20}=1.050$	$\frac{31}{30}=1.033$
	$L+1$	2	$\frac{3}{2}=1.500$	$\frac{4}{3}=1.333$	$\frac{5}{4}=1.250$	$\frac{6}{5}=1.200$	$\frac{7}{6}=1.167$
$\frac{3}{2}$	$L-\frac{3}{2}$	—	—	0	$\frac{2}{5}=0.400$	$\frac{4}{7}=0.571$	$\frac{2}{3}=0.667$
	$L-\frac{1}{2}$	—	$\frac{8}{3}=2.667$	$\frac{6}{5}=1.200$	$\frac{36}{35}=1.029$	$\frac{62}{63}=0.984$	$\frac{32}{33}=0.970$
	$L+\frac{1}{2}$	—	$\frac{26}{15}=1.733$	$\frac{48}{35}=1.371$	$\frac{26}{21}=1.238$	$\frac{116}{99}=1.172$	$\frac{162}{143}=1.133$
	$L+\frac{3}{2}$	2	$\frac{8}{5}=1.600$	$\frac{10}{7}=1.429$	$\frac{4}{3}=1.333$	$\frac{14}{11}=1.273$	$\frac{16}{13}=1.231$
2	$L-2$	—	—	$\frac{0}{0}$	0	$\frac{1}{3}=0.333$	$\frac{1}{2}=0.500$
	$L-1$	—	—	$\frac{3}{2}=1.500$	1	$\frac{11}{12}=0.917$	$\frac{9}{10}=0.900$
	L	—	$\frac{5}{2}=2.500$	$\frac{3}{2}=1.500$	$\frac{5}{4}=1.250$	$\frac{23}{20}=1.150$	$\frac{11}{10}=1.100$
	$L+1$	—	$\frac{11}{6}=1.833$	$\frac{3}{2}=1.500$	$\frac{27}{20}=1.350$	$\frac{19}{15}=1.267$	$\frac{17}{14}=1.214$
	$L+2$	2	$\frac{5}{3}=1.667$	$\frac{3}{2}=1.500$	$\frac{7}{5}=1.400$	$\frac{4}{3}=1.333$	$\frac{9}{7}=1.286$
$\frac{5}{2}$	$L-\frac{5}{2}$	—	—	—	$-\frac{2}{3}=-0.667$	0	$\frac{2}{7}=0.286$
	$L-\frac{3}{2}$	—	—	$\frac{10}{3}=3.333$	$\frac{16}{15}=1.067$	$\frac{6}{7}=0.857$	$\frac{52}{63}=0.825$
	$L-\frac{1}{2}$	—	—	$\frac{28}{15}=1.867$	$\frac{46}{35}=1.314$	$\frac{8}{7}=1.143$	$\frac{106}{99}=1.071$
	$L+\frac{1}{2}$	—	$\frac{12}{5}=2.400$	$\frac{58}{35}=1.657$	$\frac{88}{63}=1.397$	$\frac{14}{11}=1.273$	$\frac{172}{143}=1.203$
	$L+\frac{3}{2}$	—	$\frac{66}{35}=1.886$	$\frac{100}{63}=1.587$	$\frac{142}{99}=1.434$	$\frac{192}{143}=1.343$	$\frac{50}{39}=1.282$
	$L+\frac{5}{2}$	2	$\frac{12}{7}=1.714$	$\frac{14}{9}=1.556$	$\frac{16}{11}=1.455$	$\frac{18}{13}=1.385$	$\frac{4}{3}=1.333$
3	$L-3$	—	—	—	$\frac{0}{0}$	$-\frac{1}{2}=-0.500$	0
	$L-2$	—	—	—	$\frac{3}{2}=1.500$	$\frac{5}{6}=0.833$	$\frac{3}{4}=0.750$
	$L-1$	—	—	3	$\frac{3}{2}=1.500$	$\frac{7}{6}=1.167$	$\frac{21}{20}=1.050$
	L	—	—	2	$\frac{3}{2}=1.500$	$\frac{13}{10}=1.300$	$\frac{6}{5}=1.200$
	$L+1$	—	$\frac{7}{3}=2.333$	$\frac{7}{4}=1.750$	$\frac{3}{2}=1.500$	$\frac{41}{30}=1.367$	$\frac{9}{7}=1.286$
	$L+2$	—	$\frac{23}{12}=1.917$	$\frac{33}{20}=1.650$	$\frac{3}{2}=1.500$	$\frac{59}{42}=1.405$	$\frac{75}{56}=1.339$
	$L+3$	2	$\frac{7}{4}=1.750$	$\frac{8}{5}=1.600$	$\frac{3}{2}=1.500$	$\frac{10}{7}=1.429$	$\frac{11}{8}=1.375$

In the vector-coupling theory one regarded $\boldsymbol{S}$ and $\boldsymbol{L}$ simply as two classical angular momenta whose sum is $\boldsymbol{J}$. The simplest cases are $J=L+S$, in which $\boldsymbol{L}$ and $\boldsymbol{S}$ have to be parallel vectors, and $J=L-S$, in which $\boldsymbol{L}$ and $\boldsymbol{S}$ are anti-parallel. For $J=L+S$ we find

$$g=\frac{J+S}{J}=\frac{L+2S}{L+S},$$

which is just the ratio to be expected from the fact that the spin angular momentum counts double in magnetic effect. The value for $J = L - S$, however, does not correspond to the expectation based on this simple view, but is

$$1 - \frac{S}{J+1} = \frac{L - 2S + 1}{L - S + 1}.$$

Let us now consider the pattern to be expected in the transition between two levels. Using a prime to refer to the initial and a double prime to refer to the final level, we have

$$\sigma = \sigma_0 + \frac{o}{2\pi c}(g'M' - g''M'')$$

$$= \sigma_0 + \frac{o}{2\pi c}[(g' - g'')M' - g''\Delta M],$$

where $\Delta M = M'' - M'$. By the selection rule $\Delta M = 0$ for the components showing parallel polarization, while $\Delta M = \pm 1$ for those showing perpendicular polarization. The pattern is readily seen to be symmetrical about σ_0 and is conveniently characterized by reducing the values of the factor in brackets to a common denominator. Following Back and Landé the numerical values of the numerator which correspond to parallel polarization are put in parenthesis. Thus one may readily calculate that the values of the bracket factor for ${}^2P_{\frac{3}{2}}$ to ${}^2D_{\frac{3}{2}}$ are

$$\frac{(4)\,8\,(12)\,16\,24}{15},$$

which means that there are π components at $\pm\frac{4}{15}$ and $\pm\frac{12}{15}$ of $o/2\pi c$ from the undisplaced line and σ components at $\pm\frac{8}{15}$, $\pm\frac{16}{15}$, and $\pm\frac{24}{15}$. In this notation, the pattern is independent of which is the initial and which the final level.

The different Zeeman components are by no means of equal intensity, as we

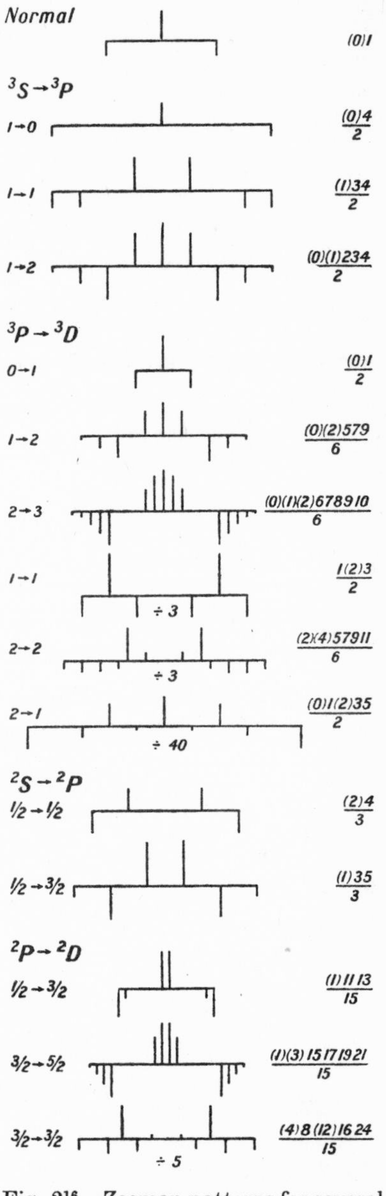

Fig. 2[16]. Zeeman patterns for several common multiplets. The relative strengths of the components of each multiplet are indicated by the lengths of the bars (divided by the indicated factor in the case of the weaker satellites).

shall see in § 4^{16}. Experimentally both the Zeeman pattern and the distribution of intensity provide important aids to the recognition of the character of the energy levels involved in a spectral line. The relative intensities are readily indicated graphically by drawing ordinates at the positions of the lines whose lengths are proportional to the intensity. Fig. 2^{16} contains such diagrams, and indicates clearly the diversity of the patterns for some common transitions.

For the Zeeman pattern it is convenient to make a double-entry table similar to those already used for multiplets. The method will be illustrated by means of the pattern for $d^6\,4p\;{}^4P_{\frac{5}{2}} \rightarrow d^6\,4s\;{}^4D_{\frac{7}{2}}$ in Mn as studied by Back.* Table 2^{16} has a row for each state of the ${}^4P_{\frac{5}{2}}$ level and a column for each state of the ${}^4D_{\frac{7}{2}}$ level. It is seen that the pattern is symmetrical around the position of the line in zero field. The upper figure in each cell is the theoretical value of the displacement, the lower is that observed by Back, the unit being $\frac{1}{35}$ of the normal Lorentz splitting.

TABLE 2^{16}.

${}^4D_{\frac{7}{2}}$ $g=\frac{10}{7}$ Unit $=\frac{1}{35}\frac{o}{2\pi c}$ cm^{-1}

			$3\frac{1}{2}$	$2\frac{1}{2}$	$1\frac{1}{2}$	$\frac{1}{2}$	$-\frac{1}{2}$	$-1\frac{1}{2}$	$-2\frac{1}{2}$	$-3\frac{1}{2}$
			175	125	75	25	−25	−75	−125	−175
	$2\frac{1}{2}$	140	−35 −34·8	+15 +15·12	+65 —					
	$1\frac{1}{2}$	84		−41 −40·8	+9 +9·07	+59 —				
${}^4P_{\frac{5}{2}}$	$\frac{1}{2}$	28			−47 −47·0	+3 +3·02	+53 +53·0			
$g=\frac{8}{5}$	$-\frac{1}{2}$	−28				−53 −53·0	−3 −3·02	+47 +47·0		
	$-1\frac{1}{2}$	−84					−59 —	−9 −9·07	+41 +40·8	
	$-2\frac{1}{2}$	−140						−65 —	−15 −15·12	+35 +34·8

3. Weak fields: general case.

In the preceding section we have considered the highly important case of the Zeeman effect for Russell-Saunders terms. Let us now consider the nature of the effect for an energy level of any type in the weak-field case. An energy level of a free atom is always rigorously characterized by a quantum number J of resultant angular momentum. The states resulting from the

* BACK, Zeits. für Phys. 15, 206 (1923).

application of a weak field will be characterized by J and M. The vector $\boldsymbol{S}$ of resultant spin is of the type considered in § 9[3], and hence

$$(\alpha\, JM|S_z|\alpha\, JM) = (\alpha\, J\vdots S\vdots \alpha\, J)\, M.$$

Therefore the magnetic alteration of the energy level is, in general,

$$(\alpha\, JM|H^M|\alpha\, JM) = o\hbar g M \tag{1}$$

as in 2[16]1, where we have, in place of 2[16]2,

$$g = 1 + (\alpha\, J\vdots S\vdots \alpha\, J). \tag{2}$$

(The special result of the preceding section is obtained if $\alpha \equiv \gamma\, SL$.) In other words all energy levels are split into $2J+1$ equally spaced levels symmetrically distributed around the unperturbed level. The amount of the splitting is governed by a Landé g factor as in Russell-Saunders levels, but now this factor is given by the general equation (2). The whole theory of the weak field effect for any term is thus reduced to an evaluation of $(\alpha\, J\vdots S\vdots \alpha\, J)$.

If we express the state $\Psi(\alpha\, JM)$ in terms of the LS-coupling scheme, we have

$$\Psi(\alpha\, JM) = \sum_{\gamma SL} \Psi(\gamma\, SL\, JM)\, (\gamma\, SL\, J|\alpha\, J).$$

Hence the matrix component that we need is

$$\bar{\Psi}(\alpha\, JM)\,(J_z + S_z)\,\Psi(\alpha\, JM)$$
$$= \sum_{\gamma SL} \bar{\Psi}(\gamma\, SL\, JM)\,(J_z + S_z)\,\Psi(\gamma SL\, JM)\, |(\gamma\, SL\, J|\alpha\, J)|^2.$$

Since $J_z + S_z$ $(= L_z + 2S_z)$ is diagonal with regard to γ, S, and L in this scheme.

The $\bar{\Psi}\,(J_z + S_z)\,\Psi$ combination occurring after the summation sign is, however, simply $g(SL\, J)\, M\hbar$, whereas the quantity on the left is $g(\alpha\, J)\, M\hbar$; hence we have the result

$$g(\alpha\, J) = \sum_{\gamma SL} g(SL\, J)\, |(\gamma\, SL\, J|\alpha\, J)|^2, \tag{3}$$

which expresses the g factor for an arbitrary level in terms of the ordinary Landé factors of the Russell-Saunders scheme and the transformation coefficients $(\gamma\, SL\, J|\alpha\, J)$.

If we regard α as a variable label running over all the levels which have a particular J value, and sum (3) over α, we obtain

$$\sum_{\alpha} g(\alpha\, J) = \sum_{\gamma SL} g(SL\, J), \tag{4}$$

since the coefficient $\sum_{\alpha} |(\gamma\, SL\, J|\alpha\, J)|^2$ which multiplies $g(SL\, J)$ on the right-hand side of the summed equation is equal to unity because the transformation is unitary. The result contained in (4) is known as the *g-sum rule.*

Actually of course in an atom there are an infinite number of levels associated with each value of J that occurs at all, so (4) merely becomes an uninteresting $\infty = \infty$ if applied to all the levels of the atom. Its importance lies in the fact that the summation over γ which expresses an actual atomic state in terms of states of the Russell-Saunders scheme is finite to a good approximation. Thus in atoms where the states $\Psi(\alpha J)$ may be assigned electron configurations, γSL are restricted to those values associated with the same electron configuration as α. In this case (4) tells us, if we let α run only over levels of a given configuration, that although the individual g values are altered by a breakdown of Russell-Saunders coupling, the sum of the g values for all the levels of a given J in the configuration is the same as in Russell-Saunders coupling.

This is exemplified by the measurements of Paschen on the Zeeman effect in neon. The configuration $2p^5\,3s$ gives rise to one level with $J=0$, two with $J=1$ and one with $J=2$. The coupling is as we have seen far from the Russell-Saunders case (§ 5^{13}) in which the levels would be labelled ${}^3P_0$, ${}^1P_1$, ${}^3P_1$ and ${}^3P_2$.

The level of $J=2$ is ${}^3P_2$ regardless of coupling so we do not need the g-sum rule to tell us that the g value for the state of $J=2$ should be $g({}^3P_2)$. This is $\frac{3}{2}$ and Paschen found 1·503 which is good agreement. For the two states with $J=1$ Paschen found $g=1{\cdot}034$ and $g=1{\cdot}464$ so $\Sigma g=2{\cdot}498$, whereas $g({}^1P_1)+g({}^3P_1)=\frac{5}{2}$, also a good agreement.

Similarly the g-sum rule may be applied to Paschen's values for the levels in the $2p^5\,3p$ configuration. Here $J=1$ is represented by four levels, which in the LS scheme are ${}^3S_1$, ${}^1P_1$, ${}^3P_1$ and ${}^3D_1$. The corresponding Russell-Saunders g values are 2, $\frac{3}{2}$, 1, $\frac{1}{2}$ with a sum of 5. The experimental values for the four levels of $J=1$ are 1·984, 1·340, 0·999, and 0·699, with sum equal to 5·022 providing a good sum-rule check although the individual values are far from the Landé values. For $J=2$ there are three states with experimental g values 1·301, 1·229, and 1·137; sum 3·667. The corresponding states in the LS scheme are ${}^3P_2$, ${}^1D_2$ and ${}^3D_2$ which have a g sum of $\frac{11}{3}$ in good agreement with experiment.

4. Intensities in the Zeeman pattern: weak fields.

The components arising from a given spectral line when the source is in a magnetic field will be said to form a Zeeman pattern. The different components, as we have just seen, are connected with the different changes in M. The relative strengths of the lines are therefore given by application of formulas 9^311 as noted in § 7^4, where the necessity arose of summing over all these separate transitions to find the total strengths of the lines of an unperturbed atom.

In the early work on the Zeeman effect, observation was made of the longitudinal effect (radiation along $\pm\mathscr{H}$) and of the transverse effect (radiation emitted in any direction perpendicular to $\mathscr{H}$). This revealed quite generally the phenomenon of the circular polarization in longitudinal observation of the lines which appear polarized perpendicular to $\mathscr{H}$ in transverse observation, and also the absence in longitudinal observation of the lines which are polarized parallel to $\mathscr{H}$ in transverse observation. Since transverse observation gives the complete pattern and is more convenient experimentally, that is the arrangement which is always used in modern work.

For the Zeeman pattern of any line in which there is no change in J, the strengths of the components in transverse observation ($\theta=\pi/2$ in 7^44) are proportional to

$$\left.\begin{aligned} |(\alpha J M|\boldsymbol{P}|\alpha' J M)|^2 &= |(\alpha J\vdots P\vdots\alpha' J)|^2 M^2 & (\pi)\\ \tfrac{1}{2}|(\alpha J M|\boldsymbol{P}|\alpha' J M\mp 1)|^2 &= |(\alpha J\vdots P\vdots\alpha' J)|^2 \tfrac{1}{4}(J\pm M)(J\mp M+1). & (\sigma)\end{aligned}\right\}\quad (1)$$

Since the component $-M\rightarrow -M+1$ has equal and opposite displacement from the original line to that of $M\rightarrow M-1$, it is evident that the pattern of strengths given by these formulas is symmetrical about the position of the unperturbed line.

Similarly we may write two more sets of formulas giving the relative strengths of the components of the Zeeman pattern for $J\rightarrow J+1$ transitions and $J\rightarrow J-1$ transitions in transverse observation. They are

$$\left.\begin{aligned} |(\alpha J M|\boldsymbol{P}|\alpha' J{+}1\, M)|^2 &= |(\alpha J\vdots P\vdots\alpha' J{+}1)|^2[(J+1)^2-M^2] & (\pi)\\ \tfrac{1}{2}|(\alpha J M|\boldsymbol{P}|\alpha' J{+}1\, M\mp 1)|^2 &= |(\alpha J\vdots P\vdots\alpha' J{+}1)|^2 \tfrac{1}{4}(J\mp M+1)(J\mp M+2) & (\sigma)\end{aligned}\right\}\quad (2)$$

and

$$\left.\begin{aligned} |(\alpha J M|\boldsymbol{P}|\alpha' J{-}1\, M)|^2 &= |(\alpha J\vdots P\vdots\alpha' J{-}1)|^2(J^2-M^2) & (\pi)\\ \tfrac{1}{2}|(\alpha J M|\boldsymbol{P}|\alpha' J{-}1\, M\mp 1)|^2 &= |(\alpha J\vdots P\vdots\alpha' J{-}1)|^2 \tfrac{1}{4}(J\pm M)(J\pm M-1). & (\sigma)\end{aligned}\right\}\quad (3)$$

These also give patterns that are symmetrical around the undisturbed line.

It should be explicitly emphasized that these formulas are valid in any coupling scheme, that is, are independent of the nature of the quantum numbers symbolized by α and α'. The positions of the components of the Zeeman pattern depend, as we have seen, on the other quantum numbers, but the relative strengths do not.

These formulas were first obtained empirically by Ornstein and Burger* and later derived in terms of the correspondence principle by Kronig and Goudsmit, and Hönl.†

As an example of the experimental test of these formulas we give some measurements by van Geel‡ on a plate, taken by Back, which gave good Zeeman patterns for $a\,^8S_{\frac{7}{2}}\rightarrow z\,^8P_{\frac{5}{2}}$ (λ4852) and $a\,^8S_{\frac{7}{2}}\rightarrow z\,^8P_{\frac{9}{2}}$ (λ4752) in Mn I. The theoretical pattern for λ4852 is given in Fig. 3^{16}. Van Geel used the theoretical values of the intensities of the perpendicular components,

* Ornstein and Burger, Zeits. für Phys. **29**, 241 (1924).

† Kronig and Goudsmit, Naturwiss. **13**, 90 (1925); Zeits. für Phys. **31**, 885 (1925); Hönl, Zeits. für Phys. **31**, 340 (1925).

‡ Van Geel, Zeits. für Phys. **33**, 836 (1925).

together with the observed blackening of the plate for these lines, as a means of finding the blackening curve of the plate. Using this blackening curve he found as the experimental values of the relative intensities of the parallel components

$$41:36:27{\cdot}5:15,$$

whereas the theoretical values are

$$40:36:28:16.$$

As the line $\lambda 4752$ was on the same plate, the same calibration was applicable to this. The relative intensities of π- and σ-components, independently, were:

	π-components	σ-components
Observed	57:46:27	70:56:37:22:12:?
Theory	54:45:25	73:52:35:21:10:3

These are evidently excellent agreements. The paper of van Geel includes several more examples of this sort, all of which agree well with theory.

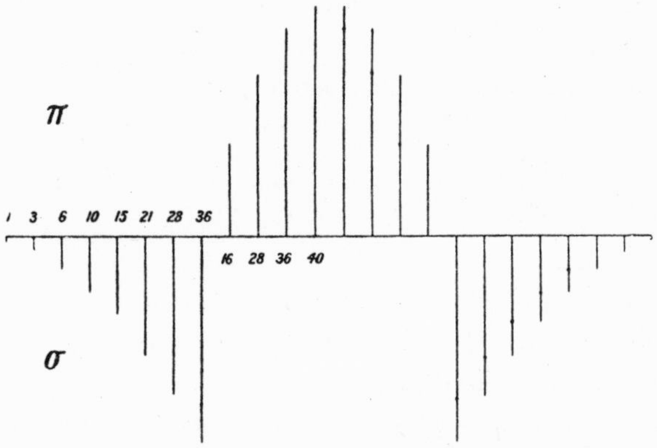

Fig. 3^{16}. Theoretical relative intensities in the ${}^8S_{\frac{7}{2}} \rightarrow {}^8P_{\frac{5}{2}}$ Zeeman pattern.

5. The Paschen–Back effect.

As already mentioned, the departure from the above theory of weak-field Zeeman effect which occurs with a magnetic field strong enough to produce splitting comparable with the interval between levels of a term is called the Paschen-Back effect. The theory has already been presented for one-electron spectra in § 10^5. Now we may consider the effect in general.

The magnetic perturbation H^M of $1^{16}1$ is evidently diagonal in any scheme of states in which M_L and M_S appear as quantum numbers, the value of the diagonal matrix element being

$$o\hbar\,(M_L + 2M_S). \qquad (1)$$

Suppose we start, as in Chapter VI, with a set of states based on a complete set of N individual sets of quantum numbers. We have seen in Chapter VII that the inclusion of the electrostatic interaction of the electrons necessitates a transformation from this scheme to an LS-coupling scheme in order to obtain states in which this part of the Hamiltonian is in diagonal form. In § 1[7] we saw that M_S and M_L can be retained as quantum numbers in an LS-coupling scheme—it is only when the spin-orbit interaction is included that it is necessary to pass to a scheme in which J and M are quantum numbers in order to obtain a scheme of states in which the complete Hamiltonian is in diagonal form.

On the other hand H^M is not diagonal in a scheme in which J and M are quantum numbers. Hence we see the place of the Paschen-Back effect in the theory of the Russell-Saunders case. For values of $\mathscr{H}$ such that the H^M matrix components are small compared to the spin-orbit-interaction matrix components the eigenstates of energy will be nearly those in which J and M are quantum numbers. This is the weak-field case of the preceding sections. For values of $\mathscr{H}$ such that H^M is large compared to the energy of spin-orbit interaction, the magnetic term will dominate and the energy eigenstates will be nearly those in which M_S and M_L are quantum numbers. For a variation of $\mathscr{H}$ from zero to such strong values there will be a continuous change in the character of the eigenstates from one of these limiting types to the other.

Predominance of the magnetic-field energy therefore draws the eigenstates toward those in which M_S and M_L are quantum numbers. In view of the fact that H^M commutes with $\boldsymbol{L}^2$ and $\boldsymbol{S}^2$, and that we can have a scheme of states in which SLM_SM_L are quantum numbers, we see that there is no tendency of the magnetic field to break down the coupling of the individual orbital or spin angular momenta of the electrons into a resultant L and S.* In the Russell-Saunders case therefore the Paschen-Back effect results from a competition between the spin-orbit interaction which works for the JM scheme and the magnetic field which works for the M_SM_L scheme. Since $M = M_S + M_L$ is a quantum number in both schemes, we see that the groups of states having different values of M may be treated independently. In dealing with a particular M group we may set up the secular equation for the energy values either in the M_SM_L scheme or in the JM scheme. For the former we need to have the non-diagonal matrix components of the spin-orbit interaction and for the latter we need the non-diagonal matrix components of the magnetic energy. The matrix in the M_SM_L scheme is more nearly diagonal at the outset for strong fields, the matrix in the JM scheme

* This is true in so far as we treat the magnetic perturbation as of the form of (1). Rigorously however there are other terms proportional to the square of the vector potential as mentioned in 10[5]2. They do not commute with the orbital angular momentum, and so at fields strong enough to make them important L and the individual l's of the electrons are no longer quantum numbers. However, this is of purely theoretical interest because the fields necessary are much greater than any attainable ones.

for weak fields. These matrices may be obtained from the calculations already given in other chapters so that the secular equations can be written down for any special case and solved by the usual methods. Except for the case of one-electron spectra, which has been already fully treated in § 10^5, this procedure will not be here carried out in detail, for it can lead into very lengthy calculations which have very little applicability to spectroscopy since the fields needed to produce large Paschen-Back effect are seldom obtainable. The best experimental illustrations of the Paschen-Back effect are in connection with hyperfine structure (§ 5^{18}) where attainable fields can effect a complete transition from one scheme of states to the other.

The Paschen-Back effect was important in the pre-quantum-mechanical theories of atomic spectra for the information it gave about the coupling relations, and was studied in this connection by Heisenberg and by Pauli.* Sommerfeld† showed the relation of an old classical coupling theory of Voigt to the effect. The quantum-mechanical treatment was first given by Heisenberg and Jordan and detailed cases were discussed by Darwin.‡

Although the preceding remarks have, for definiteness, been made for the Russell-Saunders case, it is evident that they hold with appropriate minor modifications in the case of any other coupling for the unperturbed atom.

PROBLEM.

Show that for all field strengths the sum of the magnetic changes in energy of all states of the same M is equal to $o\mathcal{H}M\Sigma g$, where Σg is the sum of the Landé g factors for the corresponding states when in a weak field. This is known as Pauli's *g-permanence rule.*

6. The Paschen-Back effect: illustrative examples.

As an illustrative example we take the $4s\,4d\,{}^3D \rightarrow 4s\,4p\,{}^3P$ multiplet in Zn I which has been studied by Paschen and Back and later by van Geel.§ The first reference includes a number of other cases, the emphasis being on the fact that strong magnetic fields bring out lines which violate the ordinary selection rule on J so that in the magnetic field the ${}^3D \rightarrow {}^3P$ multiplet is completed to all nine lines instead of the usual six. This is, of course, a simple consequence of the fact that the states in a strong magnetic field are not characterized by precise J values. The second reference provides accurate intensity measurements on the forbidden lines for various field strengths.

* Heisenberg, Zeits. für Phys. **8**, 273 (1922);
Pauli, *ibid.* **16**, 155 (1923); **31**, 765 (1925).
† Sommerfeld, Zeits. für Phys. **8**, 257 (1922).
‡ Heisenberg and Jordan, Zeits. für Phys. **37**, 263 (1926);
C. G. Darwin, Proc. Roy. Soc. **A115**, 1 (1927);
K. Darwin, *ibid.* **A118**, 264 (1928).
§ Paschen and Back, Physica **1**, 261 (1921);
van Geel, Zeits. für Phys. **51**, 51 (1928).

The matrix components of H^M in the $SLJM$ scheme are obtained from 9³11 and 10³2. They are

$$(SLJM|H^M|SLJM) = o\hbar M\, g(SLJ)$$

$$(SLJM|H^M|SLJ-1\,M) = o\hbar\sqrt{\frac{(J-L+S)(J+L-S)(J+L+S+1)(L+S+1-J)}{4J^2(2J-1)(2J+1)}}\sqrt{J^2-M^2}, \quad (1)$$

where $g(SLJ)$ is the Landé factor of § 2[16].

The Zn I $4s\,4p\,{}^3P$ intervals are 190 and 389 cm^{-1}, so the weak-field theory is adequate for them. The ${}^3D$ intervals, however, are only 3·28 and 4·94 cm^{-1}, so the quadratic term in the perturbation theory is important here. The interval ratio shows that we are fairly close to Russell-Saunders coupling. The $4s\,4d\,{}^1D$ is 300 cm^{-1} *below* the ${}^3D$, indicating a strong perturbation of it by $4p^2\,{}^1D$ as discussed in § 1[15].

Applying the perturbation theory to ${}^3D$ we find for $M=\pm 3$ a linear variation with the field of the energy of ${}^3D_2^{\pm 3}$. For $M=\pm 2$ there is an interaction between the corresponding states of ${}^3D_3$ and ${}^3D_2$ and for $M=\pm 1, 0$ there is an interaction between these states of all three levels of ${}^3D$. For the fields used the second-order perturbation is adequate. Using the matrix components (1) and the unperturbed level separations and writing η for $(o\hbar/hc)$, we readily find for the quadratic perturbations:

	Coefficient of η^2 in quadratic perturbation of		
	${}^3D_1$	${}^3D_2$	${}^3D_3$
$M=\pm 2$	—	−0·043	+0·043
±1	−0·1372	+0·0655	+0·0717
0	−0·1832	+0·1023	+0·0809

The numerical value of η is $4{\cdot}674\times 10^{-5}\mathscr{H}$ where $\mathscr{H}$ is in gauss.

Paschen and Back publish data on the line ${}^3D_2\rightarrow{}^3P_0$ for $\mathscr{H}=39340$ gauss. Since ${}^3P_0$ does not split this gives us directly the magnetic perturbation of ${}^3D_2$. For this field $\eta=1{\cdot}84$, so the quadratic displacements of the σ components which involve the $M=\pm 1$ states should be 0·22 cm^{-1} whereas they find 0·15 and 0·40 for the $M=-1$ and $+1$ components respectively. The π component involves $M=0$ for which the shift should be 0·35 and they report 0·32. This agrees within the accuracy of the data.

Van Geel's measurements on these lines are given in his Fig. 5 which we reproduce as Fig. 4[16] with the addition of a small mark on the wave-length scale at 3281·927, which is the position of unperturbed ${}^3D_2\rightarrow{}^3P_0$ given from the energy values of the levels. The broken lines show the splittings as given by the weak-field theory, the heavy lines are drawn through the experimental points. As drawn the departure for ${}^3D_2\rightarrow{}^3P_0$ appears to be linear and too large. That is because the broken lines have been drawn to converge at

3281·95 instead of 3281·927. When the change is made it is seen that the uncertainty of the points is too great for a satisfactory comparison with theory. That is probably because this is a line which is forbidden in weak fields and is weak even at the fields used and so is difficult to measure.

The data on ${}^3D_1 \to {}^3P_0$ are probably more accurate since this is a stronger line. For this line he finds the quadratic effect at $\mathscr{H} = 30{,}000$ to be $-0{\cdot}185$ and $-0{\cdot}334$ cm^{-1} for the components involving $M = \pm 1$, whereas the theoretical value is $-0{\cdot}268$; and for the component involving $M = 0$ he finds a shift $-0{\cdot}297$ as compared with a theoretical $-0{\cdot}360$.

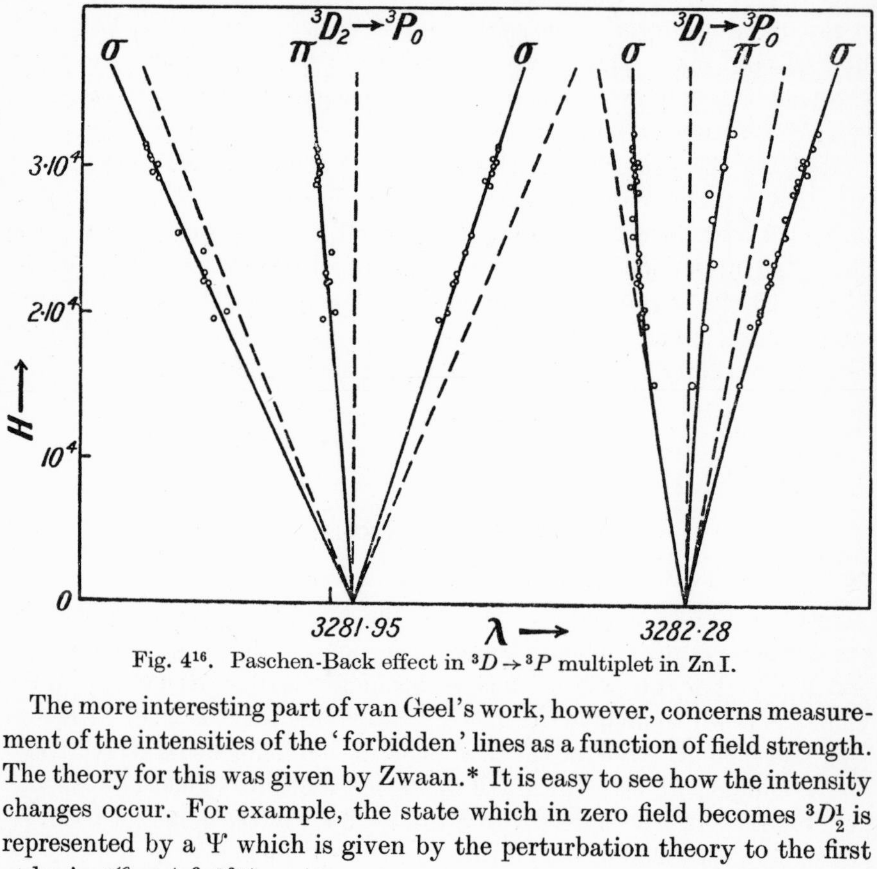

Fig. 4^{16}. Paschen-Back effect in ${}^3D \to {}^3P$ multiplet in Zn I.

The more interesting part of van Geel's work, however, concerns measurement of the intensities of the 'forbidden' lines as a function of field strength. The theory for this was given by Zwaan.* It is easy to see how the intensity changes occur. For example, the state which in zero field becomes ${}^3D_2^1$ is represented by a Ψ which is given by the perturbation theory to the first order in $\mathscr{H}$ as (cf. 9^25)

$$\Psi(A) = \Psi({}^3D_2^1) + \Psi({}^3D_1^1)\frac{({}^3D_1^1|H^M|{}^3D_2^1)}{3{\cdot}28} + \Psi({}^3D_3^1)\frac{({}^3D_3^1|H^M|{}^3D_2^1)}{-4{\cdot}94},$$

where the matrix components, given by (1), are the same as we have already used. The state that grows out of ${}^3D_2^1$ therefore acquires some of the

* Zwaan, Zeits. für Phys. **51**, 62 (1928).

characteristics of $J=1$ and $J=3$ in addition to the $J=2$ which it has, strictly, when unperturbed. If therefore we calculate the matrix component of electric moment for transition to ${}^3P_0^0$, for instance, we find

$$({}^3P_0^0|\boldsymbol{P}|A) = ({}^3P_0^0|\boldsymbol{P}|{}^3D_1^1)\frac{({}^3D_1^1|H^M|{}^3D_2^1)}{3\cdot 28},$$

the other two terms vanishing by the selection rule on J. The magnetic perturbation term is proportional to $\mathscr{H}$, so the quantity $|({}^3P_0^0|\boldsymbol{P}|A)|^2$ which measures the radiation intensity will vary as $\mathscr{H}^2$.

The intensity of all the ordinarily forbidden components as a function of field strength was worked out in this way by Zwaan. The components of the line ${}^3D_3 \to {}^3P_0$ for which $\Delta J = 3$ were found to vary as $\mathscr{H}^4$, while the two lines ${}^3D_2 \to {}^3P_0$ and ${}^3D_3 \to {}^3P_1$ for which $\Delta J = 2$ varied as $\mathscr{H}^2$ for weaker fields.

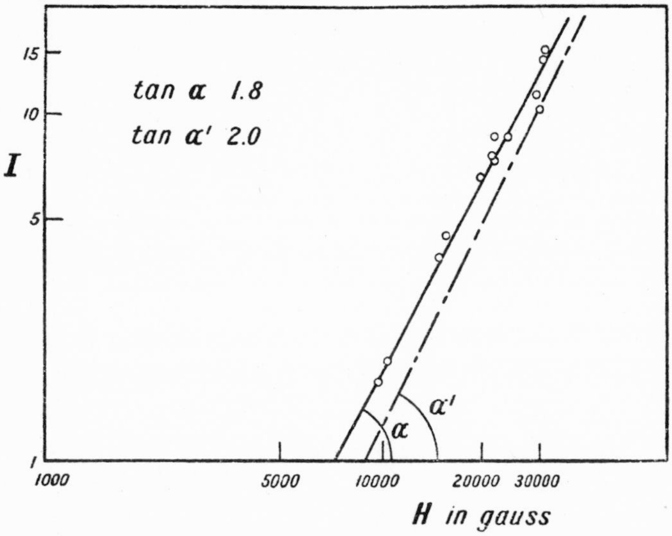

Fig. 5^{16}. Dependence of 'forbidden-line' strength on field in Paschen-Back effect for ${}^3D_2 \to {}^3P_0$ in Zn I. (I is the strength measured with $\frac{1}{100}$ of the strength of the line ${}^3D_1 \to {}^3P_0$ as unit.)

This $\mathscr{H}^2$ variation is simply the leading term of a development in powers of $\mathscr{H}$; he found as well the coefficients of $\mathscr{H}^3$ for the lines.

The results for the relative strengths of the lines ${}^3D_2 \to {}^3P_0$ and ${}^3D_1 \to {}^3P_0$ as found by Zwaan are ($\rho = 3\cdot 0 \times 10^{-5}\mathscr{H}$):

Component:	$1 \to 0$	$0 \to 0$	$-1 \to 0$
${}^3D_2 \to {}^3P_0$	$\frac{3}{16}\rho^2 - \frac{47}{432}\rho^3$	$\frac{1}{2}\rho^2$	$\frac{3}{16}\rho^2 + \frac{39}{432}\rho^3$
${}^3D_1 \to {}^3P_0$	$\frac{5}{3}$	$\frac{10}{3}$	$\frac{5}{3}$

Van Geel measured the sum of the intensities of ${}^3D_2 \to {}^3P_0$ relative to ${}^3D_1 \to {}^3P_0$ as a function of the field $\mathscr{H}$ between 10,000 and 30,000 gauss. The agreement is shown in Fig. 5^{16} (van Geel's Fig. 1) where the log of the intensity ratio is plotted against log $\mathscr{H}$. The broken line is that given by the theory. This is a very satisfactory agreement. He also studied the departure of the intensity of the $1 \to 0$ and $-1 \to 0$ components from the ρ^2 law as given by the ρ^3 terms and found good agreement with theory.

The paper by Paschen and Back contains a number of other examples of the appearance of forbidden lines in the magnetic field, but van Geel's are the only quantitative intensity data in this field.

We turn now to a brief discussion of some old data on the Zeeman effect in Hg I which provide an illustration of a Paschen-Back interaction between a singlet and a triplet term made possible because of the breakdown of Russell-Saunders coupling. The data are due to Gmelin.* He worked with low resolving power and reported all the Zeeman patterns as triplets. The splitting of $\lambda 4916$ ($6s\,8s\,{}^1S_0 \to 6s\,6p\,{}^1P_1$) he found to be $4 \cdot 72 \times 10^{-5}$ cm^{-1}/gauss, which is within 1 per cent. of the modern value of the normal splitting. From it we may infer that $g = 1 \cdot 01$ for this ${}^1P$. For $\lambda 5769$ ($6s\,6d\,{}^3D_2 \to {}^1P_1$) he found a specific splitting of 5·36, and for $\lambda 5790$ ($6s\,6d\,{}^1D_2 \to {}^1P_1$) he found 4·95.

If we assume that the relative intensities of the pattern components are given by theory, it is easy to calculate that the centre of gravity of intensity† of an unresolved blend would come at an apparent specific splitting $(\frac{3}{2}g_D - \frac{1}{2}g_P)$ times the normal, where g_D and g_P are the g factors for the D_2 and P_1 levels respectively. In this way Gmelin's data leads to observed g values of 1·04 and 1·10 for ${}^1D_2$ and ${}^3D_2$ respectively. The sum is 2·14, whereas the sum of the Landé values is $1 \cdot 00 + 1 \cdot 17 = 2 \cdot 17$, a fair agreement for such data.

From the departure of the g values from normal we can get, by the method of § 3^{16}, a rough estimate of the departure from Russell-Saunders coupling. Calling the quasi-singlet level A and the quasi-triplet B we have

$$g(A) = g({}^1D_2)\,|({}^1D_2|A)|^2 + g({}^3D_2)\,|({}^3D_2|A)|^2$$
$$g(B) = g({}^1D_2)\,|({}^1D_2|B)|^2 + g({}^3D_2)\,|({}^3D_2|B)|^2$$

from which the observed values of $g(A)$ and $g(B)$ together with the Landé values of $g({}^1D_2)$ and $g({}^3D_2)$ lead to

$$|({}^1D_2|A)|^2 = 0 \cdot 68 \qquad |({}^3D_2|A)|^2 = 0 \cdot 32,$$

which is a rather large departure from LS coupling. This rough result is borne out by intensity measurements by Bouma,‡ who found $\lambda 5769$ and $\lambda 5790$ of equal intensity, which would indicate

$$|({}^1D_2|A)|^2 = |({}^3D_2|A)|^2 = 0 \cdot 50.$$

* Gmelin, Ann. der Phys. **28**, 1079 (1909); Phys. Zeits. **9**, 212 (1908); **11**, 1193 (1910);
Green and Loring, Phys. Rev. **46**, 888 (1934), have since made much better measurements on these lines and a more detailed comparison with the intermediate-coupling theory than that given here.

† Shenstone and Blair, Phil. Mag. **8**, 765 (1929), have made good use of this method of treating unresolved Zeeman patterns.

‡ Bouma, Zeits. für Phys. **33**, 658 (1925).

So much for the first-order effect. Zeeman* found a quadratic displacement of λ5790 which was later studied by Gmelin. If we had pure LS coupling there could be no such effect because the magnetic perturbation has no matrix components connecting different singlet states. In view of the fact that the quasi-singlet A has a large component of $^3D_2$ in its eigenstate, it is able to show a Paschen-Back interaction with $^3D_1$ and $^3D_3$. It happens that A is only 3·2 cm^{-1} below $^3D_1$; hence its states will show a quadratic effect downward equal to

$$-\frac{(^3D_2\,M|H^M|^3D_1\,M)^2\,|(^3D_2|A)|^2}{3\cdot 2}.$$

Working out the details we find that the centres of gravity of the σ and π components should show respectively displacements of $-1\cdot 33\times 10^{-10}a\mathscr{H}^2$ and $-3\cdot 48\times 10^{-10}a\mathscr{H}^2$, where a is written for $|(^3D_2|A)|^2$. What Gmelin observed was no shift for the σ components and a quadratic effect of $-1\cdot 41\times 10^{-10}\mathscr{H}^2$ for the π components. This corresponds to $a=0\cdot 40$, in good agreement with the value 0·32 derived from the departure of the g's from the normal Landé values.

It thus appears that the existing data on λ5790 is in accord with the theory and exemplifies a Paschen-Back displacement of a quasi-singlet line arising from departure from Russell-Saunders coupling. More accurate measurements on this line are desirable for a more definite check. This interpretation of Gmelin's data disagrees with that given by Back,† who tries to relate the matter to hyperfine structure.

7. Quadrupole lines.

The theory of the Zeeman effect for quadrupole lines has been developed incidentally to other work in preceding sections, so brief comment relating it to the experimental work is all we need here. The splitting of the energy levels, being a property of the levels and not of the transitions, is given by the discussion of other sections of the chapter. The details of the picture concerning angular intensity and polarization, associated with different changes in M, have already been worked out in § 6^4.

The first experimental work on this point is that of McLennan and Shrum‡ on the green auroral line. As discussed in § 5^{11}, this is the $^1S_0\rightarrow{}^1D_2$ transition in the normal $2p^4$ configuration of neutral oxygen. The first Zeeman effect observations§ were taken in the longitudinal direction. From the results of § 6^4 we see that in this direction the $\Delta M=\pm 2$ and $\Delta M=0$ components have vanishing intensity, so the pattern is like that of a dipole

* Zeeman, Phys. Zeits. **10**, 217 (1909).
† Back, Handbuch der Experimentalphysik **22**, 164 (1929).
‡ McLennan and Shrum, Proc. Roy. Soc. **A106**, 138 (1924); **A108**, 501 (1925).
§ McLennan, Proc. Roy. Soc. **A120**, 327 (1928);
Sommer, Zeits. für Phys. **51**, 451 (1928).

line. The measured displacements gave $g=1$ from which the singlet character of the line was inferred. The transverse effect for this line was studied by Frerichs and Campbell* who found the σ components at twice the normal separation, corresponding to $\Delta M = \pm 2$, as well as the π components at normal separation for $\Delta M = \pm 1$, all of equal intensity, in agreement with the theory.

Segré and Bakker† studied the Zeeman effect of the $^2D \rightarrow {}^2S$ doublets in sodium and potassium. This work provides a detailed verification of the theory in several respects. For sodium the ^2D interval is so small that all fields giving a measurable splitting showed a complete Paschen-Back effect in this term. For potassium the doublet interval is great enough ($2{\cdot}32$ cm^{-1}) that the two lines can be studied separately. This was done in the oblique direction at 45 degrees, as well as transversely, and the theoretical expectations as to relative intensity and polarization of the components verified.

For higher fields the Paschen-Back effect begins to alter the intensities and permits new components to appear. The theory is exactly like that for dipole lines—we have to calculate the matrix components of the quadrupole moment with respect to the perturbed eigenfunctions. In analogy with the discussion of the preceding section the perturbed matrix components can be expressed in terms of the unperturbed matrices by means of the transformation matrix from the unperturbed to the perturbed eigenfunctions. Such calculations have been made explicitly by Milianczuk‡ and the results found to agree well with the experimental results of Segré and Bakker.

* Frerichs and Campbell, Phys. Rev. **36**, 151, 1460 (1930).
† Segré and Bakker, Zeits. für Phys. **72**, 724 (1931).
‡ Milianczuk, Zeits. für Phys. **74**, 825 (1932).

CHAPTER XVII

THE STARK EFFECT

The influence of an external electric field on atomic spectra was discovered by Stark in 1913 and is known as the Stark effect. It was natural to expect such an effect as soon as the Zeeman effect had been discovered and interpreted theoretically in terms of the electron theory. Voigt gave attention to the electric perturbation of atoms as early as 1899 and came to the conclusion that the effects would be very small. He also made attempts to discover an effect on the *D*-lines of sodium but without success.*

Stark's discovery was made on the Balmer series of hydrogen. It came in the same year as the Bohr theory of the hydrogen atom, so that it was an important additional success of that theory when, independently, Epstein and Schwarzschild† calculated the effect in terms of quantized orbits and predicted the observed line patterns exactly. Later Kramers‡ carried the theory further by using the correspondence principle to provide estimates of the relative intensities of the lines in a pattern.

It soon turned out that hydrogen occupies a somewhat exceptional position. Generally the splitting of a spectral line is much smaller than in hydrogen although there are special cases in other spectra where the effect is comparable in magnitude with that of hydrogen. We shall see that this is connected with the accidental degeneracy in hydrogen whereby terms of the same n but different l have the same energy, except for the small relativity fine structure. We shall also see that the cases of large Stark effect in other spectra are those in which two terms which may combine according to the optical selection rules lie close together.

The theory of the Stark effect in hydrogen was the first application of the perturbation theory in quantum mechanics. It turns out that the positions of the components in hydrogen as far as the first order effect (that proportional to field strength) is concerned is the same as on the quantized orbit theory. But the effects of second and third order differ in quantum mechanics from the results of the older theory. New experimental work which carries the study up to high fields has shown that the effect is in agreement with the new theory. At very high fields another aspect of the effect becomes important, namely, the destruction of quantized levels by the field. This has the effect of shifting the limit of the Balmer series toward the red by allowing

* Voigt, Ann. der Phys. **69**, 297 (1899); **4**, 197 (1901).

† Epstein, Ann. der Phys. **50**, 489 (1916);
Schwarzschild, Sitzber. Berliner Akad. 1916, p. 548.

‡ Kramers, Danske Vidensk. Selsk. Skrifter (8), iii, 3, 287, and Zeits. für Phys. **3**, 169 (1920).

the continuous spectrum to encroach on the part of the frequency scale which is reserved to the line spectrum in the absence of a field.

1. Hydrogen.

We consider first the application of the ordinary perturbation theory to atomic hydrogen to find the effect of weak fields on the energy levels. Suppose the electric field of strength $\mathscr{E}$ in the *negative* z-direction so the electron is acted on by a *force* in the *positive* z-direction. Its potential energy in the field of the nucleus is then

$$U = -\frac{e^2}{r} - e\mathscr{E}z. \tag{1}$$

Because of the fact that the potential energy tends to $-\infty$ for large positive values of z, all energy levels are allowed. But the spectrum has a quasi-discrete structure which can be investigated by treating $-e\mathscr{E}z$ as a perturbation in the usual way.

Owing to the degeneracy of the unperturbed states of hydrogen it is necessary to find the particular choice of unperturbed eigenfunctions with respect to which $e\mathscr{E}z$ is a diagonal matrix with regard to the sets of initially degenerate states. This is accomplished, as Schrödinger and Epstein* have shown, by solving the hydrogen-atom problem by separation of variables in parabolic coordinates. We write

$$x = \sqrt{\xi\eta}\cos\varphi, \quad y = \sqrt{\xi\eta}\sin\varphi, \quad z = \tfrac{1}{2}(\xi - \eta). \tag{2}$$

The coordinate surfaces so defined are paraboloids of revolution for ξ and η and meridian planes for φ. The paraboloids of both families have their foci at the origin, those with $\xi =$ constant extend to $z = -\infty$, those with $\eta =$ constant to $z = +\infty$. For these coordinates we have

$$r = \tfrac{1}{2}(\xi + \eta), \qquad d\tau = \tfrac{1}{4}(\xi + \eta)\, d\xi\, d\eta\, d\varphi,$$

and the Schrödinger equation for the perturbed atom becomes

$$\frac{\partial}{\partial\xi}\left(\xi\frac{\partial\psi}{\partial\xi}\right) + \frac{\partial}{\partial\eta}\left(\eta\frac{\partial\psi}{\partial\eta}\right) + \frac{1}{4}\left(\frac{1}{\xi} + \frac{1}{\eta}\right)\frac{\partial^2\psi}{\partial\varphi^2} + \frac{\mu}{2\hbar^2}[W(\xi+\eta) + 2e^2 + \tfrac{1}{2}e\mathscr{E}(\xi^2 - \eta^2)]\psi = 0. \tag{3}$$

This permits a separation of variables

$$\psi = F(\xi)\, G(\eta)\, e^{im\varphi}/\sqrt{2\pi}$$

leading to the equations for F and G:

$$\left.\begin{aligned} &\frac{\partial}{\partial\xi}\left(\xi\frac{\partial F}{\partial\xi}\right) + \frac{\mu}{2\hbar^2}\left[W\xi + 2e^2\beta_1 - \frac{m^2\hbar^2}{2\mu}\frac{1}{\xi} + \tfrac{1}{4}e\mathscr{E}\xi^2\right]F = 0, \\ &\frac{\partial}{\partial\eta}\left(\eta\frac{\partial G}{\partial\eta}\right) + \frac{\mu}{2\hbar^2}\left[W\eta + 2e^2\beta_2 - \frac{m^2\hbar^2}{2\mu}\frac{1}{\eta} - \tfrac{1}{2}e\mathscr{E}\eta^2\right]G = 0, \\ &\qquad\qquad\qquad\qquad \beta_1 + \beta_2 = 1. \end{aligned}\right\} \tag{4}$$

* SCHRÖDINGER, Ann. der Phys. **80**, 457 (1926); EPSTEIN, Phys. Rev. **28**, 695 (1926).

A discussion of these equations for the case $\mathscr{E}=0$ in the usual way shows that they can be solved in terms of Laguerre polynomials. The solutions of these equations are

$$F(u)=u^{|m|/2}e^{-u/2}L^{|m|}_{k_1+|m|}(u)$$
$$G(v)=v^{|m|/2}e^{-v/2}L^{|m|}_{k_2+|m|}(v), \tag{5}$$

where $$u=\xi/n\mathrm{a}, \quad v=\eta/n\mathrm{a},$$

and $$n=k_1+k_2+|m|+1,$$

k_1 and k_2 being positive integers or zero. The energy levels are, of course, the same as are obtained by solving in spherical coordinates (§ 2[5]). The values of β_1 or β_2 in terms of k_1 or k_2 are given by

$$\beta_i=\frac{k_i+\frac{1}{2}|m|+\frac{1}{2}}{n}. \tag{6}$$

The final result for the normalized wave function of a state characterized by the quantum numbers $n\,k\,m$ where k is written for k_1 is

$$\psi(n\,k\,m)=\sqrt{\frac{2\,k_1!\,k_2!}{\mathrm{a}^3n^4[(k_1+|m|)!\,(k_2+|m|)!]^3}}\,F(u)\,G(v)\,e^{im\varphi}/\sqrt{2\pi}. \tag{7}$$

For a first-order calculation of the perturbation of the atom due to an electric field we need the matrix of z with regard to the states of the same n. This may be calculated from known properties of Laguerre polynomials and found to be diagonal in k and m. The diagonal elements have the value

$$(n\,k\,m|z|n\,k\,m)=\tfrac{3}{2}n\,(k_1-k_2)\,\mathrm{a}, \tag{8}$$

so that states for which $k_1>k_2$ are those which give a greater probability for the electron to be at $z>0$. From this result it follows at once that the first-order alteration of the energy of the state $n\,k\,m$ is

$$-\tfrac{3}{2}n\,(k_1-k_2)\,e\mathscr{E}\mathrm{a}, \tag{9}$$

that is, the states in which the electron is more likely to be in the $z>0$ region are lowered in energy. This calculation neglects the relativity-spin fine structure and so is applicable only for fields producing perturbations large compared with this fine structure.*

We shall not give the details of the calculation of the second- and third-order effects† but shall pass now to a comparison of the theory with experi-

* The calculation including the fine structure was carried out independently by ROJANSKY, Phys. Rev. **33**, 1 (1929), and SCHLAPP, Proc. Roy. Soc. **A119**, 313 (1928). It turns out that there is a linear effect for very weak fields owing to the degeneracy of the levels ${}^2L_{L+\frac{1}{2}}$ and ${}^2(L+1)_{L+\frac{1}{2}}$ in the hydrogen fine structure (§ 4[5]). As the Stark displacement becomes comparable with the fine structure a non-linear effect sets in, which for larger fields goes asymptotically into that given in the text.

† The formulas (11) and (12) for the coefficients of $\mathscr{E}^2$ and $\mathscr{E}^3$ in the Stark-effect displacements were not obtained by using the elements of $-e\mathscr{E}z$ which are non-diagonal in n in the usual second- and third-order perturbation theory (this would involve a very inconvenient summation over all the levels of the discrete and continuous spectrum), but by writing F, G, W, β_1, and β_2 in (4) each as a power series in $\mathscr{E}$ and then solving in succession the equations obtained by equating the coefficient of each power of $\mathscr{E}$ to zero.

ment. If we measure $\mathscr{E}$ in kilovolt/cm and write $\nu_1(n\,k_1\,k_2)$ for the first-order alteration in cm^{-1}, (9) becomes

$$\nu_1(n\,k_1\,k_2) = 0{\cdot}0642\,n\,(k_1 - k_2)\,\mathscr{E}, \tag{10}$$

where the numerical value of the coefficient is calculated from the theoretical expression. The original absolute measurements of Stark gave an experimental value of 0·068 but recent precision measurements by Sjögren and by Kassner* agree within 1 per cent. with the theoretical value. Nearly all of the experimental data refer to the Balmer lines.† For fields up to 100 kilovolt/cm the quadratic effect is negligible in comparison with the linear effect.

From § 5^4 we know that the $\Delta m = 0$ transitions will show parallel polarization in transverse observation and will be absent in longitudinal observation. Also for transverse observation the transitions $\Delta m = \pm 1$ give radiation polarized perpendicular to the field. So far, this is just like the Zeeman effect. There is a difference in the longitudinal observation. In the Zeeman effect, the lines due to transitions where m changes show circular polarization. Here the line due to a transition $(m) \to (m \pm 1)$ is at the same frequency as $(-m) \to (-m \mp 1)$. Theoretically each of these gives circularly polarized light but in the source there is an incoherent superposition of the two contributions giving no polarization in longitudinal observation. This is also true for fields where second- or third-order effects are appreciable. These predictions concerning polarization have been checked experimentally‡ so transverse observation is used to get complete patterns.

Consideration of the range of quantum numbers shows that the pattern will be symmetric around the position of the unperturbed line. The components in a pattern differ in intensity because (*a*) individual transition probabilities differ, (*b*) some components are the sum of several transitions, (*c*) excitation conditions may put unequal numbers of atoms into various states of the initial level. As always, this latter effect is most difficult to control and is probably the principal source of discrepancies between theory and observation on relative intensities.

Theoretical calculations of relative intensities were made by Schrödinger and by Epstein (*loc. cit.*). These agree when a correction in Epstein's work is made.§ In Fig. 1^{17} the theoretical patterns for the first four lines of the Balmer series are shown, the unit of abscissa being the basic splitting $0{\cdot}0642\mathscr{E}\ \text{cm}^{-1}$ and the ordinates being proportional to the relative intensities in the pattern for any one line. The cross-marks indicate measured relative

* Sjögren, Zeits. für Phys. **77**, 290 (1932);
Kassner, *ibid.* **81**, 346 (1933).

† Frerichs, Ann. der Phys. **19**, 1 (1934), has succeeded in obtaining the Stark effect of the Lyman series. Here the patterns are entirely due to the splitting of the initial levels since the final $1s$ level is unaffected by an electric field.

‡ Stark and Wendt, Ann. der Phys. **43**, 991 (1914).

§ Gordon and Minkowski, Naturwiss. **17**, 368 (1929).

intensities. The general agreement is good but attempts to make an accurate test of the theory encounter many difficulties. When the hydrogen atoms are moving rapidly as in canal-ray sources, the relative intensity of the different lines is disturbed by what appears to be due to differential destruction of excited atoms by collision from the different states of the initial level. It has been shown* that when the atoms are moving in the direction of the

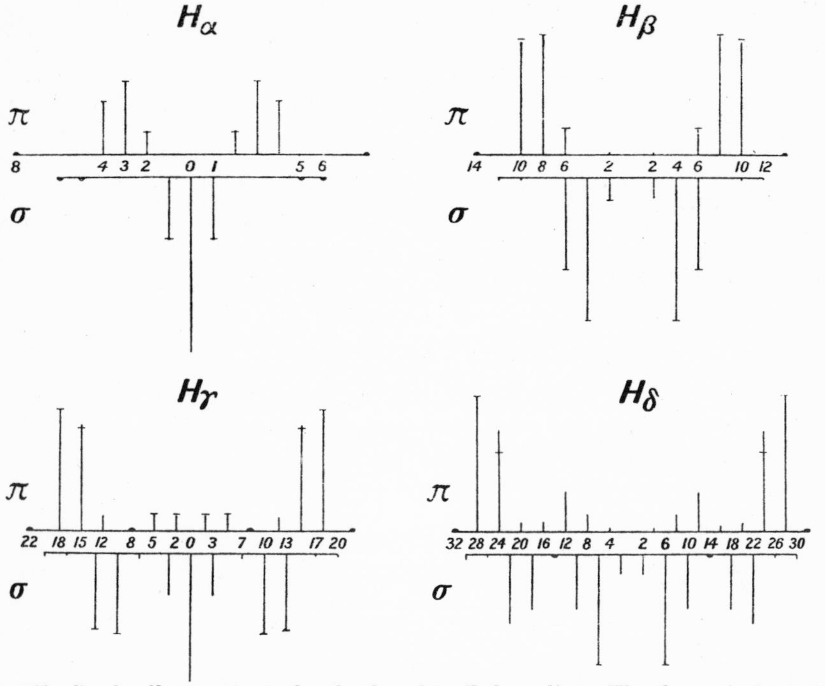

Fig. 1^{17}. Stark-effect patterns for the first four Balmer lines. The theoretical relative strengths of the components of each line are shown by the lengths of the bars, with very weak components indicated by half circles. The cross-marks indicate the measured relative intensities (principally from Foster and Chalk) of certain components of the same polarization.

field the high frequency components of the Stark pattern are weaker than the others (and are stronger when the atoms move in the direction opposite to that of the field). Since the high frequency components arise from states of higher energy in the initial state, the experimental result indicates that these states (for which the larger values of $\psi\bar{\psi}$ are on the forward, 'exposed,' side of the atom) are more subject to quenching by collision than those of lower energy.† Wierl‡ has shown that the intensity dissymmetry is not

* Stark and Kirschbaum, Ann. der Phys. **43**, 1006 (1914);
Wilsar, Gött. Nachr. 1914, p. 85;
Luneland, Ann. der Phys. **45**, 517 (1914).

† Bohr, Phil. Mag. **30**, 394 (1915);
Slack, Ann. der Phys. **82**, 576 (1927); Phys. Rev. **35**, 1170 (1930).

‡ Wierl, Ann. der Phys. **82**, 563 (1927).

present when the field is perpendicular to the motion of the atoms, nor when the canal rays go into a high vacuum.

The intensity measurements by Foster and Chalk, Mark and Wierl, and others* do not agree in detail with the theory, but provisionally we shall regard this as due to the large number of variations in physical conditions in the experimental work.

An interesting experiment has been performed by Wien† by sending a hydrogen canal ray beam through a magnetic field $\mathscr{H}$ so that the atom's velocity $\boldsymbol{v}$ is perpendicular to $\mathscr{H}$. From the standpoint of a coordinate system moving with the atom this gives rise to an electric field of amount $\boldsymbol{v} \times \mathscr{H}/c$, which can produce a Stark effect. Of course the magnetic field gives rise to a Zeeman effect, but this can be made smaller than the electrodynamic Stark effect. The ratio of the basic energy change $\frac{3}{2}e\mathscr{E}\mathrm{a}$ to the $eh\mathscr{H}/4\pi mc$ of the Zeeman effect is $(3/\alpha)(v/c)$ where α is the fine structure constant. For $v \sim 10^8$ cm/sec the ratio is 1·4, but as the strongest lines in the $\mathrm{H}\gamma$ pattern are displaced 15 to 20 times the basic unit we see that at such speeds the electric effect is about 20 times the Zeeman effect. Such speeds were used by Wien. The experiment is extremely difficult and the pictures obtained are quite unsharp. The observed effects agree with theory to within the rather low accuracy attained.

We turn now to a discussion of the experimental results for the quadratic and cubic Stark effect in hydrogen. This has been studied up to fields of 10^6 volt/cm. The theoretical energy change in the quadratic effect has been found‡ to be

$$-\frac{\mathrm{a}^3\mathscr{E}^2}{16}n^4[17n^2-3(k_1-k_2)^2-9m^2+19]. \tag{11}$$

This differs from the quadratic effect in the classical theory by the appearance of the additive $+19$ in the brackets and by the fact that the m values are all one unit lower than in the older theory. Similarly the theory for the third-order effect has been worked out by Doi§ who finds for it the energy change

$$\frac{3}{32}\frac{\mathrm{a}^5\mathscr{E}^3}{e}n^7(k_1-k_2)[23n^2-(k_1-k_2)^2+11m^2-71]. \tag{12}$$

* Foster and Chalk, Proc. Roy. Soc. **A123**, 108 (1929);
Mark and Wierl, Zeits. für Phys. **53**, 526 (1929); **55**, 156 (1929);
Kiuti, Jap. Jour. Phys. **4**, 13 (1925);
Ishida and Hiyama, Sci. Pap. Inst. Phys. Chem. Res. Tokyo **9**, 1 (1928);
Stark, Ann. der Phys. **1**, 1009 (1920); **48**, 193 (1915).

† Wien, Ann. der Phys. **49**, 842 (1916).

‡ Epstein, Phys. Rev. **28**, 695 (1926);
Wentzel, Zeits. für Phys. **38**, 527 (1926);
Waller, *ibid.* **38**, 640 (1926);
Van Vleck, Proc. Nat. Acad. Sci. **12**, 662 (1926).

§ Result only published in Ishida and Hiyama (*loc. cit.*).

It is convenient to express the shift of a particular component of a Balmer line in the form

$$\Delta(\mathrm{cm}^{-1}) = \pm a\mathscr{E} - b\mathscr{E}^2 \pm c\mathscr{E}^3, \qquad (13)$$

measuring $\mathscr{E}$ in million volt/cm in order that the coefficients be of comparable order of magnitude. Ishida and Hiyama have given a table, which we reproduce as Table 1[17] of the coefficients a, b, c, for the strong components of H α, H β and H γ. There are two columns for b and c corresponding to the quantum-mechanical theory and the old quantized-orbit theory predictions. The column headed 'transition' gives the quantum numbers $k_1 k_2 m$ of initial and final states for the transition in which the linear effect is positive, that headed 'label' gives the linear effect in terms of the basic splitting, and the polarization of the component.

TABLE 1[17]. *Coefficients of Stark displacements in* (13).

Line	Transition	Label	a	b New	b Old	c New	c Old
H α	110→010	2 π	128·78	6·715	5·707	0·003	0·002
	101 001	3 π	193·17	6·207	4·622	0·088	0·086
	200 100	4 π	257·56	6·309	5·395	0·164	0·145
	002 001	0 σ	0	5·177	2·819	0	0
	110 001	0 σ	0	6·705	5·707	0	0
	101 100	1 σ	64·39	6·156	4·419	0·085	0·083
H β	210→010	6 π	386·34	38·36	34·80	1·045	0·97
	201 001	8 π	515·12	35·97	30·12	2·125	2·11
	300 100	10 π	643·90	35·10	31·54	3·065	2·855
	111 010	2 σ	128·78	37·54	31·54	0·003	0·002
	102 001	4 σ	257·56	33·53	25·23	1·155	1·20
	210 001	4 σ	257·56	38·41	35·00	1·042	0·97
	201 100	6 σ	386·34	35·92	30·15	2·125	2·11
H γ	220→010	2 π	128·78	146·3	137·3	0·003	0·002
	211 001	5 π	321·95	142·2	127·6	7·65	7·53
	310 010	12 π	772·68	142·5	133·0	14·92	14·29
	301 001	15 π	965·85	134·2	119·6	22·64	22·41
	400 100	18 π	1159·02	130·5	121·3	29·30	28·00
	112 001	0 σ	0	134·1	113·5	0	0
	220 001	0 σ	0	146·3	137·2	0	0
	211 100	3 σ	193·17	142·2	127·0	7·65	7·58
	202 001	10 σ	643·90	130·5	109·8	16·00	16·40
	310 001	10 σ	643·90	142·5	127·1	14·93	14·30
	301 100	13 σ	837·07	134·3	113·3	22·64	22·40

The quadratic effect was discovered by Takamine and first studied carefully by Kiuti and by Rausch von Traubenberg and Gebauer.* The cubic effect is studied on the components 18 π and 15 π of H γ in a later paper by Gebauer and Rausch von Traubenberg.† We shall content ourselves with a report of this work. By reversion of the series (13) we see that the field strength can be calculated from the observed shift Δ by the formula

$$\mathscr{E} = \frac{\Delta}{a}\left[1 + \frac{b}{a}\frac{\Delta}{a} + \left(2\frac{b^2}{a^2} - \frac{c}{a}\right)\frac{\Delta^2}{a^2}\right].$$

* KIUTI, Zeits. für Phys. **57**, 658 (1929);
RAUSCH VON TRAUBENBERG and GEBAUER, *ibid.* **54**, 307 (1929); **56**, 254 (1929).

† GEBAUER and RAUSCH VON TRAUBENBERG, Zeits. für Phys. **62**, 289 (1930).

If the theory is correct this formula gives the same value of the field strength on a particular plate when applied to the various components of the patterns. This is the case to an accuracy of about 0·1 per cent. in the experimental work. The procedure was to calculate $\mathscr{E}$ from the observed effect on $H\beta$ and then to compare the observed effect on $H\gamma$ $+15\pi$ and $+18\pi$ with the theoretical values as calculated with the values of $\mathscr{E}$ determined from $H\beta$.

We may regard the theory of the linear effect as already adequately tested in other experiments and subtract its theoretical amount from the observed shift to obtain the observed amounts of the shift due to the higher order effects. We have prepared Fig. 2^{17} in this way from the data on the 18π violet component of $H\gamma$. The points give the observed data and, for comparison, are given theoretical curves for the amount of the quadratic effect

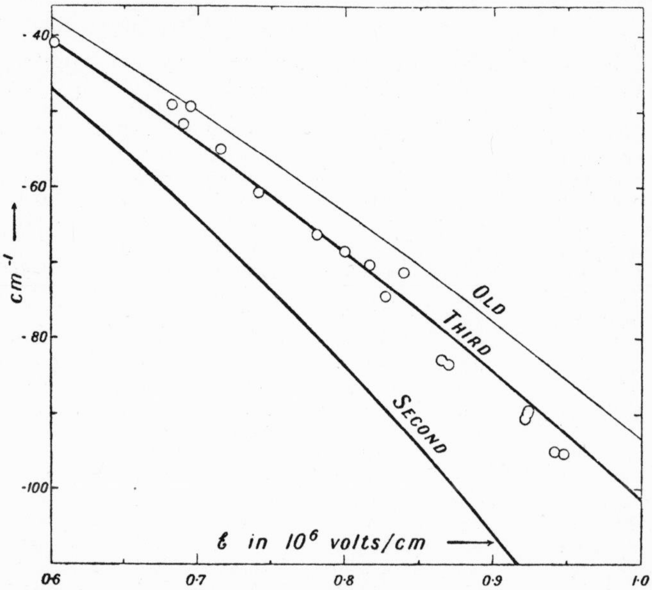

Fig. 2^{17}. The second- and third-order Stark effect on the 18π violet component of $H\gamma$.

alone, the quadratic plus cubic effect, and the quadratic plus cubic effect as calculated on the old theory. It is clear that the data are best represented by the curve representing quadratic plus cubic effect on the quantum-mechanical theory. There appears to be a small systematic departure at the higher field strengths which is perhaps due to the fourth-order effect. The data are evidently not well represented by the old theory.

2. Stark effect at the series limit.

As already mentioned in the preceding section, the potential energy of the electron in the applied field tends to $-\infty$ as z tends to large positive values. This has as a consequence that the spectrum of allowed values is really continuous, there being admissible wave functions for all values of the energy. Nevertheless we have seen that the spectrum gives sharp line patterns in Stark-effect experiments. We wish now to examine the situation more carefully.

The potential energy U of $1^{17}1$ has an extremum at the point $x=0$, $y=0$, $z=d=\sqrt{e/\mathscr{E}}$, where the electric force due to the nucleus is just balanced by that of the external field. At this point the potential energy has the value

$$U_0=-2\sqrt{e^3\mathscr{E}}. \qquad (1)$$

It is natural to expect that this energy value and the distance d play an important role in the Stark effect. This is in fact the case. Examining the formulas $1^{17}9$ and $1^{17}11$ we see that the coefficients of the quantum numbers can be written $-\frac{3}{4}U_0\,(\mathrm{a}/d)$ and $\frac{1}{32}U_0\,(\mathrm{a}/d)^3$ respectively, and similarly for the cubic term of $1^{17}12$. In a general way this might lead us to suspect that the developments have a validity only for small (a/d) and, as the region of space occupied effectively by an electron in the nth state has linear dimensions of the order $n^2\mathrm{a}$, we may expect the perturbation theory to fail for the nth state when $d\sim n^2\mathrm{a}$.

The first theoretical discussion of the special features of the problem in this region was given by Robertson and Dewey* from the standpoint of the classical quantized orbits. In that theory only those orbits can be quantized which are conditionally periodic, so the upper limit of the discrete energy spectrum is the maximum value of the energy for which conditionally periodic orbits exist. They found that there exist no quantizable orbits whose total energy is greater than

$$W_1=-\frac{3}{2}\left(\frac{e\mathscr{E}L_z}{\mu^{\frac{1}{2}}}\right)^{\frac{2}{3}},$$

and quantizable orbits in the neighbourhood of this limit only exist provided that

$$|L_z|<\frac{8}{9}\left(\frac{\mu e^3}{\sqrt{3e\mathscr{E}}}\right)^{\frac{1}{2}}.$$

In the classical mechanics the orbits for which $L_z=0$ are ruled out because they lead to collisions with the nucleus. Hence the least admissible value of L_z is $\hbar$. This value of L_z leads to an upper limit of the discrete spectrum which in wave numbers is given by

$$\sigma_1=-1110\mathscr{E}^{\frac{2}{3}},$$

whereas the value of the extremum of potential energy U_0 is given by

$$\sigma_0=-6271\mathscr{E}^{\frac{1}{2}},$$

where $\mathscr{E}$ is measured in 10^6 volt/cm in both formulas. These two expressions

* Robertson and Dewey, Phys. Rev. **31**, 973 (1928).

are plotted as functions of $\mathscr{E}$ in Fig. 3^{17}. From this we see that the discrete spectrum may extend up to higher energy values than the extremum of the potential energy function. At first sight this may seem a surprising result. If the particle has enough energy to get to the configuration where the field force balances the nuclear attraction we are apt to think that it will surely do so. But this is not the case. A simple mechanical example is that of a ball rolling under gravity inside of a spherical bowl, one side of which is partly broken away (as with the Mad Hatter's tea-cup). The ball may perform oscillations in a vertical plane which does not pass through the broken wall, whose total energy exceeds the energy limit where the non-periodic motions begin, that is, at the energy corresponding to the lowest place in the rim of the bowl.

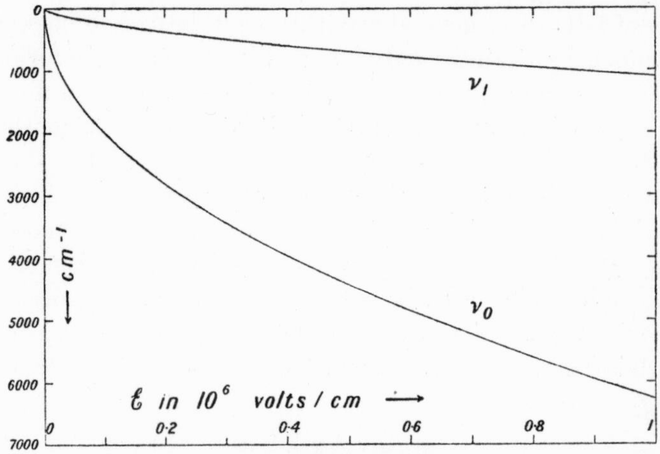

Fig. 3^{17}. Classical-dynamical energy limits for quantized orbits in the hydrogen atom in an electric field.

So much for the classical-mechanical theory of the problem. The quantum-mechanical theory has been treated by Lanczos* and is based on a direct discussion of equations $1^{17}4$ for the case $\mathscr{E} \neq 0$. The problem is somewhat complicated by the peculiar dynamical interaction of the motion in the ξ and η coordinates, but the essential point is the coupling of discrete and continuous energy eigenstates of the same energy value. This feature of quantum mechanics was first discussed by Oppenheimer,† who applied it to a calculation of the probability of ionization of hydrogen in the normal state by the application of a steady field and showed how this mechanism could account for the emission of electrons by cold metals in strong electric fields. It has become well known through its application by Gurney and Condon

* Lanczos, Zeits. für Phys. **62**, 518 (1930); **65**, 431 (1930); **68**, 204 (1931).
† Oppenheimer, Phys. Rev. **31**, 66 (1928).

and by Gamow* to the problem of α-particle disintegration of radioactive elements.

The effective potential energy functions in the two equations are

$$U_1(\xi) = -\frac{2e^2\beta_1}{\xi} + \frac{m^2\hbar^2}{2\mu}\frac{1}{\xi^2} - \tfrac{1}{2}e\mathscr{E}\xi,$$

$$U_2(\eta) = -\frac{2e^2\beta_2}{\eta} + \frac{m^2\hbar^2}{2\mu}\frac{1}{\eta^2} + \tfrac{1}{2}e\mathscr{E}\eta,$$

which are of the forms indicated in Fig. 4[17]. The wave equations cannot be solved exactly but have been treated by the Wentzel-Brillouin-Kramers method. The second of them leads to discrete energy levels which depend parametrically on β_2, m and the quantum number k_2 which gives the number of nodes in $G(\eta)$. We shall call this relation $W = W_2(\beta_2, k_2, m)$. The first equation does not lead to discrete values of W for a fixed β_1; instead we are presented with the phenomenon of the leaky potential barrier. We are

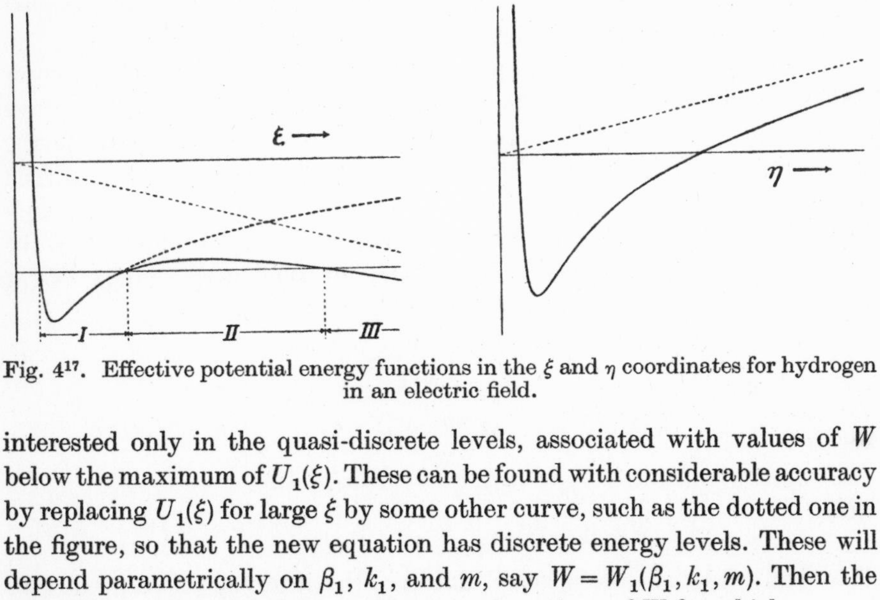

Fig. 4[17]. Effective potential energy functions in the ξ and η coordinates for hydrogen in an electric field.

interested only in the quasi-discrete levels, associated with values of W below the maximum of $U_1(\xi)$. These can be found with considerable accuracy by replacing $U_1(\xi)$ for large ξ by some other curve, such as the dotted one in the figure, so that the new equation has discrete energy levels. These will depend parametrically on β_1, k_1, and m, say $W = W_1(\beta_1, k_1, m)$. Then the allowed energy levels of the problem are the values of W for which

$$W_1(\beta_1, k_1, m) = W_2(1-\beta_1, k_2, m),$$

where the relation $\beta_1 + \beta_2 = 1$ has already been taken into account.

Now let us consider the fact that the true $U_1(\xi)$ slopes downward for large ξ, instead of having a horizontal asymptote as in the dotted curve. For a value of $W = W_1(\beta_1, k_1, m)$ the wave function will oscillate and have k_1

* Gurney and Condon, Nature **122**, 439 (1928); Phys. Rev. **33**, 127 (1929); Gamow, Zeits. für Phys. **31**, 204 (1928).

nodes in the region I. In the region II it will approach zero in a quasi-exponential way and in the region III it will again assume an oscillatory character but will have a greatly reduced amplitude relative to the amplitude in I. For a value of W differing slightly from $W_1(\beta_1, k_1, m)$, the behaviour in I will be quite similar, but owing to the slight difference in the wave-length of the de Broglie waves it will increase quasi-exponentially in region II so that the oscillatory portion in region III will have a much larger amplitude relatively than before. This change in character of the eigenfunction from 'large inside—small outside' to 'small inside—large outside' takes place as a continuous function of W but occurs in an extremely small range of energy values near the value $W_1(\beta_1, k_1, m)$. We shall want the wave functions normalized so that the integral of their square out to some large value of ξ is equal to unity. (This suffices to give the correct relative normalization, for the behaviour of the oscillatory part of each wave function from a large value of ξ out to $\xi = \infty$ is essentially the same, and thus we avoid dealing with divergent integrals and the special technique for normalizing continuous spectrum wave functions.)

In computing the transition probability from any energy W to a lower state which shows a much smaller perturbation it is only the part of the wave function in region I which plays an essential role, for in calculating the electric-moment matrix component the wave function of the less-perturbed state will be essentially zero outside of this region. By applying the Wentzel-Brillouin-Kramers approximation, Lanczos was able to show that the intensity of radiation from initial states in the energy range $d\omega$ at $W_1(\beta_1, k_1, m) + \omega$ is proportional to

$$\frac{1}{\pi}\frac{d\omega/X}{1+(\omega^2/X^2)},$$

where

$$X = \frac{h\nu}{2\pi}e^{-2I},$$

with

$$\nu = \frac{1}{2}\left[\int_{\text{I}} \frac{d\xi}{\sqrt{2(W-U)/\mu}}\right]^{-1}, \quad I = \frac{1}{\hbar}\int_{\text{II}} \sqrt{2\mu(U-W)}\,d\xi.$$

The 'half-width' of the spectral line (i.e. the interval between the two energies at which the radiation intensity has sunk to half value) is evidently equal to $2X$. The value of X depends both on ν, which is the classical frequency of vibration in the ξ coordinate for the state in question, and I, which measures the penetrability of the barrier between regions I and III.

If at $t = 0$ we are sure that the atom has an energy about equal to that of the energy level in question and that the electron is in the region I, we represent the state by a wave packet built to give constructive interference of the waves in region I and destructive interference in III. Such a packet can be built by superposition of states in the energy range of the order of X. As time goes on the constituent waves in the packet will get out of phase, so the amplitude of the packet will decay in such a way that the probability of the electron being in I contains a factor $e^{-\delta t}$ where $\delta = 2\nu e^{-I}/\pi$. This diminution in region I is compensated by an increase in region III, so the process corresponds to an ionization of the atoms by the steady field, or a radiationless transition of electrons from the bound state of region I to the free state of region III. Suppose α is the probability per unit time of leaving region I by radiation. Then the total probability of leaving region I in unit time is $(\alpha+\delta)$, so the fraction of excited atoms which leave by radiation is $\alpha/(\alpha+\delta)$. The total intensity of the radiation from this level is therefore this fraction of the value it would have been in the absence of the field ionization.

Thus the effect of the field is to give an unsharp line whose width is $h\nu e^{-2I}/\pi$ and whose total intensity is diminished by the field-ionization process. The actual calculation of the quantities ν and I for high fields is somewhat laborious although the values can be expressed in terms of elliptic integrals. Lanczos has made some calculations for the 18π violet and red components of $H\gamma$ which are in general agreement with the experimental work. The rather surprising result is that the violet component remains sharp and strong to much higher field strengths than the red component which fades out at about 0·72 million volt/cm. As the violet component arises from an initial state of higher energy than the red component we might expect the opposite, for the high state has a more penetrable barrier to go through. The explanation is that the initial state of high energy corresponds to the electron being mostly on the opposite side of the nucleus from the low place in the potential where the leaking takes place. In the violet component's initial state the electron is cautious and conservative, avoiding the leaky barrier; in the red state the electron is reckless and wastrel, hurling itself against a less leaky barrier often enough to rob itself of the chance to shine!

3. General theory for non-hydrogenic atoms.

In this section we shall sketch the general results obtained by calculating the electric perturbation of any atom by the perturbation theory. If the field $\mathscr{E}$ is in the *negative* z direction and P_z is the z-component of electric moment of the atom, then the Hamiltonian of the atom in the field is

$$H = H_0 + \mathscr{E} P_z, \tag{1}$$

where H_0 is written for the Hamiltonian of the atom in zero field. The perturbation is therefore determined entirely by the matrix components of the z-component of the electric moment, a quantity with which we are already quite familiar because it measures the strengths of spectral lines.

If for a particular atom we already have a rather complete knowledge of the energy levels and corresponding eigenfunctions, we may calculate the matrix of P_z and proceed to apply the perturbation theory. The details of the results so obtained naturally will depend a great deal on the mode of coupling of the angular-momentum vectors and the amount of configuration interaction. But we can, nevertheless, make some remarks which hold quite generally for all atoms before treating the separate special cases that arise.

In the first place P_z has matrix components only between states of opposite parity. Second, for a state of given J value it has matrix components with other states whose J values are $J-1$, J and $J+1$. Third, the order of magnitude of the non-vanishing matrix components involving electrons in states of total quantum number n will be the same as computed with hydrogenic wave functions. This we shall see is of the order $6{\cdot}4\,n^2\,\text{cm}^{-1}$ for a field of 100 kilovolt/cm. Because of this the second-order perturbation between two states whose energy difference is more than $10\,.\,(6{\cdot}4)^2\,n^4 \sim 400\,n^4$ cm^{-1} is almost sure to be less than $0{\cdot}1\ \text{cm}^{-1}$ even at 10^5 volt/cm. As most of the experimental work has $0{\cdot}1\,\text{cm}^{-1}$ as the limit of accuracy and is carried out at smaller fields, we see that the interaction of two terms several thousand wave numbers apart is negligible. The combined effect of an entire series of terms on a given term may be appreciable however. From our previous study (§ 10[2]) of the general features of perturbations we know that the effects will be given quite accurately by the second-order formula provided the initial separation of two interacting states is large compared with the energy of interaction between them. So for terms separated by more than about $6{\cdot}4\,n^2$ it will be sufficient to use the second-order formula. This qualitative restriction coupled with the parity and J selection rules suffices to make most cases tractable in this way. In such cases the displacement of the terms is small compared to the effect in hydrogen and is strictly proportional to the square of the field strength.

For a group of states lying close to each other a more accurate calculation is needed. This can be made by forming a secular equation from the matrix of the perturbed Hamiltonian by including just those rows and columns which refer to the small community of close, interacting states. Solution of this finite-matrix transformation problem will determine the mutual perturbation of these states and the transformation to a new set of states with regard to which this part of the Hamiltonian is diagonal. This transformation also enables us to calculate the values of the perturbation matrix components connecting these perturbed states with the distant states and thus to know how the second-order action due to them is distributed over the community. This procedure is exemplified in the next section.

There is another qualitative remark of importance. Generally the electric displacement of excited states is much greater than that of low-lying energy levels so that in most cases the displacements of the final state of a line are negligible, the observed line pattern being a direct picture of the perturbation of the initial state.

Let us next consider quite generally the kind of pattern to be observed in the quadratic Stark effect of a state described by the quantum numbers $\alpha J M$. Using $9^3 11$ in the second-order formula of $9^2 4$ to express the dependence on M of the matrix components, we may write for the displacement

$$\Delta(\alpha J M) = \frac{\mathscr{E}^2}{hc}[J^2 a_- + (J+1)^2 a_+ - (a_+ + a_- - a_0) M^2]. \tag{2}$$

Here
$$a_- = \sum_\beta \frac{|(\alpha J \vdots P \vdots \beta J-1)|^2}{E_{\alpha J} - E_{\beta J-1}}$$

represents the perturbation by the levels for which $J' = J-1$; a_0 and a_+ are similar expressions for the perturbation by levels of $J' = J$ and $J+1$ respectively.

The relative strength of the different components of the line pattern for transverse observation resulting when these states combine with an unperturbed level $\gamma J'$ is likewise given by $9^3 11$ just as in the Zeeman effect (cf. § 4^{16}). Writing (2) in the form

$$\Delta(\alpha J M) = A - BM^2, \tag{3}$$

let us consider the transitions from the states of the level αJ to a level γJ of the same J value. The π component whose displacement from the position of the unperturbed line is $A - BM^2$ is produced by the transitions $M \to M$ and $-M \to -M$ so its whole strength is proportional to $2M^2$, assuming equal population of the various initial states. The σ component line at the same place is produced by $M \to M \pm 1$ and $-M \to M \mp 1$, so its whole strength is proportional to $[J(J+1) - M^2]$ for $M \neq 0$ while for $M = 0$ the strength is proportional to $\frac{1}{2}J(J+1)$. In the same way the relative strengths of the lines in the patterns for a $J \to J-1$ and a $J \to J+1$ transition may be worked out. The results are summarized in the table:

Relative strength of line component at $A - BM^2$

Transition	Polarization π	Polarization σ
$\alpha J \to \gamma J$	$2M^2$	$\epsilon J(J+1) - M^2$
$\alpha J \to \gamma J-1$	$2(\epsilon J^2 - M^2)$	$\epsilon J(J-1) + M^2$
$\alpha J \to \gamma J+1$	$2[\epsilon(J+1)^2 - M^2]$	$\epsilon(J+1)(J+2) + M^2$
$\epsilon = 1$ for $M \neq 0$ and $\frac{1}{2}$ for $M = 0$		

These relative strengths are the analogues of the weak-field strengths of § 4[16] in the Zeeman effect since they are calculated with the unperturbed eigenfunctions.

We can also obtain some general results with regard to the appearance of 'forbidden' lines in the Stark effect. The theory, of course, is analogous to the corresponding case in the Zeeman effect as treated by Zwaan (§ 6[16]). The states $\psi(\beta J' M)$ which perturb the state $\psi(\alpha J M)$ (where $J' = J$ or $J \pm 1$) are of opposite parity and therefore will not give dipole radiation in combination with the states which combine with $\psi(\alpha J M)$. But in the electric field the eigenfunctions become perturbed so that in the perturbed eigenfunction arising from $(\beta J' M)$ there appears some of the eigenfunction of the states $(\alpha J M)$. By the results of § 9[2] we have, to the first power in $\mathscr{E}$,

$$\begin{aligned}\psi(\beta J' M) = \psi_0(\beta J' M) &- \mathscr{E} \sum_\alpha \frac{(\beta J' \vdots P \vdots \alpha J'-1)\sqrt{J^2 - M^2}}{E_{\beta J'} - E_{\alpha J'-1}} \psi_0(\alpha J'-1\, M) \\ &- \mathscr{E} \sum_\alpha \frac{(\beta J' \vdots P \vdots \alpha J')\, M}{E_{\beta J'} - E_{\alpha J'}} \psi_0(\alpha J' M) \\ &- \mathscr{E} \sum_\alpha \frac{(\beta J' \vdots P \vdots \alpha J'+1)\sqrt{(J+1)^2 - M^2}}{E_{\beta J'} - E_{\alpha J'+1}} \psi_0(\alpha J'+1\, M),\end{aligned} \tag{4}$$

where the subscript 0 refers to the unperturbed eigenfunctions. Hence, if $(\gamma J'' M'')$ is a final state, whose perturbation by the field we may neglect, of the same parity as β (with which it therefore does not combine in the absence of a perturbation), the matrix component of electric moment connecting the two states is given by

$$\begin{aligned}(\gamma J'' M''|\boldsymbol{P}|\beta J' M) &= -\mathscr{E} \left\{ \sum_\alpha \frac{(\gamma J'' M''|\boldsymbol{P}|\alpha J'-1\, M)\,(\beta J' \vdots P \vdots \alpha J'-1)\sqrt{J^2 - M^2}}{E_{\beta J'} - E_{\alpha J'-1}} \right. \\ &+ \sum_\alpha \frac{(\gamma J'' M''|\boldsymbol{P}|\alpha J' M)\,(\beta J' \vdots P \vdots \alpha J')\, M}{E_{\beta J'} - E_{\alpha J'}} \\ &+ \left. \sum_\alpha \frac{(\gamma J'' M''|\boldsymbol{P}|\alpha J'+1\, M)\,(\beta J' \vdots P \vdots \alpha J'+1)\sqrt{(J+1)^2 - M^2}}{E_{\beta J'} - E_{\alpha J'+1}} \right\}.\end{aligned} \tag{5}$$

This expression can be somewhat reduced by using 9[3]11 to express the dependence on M'' and M of the first factor of the numerator in each summand. The squared magnitude of (5) measures the strength of the component produced by the transition $\gamma J'' M'' \to \beta J' M$. It is evidently proportional to the square of the applied field.

It will often happen that only one term in one of the three sums is appreciable. The strength of the forbidden pattern is then given by the square of that term alone. This is the case when the perturbation is principally due to the interaction of two neighbouring terms. From (5) we see that the relative intensity distribution in the pattern of the forbidden line is not the same as in the case of a non-forbidden line.

To make this more explicit suppose that the important term in (5) in a given case is one of those on the first line for a particular value of α. Then the intensities are proportional to

$$\left.\begin{matrix}1\cdot\\ \tfrac{1}{2}\cdot\end{matrix}\right\} |(\gamma J'' M''|\boldsymbol{P}|\alpha J'-1\, M)|^2 \frac{\Delta(\beta J' M)}{E_{\beta J'}-E_{\alpha J'-1}}. \qquad \left.\begin{matrix}(\pi)\\ (\sigma)\end{matrix}\right\} \qquad (6)$$

The pattern to be expected is therefore that which applies for a combination with the *perturbing* (not the *perturbed*) level modified by the factor

$$\Delta(\beta J' M)/(E_{\beta J'}-E_{\alpha J'-1}),$$

which is the ratio of the perturbation of the state in question to the unperturbed interval between the interacting levels.

This is about all one can say without consideration of special cases. It is evident that all of the knowledge concerning the matrix components of $\boldsymbol{P}$ which we have acquired in connection with the intensity problem is available for application here. Thus in the Russell-Saunders case we can use the results of § 10[3] to write down formulas for the electric-field interaction of adjacent Russell-Saunders terms. Likewise knowledge of the eigenfunctions in configurations showing intermediate coupling permits us to make corresponding predictions concerning the Stark effect. The method is so straightforward that there is no point in writing down explicitly formulas for the various special cases. We shall turn therefore to the application of the theory to typical cases in the experimental data.

4. Helium.

The first application of quantum mechanics to a non-hydrogenic Stark effect was due to Foster,* who carried out the calculations and applied them to his own experimental data on helium. He uses matrix components given by use of hydrogenic wave functions, for which

$$(n\,S\,L\vdots P\vdots n\,S\,L-1) = -\tfrac{3}{2}ean\sqrt{\frac{n^2-L^2}{4L^2-1}} \qquad (1)$$

(cf. 6[5]2, 3). In this spectrum the triplet intervals are so small that they may be neglected. The lines for which the Stark effect is studied are the combinations of the states with $n=4$ and 5 with the $2\,S$ and $2\,P$ terms.

First let us estimate the Stark displacement of the final terms to see that it is negligible. $2\,^1S$ lies 5857 cm^{-1} below $2\,^1P$, so by 3[17]2 the mutual perturbation of the $M=0$ states amounts to $0{\cdot}0285\mathscr{E}^2$ cm^{-1}, where $\mathscr{E}$ is measured in

* Foster, Proc. Roy. Soc. **A117**, 137 (1927).

10^5 volt/cm. The $M=1$ state of $2\,^1P$ is not perturbed by $2\,^1S$. Similarly $2\,^3S$ lies 9231 cm^{-1} below $2\,^3P$, so the mutual interaction here is even smaller in the ratio $\frac{5857}{9231}$. These effects are quite negligible alongside of the interactions in the $n=4$ and $n=5$ groups.

Using the observed intervals between $4\,^1S$, $4\,^1P$, $4\,^1D$, $4\,^1F$ and the hydrogenic matrix components (1) together with their dependence on M and J as given by $9^3 11$ and $11^3 8$, Foster set up the secular equations for the cases $M=0, 1, 2$ and calculated the roots for fields up to 10^5 volt/cm. The results are shown in Fig. 5^{17}, where the ordinates are wave numbers measured from

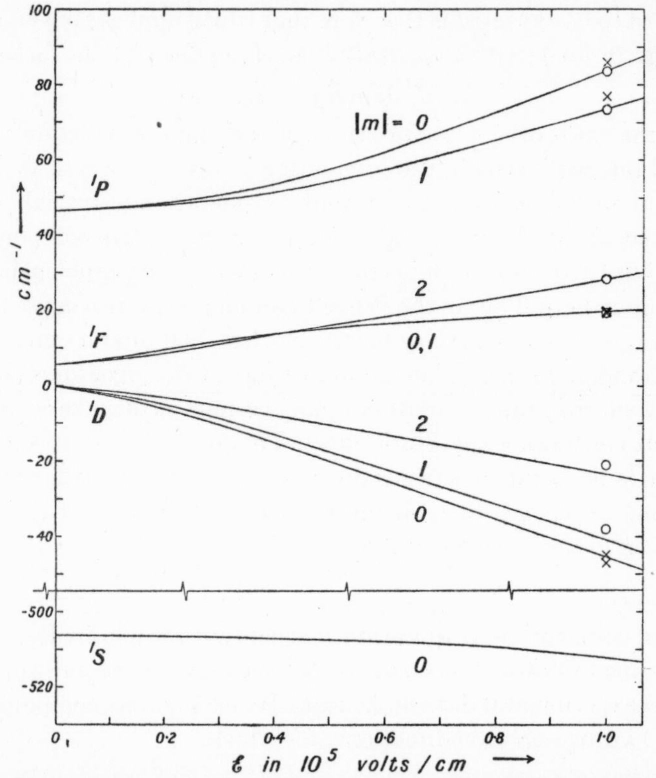

Fig. 5^{17}. Stark-effect interaction of the $4\,^1S$, $4\,^1P$, $4\,^1D$ and $4\,^1F$ levels in helium.

the unperturbed D level and the abscissas are $\mathscr{E}$ in 10^5 volt/cm. They form an excellent illustration of the perturbation theory. The S term is so far away that it is very slightly perturbed, but even so its perturbation is large compared with that in the $2\,S$ or $2\,P$ states. The tendency of the terms to repel each other is nicely brought out, especially in regard to the F terms. There the components with $M=0$ and 1 start out with larger displacements than the $M=2$ state but are prevented from reaching as large displacement as the $M=2$ state because of repulsion by the $M=0$, 1 components of the P level above. The $M=0$, 1 components of the F level are really not identical but coincide within the accuracy of the drawing. The accuracy with which the data is represented by the theory is shown by the experimental points for 10^5 volt/cm as given by Foster, the o's representing data from the $2\,^1P-4\,^1P$, ^1D, ^1F group and the ×'s from the $2\,^1S-4\,^1P$, ^1D, ^1F group.

Foster also calculates the variation of the relative strengths of lines in the patterns with the field-strength. This is done by finding the transformation to states in the groups which makes the perturbed Hamiltonian diagonal and calculating their matrix components of electric moment with $2\,^1S$ and $2\,^1P$ as final terms. The observations do not include intensity measurements, so it can only be said that the observed intensities agree qualitatively with the theory. An interesting feature of the intensity theory is that certain components may fade out at a certain field-strength and reappear again at a larger value. This was checked experimentally.

Ishida and Kamijima* have obtained plates of the effect in helium up to 5×10^5 volt/cm and find their data in good agreement with the theory.

A study of the relative intensities of the lines in the patterns has also been made by Dewey.† Her calculations were based on an application of the Kramers-Heisenberg dispersion formula to the Stark effect as suggested by Pauli.‡ This formulation of the problem, reached through the correspondence principle, agrees with the quantum-mechanical theory.

Foster § has studied the perturbation of certain helium lines in the presence of both electric and magnetic fields.

5. Alkali metals.

We shall consider next the effect on the one-electron spectra of the alkalis. Passing over the older theoretical work, || the first quantum-mechanical calculations are those of Unsöld,¶ who simply considered the interaction in a weak field of the terms of the same n but different l, using the hydrogenic matrix components and the empirical values of the separations as in Foster's helium calculations. A discussion of this formula in comparison with experimental data has been given by Ladenburg.** It gives fairly good values in all cases where it is applicable. This type of calculation has been extended to stronger fields by Rojansky.††

Consideration of the effect in the alkalis falls naturally into two parts: the very small quadratic effect on lines whose initial and final states are very

* Ishida and Kamijima, Sci. Papers, Inst. Phys. Chem. Res. Tokyo **9**, 117 (1928).
† Dewey, Phys. Rev. **28**, 1108 (1926).
‡ Pauli, Kgl. Danske Vid. Selskab, Math.-fys. **7**, No. 3 (1925).
§ Foster, Proc. Roy. Soc. **A122**, 599 (1928); Nature **123**, 414 (1929).
|| Becker, Zeits. für Phys. **9**, 332 (1922).
¶ Unsöld, Ann. der Phys. **82**, 355 (1927).
** Ladenburg, Phys. Zeits. **30**, 369 (1929).
†† Rojansky, Phys. Rev. **35**, 782 (1930).

slightly perturbed, and the transition from quadratic to linear effect, as in helium, when the interacting terms lie very close together. As an example of the first kind we consider the data of Grotrian and Ramsauer* on the principal series of potassium. The data were obtained in absorption for the second, third and fourth lines.

According to § 3^{17} the ^2P terms will be perturbed by the ^2S and ^2D terms. As remarked at the end of that section we may use formulas of § 10^3 in the Russell-Saunders case to find the relative values of the perturbations on the components of a term. If this be done it will readily be found that the perturbation of a ^2P term in a one-electron spectrum can be written

$$\begin{aligned}\Delta(^2P^{\frac{1}{2}}_{\frac{3}{2}}) &= 2A_s + \tfrac{11}{3}A_d\\ \Delta(^2P^{\frac{3}{2}}_{\frac{3}{2}}) &= \qquad 3A_d\\ \Delta(^2P^{\frac{1}{2}}_{\frac{1}{2}}) &= A_s + \tfrac{10}{3}A_d,\end{aligned} \tag{1}$$

in which A_s and A_d are coefficients measuring the perturbation by the S and D series respectively,

$$A_s = \frac{\mathscr{E}^2}{hc}\sum_{n'}\frac{|(n\,^2P\vdots P\vdots n'\,^2S)|^2}{E_{np}-E_{n's}}, \quad A_d = \frac{\mathscr{E}^2}{hc}\sum_{n'}\frac{|(n\,^2P\vdots P\vdots n'\,^2D)|^2}{E_{np}-E_{n'd}}.$$

These formulas are based on the supposition that the ^2D interval is negligible and the ^2P interval large compared with the amount of the effect. For $5\,^2P$ in potassium the experimental values of the Δ's are $-0{\cdot}44\mathscr{E}^2$, $-0{\cdot}21\mathscr{E}^2$, and $-0{\cdot}37\mathscr{E}^2$ respectively (unit of $\mathscr{E} = 10^5$ volt/cm) which are represented by taking $A_s = 0{\cdot}13$ and $A_d = 0{\cdot}07$ probably within the accuracy of the data.

The kind of pattern to be expected and the relative intensities are readily worked out from the results of § 3^{17}. The prediction is that the line from $^2P_{\frac{1}{2}}$ is displaced by the amount $\Delta(^2P^{\frac{1}{2}}_{\frac{1}{2}})$, the single component appearing of relative strength 2 and 2 in π and σ polarization, while from $^2P_{\frac{3}{2}}$ we have a component displaced by $\Delta(^2P^{\frac{1}{2}}_{\frac{3}{2}})$ which has strength 4 in π and 1 in σ polarization and a component displaced by $\Delta(^2P^{\frac{3}{2}}_{\frac{3}{2}})$ of strength 3 purely in σ polarization. This accords with the data except that Grotrian and Ramsauer could not detect the theoretically weak $\Delta(^2P^{\frac{1}{2}}_{\frac{3}{2}})$ component in σ polarization.

It is hard to say anything definite about the absolute value of the coefficients. They are small both because the nearest perturbing terms are about 3000 cm^{-1} away and also because the nearest S and D terms below $5\,^2P$ are almost as close as the nearest S and D terms above, so there is a tendency toward cancellation of effects. The quadratic effect on the sodium D lines has been studied by Ladenburg.†

* Grotrian and Ramsauer, Phys. Zeits. **28**, 846 (1927).
† Ladenburg, Zeits. für Phys. **28**, 51 (1925).

As an illustration of a more hydrogen-like alkali spectrum we shall mention the effect in lithium. This has been discussed by Condon* in a manner analogous to Foster's work on helium. The $4\,D$ and $4\,F$ terms are close enough that a linear effect sets in at fields of the order 30 000 volt/cm, while the $5\,D$, $5\,F$ and $5\,G$ are so close together that their interaction gives a linear effect from the outset. Ishida and Fukushima† have published observations on the $2\,P-4\,PDF$ and $2\,P-5\,PDFG$ patterns which show in all respects agreement with the qualitative predictions of the theory, and in which the displacements are quite accurately given by calculating the secular equation with hydrogenic matrix components.

* Condon, Phys. Rev. **43**, 648 (1933).
† Ishida and Fukushima, Sci. Pap. Inst. Phys. Chem. Res. Tokyo **9**, 141 (1928).

CHAPTER XVIII

THE NUCLEUS IN ATOMIC SPECTRA

Except for brief consideration of the finite mass of the proton in Chapter v, in connection with the theory for atomic hydrogen, we have treated the nucleus so far not as a dynamical particle but as a fixed centre of Coulomb force characterized solely by the atomic number Z. In this chapter we shall consider the way in which the nucleus affects the structure of atomic spectra. The very fact that this topic can be put off so long indicates that the effects are small. Nevertheless they are of great importance and afford a tool for studying atomic nuclei. The most obvious feature to be considered is the finite mass of the nucleus, as a consequence of which the nucleus has some kinetic energy. The effect of this on the atomic energy levels we consider in § 1. But more interesting is the fact that some spectra show a fine structure of the lines, finer than the ordinary multiplet structure (on a scale of 0·1 to 1·0 cm^{-1}). This is known usually as 'hyperfine' structure and, following Pauli, is to be associated with quantum numbers specifying a degree of freedom for angular momentum of the nucleus. The theory of the energy levels resulting from this picture has received a great deal of attention in the last few years and is now fairly well understood, although much remains to be done.

1. Effect of finite mass.

The theory for a nucleus of finite mass in an N-electron problem has been considered by Hughes and Eckart* and also by Bartlett and Gibbons.† The kinetic energy of a system of N electrons, each of mass μ, and a nucleus of mass M is given in terms of the velocities by

$$T = \tfrac{1}{2}\mu \sum \dot{\boldsymbol{r}}_i^2 + \tfrac{1}{2} M \dot{\boldsymbol{r}}^2,$$

where $\boldsymbol{r}_i$ and $\boldsymbol{r}$ are the position vectors of the electrons and nucleus relative to a fixed origin. We introduce the position vector $\boldsymbol{R}$ of the centre of mass,

$$(M + N\mu)\boldsymbol{R} = M\boldsymbol{r} + \mu \Sigma \boldsymbol{r}_i,$$

and the position vectors $\boldsymbol{s}_i$ which give the location of each electron relative to the nucleus,

$$\boldsymbol{s}_i = \boldsymbol{r}_i - \boldsymbol{r}.$$

Then the kinetic energy becomes, when expressed as a function of the $\boldsymbol{p}_i$, conjugate to $\boldsymbol{s}_i$, and $\boldsymbol{P}$, conjugate to $\boldsymbol{R}$,

$$T = \frac{1}{2\mu'} \sum_i \boldsymbol{p}_i^2 + \frac{1}{M} \sum_{i \neq j} \boldsymbol{p}_i \cdot \boldsymbol{p}_j + \frac{1}{2(M + N\mu)} \boldsymbol{P}^2,$$

* Hughes and Eckart, Phys. Rev. **36**, 694 (1930).
† Bartlett and Gibbons, Phys. Rev. **44**, 538 (1933).

where μ' is the reduced mass $\mu M/(\mu+M)$. Since the translational energy of the centre of mass cannot change appreciably in a radiation process and since $\boldsymbol{P}$ is a constant of the motion, we may as well confine ourselves to states for which $\boldsymbol{P}=0$. Then we see that the finite nuclear mass has altered the kinetic energy part of the Hamiltonian in two ways, by the substitution of μ' for μ in the first term and by the appearance of the second term which we shall call S:

$$S=\frac{1}{M}\sum_{i\neq j}\boldsymbol{p}_i\cdot\boldsymbol{p}_j.$$

This is evidently a quantity of type G, as considered in § 7^6.

Suppose we know the energy levels for the case of infinite nuclear mass. Then for finite mass the effect of the change in the first term (neglecting spin-orbit interaction) is to multiply these levels by $\mu'/\mu=[1+\mu/M]^{-1}$. This is what we found in Chapter V for hydrogen and hydrogen-like ions. For any element the result readily follows from the fact that the potential-energy function is homogeneous of degree -1 in all the positional coordinates. This part has been called the normal effect.

The specific effect is that due to the matrix components of S, which we treat as a perturbation. In the zero-order scheme, we have from $7^6 7$ the diagonal element

$$(A|S|A)=\frac{1}{M}\sum_{k>t}\{(a^k|\boldsymbol{p}|a^k)\cdot(a^t|\boldsymbol{p}|a^t)-(a^k|\boldsymbol{p}|a^t)\cdot(a^t|\boldsymbol{p}|a^k)\},$$

the reduction to individual one-electron matrix components being possible because $\boldsymbol{p}_i\cdot\boldsymbol{p}_j$ factorizes. Now $\boldsymbol{p}$ is a vector which anticommutes with the parity operator $\mathscr{P}$ (§ 11^6), hence all of the diagonal matrix components in the first term vanish. Moreover the second term will vanish unless a^k and a^t refer to individual sets of opposite parity. Similar remarks can be made about the non-diagonal matrix components given by § 7^6. With the main features of the S matrix known the detailed calculation of the energies due to this term may be made in any special case by the same methods as for other types of perturbation.

Bartlett and Gibbons have carried out the calculations for the $2p^5\,3p\rightarrow 2p^5\,3s$ lines in neon using the Hartree field found by Brown.* They calculate the matrix components in the zero-order scheme and from these the perturbations of the Russell-Saunders energy levels are obtained with the diagonal-sum rule (exactly as in the calculation of electrostatic energies in Chapter VII). In the case of the p^5p configuration all the electrons outside closed shells have the same parity, so the diagonal matrix components of S contain only the contributions due to interaction of the s closed shells with the p electrons; these are the same for all the states of the configuration. In the

* Brown, Phys. Rev. **44**, 214 (1933).

$p^5 s$ configuration this is not the case, and it turns out there that the ^3P term is displaced by $3k$, while the ^1P term's displacement is k, where k is a matrix component connecting the $2p$ and $3s$ states. Calculating the specific shifts for the Ne^{22} and Ne^{20} isotopes and taking the difference (heavier minus lighter), they calculate net shifts of 0·0136 cm^{-1} for the $p^5 p$ levels, of 0·0098 cm^{-1} for the $p^5 s\,^1P$ level, and 0·0293 cm^{-1} for the $p^5 s\,^3P$ levels. Hence in the $p^5 p \rightarrow p^5 s$ array, the singlet lines will be displaced 0·0195 cm^{-1} more than the triplet lines. This is in accord with the observations of Nagaoka and Mishima,* although the theoretical value of the absolute shift is much too small.

2. Local nuclear fields.

Unquestionably the nucleus is not truly a point centre of force, so that the Coulomb potential, $-Ze^2/r$, cannot be correct in the limit as $r \rightarrow 0$. Experiments on scattering of α-particles by nuclei have shown that there are departures from the Coulomb law in the neighbourhood of $r \sim 10^{-12}$ cm, and various theories of nuclear structure agree in assigning to the size of the nucleus a linear dimension of this order of magnitude. Calculations of the effect on atomic spectra of such departures from the Coulomb law have been made by Racah, Breit, and Rosenthal.†

Nothing very definite is known about the nature of the departure from the Coulomb law. Racah makes the simple assumption that the nucleus is spherical in shape and that the potential inside the nucleus is constant and continuous with the value at the boundary of the nucleus. Rosenthal and Breit work with a model having a discontinuity at the nuclear boundary corresponding to the potential barrier model used in current theories of α-particle disintegration. The calculations of Rosenthal and Breit are carried through using the Dirac equations of the relativistic theory (§ 5^5). In the first work it turned out, on assuming that the radius of isotopic nuclei varies as the cube-root of the atomic weight (constant nuclear density), that the theoretical values of the isotope shifts of spectra of thallium, lead, and mercury were considerably larger than the experimental values. One of the most uncertain elements entering into the calculations is the value of the atomic eigenfunctions at the nucleus. In the later paper, Breit shows that much of the discrepancy can be removed if a semi-empirical formula for $\psi^2(0)$ due to Goudsmit is used, instead of the values used in the first paper.

Data on differences in spectra associated with different isotopes of heavy

* Nagaoka and Mishima, Sci. Pap. Inst. Phys. Chem. Res. Tokyo **13**, 293 (1930).

† Racah, Nature **129**, 723 (1932);
Rosenthal and Breit, Phys. Rev. **41**, 459 (1932);
Breit, Phys. Rev. **42**, 348 (1932).

elements is being accumulated rapidly at present and may in the future prove an important source of knowledge about the nucleus, but at present not much more can be said than that reasonable assumed departures from the Coulomb law in the neighbourhood of the nucleus can account in order of magnitude for the observed effects.

3. Nuclear spin in one-electron spectra.

The effects considered in the two preceding sections do not produce a splitting of the levels of a single atomic species and so can only give rise to a hyperfine structure when several isotopes are present. But such structure is observed in atoms having no isotopes, notably bismuth, so some additional hypothesis is needed for description of this structure. This was supplied in 1924 by Pauli,* who postulated that the nucleus itself may have a spin angular momentum and an associated magnetic moment. It is supposed that a nucleus of given Z and M always has the same spin, denoted by $\boldsymbol{I}$, but that different kinds of nuclei have different spins. This hypothesis of nuclear spin has also proven of great importance in the theory of molecular spectra, so that it is now an indispensable part of atomic theory.

With nuclear spin postulated, we need in addition to know the term in the Hamiltonian corresponding to the interaction of the nuclear spin with the electronic structure. This was first obtained from the classical picture of a nuclear magnetic moment whose energy in the magnetic field produced by the electronic structure depends on the orientation of the nucleus relative to that field. This brings up the question of the magnitude of the nuclear magnetic moment. It is difficult in the present state of knowledge of nuclear structure to say anything about this that is very definite. For an electron the magnetic moment is $e/\mu c$ times the angular momentum, so it was natural to suppose that for a proton the magneto-mechanical ratio would be e/Mc or $\mu/M = 1/1838$ times as great. But this appears not to be the case according to the recent experiments of Stern, Frisch, and Estermann,† in which the magnetic moment of the proton was measured by a molecular beam method on molecular hydrogen. The result was a magnetic moment 2·5 times larger than $(e/Mc)(\frac{1}{2}\hbar)$. That the spin of the proton is $\frac{1}{2}\hbar$ is known from the band spectra of H_2. This unexpected result for the proton is made certain by a more direct method, developed by Rabi, Kellogg, and Zacharias,‡ which uses atomic instead of molecular hydrogen.

Somewhat surprising is the fact that those nuclei which are supposed to contain an odd number of electrons also have magnetic moments of this

* Pauli, Naturwiss. **12**, 741 (1924).

† Frisch and Stern, Zeits. für Phys. **85**, 4 (1933); Estermann and Stern, *ibid.* **85**, 17 (1933).

‡ Rabi, Kellogg, and Zacharias, Phys. Rev. **46**, 157 (1934).

small size; this makes it appear that a nuclear electron is quite different from a free electron or one in the extra-nuclear structure. This is just one of the properties of nuclei which lends support to the newer view that nuclei are composed wholly of protons and neutrons and do not contain any electrons.

The first quantitative theory of the interaction of the nuclear magnetic moment and the outer electrons is due to Fermi and Hargreaves.* The magnetic moment **M** of the nucleus gives rise to a field described by the vector potential

$$\boldsymbol{A}=\frac{1}{r^3}\mathbf{M}\times\boldsymbol{r}. \tag{1}$$

Since the magnitude of the nuclear spin is constant in all states of the system, it does not need to be explicitly mentioned but we do need to introduce M_I, the z-component of nuclear spin, as a coordinate and quantum number. Then, since **M** is proportional to the nuclear angular momentum $\boldsymbol{I}$, that is

$$\mathbf{M}=g_N\frac{e}{2Mc}\boldsymbol{I}, \tag{2}$$

where g_N is analogous to the Landé g-factor in the theory of the Zeeman effect, it follows that the components of **M** are represented by three non-commuting matrices whose rows and columns are labelled by the values of M_I. Fermi handles the one-electron problem with Dirac's equations (§ 5^5), including the interaction with the nucleus through the vector potential given by (1).

We shall not reproduce the calculations in detail. It is easy to see that we have here another case of vector coupling. We start with a scheme in which the $\boldsymbol{J}$ of the outer electron and the $\boldsymbol{I}$ of the nucleus is known. The resultant angular momentum $\boldsymbol{F}$ is the vector sum of these two and we have to transform from a scheme of states in which $\boldsymbol{J}^2$ and $\boldsymbol{I}^2$ are diagonal to one in which $\boldsymbol{F}^2$ is diagonal. All this is exactly what we have been through in studying the coupling of $\boldsymbol{S}$ and $\boldsymbol{L}$ to form $\boldsymbol{J}$.

Fermi finds that the ${}^2S_{\frac{1}{2}}$ terms of the one-electron ns configurations are split into two levels, the displacements from the unperturbed positions being

$$\tfrac{8}{3}\mu_0 \mathrm{M}\pi\,\psi^2(0)\left(1,\,-\frac{I+1}{I}\right). \tag{3}$$

Here $\mu_0=e\hbar/2\mu c$ is the magnetic moment of the Bohr magneton, M the actual magnetic moment of the nucleus and $\psi^2(0)$ is the value of the normalized s-eigenfunction at the nucleus. The factor 1 belongs with an F value of $I+\frac{1}{2}$ and the other with $F=I-\frac{1}{2}$. For positive M, the level with $F=I-\frac{1}{2}$

* Fermi, Zeits. für Phys. **60**, 320 (1930);
Hargreaves, Proc. Roy. Soc. **A124**, 568 (1929); **A127**, 141, 407 (1930).

is lower in energy than the other. Positive M corresponds to the magnetic moment parallel to the angular momentum as if the magnetic moment were produced by rotation of the positive charge distribution of the nucleus. Likewise Fermi finds that the $^2P_{\frac{1}{2}}$ level is split into two levels in the same way, the amount of the perturbation being given by (3) with $\overline{r^{-3}}$ in place of $\pi\psi^2(0)$.

The $^2P_{\frac{3}{2}}$ level is split in general into four levels whose F values are $I-\frac{3}{2}$, $I-\frac{1}{2}$, $I+\frac{1}{2}$, and $I+\frac{3}{2}$, the displacements in energy being

$$\tfrac{8}{5}\mu_0 \mathrm{M}\overline{r^{-3}}\left(-\frac{I+1}{I}, -\frac{I+4}{3I}, \frac{I-3}{3I}, 1\right) \tag{4}$$

respectively. (For $I=1$ there are but three levels, as $I-\frac{3}{2}$ is impossible, and for $I=\frac{1}{2}$ there are only two.) The four levels have the same kind of interval rule as with Russell-Saunders terms: the interval between adjacent levels is proportional to the larger F value of the pair.

Let us now consider the application of these results to the alkali metals. Schüler, and Dobrezov and Terenin* find in Na I that each of the D-lines consists of two components with a separation of 0·022Å. For potassium the corresponding structure is either absent or very much narrower.† In Rb I the second line of the principal series shows doubling as in sodium with a component distance of 0·020Å.‡ Jackson § studied the first three lines of the principal series of Cs I and found that each component in all three lines showed a separation of 0·300 cm^{-1} and that the two hyperfine components were of about equal intensity.

These observations are in accord with the view that the $^2S_{\frac{1}{2}}$ level has been doubled by interaction with the nucleus and that a splitting of the ^2P levels is negligible. From the splitting one cannot find the value of I, since in any case the $^2S_{\frac{1}{2}}$ level would be split into two levels.

That the nuclear spin of Na I is equal to $\frac{3}{2}$ is shown by the measurements of Rabi and Cohen by a modification of the Stern-Gerlach experiment, by Joffe and Urey from alternating intensities in the Na_2 bands, and by the recent accurate measurements of the hyperfine structure intensities by Granath and Van Atta. || We thus have three different methods leading to the same value in the case of this nucleus. The calculation of the nuclear magnetic moment from the observed splittings is of course a much more difficult and uncertain matter. The value of I is inferred from hyperfine

* Schüler, Naturwiss. **16**, 512 (1928);
Dobrezov and Terenin, *ibid.* **16**, 658 (1928).
† Schüler and Brück, Zeits. für Phys. **58**, 735 (1929).
‡ Filippov and Gross, Naturwiss. **17**, 121 (1929).
§ Jackson, Proc. Roy. Soc. **A121**, 432 (1928).
|| Rabi and Cohen, Phys. Rev. **43**, 582 (1933); **46**, 707 (1934);
Joffe and Urey, *ibid.* **43**, 761 (1933);
Granath and Van Atta, *ibid.* **44**, 935 (1933).

structure intensities by an application of the sum rules, for evidently the theory of Russell-Saunders multiplet intensities (§ 2^9) is applicable here if we write F for J, J for S, and I for L. Thus in the splitting of a line ending on the ^2S level, the two components have an intensity ratio of $I:(I+1)$, this being the ratio of the two values of $2F+1$ for $F=I-\frac{1}{2}$ and $F=I+\frac{1}{2}$.

We shall not undertake a detailed review of the theory by which the nuclear magnetic moment is calculated from the data. For one-electron spectra in particular this has been most carefully discussed in a paper by Fermi and Segré.* This paper includes a compilation of known values of nuclear moments, as does an independent paper by Goudsmit.†

Table 1^{18} is based on these two compilations and gives the values of the magnetic moment in units $e\hbar/2Mc$ from each paper, so that the reader may judge for himself the degree of consistency obtainable by different workers in the present stage of development of the theory.

TABLE 1^{18}. *Nuclear spins and magnetic moments.*

Element	Z	Atomic weight	I	Magnetic moment: Goudsmit	Magnetic moment: Fermi and Segré
Li	3	7	$\frac{3}{2}$	3·29	3·2
Na	11	23	$\frac{3}{2}$	2·1	2·7‡
Al	13	27	$\frac{1}{2}$	2·1	
Cu	29	63, 65	$\frac{3}{2}$	2·5	2·36
Ga	31	69	$\frac{3}{2}$	2·01	2·14
Ga	31	71	$\frac{3}{2}$	2·55	2·75
As	33	75	$\frac{3}{2}$	0·9	
Rb	37	85	$\frac{5}{2}$	1·3	1·36
Rb	37	87	$\frac{3}{2}$	2·7	2·8
Cd	48	111, 113	$\frac{1}{2}$	−0·67	−0·53
In	49	115	$\frac{9}{2}$	5·4	5·3
Sb	51	121	$\frac{5}{2}$	2·7	
Sb	51	123	$\frac{7}{2}$	2·1	
Cs	55	133	$\frac{7}{2}$	—	2·6
Ba	56	137	$\frac{5}{2}$	—	1·05
Au	79	197	$\frac{3}{2}$?	—	1·8?
Hg	80	199	$\frac{1}{2}$	0·55	0·46
Hg	80	201	$\frac{3}{2}$	−0·62	−0·51
Tl	81	203, 205	$\frac{1}{2}$	1·8	1·4
Pb	82	207	$\frac{1}{2}$	0·60	0·53
Bi	83	209	$\frac{9}{2}$	4·0	3·54

4. The hyperfine structure of two-electron spectra.

The most thoroughly worked out case of hyperfine structure where there is more than one electron outside closed shells is that of the $1s\,2s\,^3S$ and

* FERMI and SEGRÉ, Zeits. für Phys. **82**, 729 (1933).
† GOUDSMIT, Phys. Rev. **43**, 636 (1933).
‡ From GRANATH and VAN ATTA (*loc. cit.*).

$1s\,2p\,^3P$ terms of Li II.* The hyperfine structure of the multiplet due to the combination of these terms is of particular interest because it is comparable in magnitude with the ordinary level structure of Li II due to spin-orbit interaction. The theory is in good accord with experiment† when the value of I for Li^7, the most abundant isotope, is taken to be $\frac{3}{2}$. This value is in accord with that from the spectrum of the molecule Li_2.‡

We shall not give the details of the calculations. These are based on the use of a law of interaction of the electrons with the nucleus as used by Fermi for the one-electron problem, together with use of appropriate two-electron wave functions. For the $^3S_1$ level, Breit and Doermann find that the perturbation due to interaction with the nucleus is

$$1{\cdot}06\frac{8\pi}{3}\mu_0 M\psi^2(0)\left(-\tfrac{5}{3}, -\tfrac{2}{3}, 1\right) \qquad (1)$$

for the three components $F=\frac{1}{2}, \frac{3}{2}$, and $\frac{5}{2}$ respectively. This counts only the interaction of the $1s$ electron with the nucleus, $\psi(0)$ being here the value at the origin of the $1s$ wave function.

The ^3P levels are split according to the laws of vector addition of angular momenta: $^3P_0$ is not split, $^3P_1$ becomes a group of three levels with $F=\frac{1}{2}, \frac{3}{2}$ and $\frac{5}{2}$ and $^3P_2$ a group of four levels with $F=\frac{1}{2}, \frac{3}{2}, \frac{5}{2}$ and $\frac{7}{2}$. An energy level diagram (after Güttinger and Pauli) is given in Fig. 1^{18} which shows how the magnitude of the hyperfine structure in ^3P is related to the ordinary level intervals. The fact that they are comparable implies that the non-diagonal matrix components of the nuclear interaction which connect different levels of ^3P are important. These were considered in the calculations of Güttinger and Pauli.

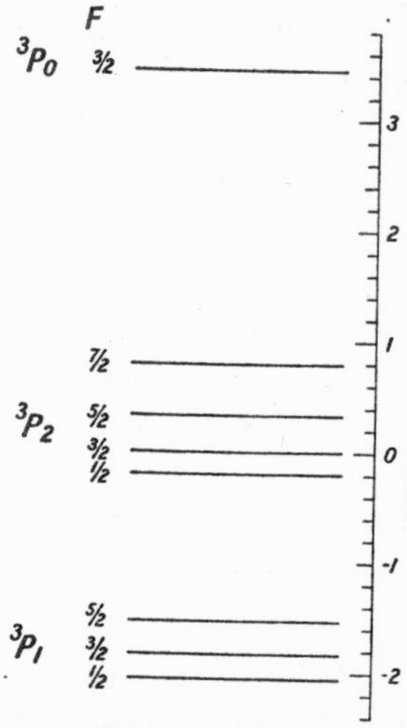

Fig. 1^{18}. Hyperfine structure of the $2\,^3P$ term in Li I.

The experimental data are in good general agreement with the theoretical calculations, although there are discrepancies in the relative intensities.

The detailed interpretation of the observed hyperfine structure for the case of more than one electron outside closed shells is complicated by the necessity of considering the state in intermediate coupling (Chapter XI) and

* Güttinger, Zeits. für Phys. **64**, 749 (1930);
Güttinger and Pauli, *ibid.* **67**, 743 (1931);
Breit and Doermann, Phys. Rev. **36**, 1262, 1732 (1930).

† Schüler, Zeits. für Phys. **42**, 487 (1927);
Granath, Phys. Rev. **36**, 1018 (1930).

‡ Wurm, Zeits. für Phys. **58**, 562 (1929);
Harvey and Jenkins, Phys. Rev. **34**, 1286 (1929); **35**, 789 (1930).

also the influence of configuration interaction (Chapter XV). The effect of the intermediate coupling has been considered in detail by Breit and Wills,* while the importance of considering the interaction of configurations in this connection has been stressed by Fermi and Segré.† At present not enough is known about configuration interaction to make possible quantitative allowance for its effects, but Fermi and Segré show that qualitatively its effects account for a number of observed features of the hyperfine structure which are otherwise without explanation.

The work of Breit and Wills is in extension of that of Goudsmit,‡ who first gave a general discussion of the interaction with the nucleus in the case of complex spectra by vector-coupling methods. Their developments agree fairly well with experiment but there are many outstanding discrepancies, indicating the need for further refinements of the theory.

In Hg and Al II cases occur in which the hyperfine structure is comparable with the ordinary spin-orbit and electrostatic separations. This makes necessary more accurate calculations of the nuclear perturbation of the levels. Empirically these effects and their nature were recognized by Paschen.§ They have been discussed in detail by Goudsmit and Bacher|| with satisfactory results.

5. Zeeman effect of hyperfine structure.

We shall not give a detailed account of the results which have been obtained for the perturbation by a magnetic field of the structure due to nuclear interaction. The work is extremely important as indicating the essential correctness of the nuclear-magnetic-moment picture, but presents no new theoretical problems. In the ordinary theory of the Zeeman effect (§ 1^{16}) the departures from the simple Lorentz theory arise because of the different values of the magneto-mechanical ratio associated with the vectors $\boldsymbol{L}$ and $\boldsymbol{S}$. We have a similar situation here. For fields such that the ordinary Paschen-Back effect is negligible we have a magneto-mechanical ratio of the electronic structure represented by $g(J)$ where this is the Landé g factor of the level in question. Associated with the nucleus is a magneto-mechanical ratio about 10^{-3} as great so that the direct interaction of the nucleus with the external magnetic field is negligible.

A direct application of the formulas of Chapter III then shows the splitting of a given hyperfine level to be governed by $g(J)$ of the ordinary level to which the level belongs multiplied by the factor ($10^3$2a) which gives the

* Breit and Wills, Phys. Rev. **44**, 470 (1933).
† Fermi and Segré, Zeits. für Phys. **82**, 729 (1933).
‡ Goudsmit, Phys. Rev. **37**, 663 (1931).
§ Paschen, Sitzber. Preuss. Akad. 1932, p. 502.
|| Goudsmit and Bacher, Phys. Rev. **43**, 894 (1933).

matrix component of J_z in a state labelled by precise values of $\boldsymbol{F}^2$ and F_z. That is, the g value for a state of resultant angular momentum F composed of electronic J and nuclear I is

$$g(F)=g(J)\frac{F(F+1)+J(J+1)-I(I+1)}{2F(F+1)}.$$

This result was first obtained by Goudsmit and Bacher* from vector-model considerations.

Just as in the discussion of § 5^{16}, the difference in the magneto-mechanical ratio of the vectors $\boldsymbol{J}$ and $\boldsymbol{I}$ means that strong magnetic fields can produce a breakdown of the coupling of $\boldsymbol{J}$ and $\boldsymbol{I}$. This is the Paschen-Back effect of the hyperfine structure. Experimentally and theoretically this phenomenon has been carefully studied by Back and Goudsmit† for Bi. They find good agreement for $I=\frac{9}{2}$ at field strengths corresponding to various stages of the transformation of coupling schemes. The magnetic transformation of the hyperfine structure in thallium has been investigated thoroughly by Back and Wulff.‡

* GOUDSMIT and BACHER, Zeits. für Phys. **66**, 13 (1930).
† BACK and GOUDSMIT, Zeits. für Phys. **47**, 174 (1928); ZEEMAN, BACK, and GOUDSMIT, *ibid.* **66**, 1 (1930).
‡ BACK and WULFF, Zeits. für Phys. **66**, 31 (1930).

APPENDIX

UNIVERSAL CONSTANTS AND NATURAL ATOMIC UNITS

Measurements in physics are statements of relation of the quantity measured to quantities of like kind which are called units. It is customary to build up the system in such a way that the unit of any kind of physical quantity is defined in terms of three conventional units of mass, length and time. The choice of the basic units for these quantities is wholly arbitrary, the general order of magnitude in the centimetre-gram-second system being such that the numerical measure of quantities occurring in ordinary laboratory experiments is of the general order of unity. Thus the velocity of light in the c g s system is 3×10^{10} cm sec^{-1}. The centimetre and second being so chosen that 1 cm sec^{-1} is of the order of velocities of common experience, the bigness of the number measuring velocity of light on this system is simply a statement that velocities of common experience are very small compared with that of light.

There is, therefore, nothing especially fundamental about the c g s basis. Its basic units are of convenient magnitude for common laboratory apparatus, so ultimately the foundation is anthropomorphic since laboratory apparatus is built and designed on a scale convenient for manipulation and observation by a human observer. To recognize this fact is not to deplore it. Certainly the c g s system is convenient for description of the macroscopic apparatus which provides the refined sense-data of physics. But the fact shows us clearly that a metric resting on such a basis will probably not provide units of convenient size for dealing with another branch of physics like the theory of atomic structure.

This is in fact the case. The basic universal constants of atomic theory have values which are very large or very small compared with unity. For example*:

Electron charge:	$-e = -4{\cdot}770 \times 10^{-10}\ \mathrm{g}^{\frac{1}{2}}\ \mathrm{cm}^{\frac{3}{2}}\ \mathrm{sec}^{-1}$;
Quantum constant:	$\hbar = 1{\cdot}043 \times 10^{-27}\ \mathrm{g\ cm^2\ sec^{-1}}$;
Electron mass:	$\mu = 9{\cdot}035 \times 10^{-28}\ \mathrm{g}$;
Light velocity:	$c = 2{\cdot}99796 \times 10^{10}\ \mathrm{cm\ sec^{-1}}$.

These great powers of ten are rather inconvenient in theoretical calculations. There are two ways of avoiding them which might be adopted. One is the way which the metric system has already adopted for extending itself to larger

* We use the values of the universal constants as given by Birge after a critical survey of all the relevant data: Rev. Mod. Phys. **1**, 1 (1929); Phys. Rev. **40**, 228 (1932). For later modifications see Birge, Phys. Rev. **43**, 211 (1933); Michelson, Pease, and Pearson, Science **81**, 100 (1935).

or smaller units, namely to choose another system of fundamental units related to the metric system by conversion factors which are powers of 10. By taking as our units of mass, length and time

$$10^{-28}\,\text{g}, \quad 10^{-8}\,\text{cm}, \quad \text{and} \quad 10^{-17}\,\text{sec},$$

the new electrostatic unit of charge becomes $10^{-9}\,\text{g}^{\frac{1}{2}}\,\text{cm}^{\frac{3}{2}}\,\text{sec}^{-1}$, so in this system the electronic charge has the value $-0{\cdot}4770$. Similarly the new unit of velocity is $10^{9}\,\text{cm sec}^{-1}$, and of angular momentum $10^{-27}\,\text{g cm}^2\,\text{sec}^{-1}$, so the velocity of light and the quantum constant have the numerical values 29·9796 and 1·043 respectively.

For theoretical purposes, however, it is still more convenient to introduce a system of units in which certain of these universal constants are set equal to unity. From this standpoint the numerical values of the universal constants become the numerical conversion factors which connect this system with the ordinary c g s units. Quite a variety of such systems have an equal claim to use so far as general convenience is concerned. Thus it is a matter of arbitrary choice whether we set h or $\hbar$ equal to unity. This kind of arbitrary choice merely alters the place where the pure number 2π appears in the calculations and is analogous to the difference between the Heaviside-Lorentz units and the older set of electromagnetic units. Another arbitrary element lies in the fact that there are only three fundamental units at our disposal, so that it is not possible to assign the numerical values of more than three of the universal constants—different systems arise according to which choice is made in this respect. There is not much point in debating which of the choices is most convenient. As it happens one particular choice, recommended by Hartree,* has already been quite generally employed in theoretical work, so we adopt that one. Hartree's atomic units are such that e, μ and $\hbar$ have each the numerical value unity. Denoting by a and τ the Hartree units of length and time respectively this means that

$$\begin{aligned} e &= \mu^{\frac{1}{2}}\mathsf{a}^{\frac{3}{2}}\tau^{-1} \\ \hbar &= \mu\mathsf{a}^2\tau^{-1} \end{aligned} \quad \text{so} \quad \begin{aligned} \mathsf{a} &= \hbar^2\mu^{-1}e^{-2}, \\ \tau &= \hbar^3\mu^{-1}e^{-4}. \end{aligned}$$

From this standpoint the numerical values of the universal constants e, $\hbar$ and μ given in the c g s system are the conversion factors by means of which quantities expressed in Hartree's system are to be expressed in the c g s system. Thus an angular momentum expressed as $x\hbar$ in the Hartree system, where x is a pure number, is expressed as $(1{\cdot}043 \times 10^{-27})\,x\,\text{g cm}^2\,\text{sec}^{-1}$ in the c g s system. Therefore the length expressed as $1\mathsf{a}$ in the Hartree system is expressed as $(1{\cdot}043 \times 10^{-27})^2\,(9{\cdot}035 \times 10^{-28})^{-1}\,(4{\cdot}770 \times 10^{-10})^{-2}$ cm, which works out to be

$$1\mathsf{a} = 0{\cdot}528 \times 10^{-8}\,\text{cm}.$$

* Hartree, Proc. Camb. Phil. Soc. **24**, 89 (1926).

Similarly the value of τ in the c g s system is

$$1\tau = 2{\cdot}419 \times 10^{-17}\,\text{sec.}$$

From these basic conversion factors the value of the Hartree unit of any derived quantity in the c g s system is readily found. Thus unit velocity is

$$1\mathrm{a}\tau^{-1} = 2{\cdot}18 \times 10^{8}\,\text{cm sec}^{-1};$$

hence the velocity of light has the value

$$c = 137{\cdot}29\,\mathrm{a}\tau^{-1}.$$

The reciprocal of the pure number which gives the value of the velocity of light in these units is known in the theory as the fine structure constant. This is usually defined by the equation

$$\alpha = e^2/c\hbar.$$

It appears, for example, as a parameter of the relativity-spin structure of the hydrogen spectrum in Chapter V. Another important parameter of atomic theory is the mass of the proton, which is

$$M = 1838{\cdot}3\,\mu.$$

The fact that the numerical magnitudes of c^2 and of M in this system of units are both large compared with unity is of fundamental importance for the theory. The largeness of c^2 is what makes the relativistic corrections small and the largeness of M makes it a good approximation to treat the nucleus as a fixed centre of force.

Aside from the fact that e, $\hbar$, and μ have simple numerical magnitudes in this system of units, the other units have simple physical interpretations in the theory. Thus unit length is the radius of the first orbit in Bohr's theory of hydrogen for an infinitely massive nucleus, and unit velocity is the velocity of the electron in this first Bohr orbit. Unit energy is the potential energy of the electron in the first Bohr orbit and hence the ionization energy is half a Hartree unit. Numerous other examples are to be found in the table in this appendix.

We shall not discuss the precision with which these quantities are known in terms of the c g s system other than to say that it is quite generally that associated with a probable error somewhat less than 0·1 per cent. The quantity whose relation to the c g s system is known most precisely is the combination $\alpha\mathrm{a}^{-1}$. The combination $\alpha\mathrm{a}^{-1}/4\pi$ is known as the Rydberg constant, R. Its value is

$$\mathsf{R} = \frac{\alpha}{4\pi}\mathrm{a}^{-1} = 109737{\cdot}42 \pm 0{\cdot}06\,\text{cm}^{-1}.$$

Its relation to the c g s system is thus known to an accuracy of about 6 parts in 10 million, although the accuracy with which the values of α and of a separately are known is much less. The theoretical energy levels of atoms are given in terms of the Hartree energy unit and the parameters like α, μ/M, the atomic number Z, and the quantum numbers labelling the particular level in question. Of these Z and the quantum numbers are integers exactly, so there is no uncertainty in their numerical values. The effects of α and μ/M on the energy are usually in the form of small corrections, so we are not greatly hampered by the uncertainty in their values in making comparisons between theory and experiment. In using the energy levels to predict the wave-numbers of spectral lines, the wave-number equivalent of the energy given by Bohr's relation, $E = hc\sigma$, is what enters. As the wave number equivalent to one Hartree unit of energy is $2\mathsf{R}$, it follows that the principal factor needed for passing from Hartree units to c g s units is the one known with the greatest precision. The precision of knowledge in R is much greater than the accuracy attained in the perturbation theory calculations at present, so that the comparisons with experiment are not at all hampered by the inaccuracy in the Rydberg constant. For that reason we may express the results of the theory at once in terms of cm^{-1} without paying attention to the uncertainties in R.

The only other universal constant of interest for spectroscopic theory is the Boltzmann constant, k, which measures the relation between thermal energy and the thermodynamic temperature scale measured in conventional Centigrade degrees. As we nearly always express energy in terms of equivalent wave-numbers in cm^{-1}, the important conversion factor is not k itself but k/hc, which when multiplied by T in Centigrade degrees gives directly the wave-number equivalent of the energy kT. This factor has the value

$$\frac{k}{hc} = 0{\cdot}698\,\text{cm}^{-1}\,\text{deg}^{-1}.$$

This is the reciprocal of the quantity often denoted by c_2 in classical radiation theory.

Values of some Important Physical Quantities

Kind of quantity	Value in atomic units	Value in c g s units
Length:		
Radius of first Bohr orbit, $\mathrm{a} = \hbar^2/\mu e^2$	1 a	$(0\cdot5282 \pm 0\cdot0004) \times 10^{-8}$ cm
Shift in X-ray wave-length by Compton scattering through $\pi/2$, $2\pi\alpha\mathrm{a}$ (This is also the wave-length of γ-radiation associated with annihilation of an electron)	0·0458 a	$24\cdot2 \times 10^{-11}$ cm
Electromagnetic radius of electron, $e^2/\mu c^2 = \alpha^2\mathrm{a}$	$5\cdot31 \times 10^{-5}$ a	$2\cdot80 \times 10^{-13}$ cm
Wave-length of limit of Lyman series, $4\pi\alpha^{-1}\mathrm{a}_{\mathrm{H}}$	1722 a	910×10^{-8} cm
Wave-number:		
Atomic unit, a^{-1}	$1\ \mathrm{a}^{-1}$	$1\cdot893 \times 10^{8}\ \mathrm{cm}^{-1}$
Rydberg constant, $\mathrm{R} = \alpha\mathrm{a}^{-1}/4\pi$	$5\cdot81 \times 10^{-4}\ \mathrm{a}^{-1}$	$109737\cdot42 \pm 0\cdot06\ \mathrm{cm}^{-1}$
Wave-number of first Balmer line, $\frac{5}{36}\mathrm{R}_{\mathrm{H}}$	$0\cdot807 \times 10^{-4}\ \mathrm{a}^{-1}$	$15233\ \mathrm{cm}^{-1}$
Hydrogen doublet constant, $\alpha^2\mathrm{R}_{\mathrm{H}}/16$	$1\cdot93 \times 10^{-9}\ \mathrm{a}^{-1}$	$0\cdot3636 \pm 0\cdot0006\ \mathrm{cm}^{-1}$
Wave-number associated with one electron volt	$4\cdot281 \times 10^{-5}\ \mathrm{a}^{-1}$	$8106 \pm 3\ \mathrm{cm}^{-1}$
Mass:		
Mass of electron, μ	$1\ \mu$	$(9\cdot035 \pm 0\cdot010) \times 10^{-28}$ g
Mass of proton, M	$(1838\cdot3 \pm 1\cdot0)\ \mu$	$1\cdot661 \times 10^{-24}$ g
Time:		
Time for electron to go $(2\pi)^{-1}$ revolutions in first Bohr orbit, τ	$1\ \tau$	$2\cdot419 \times 10^{-17}$ sec
Mean life for $2p \rightarrow 1s$ transition in hydrogen	$6\cdot62 \times 10^{7}\ \tau$	$1\cdot6 \times 10^{-9}$ sec
Velocity:		
Speed of electron in first Bohr orbit, $\mathrm{a}\tau^{-1}$	$1\ \mathrm{a}\tau^{-1}$	$2\cdot18 \times 10^{8}\ \mathrm{cm\ sec}^{-1}$
Speed of light, $\alpha^{-1}\mathrm{a}\tau^{-1}$	$(137\cdot29 \pm 0\cdot11)\ \mathrm{a}\tau^{-1}$	$(2\cdot99796 \pm 0\cdot00004) \times 10^{10}\ \mathrm{cm\ sec}^{-1}$
Momentum:		
Of electron in first Bohr orbit, $\mu\mathrm{a}\tau^{-1}$	$1\ \mu\mathrm{a}\tau^{-1}$	$1\cdot966 \times 10^{-19}\ \mathrm{g\ cm\ sec}^{-1}$
Basic quantity μc of relativistic theory	$137\cdot29\ \mu\mathrm{a}\tau^{-1}$	$2\cdot70 \times 10^{-17}\ \mathrm{g\ cm\ sec}^{-1}$
Angular momentum:		
Basic quantum unit, $\hbar = \mu\mathrm{a}^2\tau^{-1}$	$1\ \mu\mathrm{a}^2\tau^{-1}$	$1\cdot043 \times 10^{-27}\ \mathrm{g\ cm}^2\ \mathrm{sec}^{-1}$
Planck constant, $2\pi\hbar = h$	$2\pi\ \mu\mathrm{a}^2\tau^{-1}$	$6\cdot547 \times 10^{-27}\ \mathrm{g\ cm}^2\ \mathrm{sec}^{-1}$
Energy:		
Atomic unit, $\mu\mathrm{a}^2\tau^{-2} = \mu e^4/\hbar^2 = 2\mathrm{R}hc$, twice the ionization energy of hydrogen with infinite nuclear mass	$1\ \mu\mathrm{a}^2\tau^{-2}$	$4\cdot304 \times 10^{-11}$ erg $\simeq 219\,474\cdot84\ \mathrm{cm}^{-1}$ $\simeq 27\cdot07$ electron volt
Energy equivalent of electron mass, $\mu c^2 = \alpha^{-2}\mu\mathrm{a}^2\tau^{-2}$	$18859\ \mu\mathrm{a}^2\tau^{-2}$	$8\cdot10 \times 10^{-7}$ erg
Force:		
Force of attraction toward nucleus on electron in first Bohr orbit, e^2/a^2	$1\ \mu\mathrm{a}\tau^{-2}$	$8\cdot19 \times 10^{-3}$ dyne
Electric charge:		
Atomic unit, e, negative of charge on electron	$1\ \mu^{\frac{1}{2}}\mathrm{a}^{\frac{3}{2}}\tau^{-1}$	$4\cdot770 \times 10^{-10}$ e s u
Potential:		
Potential of electron's field at atomic unit distance, e/a	$1\ \mu^{\frac{1}{2}}\mathrm{a}^{\frac{1}{2}}\tau^{-1}$	$9\cdot03 \times 10^{-2}\ \mathrm{e\ s\ u\ cm}^{-1}$ 27·2 volts
Electric field:		
Field strength at atomic unit distance from electron, e/a^2	$1\ \mu^{\frac{1}{2}}\mathrm{a}^{-\frac{1}{2}}\tau^{-1}$	$17\cdot1 \times 10^{6}\ \mathrm{g}^{\frac{1}{2}}\ \mathrm{cm}^{-\frac{1}{2}}\ \mathrm{sec}^{-1}$ $51\cdot3 \times 10^{8}\ \mathrm{volt\ cm}^{-1}$
Electric moment:		
Moment of dipole formed by hydrogen atom in first Bohr orbit, $e\mathrm{a}$	$1\ \mu^{\frac{1}{2}}\mathrm{a}^{\frac{5}{2}}\tau^{-1}$	$2\cdot52 \times 10^{-18}$ e s u cm
Magnetic field:		
Atomic unit, $\mu^{\frac{1}{2}}\mathrm{a}^{-\frac{1}{2}}\tau^{-1}$	$1\ \mu^{\frac{1}{2}}\mathrm{a}^{-\frac{1}{2}}\tau^{-1}$	$17\cdot1 \times 10^{6}$ gauss
Field at nucleus due to motion of electron in first Bohr orbit, $\alpha\mu^{\frac{1}{2}}\mathrm{a}^{-\frac{1}{2}}\tau^{-1}$	$7\cdot283 \times 10^{-3}\ \mu^{\frac{1}{2}}\mathrm{a}^{-\frac{1}{2}}\tau^{-1}$	$1\cdot245 \times 10^{5}$ gauss
Magnetic moment:		
Atomic unit, $\mu^{\frac{1}{2}}\mathrm{a}^{\frac{5}{2}}\tau^{-1}$	$1\ \mu^{\frac{1}{2}}\mathrm{a}^{\frac{5}{2}}\tau^{-1}$	$2\cdot52 \times 10^{-18}\ \mathrm{erg\ gauss}^{-1}$
Bohr magneton, $e\hbar/2\mu c = \frac{1}{2}\alpha\mu^{\frac{1}{2}}\mathrm{a}^{\frac{5}{2}}\tau^{-1}$	$3\cdot642 \times 10^{-3}\ \mu^{\frac{1}{2}}\mathrm{a}^{\frac{5}{2}}\tau^{-1}$	$0\cdot9174 \times 10^{-20}\ \mathrm{erg\ gauss}^{-1}$

LIST OF PRINCIPAL TABLES

INDEX OF SUBJECTS

INDEX OF NAMES

LIST OF PRINCIPAL TABLES

INDEX OF SUBJECTS

INDEX OF NAMES